T0299179

Key Technologies for 5G Wireless Systems

Gain a detailed understanding of the protocols, network architectures, and techniques being considered for 5G wireless networks with this authoritative guide to the state of the art.

- Get up to speed with key topics such as cloud radio access networks, mobile edge computing, full duplexing, massive MIMO, mmWave, NOMA, the Internet of Things, M2M communications, D2D communications, mobile data offloading, interference mitigation techniques, radio resource management, visible light communication, and smart data pricing.
- Learn from leading researchers in academia and industry about the most recent theoretical developments in the field.
- Discover how each potential technology can increase the capacity, spectral efficiency, and energy efficiency of wireless systems.

Providing the most comprehensive overview of 5G technologies to date, this is an essential reference for researchers, practicing engineers, and graduate students working in wireless communications and networking.

Vincent W. S. Wong is a Professor in the Department of Electrical and Computer Engineering at the University of British Columbia, Canada, and a Fellow of the IEEE.

Robert Schober is an Alexander von Humboldt Professor and the Chair for Digital Communication at the Friedrich-Alexander University of Erlangen-Nuremberg, Germany. He is a Fellow of the IEEE, the Canadian Academy of Engineering, and the Engineering Institute of Canada.

Derrick Wing Kwan Ng is a Lecturer in the School of Electrical Engineering and Telecommunications at the University of New South Wales, Australia. He is an Associate Editor of *IEEE Communications Letters*.

Li-Chun Wang is a Professor in the Department of Electrical and Computer Engineering at National Chiao Tung University, Taiwan, and a Fellow of the IEEE.

Key Technologies for 5G Wireless Systems

VINCENT W. S. WONG
University of British Columbia

ROBERT SCHOBER
University of Erlangen-Nuremberg

DERRICK WING KWAN NG
University of New South Wales

LI-CHUN WANG
National Chiao-Tung University

CAMBRIDGE
UNIVERSITY PRESS

CAMBRIDGE
UNIVERSITY PRESS

University Printing House, Cambridge CB2 8BS, United Kingdom

One Liberty Plaza, 20th Floor, New York, NY 10006, USA

477 Williamstown Road, Port Melbourne, VIC 3207, Australia

314-321, 3rd Floor, Plot 3, Splendor Forum, Jasola District Centre, New Delhi-110025, India

79 Anson Road, #06-04/06, Singapore 079906

Cambridge University Press is part of the University of Cambridge.

It furthers the University's mission by disseminating knowledge in the pursuit of
education, learning and research at the highest international levels of excellence.

www.cambridge.org
Information on this title: www.cambridge.org/9781107172418
10.1017/9781316771655

First published 2017

A catalogue record for this publication is available from the British Library

Library of Congress Cataloging in Publication data
Names: Wong, Vincent W. S., editor.
Title: Key technologies for 5G wireless systems / edited by Vincent W.S. Wong
[and 3 others]. Other titles: Key technologies for five G wireless systems
Description: Cambridge ; New York, NY : Cambridge University Press, 2017.
Identifiers: LCCN 2016045220 | ISBN 9781107172418 (hardback)
Subjects: LCSH: Wireless communication systems. | Machine-to-machine
communications. | Internet of things.
Classification: LCC TK5103.2.K49 2017 | DDC 621.3845/6–dc23
LC record available at https://lccn.loc.gov/2016045220

ISBN 978-1-107-17241-8 Hardback

Contents

List of Contributors   *page* xvi

Preface   xxi

1   Overview of New Technologies for 5G Systems   1
Vincent W. S. Wong, Robert Schober, Derrick Wing Kwan Ng, and Li-Chun Wang
    1.1   Introduction   1
    1.2   Cloud Radio Access Networks   3
    1.3   Cloud Computing and Fog Computing   4
    1.4   Non-orthogonal Multiple Access   4
    1.5   Flexible Physical Layer Design   6
    1.6   Massive MIMO   7
    1.7   Full-Duplex Communications   9
    1.8   Millimeter Wave   12
    1.9   Mobile Data Offloading, LTE-Unlicensed, and Smart Data Pricing   13
    1.10   IoT, M2M, and D2D   14
    1.11   Radio Resource Management, Interference Mitigation, and Caching   16
    1.12   Energy Harvesting Communications   17
    1.13   Visible Light Communication   19
    Acknowledgments   20
    References   20

Part I Communication Network Architectures for 5G Systems   25

2   Cloud Radio Access Networks for 5G Systems   27
Chih-Lin I, Jinri Huang, Xueyan Huang, Rongwei Ren, and Yami Chen
    2.1   Rethinking the Fundamentals for 5G Systems   27
    2.2   User-Centric Networks   29
    2.3   C-RAN Basics   29
        2.3.1   C-RAN Challenges Toward 5G   30
    2.4   Next Generation Fronthaul Interface (NGFI): The FH Solution
        for 5G C-RAN   31
        2.4.1   Proof-of-Concept Development of NGFI   33

2.5	Proof-of-Concept Verification of Virtualized C-RAN	35
	2.5.1 Data Packets	37
	2.5.2 Test Procedure	38
	2.5.3 Test Results	39
2.6	Rethinking the Protocol Stack for C-RAN	40
	2.6.1 Motivation	40
	2.6.2 Multilevel Centralized and Distributed Protocol Stack	40
2.7	Conclusion	45
	Acknowledgments	46
	References	46

3 Fronthaul-Aware Design for Cloud Radio Access Networks — 48
Liang Liu, Wei Yu, and Osvaldo Simeone

3.1	Introduction	48
3.2	Fronthaul-Aware Cooperative Transmission and Reception	49
	3.2.1 Uplink	51
	3.2.2 Downlink	57
3.3	Fronthaul-Aware Data Link and Physical Layers	61
	3.3.1 Uplink	63
	3.3.2 Downlink	69
3.4	Conclusion	73
	Acknowledgments	74
	References	74

4 Mobile Edge Computing — 76
Ben Liang

4.1	Introduction	76
4.2	Mobile Edge Computing	77
4.3	Reference Architecture	79
4.4	Benefits and Application Scenarios	80
	4.4.1 User-Oriented Use Cases	80
	4.4.2 Operator-Oriented Use Cases	81
4.5	Research Challenges	82
	4.5.1 Computation Offloading	82
	4.5.2 Communication Access to Computational Resources	83
	4.5.3 Multi-resource Scheduling	84
	4.5.4 Mobility Management	85
	4.5.5 Resource Allocation and Pricing	85
	4.5.6 Network Functions Virtualization	86
	4.5.7 Security and Privacy	86
	4.5.8 Integration with Emerging Technologies	87
4.6	Conclusion	88
	References	88

5   **Decentralized Radio Resource Management for Dense Heterogeneous**
          Wireless Networks   92
          Abolfazl Mehbodniya and Fumiyuki Adachi
          5.1   Introduction   92
          5.2   System Model   93
                 5.2.1   SINR Expression   95
                 5.2.2   Load and Cost Function Expressions   95
          5.3   Joint BSCSA/UECSA ON/OFF Switching Scheme   96
                 5.3.1   Strategy Selection and Beacon Transmission   96
                 5.3.2   UE Association   96
                 5.3.3   Proposed Channel Segregation Algorithms   98
                 5.3.4   Mixed-Strategy Update   100
          5.4   Computer Simulation   101
          5.5   Conclusion   104
          Acknowledgments   104
          References   105

Part II Physical Layer Communication Techniques   107

6   **Non-Orthogonal Multiple Access (NOMA) for 5G Systems**   109
          Wei Liang, Zhiguo Ding, and H. Vincent Poor
          6.1   Introduction   110
          6.2   NOMA in Single-Input Single-Output (SISO) Systems   112
                 6.2.1   The Basics of NOMA   112
                 6.2.2   Impact of User Pairing on NOMA   113
                 6.2.3   Cognitive Radio Inspired NOMA   116
          6.3   NOMA in MIMO Systems   120
                 6.3.1   System Model for MIMO-NOMA Schemes   121
                 6.3.2   Design of Precoding and Detection Matrices with Limited CSIT 123
                 6.3.3   Design of Precoding and Detection Matrices with Perfect CSIT 126
          6.4   Summary and Future Directions   128
          References   130

7   **Flexible Physical Layer Design**   133
          Maximilian Matthé, Martin Danneberg, Dan Zhang, and Gerhard Fettweis
          7.1   Introduction   133
          7.2   Generalized Frequency Division Multiplexing   135
          7.3   Software-Defined Waveform   137
                 7.3.1   Time Domain Processing   138
                 7.3.2   Implementation Architecture   138
          7.4   GFDM Receiver Design   141
                 7.4.1   Synchronization Unit   142
                 7.4.2   Channel Estimation Unit   144
                 7.4.3   MIMO-GFDM Detection Unit   145

7.5   Summary and Outlook   147
Acknowledgments   148
References   148

8   Distributed Massive MIMO in Cellular Networks   151
Michail Matthaiou and Shi Jin
8.1   Introduction   151
8.2   Massive MIMO: Basic Principles   152
       8.2.1   Uplink/Downlink Channel Models   153
       8.2.2   Favorable Propagation   154
8.3   Performance of Linear Receivers in a Massive MIMO Uplink   154
8.4   Performance of Linear Precoders in a Massive MIMO Downlink   157
8.5   Channel Estimation in Massive MIMO Systems   158
       8.5.1   Uplink Transmission   159
       8.5.2   Downlink Transmission   160
8.6   Applications of Massive MIMO Technology   161
       8.6.1   Full-Duplex Relaying with Massive Antenna Arrays   161
       8.6.2   Joint Wireless Information Transfer and Energy Transfer for
                Distributed Massive MIMO   163
8.7   Open Future Research Directions   167
8.8   Conclusion   168
References   169

9   Full-Duplex Protocol Design for 5G Networks   172
Taneli Riihonen and Risto Wichman
9.1   Introduction   172
9.2   Basics of Full-Duplex Systems   173
       9.2.1   In-Band Full-Duplex Operation Mode   173
       9.2.2   Self-Interference and Co-channel Interference   174
       9.2.3   Full-Duplex Transceivers in Communication Links   175
       9.2.4   Other Applications of Full-Duplex Transceivers   178
9.3   Design of Full-Duplex Protocols   179
       9.3.1   Challenges and Opportunities in Full-Duplex Operation   179
       9.3.2   Full-Duplex Communication Scenarios in 5G Networks   180
9.4   Analysis of Full-Duplex Protocols   182
       9.4.1   Operation Modes in Wideband Fading Channels   182
       9.4.2   Full-Duplex Versus Half-Duplex in Wideband Transmission 184
9.5   Conclusion   184
       9.5.1   Prospective Scientific Research Directions   184
       9.5.2   Full-Duplex in Commercial 5G Networks   185
References   186

10   Millimeter Wave Communications for 5G Networks   188
Jiho Song, Miguel R. Castellanos, and David J. Love

10.1   Motivations and Opportunities   188
10.2   Millimeter Wave Radio Propagation   189
        10.2.1   Radio Attenuation   189
        10.2.2   Free-Space Path Loss   191
        10.2.3   Severe Shadowing   193
        10.2.4   Millimeter Wave Channel Model   193
        10.2.5   Link Budget Analysis   194
10.3   Beamforming Architectures   195
        10.3.1   Analog Beamforming Solutions   196
        10.3.2   Hybrid Beamforming Solutions   200
        10.3.3   Low-Resolution Receiver Architecture   201
10.4   Channel Acquisition Techniques   201
        10.4.1   Subspace Sampling for Beam Alignment   202
        10.4.2   Compressed Channel Estimation Techniques   205
10.5   Deployment Challenges and Applications   207
        10.5.1   EM Exposure at Millimeter Wave Frequencies   207
        10.5.2   Heterogeneous and Small-Cell Networks   208
Acknowledgments   209
References   209

11   **Interference Mitigation Techniques for Wireless Networks**   214
      Koralia N. Pappi and George K. Karagiannidis
11.1   Introduction   214
11.2   The Interference Management Challenge in the 5G Vision   214
        11.2.1   The 5G Primary Goals and Their Impact on Interference   214
        11.2.2   Enabling Technologies for Improving Network Efficiency
                 and Mitigating Interference   216
11.3   Improving the Cell-Edge User Experience: Coordinated Multipoint   218
        11.3.1   Deployment Scenarios and Network Architecture   218
        11.3.2   CoMP Techniques for the Uplink   220
        11.3.3   CoMP Techniques for the Downlink   221
11.4   Interference Alignment: Exploiting Signal Space Dimensions   223
        11.4.1   The Concept of Linear Interference Alignment   224
        11.4.2   The Example of the X-Channel   225
        11.4.3   The K-User Interference Channel and Cellular Networks:
                 Asymptotic Interference Alignment   226
        11.4.4   Cooperative Interference Networks   227
        11.4.5   Insight from IA into the Capacity Limits of Wireless Networks 227
11.5   Compute-and-Forward Protocol: Cooperation at the Receiver
        Side for the Uplink   228
        11.5.1   Encoding and Decoding of the CoF Protocol   228
        11.5.2   Achievable-Rate Region and Integer Equation Selection   230
        11.5.3   Advantages and Challenges of the CoF Protocol   232
11.6   Conclusion   233
References   233

12   Physical Layer Caching with Limited Backhaul in 5G Systems   236
Vincent Lau, An Liu, and Wei Han
12.1   Introduction   236
12.2   What Is PHY Caching?   238
       12.2.1   Typical Physical Layer Topologies   238
       12.2.2   Basic Components of PHY Caching   240
       12.2.3   Benefits of PHY Caching   241
       12.2.4   Design Challenges and Solutions in PHY Caching   243
12.3   DoF Upper Bound for Cached Wireless Networks   245
       12.3.1   Architecture of Cached Wireless Networks   246
       12.3.2   Generic Cache Model   246
       12.3.3   Cache-Assisted PHY Transmission Model   249
       12.3.4   Upper Bound of Sum DoF for Cached Wireless Networks   251
12.4   MDS-Coded PHY Caching and the Achievable DoF   255
       12.4.1   MDS-Coded PHY Caching with Asynchronous Access   256
       12.4.2   Cache-Assisted MIMO Cooperation in the PHY   256
       12.4.3   MIMO Cooperation Probability of MDS-Coded PHY
                Caching with Asynchronous Access   258
       12.4.4   Achievable DoF for Cached Wireless Networks   260
12.5   Cache Content Placement Algorithm for DoF Maximization   261
12.6   Closed-Form DoF Analysis and Discussion   264
       12.6.1   Content Popularity Model and Definition of DoF Gain   264
       12.6.2   Asymptotic DoF Gain with Respect to the Number of Files   265
       12.6.3   Asymptotic DoF Gain with Respect to the Number of Users   267
12.7   Conclusion and Future Work   267
References   268

13   Cost-Aware Cellular Networks Powered by Smart Grids and Energy Harvesting   271
Jie Xu, Lingjie Duan, and Rui Zhang
13.1   Introduction   271
13.2   Energy Supply and Demand of Cellular Systems   274
13.3   Energy Cooperation   276
       13.3.1   Aggregator-Assisted Energy Trading   276
       13.3.2   Aggregator-Assisted Energy Sharing   277
13.4   Communication Cooperation   278
       13.4.1   Cost-Aware Traffic Offloading   278
       13.4.2   Cost-Aware Spectrum Sharing   279
       13.4.3   Cost-Aware Coordinated Multipoint   280
13.5   Joint Energy and Communication Cooperation   280
       13.5.1   Joint Energy and Spectrum Sharing   281
       13.5.2   Joint Energy Cooperation and CoMP   281
       13.5.3   A Case Study   282
13.6   Extensions and Future Directions   284
13.7   Conclusion   286
References   286

14   **Visible Light Communication in 5G**   289
        Harald Haas and Cheng Chen
        14.1 Introduction   289
        14.2 Differences between Light-Fidelity and Visible Light
              Communication   290
        14.3 LiFi LED Technologies   292
        14.4 LiFi Attocell Networks   293
              14.4.1 Optical OFDM Transmission   294
              14.4.2 Channel Model   296
              14.4.3 Light Source Output Power   302
              14.4.4 Signal Clipping   303
              14.4.5 Noise at Receiver   303
              14.4.6 Multiple Access and Spatial-Reuse Schemes   304
        14.5 Design of Key Parameters for LiFi Attocell Networks   304
              14.5.1 Co-channel Interference Minimization   305
              14.5.2 Maximization of Strength of Desired Signal   306
              14.5.3 Parameter Configurations   307
        14.6 Signal-to-Interference-Plus-Noise Ratio in LiFi Attocell
              Networks   308
              14.6.1 System Model Assumptions   309
              14.6.2 Hexagonal Cell Deployment   309
              14.6.3 PPP Cell Deployment   312
              14.6.4 SINR Statistics Results and Discussion   316
        14.7 Cell Data Rate and Outage Probability   318
        14.8 Performance of Finite Networks and Multipath Effects   322
        14.9 Practical Cell Deployment Scenarios   324
              14.9.1 Square Network   324
              14.9.2 Hard-Core Point Process Network   324
              14.9.3 Performance Comparison   325
        14.10 LiFi Attocell Networks Versus Other Small-Cell Networks   325
        14.11 Summary   328
        References   329

Part III Network Protocols, Algorithms, and Design   333

15   **Massive MIMO Scheduling Protocols**   335
        Giuseppe Caire
        15.1 Introduction   335
        15.2 Network Model and Problem Formulation   337
              15.2.1 Timescales   337
              15.2.2 Request Queues and Network Utility Maximization   338
        15.3 Dynamic Scheduling Policy   342
              15.3.1 The Drift-Plus-Penalty Expression   342
              15.3.2 Pull Congestion Control at the UEs   344

	15.3.3	Greedy Maximization of the Individual Utilities at the UEs	344
	15.3.4	PHY Rate Scheduling at the BSs	344
15.4	Policy Performance		345
15.5	Wireless System Model with Massive MU-MIMO Helpers		347
	15.5.1	PHY Rates of Massive MIMO BSs	347
	15.5.2	Transmission Scheduling with Massive MIMO BSs	350
15.6	Numerical Experiments		351
15.7	Conclusion		355
	References		356

16 Mobile Data Offloading for Heterogeneous Wireless Networks   358
Man Hon Cheung, Haoran Yu, and Jianwei Huang

16.1	Introduction		358
16.2	Current Standardization Efforts		359
	16.2.1	Access Network Discovery and Selection Function (ANDSF)	359
	16.2.2	Hotspot 2.0	360
	16.2.3	Next Generation Hotspot (NGH)	361
	16.2.4	Radio Resource Management	361
	16.2.5	Design Considerations in Data Offloading Algorithms	361
16.3	DAWN: Delay-Aware Wi-Fi Offloading and Network Selection		363
	16.3.1	System Model	363
	16.3.2	Problem Formulation	364
	16.3.3	General DAWN Algorithm	366
	16.3.4	Threshold Policy	367
	16.3.5	Performance Evaluation	369
16.4	Data Offloading Considering Energy–Delay Trade-off		370
	16.4.1	Background on Energy-Aware Data Offloading	371
	16.4.2	System Model	372
	16.4.3	Problem Formulation	374
	16.4.4	Energy-Aware Network Selection and Resource Allocation (ENSRA) Algorithm	375
	16.4.5	Performance Analysis of ENSRA	376
	16.4.6	Performance Evaluation	376
16.5	Open Problems		377
16.6	Conclusion		378
	Acknowledgment		378
	References		378

17 Cellular 5G Access for Massive Internet of Things   380
Germán Corrales Madueño, Nuno Pratas, Čedomir Stefanović, and Petar Popovski

17.1	Introduction to the Internet of Things (IoT)	380
17.2	IoT Traffic Patterns in Network Access	381
17.3	The Features of Cellular Access That Are Suitable for the IoT	386
17.4	Overview of Cellular Access Protocols	387

17.4.1   One-Stage Access   388
17.4.2   Two-Stage Access   389
17.4.3   Periodic Reporting   390
17.4.4   Case Study: LTE Connection Establishment   390
17.5   Improving the Performance of One-Stage Access for 5G Systems   392
17.6   Reliable Two-Stage Access for 5G Systems   393
17.7   Reliable Periodic Reporting Access for 5G Systems   395
17.8   Emerging Technologies for the IoT   396
17.8.1   LTE-M: LTE for Machines   397
17.8.2   Narrowband IoT (NB-IoT): A 3GPP Approach to Low-Cost IoT   397
17.8.3   Extended Coverage GSM (EC-GSM): Evolution of GSM for
         the IoT   398
17.9   Conclusion   398
References   399

18   **Medium Access Control, Resource Management, and Congestion Control
     for M2M Systems**   402
     Shao-Yu Lien and Hsiang Hsu
     18.1   Introduction   402
     18.2   Architectures for M2M Communications   403
            18.2.1   WLAN Architecture for M2M Communications   403
            18.2.2   Cellular Radio Access Network for M2M Communications   404
            18.2.3   Heterogeneous Cloud Radio Access Network for M2M
                     Communications   405
            18.2.4   FogNet Architecture for M2M Communications   407
     18.3   MAC Design for M2M Communications   408
            18.3.1   Grouping-Based M2M MAC in H-CRAN   409
            18.3.2   Access Class Barring Based M2M MAC in FogNet/WLAN 410
            18.3.3   Random-Backoff-Based M2M MAC   411
            18.3.4   Harmonized M2M MAC for Low-Power/Low-Complexity
                     Machines   412
     18.4   Congestion Control and Low-Complexity/Low-Throughput Massive
            M2M Communications   416
            18.4.1   Congestion Control in ACB-Based M2M MAC   416
            18.4.2   Massive MTC and Low-Complexity/Low-Throughput IoT
                     Communications   417
     18.5   Conclusion   419
     References   420

19   **Energy-Harvesting Based D2D Communication in Heterogeneous Networks**   423
     Howard H. Yang, Jemin Lee, and Tony Q. S. Quek
     19.1   Introduction   423
     19.2   Energy Harvesting Heterogeneous Network   425
            19.2.1   Energy Harvesting Region   426

	19.2.2	Energy Harvesting Process and UE Relay Distribution	427
	19.2.3	Transmission Mode Selection and Outage Probability	429
19.3	Numerical Analysis and Discussion		432
19.4	Conclusion		435
References			435

20   LTE-Unlicensed: Overview and Distributed Coexistence Design   438
Yunan Gu, Lin X. Cai, Lingyang Song, and Zhu Han

	20.1	Motivations		438
		20.1.1	Better Network Performance	441
		20.1.2	Enhanced User Experience	441
		20.1.3	Unified LTE Network Architecture	441
		20.1.4	Fair Coexistence with Wi-Fi	441
	20.2	Coexistence Issues in LTE-Unlicensed		441
	20.3	Distributed Resource Allocation Applications of LTE-Unlicensed		444
		20.3.1	Matching Theory Framework	444
		20.3.2	Static Resource Allocation: Student–Project Allocation Matching	446
		20.3.3	Dynamic Resource Allocation: Random Path to Matching Stability	451
	20.4	Conclusion		457
	References			458

21   Scheduling for Millimeter Wave Networks   460
Lin X. Cai, Lin Cai, Xuemin Shen, and Jon W. Mark

	21.1	Introduction		460
	21.2	Background		461
		21.2.1	Multiplexing Technologies for mmWave Networks	461
		21.2.2	Directional Antennas	461
		21.2.3	Network Architecture	462
	21.3	Exclusive Regions		462
		21.3.1	Case 1: Omni-antenna to Omni-antenna	464
		21.3.2	Case 2: Directional Antenna to Omni-antenna	465
		21.3.3	Case 3: Omni-antenna to Directional Antenna	465
		21.3.4	Case 4: Directional Antenna to Directional Antenna	465
	21.4	REX: Randomized Exclusive Region Based Scheduler		466
	21.5	Estimating the Average Number of Concurrent Transmissions Using REX		467
		21.5.1	Case 1: Omni-antenna to Omni-antenna	468
		21.5.2	Case 2: Directional Antenna to Omni-antenna	468
		21.5.3	Case 3: Omni-antenna to Directional Antenna	469
		21.5.4	Case 4: Directional Antenna to Directional Antenna	469
		21.5.5	Edge Effect	469
	21.6	Performance Evaluation		470
		21.6.1	Spatial Multiplexing Gain	470
		21.6.2	Fairness	472

21.7 Further Discussion 473

21.7.1 Fast Fading 473

21.7.2 Shadowing Effect 473

21.7.3 Three-Dimensional Networks 473

21.7.4 Distributed Medium Access 474

21.7.5 Hybrid Medium Access 474

21.7.6 Optimal Scheduling 475

21.8 Conclusion 475

References 475

22 Smart Data Pricing in 5G Systems 478

Carlee Joe-Wong, Liang Zheng, Sangtae Ha, Soumya Sen, Chee Wei Tan, and Mung Chiang

22.1 Introduction 478

22.2 Smart Data Pricing 482

22.2.1 How Should ISPs Charge for Data? 482

22.2.2 Whom Should ISPs Charge for Data? 483

22.2.3 What Should ISPs Charge For? 484

22.3 Trading Mobile Data 485

22.3.1 Related Work on Data Auctions 485

22.3.2 Modeling User and ISP Behavior 486

22.3.3 User and ISP Benefits 487

22.4 Sponsoring Mobile Data 489

22.4.1 Modeling Content Provider Behavior 489

22.4.2 Implications of Sponsored Data 490

22.5 Offloading Mobile Data 491

22.5.1 User Adoption and Example Scenarios 491

22.5.2 Optimal ISP Behavior 494

22.6 Future Directions 494

22.6.1 Capacity Expansion and New Supplementary Networks 495

22.6.2 Two-Year Contracts Versus Usage-Based Pricing 495

22.6.3 Incentivizing Fog Computing 496

22.7 Conclusion 496

References 497

Index 501

List of Contributors

Fumiyuki Adachi
Tohoku University, Japan

Lin Cai
University of Victoria, Canada

Lin X. Cai
Illinois Institute of Technology, USA

Giuseppe Caire
Technical University of Berlin, Germany

Miguel R. Castellanos
Purdue University, USA

Cheng Chen
The University of Edinburgh, United Kingdom

Yami Chen
China Mobile Research Institute

Man Hon Cheung
The Chinese University of Hong Kong, Hong Kong

Mung Chiang
Princeton University, USA

Martin Danneberg
Technische Universität Dresden, Germany

Zhiguo Ding
Lancaster University, United Kingdom

Lingjie Duan
Singapore University of Technology and Design, Singapore

Gerhard Fettweis
Technische Universität Dresden, Germany

Yunan Gu
University of Houston, USA

Sangtae Ha
University of Colorado at Boulder, USA

Harald Haas
The University of Edinburgh, United Kingdom

Wei Han
Hong Kong University of Science and Technology, Hong Kong

Zhu Han
University of Houston, USA

Hsiang Hsu
National Taiwan University, Taiwan

Jianwei Huang
The Chinese University of Hong Kong, Hong Kong

Jinri Huang
China Mobile Research Institute, China

Xueyan Huang
China Mobile Research Institute, China

Chih-Lin I
China Mobile Research Institute, China

Shi Jin
Southeast University, China

Carlee Joe-Wong
Carnegie Mellon University, USA

George K. Karagiannidis
Aristotle University of Thessaloniki, Greece

Vincent Lau
Hong Kong University of Science and Technology, Hong Kong

Jemin Lee
Daegu Gyeongbuk Institute of Science and Technology, Korea

Ben Liang
The University of Toronto, Canada

Wei Liang
Lancaster University, United Kingdom

Shao-Yu Lien
National Formosa University, Taiwan

An Liu
Hong Kong University of Science and Technology, Hong Kong

Liang Liu
University of Toronto, Canada

David J. Love
Purdue University, USA

Germán Corrales Madueño
Aalborg University, Denmark

Jon W. Mark
University of Waterloo, Canada

Michail Matthaiou
Queen's University Belfast, United Kingdom

Maximilian Matthé
Technische Universität Dresden, Germany

Abolfazl Mehbodniya
Tohoku University, Japan

Derrick Wing Kwan Ng
The University of New South Wales, Australia

Koralia N. Pappi
Aristotle University of Thessaloniki, Greece

H. Vincent Poor
Princeton University, USA

Petar Popovski
Aalborg University, Denmark

Nuno Pratas
Aalborg University, Denmark

Tony Q.S. Quek
Singapore University of Technology and Design, Singapore

Rongwei Ren
China Mobile Research Institute, China

Taneli Riihonen
Aalto University, Finland

Robert Schober
Friedrich-Alexander-University Erlangen-Nürnberg, Germany

Soumya Sen
University of Minnesota, USA

Xuemin Shen
University of Waterloo, Canada

Osvaldo Simeone
New Jersey Institute of Technology, USA

Jiho Song
Purdue University, USA

Lingyang Song
Peking University, China

Čedomir Stefanović
Aalborg University, Denmark

Chee Wei Tan
The City University of Hong Kong, Hong Kong

Li-Chun Wang
National Chiao-Tung University, Taiwan

Risto Wichman
Aalto University, Finland

Vincent W. S. Wong
The University of British Columbia, Canada

Jie Xu
Guangdong University of Technology, China, and Singapore University of Technology
and Design, Singapore

Howard H. Yang
Singapore University of Technology and Design, Singapore

Haoran Yu
The Chinese University of Hong Kong, Hong Kong

Wei Yu
University of Toronto, Canada

Dan Zhang
Technische Universität Dresden, Germany

Rui Zhang
National University of Singapore, and Institute for Infocomm Research, A*STAR,
Singapore

Liang Zheng
Princeton University, USA

Preface

Mobile devices (e.g., smartphones and tablets) have become a commodity in our daily lives. While these devices already support many different types of applications and services, there will be a continual increase in demand for mobile data traffic due to web applications, real-time and streaming video traffic, and applications related to the Internet of Things (IoT). The future fifth generation (5G) wireless cellular systems aim not only to provide a higher aggregate throughput, but also to support applications which have stringent quality of service (QoS) requirements, such as seamless mobility, ultra-low latency (e.g., Tactile Internet), and high reliability (e.g., vehicular communications). Further improvements in spectrum efficiency, energy efficiency, and cost per bit are also important. In order to meet these demands, fundamental changes to the network architecture and all layers of the protocol stack compared with fourth generation (4G) wireless systems are needed.

This book aims to provide a comprehensive treatment of the ongoing research into and state-of-the-art techniques for addressing the challenges arising from the design of 5G wireless systems. Written by leading experts on the subject, this book includes 22 chapters, which cover various aspects of 5G systems, including network architecture design, physical layer techniques, algorithms, and network protocol design. Chapter 1 serves as an introductory chapter and provides an overview of the different key technologies related to 5G systems. Each of the other chapters tackles one specific challenge for system design. The chapters can be read independently.

This book will be of interest to a readership from the communications, signal processing, and networking communities. The primary audience for this book is researchers and engineers who are interested in studying advanced communication and networking techniques, as well as state-of-the-art research on 5G systems. This book will serve as a resource for self-study and as a reference book for researchers and engineers involved in the design of wireless communication systems. It is also suitable for graduate students who are interested in 5G systems and the related communication and networking issues. It may serve as a reference book for graduate-level courses for students in electrical engineering, communication engineering, and networking.

We would like to thank all the authors for their outstanding contributions and their timeliness in completing their respective chapters. In addition, we would like to thank Elizabeth Horne and Heather Brolly from Cambridge University Press for their valuable advice throughout the production of this book. Last but not least, we would like to thank the Natural Sciences and Engineering Research Council of Canada (NSERC) for its financial support.

Vincent W. S. Wong
Robert Schober
Derrick Wing Kwan Ng
Li-Chun Wang

1 Overview of New Technologies for 5G Systems

Vincent W. S. Wong, Robert Schober, Derrick Wing Kwan Ng, and Li-Chun Wang

1.1 Introduction

In recent years, wireless service providers in different countries have deployed both the Third Generation Partnership Project (3GPP) Long Term Evolution (LTE) and LTE-Advanced systems. Despite the unprecedented data rates and quality of service (QoS) provided by these new networks, user demand is beginning to exceed their capabilities. For example, the proliferation of smartphones and tablets has caused a significant and sustained increase in mobile data traffic. In fact, in 2015 alone, global mobile data traffic grew by 74% from 2.1 to 3.7 exabytes [1]. Furthermore, the existing networks are not well suited for the exceedingly large number of devices and appliances that are expected to be connected wirelessly to the Internet in future Internet of Things (IoT) applications and machine-to-machine (M2M) communications. Moreover, to make the growth of networks and the number of connected devices economically and ecologically sustainable, energy efficiency has to be substantially improved. Also, emerging new applications such as remote surgery in healthcare, autonomous driving, and wireless control of industrial robots require ultra-low latencies in the sub-millisecond range and ultra-high reliability, giving rise to the notion of the Tactile Internet. In order to support the exponential growth of existing mobile traffic and the emergence of new wireless applications and services, researchers and standardization bodies worldwide have set out to develop a fifth generation (5G) of wireless networks [2–6]. Some of the stringent requirements for this next generation of wireless networks are listed in Table 1.1 [7].

To meet these challenging requirements, a mere evolution of the current networks is not sufficient. Instead, a true revolution of technologies in both the radio access network and the mobile core network is needed (Figure 1.1). In the radio access network, fundamentally new physical layer technologies such as massive multiple-input multiple-output (MIMO), non-orthogonal multiple access (NOMA), full-duplex (FD) communication, millimeter wave (mmWave) communication, device-to-device (D2D) communication, and visible light communication (VLC) will be deployed. Furthermore, leveraging cloud computing, the cloud radio access network (C-RAN) has emerged as a promising and cost-efficient mobile network architecture to enhance the spectrum and

Table 1.1 Requirements for 5G wireless communication systems [7].

Figure of merit	5G requirement	Comparison with 4G
Peak data rate	10 Gb/s	100 times higher
Guaranteed data rate	50 Mb/s	–
Mobile data volume	10 Tb/s/km^2	1000 times higher
End-to-end latency	Less than 1 ms	25 times lower
Number of devices	1 M/km^2	1000 times higher
Total number of human-oriented terminals	$\geq$ 20 billion	–
Total number of IoT terminals	$\geq$ 1 trillion	–
Reliability	99.999%	99.99%
Energy consumption	–	90% less
Peak mobility support	$\geq$ 500 km/h	–
Outdoor terminal location accuracy	$\leq$ 1 m	–

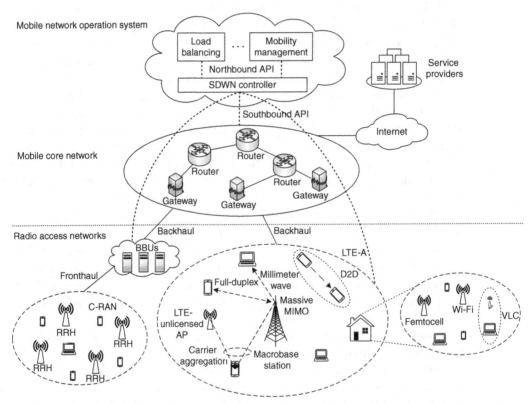

Figure 1.1 Illustration of the 5G network architecture. The radio access network includes various technologies such as C-RAN, massive MIMO, full duplexing, mmWave, femtocells, Wi-Fi, D2D, and VLC. The mobile core network can be controlled by a software-defined wireless network (SDWN) controller.

energy efficiency of 5G networks. In addition, different access technologies, including LTE and Wireless Fidelity (Wi-Fi), may be integrated to guarantee seamless coverage, and to support high data-rate transmission and data offloading.

In this chapter, we provide an overview of some of the exciting new technologies which are expected to be incorporated into 5G systems. These techniques will then be covered in detail in the subsequent chapters.

1.2   Cloud Radio Access Networks

It is reported [8] that peak traffic demand can be 10 times higher than off-peak traffic demand. However, as network resources for base stations are always provisioned for peak traffic demand, many base stations are lightly loaded or even in idle mode during off-peak hours, which leads to low utilization of the deployed cell sites. On the other hand, the energy efficiency of the lightly loaded base stations may also be low since the circuit power consumption constitutes a significant part of the total power consumption of a base station. C-RAN has recently been identified as a leading candidate for the 5G network architecture. In C-RAN, the baseband signal processing and the radio functionalities are decoupled. In general, a C-RAN consists of a baseband unit (BBU) pool placed in a cloud-based data center, and a large number of low-cost remote radio heads (RRHs) each deployed in a small cell. The BBUs and RRHs are connected through fronthaul links. The BBUs perform centralized signal processing and interference management. The RRHs retain only the radio functionality and communicate with the users over radio channels.

The C-RAN architecture has several advantages. First, C-RAN can adapt to spatial and temporal traffic fluctuations to provide on-demand services by exploiting the statistical multiplexing gain [9]. To this end, the number of BBUs required to support peak traffic demand can be reduced and the idle RRHs can be switched off to reduce power consumption, which leads to lower network capital expenditure (CAPEX) and operating expenditure (OPEX), respectively. Second, C-RAN facilitates the implementation of cooperative transmission/reception strategies, for example, enhanced intercell interference coordination (eICIC) and coordinated multipoint (CoMP) transmission [10]. With these cooperative strategies, spectrum efficiency can be significantly boosted via effective interference management among multiple RRHs. Third, C-RAN simplifies upgrading and maintenance of the network. Specifically, the virtualization of baseband signal processing on general-purpose processors or cloud servers simplifies network upgrades.

Despite the aforementioned advantages, C-RAN also poses new research challenges. First, in practice the fronthaul links have finite capacity, which can significantly degrade the performance gain achieved by C-RAN [11]. Second, to facilitate the centralized signal-processing and cooperative transmission strategies, massive amounts of accurate channel state information (CSI) are required at the BBUs [12]. In addition, user mobility leads to time-varying channels, which increases the CSI update frequency. Furthermore, owing to limited training resources and the transmission delay introduced by the fronthaul links, the CSI received by the BBUs may not be accurate, which may degrade the ability to perform effective interference management. The C-RAN architecture will be discussed in detail in Chapter 2. An information-theoretic approach to determine the achievable rates of C-RAN with fronthaul capacity constraints will be presented in Chapter 3.

1.3   Cloud Computing and Fog Computing

Ubiquitous and pervasive computing services are crucial to the processing and storing of the significant amounts of data generated in IoT systems. The limited processing capacity of IoT objects may not always provide the computational power required for IoT applications. In this case, cloud computing can provide the necessary storage and processing capabilities. The cloud servers can collect data from different IoT devices, store the data, and run software applications to process and analyze the data. Cloud platforms such as ThingWorx, OpenIoT, Google Cloud, and the Amazon Web Services (AWS) IoT platform provide computing services for IoT application developers and service providers. Finally, the IoT service providers offer a set of services to IoT end users based on the information collected from IoT objects.

For delay-sensitive IoT applications with stringent latency requirements, conventional cloud services may not be appropriate. Fog computing [13–15], which is also known as *mobile edge computing*, extends cloud computing to the edge of the network. Here, IoT devices with processing and storage capability, called fog nodes, are deployed in the system and run applications on behalf of other devices. Since fog nodes are located in close proximity, the delay performance of computing services will be improved. In addition, fog aggregation nodes, which are network edge devices (e.g., routers and smart gateways) and have computing and storage capabilities, can provide further computing services to tasks that have more relaxed latency requirements. Figure 1.2 illustrates the coexistence of fog computing and cloud computing in IoT systems. In fact, fog computing is not a substitute for cloud computing. These two computing paradigms complement each other and together can provide the computational services required for IoT and improve the scalability of IoT applications. Mobile edge computing and fog computing will be discussed in detail in Chapter 4.

1.4   Non-orthogonal Multiple Access

Wireless communication systems have to provide services to many users in the same area (e.g., the same cell) concurrently. To coordinate and guarantee services for multiple users, some form of multiple access technique is required. The first four generations of cellular systems have relied on orthogonal multiple access (OMA). In particular, the first generation (1G) systems employed frequency division multiple access (FDMA). The second generation (2G) systems, which implemented the Global System for Mobile Communications (GSM), primarily used time division multiple

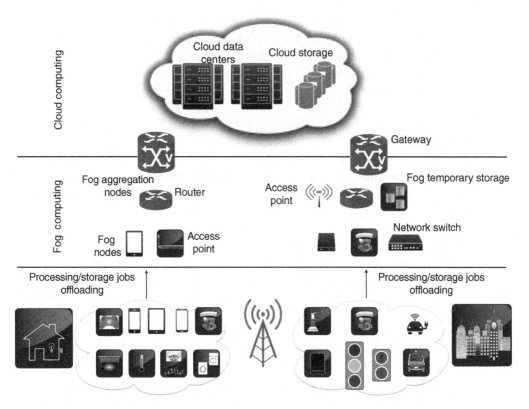

Figure 1.2 Fog computing and cloud computing can together provide computational resources for IoT objects and enable different IoT applications such as smart homes and smart cities. Fog nodes are located in close proximity to IoT objects and can respond to an emerging processing job quickly. Fog aggregation nodes, typically with higher processing power, can also support the computational requirements of IoT objects.

access (TDMA). The third generation (3G) systems, which implemented the Universal Mobile Telecommunications System (UMTS), relied on code division multiple access (CDMA). The fourth generation (4G) systems, which implemented LTE, adopted orthogonal frequency division multiple access (OFDMA). The main advantage of OMA is that under ideal conditions interuser interference is avoided, which significantly simplifies system and protocol design, including detection, channel estimation, and resource allocation. However, on the negative side, the number of users that can be supported in an OMA system is limited by the number of available orthogonal dimensions and, in practice, the orthogonality is often lost owing to effects such as frequency selectivity of the channel (in TDMA and CDMA) and phase noise and frequency offsets (in FDMA and OFDMA). Furthermore, from an information theory point of view, OMA is not optimal [16].

The shortcomings of OMA can be overcome by non-orthogonal multiple access (NOMA) techniques [17–21]. In NOMA, multiple users are scheduled on the same resource, i.e., in the same temporal, spectral, and spatial dimension. Thereby, a certain amount of interuser interference introduced by non-orthogonal transmission is tolerated and removed at the receiver via successive interference cancellation (SIC). Because of its more efficient use of resources, both academia and industry see NOMA as one of the key enabling technologies for 5G systems [17, 19]. Although there are several different forms of NOMA, the two schemes that have received the most attention so far are power domain NOMA [18] and sparse code multiple access (SCMA) [17, 22].

In power domain NOMA, multiple users are multiplexed on the same time and frequency resource and power differences are exploited at the receiver for separation of the users' signals via SIC [20]. The advantages of power domain NOMA compared with OMA in terms of achievable data rate, coverage, and reliability were demonstrated in [21]. In addition, the combination of power domain NOMA with other 5G candidate technologies such as (massive) MIMO [23, 24] and FD transmission has been investigated. Although primarily a candidate for 5G, power domain NOMA has also been considered for standardization in 3GPP for downlink LTE transmission [19].

SCMA enables non-orthogonal access by overloading the multiuser system. Thereby, at the transmitter, the coded bits of a user are mapped to a complex codeword and the codewords of different users are overlaid using sparse spreading. At the receiver, joint multiuser detection and channel decoding using a message-passing algorithm is employed, where the sparsity of the spreading code limits the computational complexity [22]. Because of the overloading, SCMA can accommodate more users than OFDMA and achieve a higher throughput and connectivity.

NOMA is currently an active area of research and many challenges have not been fully resolved yet; examples include the fundamental information-theoretical limits of NOMA, channel coding and modulation design for NOMA, the integration of NOMA and other 5G techniques, including (massive) MIMO and full-duplex, security provisioning for NOMA, resource allocation for NOMA, and hardware implementation of NOMA. A thorough introduction to NOMA is provided in Chapter 6.

1.5   Flexible Physical Layer Design

As mentioned in Section 1.1, 5G networks aim to support not only voice and mobile Internet applications but also applications such as the IoT, M2M, the Tactile Internet, and vehicular communications. These applications have different requirements in terms of delay, reliability, and power consumption. In the last decade, orthogonal frequency division multiplexing (OFDM) and OFDMA have become the dominant physical layer technologies for providing high-data-rate services for major wideband wireless communication systems such as IEEE 802.11-based wireless local area networks (WLANs), Worldwide Interoperability for Microwave Access (WiMAX), and LTE-A. However, OFDM/OFDMA systems suffer from the following drawbacks: (i) OFDM-based systems have a high peak-to-average power ratio (PAPR), which requires the use of power-inefficient linear power amplifiers at the transmitter; (ii) the rectangular pulse shape of the OFDM symbol leads to large spectral side lobes and high out-of-band radiation; and (iii) subcarrier orthogonality is sensitive to carrier frequency offsets and phase noise. Therefore, the existing OFDM technology may not be well suited for the transmission of the data of some 5G applications. Hence, several alternative non-orthogonal waveforms have been proposed and will be considered for the 5G physical layer. Promising candidates are filterbank multicarrier (FBMC), universal filtered multicarrier (UFMC), and generalized frequency division multiplexing (GFDM) [25]. These non-orthogonal signaling schemes attempt to overcome the limitations of OFDM/OFDMA by introducing new features into the signal and frame structure. For instance, GFDM is based on the modulation of independent blocks, and each block consists of a number of subcarriers and subsymbols. In GFDM, cyclic prefixes (CPs) and circular filtering are employed. In particular, GFDM exploits the "tail-biting" technique through circular filtering to decrease the length of the signal pulse tails. The circulant signal structure of GFDM also enables the use of one CP for an entire data block containing multiple GFDM symbols, which improves the spectral efficiency compared with conventional OFDM. In fact, GFDM is a flexible physical layer scheme since it covers both CP-OFDM and single-carrier frequency domain equalization (SC-DFE) as special cases. Besides, GFDM allows the time and frequency spacing of each data symbol to be adapted according to the channel properties and the type of application. The details of GFDM, including the receiver design and hardware implementation, will be presented in Chapter 7.

1.6   Massive MIMO

Multiple-antenna technology is a key element of current and future wireless communication systems. Traditionally, the number of antennas envisioned for MIMO communication systems has been limited to a comparatively small number, say 20 or less. However, recently it has been shown that multiuser MIMO communication systems exhibit several favorable properties if the number of antennas at the base station is increased to hundreds or even thousands [26], giving rise to so-called large-scale or massive MIMO systems. As a result, a large amount of theoretical and experimental research has been dedicated to massive MIMO communication systems since 2010.

It is well known that MIMO communication systems achieve substantial gains in spectral, power, and energy efficiency compared with conventional single-input single-output (SISO) systems. In fact, it has been shown that under ideal conditions the capacity of a point-to-point MIMO system with N_T transmit antennas and N_R receive antennas scales linearly with $\min\{N_T, N_R\}$, which is referred to as the multiplexing gain in the literature [27, 28]. However, point-to-point MIMO systems have several disadvantages in practice. First, the number of antennas that a mobile terminal (e.g., a smartphone) can accommodate is limited owing to size, power consumption, and cost constraints, which negatively impacts the achievable multiplexing gain. Second, the multiplexing gain may disappear altogether in the case of strong interference (e.g., at the cell edges), unfavorable channel conditions (e.g., insufficient scattering), and narrow antenna spacing mandated by the size constraints of mobile terminals. The disadvantages of point-to-point MIMO systems can be overcome by multiuser MIMO systems [29–31]. In multiuser MIMO systems, a central node with multiple antennas (e.g., a base station) serves a number of (mobile) users with a small number of antennas. Thus, the signal-processing complexity at the mobile terminals is low, especially in the case of single-antenna terminals. In addition, since the users are spatially distributed over an entire cell, the angular separation of the terminals typically exceeds the Rayleigh resolution of the array and the channels of different users can be assumed independent. However, the multiple users in the system introduce interuser interference, which has to be mitigated by appropriate processing at the transmitter and receiver for downlink (i.e., base station to users) and uplink (i.e., users to base station) transmission, respectively. The uplink channel may be classified as a classical multiple access channel, for which many suitable linear and nonlinear receiver processing techniques are known from the rich literature on CDMA systems [32]. Thereby, while being computationally more complex, nonlinear receiver structures achieve a higher performance than linear ones. The downlink channel is a broadcast channel and suitable precoding techniques at the transmitter are needed to achieve high performance. Dirty paper coding was shown to be the optimal capacity-achieving precoding technique for a Gaussian MIMO broadcast channel [29, 30]. However, it entails a high computational complexity in practical implementation. Thus, linear precoding techniques such as zero-forcing (ZF) precoding, minimum mean-square error (MMSE) precoding, and regularized ZF precoding have attracted considerable attention as a good compromise between performance and complexity [33, 34].

For both uplink and downlink transmission, the availability of CSI at the base station is crucial for exploiting the full potential of multiuser MIMO systems. This is not critical for the uplink, where the users can simply send pilot symbols along with their data packets. In the downlink, channel estimation is more challenging and the best approach depends on the type of duplexing used. For frequency division duplex (FDD) systems, where different carrier frequencies are used for uplink and downlink transmission, the uplink and downlink channels are mutually statistically independent. Thus, each base station antenna has to first transmit pilots, which enable the users to estimate their respective downlink channels. Subsequently, each user has to feed back its channel estimate to the base station. Thus, assuming K users, the required number of feedback symbols grows linearly with $N_T K$. On the other hand, for time division duplex (TDD) systems, the same carrier frequency is used for uplink and downlink. Thus, assuming a sufficiently large coherence time, the uplink and downlink channels are reciprocal and the base station can obtain the downlink channel by estimating the uplink channel based on pilots transmitted by the users. In this case, the number of pilots required grows linearly with the number of users K but is independent of the number of base station antennas.

Unlike conventional multiuser MIMO systems, which employ a comparatively small number of base station antennas (e.g., fewer than 20), massive MIMO systems are expected to employ hundreds or even thousands of base station antennas. Although such a tremendous increase in the number of antennas introduces new challenges in transceiver design and implementation, it has some interesting advantages for signal processing and communication. For example, if the number of base station antennas is much larger than the number of users in the system, simple matched-filter (MF) precoding (downlink) and MF detection (uplink) at the base station lead to close-to-optimal performance, facilitating low-complexity signal processing at both the base station and the user terminals. Nevertheless, as will be shown in Chapter 8, as the number of users increases, significant performance gains can be achieved with ZF and MMSE precoding and detection schemes. Furthermore, random impairments such as small-scale fading and noise are averaged out as the number of base station antennas grows large. To keep the signaling overhead for CSI acquisition in massive MIMO systems manageable, TDD operation is preferred, since for FDD systems the amount of CSI feedback grows with the number of base station antennas. However, a major impairment in massive MIMO systems is so-called *pilot contamination*. Pilot contamination is caused by the reuse of the same (or linearly dependent) pilot sequences in different cells. This reuse is unavoidable as, for a given pilot sequence length, the number of linearly independent pilot sequences is limited. Recently, several efficient techniques have been proposed to overcome pilot contamination [35, 36].

Massive MIMO also has the potential to significantly improve energy efficiency. It has been shown [37] that if N_T grows large and all other system parameters are assumed constant, the transmit power per user in multiuser massive MIMO systems can be reduced proportionally to $1/N_T$ and $1/\sqrt{N_T}$ for perfect and imperfect CSI knowledge respectively, at the base station, without affecting throughput and reliability. Hence, massive MIMO systems offer a simple path to more energy-efficient and "greener" communication networks. Furthermore, a major concern for future wireless communication systems is security and privacy. Massive MIMO is well suited to addressing these issues. In fact, because of the large number of spatial degrees of freedom, massive MIMO can be exploited to protect cellular systems against passive [38] and active [39] eavesdropping.

Because of its favorable properties, massive MIMO is expected to be one of the core technologies of 5G systems [3]. Nevertheless, massive MIMO still has many challenging open research problems. For example, because of the large scale of massive MIMO systems, the use of cheap hardware components is desirable. However, this in turn gives rise to hardware impairments, such as phase noise, in-phase/quadrature phase imbalance, and amplifier nonlinearities, which have to be properly dealt with to avoid performance degradation [40]. Furthermore, the channel-hardening effect induced by the large number of base station antennas necessitates the design of new resource allocation and user association algorithms. Massive MIMO systems and scheduling protocols will be discussed in detail in Chapters 8 and 15, respectively.

1.7   Full-Duplex Communications

Although advanced technologies such as C-RAN, NOMA, and massive MIMO are able to alleviate system resource shortages, the spectral resource is still underutilized. In particular, traditional wireless communication devices operate in the half-duplex (HD) mode, where downlink and uplink communications are separated orthogonally in either frequency or time, which results in a significant loss in spectral efficiency. Although researchers have proposed various techniques for minimizing/recovering the spectral-efficiency loss inherent in HD communication, such as joint dynamic uplink and downlink resource allocation [41] and two-way HD relaying [42], these schemes do not solve the problem fundamentally, since the associated protocols still operate in the HD communication mode.

Recently, full-duplex (FD) wireless communication has emerged as a candidate technique for 5G networks and has received significant attention from both industry [43, 44] and academia [45–47]. Compared with existing communication networks adopting HD transmission, FD systems simultaneously transmit and receive data signals in the same frequency band, which has the following advantages [45]. First, FD systems can provide a better utilization of time and frequency resources such that it is possible for FD systems to double the link capacity compared with existing HD systems. Second, in practice, in addition to data signals, feedback signals such as control information or CSI can be also transmitted concurrently to facilitate data communication. Thus, FD systems can reduce the feedback delay by receiving feedback signals during data transmission. Third, FD systems can improve communication security. In fact, an HD base station is unable to guarantee physical layer security in the uplink unless external helpers perform cooperative jamming to interfere with potential eavesdroppers. In contrast, an FD base station can guarantee secure uplink transmission by transmitting jamming signals in the downlink while receiving the desired uplink information signals. Last but not least, a hybrid HD and FD protocol, which retains the option to utilize one frequency band for FD communication or two orthogonal frequency bands for two parallel HD communications, can be adopted to increase the flexibility in spectrum usage.

Despite the potential benefits, the performance of FD communication systems is limited by self-interference (SI), which is caused by signal leakage from the downlink transmission to the uplink signal reception (Figure 1.3). In particular, the ratio between the SI power and the desired incoming signal power can easily exceed 100 dB [48]. The huge difference between the signal powers causes saturation in the analog-to-digital converter (ADC) at the receiver front end of FD devices which severely jeopardizes signal reception. Hence, FD communication has been considered impractical for the past 60 years. Fortunately, recent research has shown that FD communication is feasible by using spatial SI suppression, digital/radio frequency (RF) interference cancellation techniques, and transmit/receive antenna isolation [47]. Several prototypes of FD transceivers using various SI cancellation techniques have been built to demonstrate the feasibility of FD communication and the expected performance gains compared with HD communication in different physical environments [49–51].

In fact, FD technology introduces new research challenges for wireless communication engineers in both resource allocation and communication protocol design. In the following, we briefly discuss some open issues in FD communication systems. In general, strong SI is an obstacle in realizing FD communications since it increases with the transmit power of the FD devices. As a result, multiple-antenna technology has been proposed to overcome SI. In particular, by utilizing the extra spatial degrees of

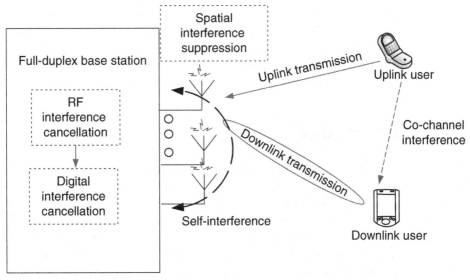

Figure 1.3 An FD communication system consisting of an FD base station serving an HD downlink user and an HD uplink user simultaneously in the same frequency band. Self-interference and co-channel interference impair uplink and downlink communication, respectively.

freedom offered by multiple antennas, the information signal can be accurately steered in the desired direction and thus the SI can be effectively suppressed. Nevertheless, accurate CSI is required to perform resource allocation, which may not be available in practical systems with time-varying wireless communication channels. Therefore, a robust FD communication system design taking into account the imperfections of the CSI is critical. On the other hand, in a cellular network with multiple FD base stations sharing a given frequency band, the transmitted signal of one FD base station may cause strong interference at other base stations, even if the SI at each base station can be controlled to a manageable level. In fact, inter-base-station interference becomes a severe problem when the base stations are close to each other, for example, in a small-cell environment. As a result, FD base station coordination may be a viable solution to reducing inter-base-station interference. Specifically, a set of FD base stations with limited cross-interference can be selected to serve a set of users to reduce the overall interference. However, interference coordination requires a significant amount of information exchange between the FD base stations, which may consume a large portion of the limited system resources. Finally, the traditional communication protocols for the physical layer, network layer, and medium access control (MAC) layer were designed for HD communication systems. For instance, many existing physical layer techniques and MAC layer protocols require a knowledge of the state of the wireless communication terminals, such as the modulation format, code rate, QoS requirements, and acknowledgment (ACK) status. For the current HD communication protocol, multiple orthogonal time slots may be needed to convey the required information overhead, which introduces unnecessary delays into the system and causes a performance bottleneck for delay-sensitive communication services. In contrast, with FD, these terminals could potentially exchange state information while simultaneously transmitting data. As a result, in order to fully exploit the potential performance gains introduced by FD technology into 5G networks, new communication protocols are needed. This issue will be discussed in detail in Chapter 9.

1.8   Millimeter Wave

A key goal of 5G systems is to increase the data rate by a factor of 1,000 compared with the previous generation [2]. A seemingly straightforward approach to increasing the data rate is to increase the bandwidth used for transmission. However, the gains in bandwidth that can be achieved by reframing or more efficient usage of the traditionally employed sub-6 GHz frequency bands are very limited. Hence, it is natural to turn to frequency bands that have seen limited or no usage so far. In this respect, the so-called mmWave frequency bands ranging from 30 GHz to 300 GHz have received considerable attention recently [2, 3].

Originally, mmWave frequencies were viewed mainly as a solution for extremely short transmission distances in wireless personal area networks (WPANs) and WLANs, with the new WiGig standard in the 60 GHz band being a prominent example [52], and for fixed wireless in the E-band at 71–76 GHz, 81–86 GHz, and 92–95 GHz [53]. The perception was that for applications which require longer transmission distances, the propagation loss in the mmWave bands would be insurmountable. As such, mmWave frequencies were ruled out for use in cellular systems.

The general attitude toward using mmWave frequencies for communications has dramatically changed in the past few years [54]. This has two main reasons. On the one hand, considering the trend toward decreasing cell sizes and the emergence of short-range techniques such as D2D communication, the required transmission distances have been considerably decreased. On the other hand, extensive channel measurement campaigns have shown that the path loss is not as severe as previously thought. In fact, the use of large antenna arrays can largely compensate for the less favorable propagation conditions compared with lower frequencies [55]. One remaining obstacle is the heavy blockage of mmWave signals, preventing the penetration of building walls and making a line of sight (LOS) mandatory for reliable communication.

Although mmWave is a rapidly maturing technology, many research challenges remain. The high sampling rates associated with transmission at such high frequencies/large bandwidths make analog-to-digital and digital-to-analog conversion challenging and power-inefficient. As such, instead of the fully digital signal processing that has been traditionally used at lower frequencies, the use of analog phase shifters and hybrid analog–digital beamforming techniques is required [56]. Owing to heavy blockage, new protocol and user association designs are needed [57]. Also, the narrow beams and imperfect hardware typical of mmWave systems may lead to noise-limited rather than interference-limited systems, which has to be taken into account in system and protocol design. To overcome the blockage problem, hybrid designs employing both sub-6 GHz and mmWave frequencies for indoor–outdoor communication may be necessary. Another important topic which has many challenges and opportunities is the combination of mmWave technology with massive MIMO [58, 59]. Despite these challenges, mmWave is seen today as a key enabler for both access and wireless backhaul in 5G communication systems [60, 61]. Physical layer and protocol design for mmWave systems will be addressed in Chapters 10 and 21, respectively.

1.9   Mobile Data Offloading, LTE-Unlicensed, and Smart Data Pricing

Mobile data offloading can help mobile network operators cope with the exponential growth of cellular traffic by delivering mobile traffic through complementary network technologies such as Wi-Fi access points and small cells (e.g., femtocells). A recent report from Cisco [1] shows that in 2015, for the first time, mobile data traffic offloaded onto Wi-Fi and femtocell networks exceeded cellular traffic. One of the benefits of mobile data offloading is that it can alleviate congestion in cellular networks. It is expected that Wi-Fi networks will be an important complementary technology in 5G systems. The early deployment of mobile data offloading has focused on user-initiated offloading, where the offloading decision is made by the users. The cellular networks and Wi-Fi networks are owned and administered by different operators. The user-initiated approach makes the cellular network operators unable to monitor and control the activities of customers. In most of the current deployments, the mobile network operators deploy their own Wi-Fi networks and provide operator-initiated offloading, where the offloading decision is made by the network operator. This facilitates seamless authentication and roaming, automatic network selection, and integrated policy and charging functions. Chapter 16 summarizes several related standards on mobile data offloading, including Access Network Discovery and Selection Function (ANDSF), Hotspot 2.0, and Next Generation Hotspot (NGH). It then presents two data offloading algorithms which address the issues of cost, delay, and energy.

Besides mobile data offloading, an alternative way to improve the aggregate network throughput of LTE-Advanced systems is to utilize both licensed and unlicensed frequency bands under a unified LTE network infrastructure [62]. This is known as LTE-Unlicensed, or LTE-U. The unlicensed frequency band includes the 2.4 GHz Industry, Scientific, and Medical (ISM) band and the 5 GHz Unlicensed National Information Infrastructure (U-NII) band [63]. From the perspective of network operators, the advantages of LTE-Unlicensed include unified authentication procedures, integrated network management, lower operational costs, and efficient use of network resources. From the perspective of the users, the benefits of LTE-Unlicensed include higher throughput, reliable and secure communications, and seamless mobility and network coverage. However, there are several technical challenges in deploying LTE-Unlicensed. One of them is harmonious coexistence between LTE-Unlicensed and the incumbent Wi-Fi systems. Since Wi-Fi systems use carrier-sense multiple access with collision avoidance (CSMA-CA) as the MAC protocol, signals from

LTE-Unlicensed can without proper design become continuous interference to Wi-Fi systems. To address the coexistence issue, an LTE-Unlicensed device needs to sense whether a channel is occupied by using techniques such as clear channel assessment and listen-before-talk. These techniques also prevent an LTE-Unlicensed device occupying a channel for a long time. Thus, the device can share the unlicensed bandwidth with Wi-Fi users in a fair and friendly manner. Furthermore, in LTE-Unlicensed, during a communication session, a licensed carrier, called the primary component carrier, and several unlicensed carriers, called secondary component carriers, are assigned to a user at one time. Control signals are always transmitted via the primary component carrier. This ensures ubiquitous coverage. The allocation of the secondary component carriers can be dynamically adjusted based on the users' traffic demand. Thus, the aggregate network throughput can be improved via either carrier aggregation or link aggregation. Licensed Assisted Access (LAA) and LTE-WLAN Aggregation (LWA) are under standardization within 3GPP [64]. LTE-Unlicensed, its coexistence issues, and the corresponding resource allocation algorithms will be discussed in detail in Chapter 20.

Most of mobile service providers today implement usage-based pricing to charge for the data usage of their mobile customers. Chapter 22 presents different types of smart data-pricing mechanisms. These mechanisms focus on the prices that the users pay for consuming data on wireless cellular networks or other types of networks. By varying the prices, the mobile service providers can influence users' demands for different types of mobile data. This helps service providers offload data traffic during congested periods. There are several benefits to the mobile service providers from implementing smart data pricing [65]. First, traded data plans enable mobile service providers to attract more users, especially high-volume users. Second, mobile service providers can obtain additional revenue by allowing both content and application providers to sponsor the mobile data of their customers. Finally, smart data pricing can reduce congestion in LTE networks by offloading the traffic to other supplementary networks. Details of smart data-pricing mechanisms will be presented in Chapter 22.

1.10   IoT, M2M, and D2D

The existing wireless networks and LTE systems are designed to support human-to-human (H2H) communications, with the aim of supporting voice and multimedia transmission with low access delay and high throughput. There are two other major trends which are reshaping the wireless industry: M2M communications and the IoT [66, 67]. An M2M system is a network which consists of a large number of devices that can communicate with little or no human intervention in order to accomplish specific tasks. M2M communications enable the implementation of the IoT, in which ubiquitous connections can be established either on demand or periodically. Some IoT objects are equipped with sensing capabilities to interact with the environment and gather data. According to Cisco's forecast [1], M2M connections will grow from 604 million in 2015 to 3.1 billion by 2020. An IoT object may be a home appliance, surveillance camera, smart meter, vehicle, or device, which can be assigned a unique identifier (ID). Applications of the IoT include vital-sign monitoring in healthcare systems, remote security sensing, freight management and tracking in intelligent transportation systems, and monitoring and control in smart cities.

Since many IoT objects are battery powered, low-power communication is essential. Technologies such as ZigBee, Bluetooth, and Wi-Fi can provide low-cost solutions for short-range communication in IoT systems. IoT communications in the licensed bands are crucial to improving the network coverage, extending IoT applications to suburban and mission-critical environments, and enhancing communication security. Both LTE and LTE-Advanced can provide reliable communication and high-speed data transfer. They can also support mobility of IoT objects. The 5G mobile network is a promising communication platform for future IoT systems. M2M and IoT applications have several distinct features, which include low mobility, low-volume and infrequent data transmissions, small data payload, and location-specific triggering. It is crucial that the current wireless systems evolve to support the IoT. The International Telecommunications Union (ITU) classified machine-type communications (MTC) into two categories. The first category is massive MTC (mMTC), which is characterized by a high connection density, i.e., a massive number of active low-cost and low-power MTC devices are deployed per unit area. The second category is ultra-reliable low-latency communications (URLLC), which requires reliable data transmission subject to stringent real-time latency constraints. 3GPP is developing a standard to support MTC and IoT communication sessions efficiently. For example, narrowband IoT (NB-IoT) has been proposed to enable low-cost, low-power, extended-coverage connectivity for MTC devices. Design issues and solutions in relation to massive IoT devices in 5G wireless cellular systems will be discussed in Chapter 17.

Owing to massive concurrent data transmission from MTC devices, network congestion may occur. Generally, MTC may cause radio access network and core network congestion. Radio access network congestion usually occurs in the coverage area of a specific cell when many MTC devices concurrently access the same base station. When a large number of MTC devices transmit messages from many base stations to a single MTC server, core network congestion may happen. Core network congestion may result in intolerable delays, packet loss, and service unavailability. In Chapter 18, M2M communication network architectures, MAC design, and congestion control will be discussed in detail.

D2D communication is a promising technique for enhancing the system capacity of the next generation wireless communication systems. D2D communication enables a direct link between two users over a short transmission distance without routing data through a macrocell base station, thereby reducing the transmission power consumed. The macrocell base station can send control signals to devices, which contain instructions regarding power control and channel allocation for D2D communication. Another advantage of D2D communication is that a D2D pair can concurrently reuse the available spectrum of macrocells. With the advantages of high spectrum efficiency, low energy consumption, and ubiquitous coverage, D2D communication has been proposed for applications such as public safety, advertising, and vehicular communication networks. In Chapter 19, energy harvesting enabled D2D communication systems are discussed. Stochastic-geometry analysis techniques are applied to investigate the effects of user density and access point density on energy harvesting enabled D2D networks, and a transmission mode selection strategy will be presented.

1.11   Radio Resource Management, Interference Mitigation, and Caching

More spectrum, higher spectrum efficiency, and denser cell sites are the keys to achieving the goal of thousand-fold capacity improvement in 5G wireless [5]. Because spatial densification can provide a significant spectrum efficiency gain, small cells have drawn considerable attention recently [4]. However, the capacity gain of small cells comes at the cost of severe intercell interference and complicated interference scenarios. In order to provide capacity enhancement and coverage extension simultaneously, 3GPP LTE-Advanced focuses on heterogeneous networks (HetNets), where the macrocells and small cells are coordinated to coexist in the same coverage area. To this end, eICIC techniques are specified for LTE-Advanced HetNets in support of two functions: range expansion and resource partition. The former allows more user terminals to benefit from small cells by introducing a cell selection bias, and the latter helps in mitigating the interference between a macrocell and several small cells through the assignment of almost-blank subframes. Specifically, almost-blank subframes are defined as empty subframes in the time domain of a macrocell. Small cells can transmit data much faster in almost-blank subframes owing to the absence of interference from the macrocell. Despite the capacity loss for macrocellular users, because of the capacity gain of the small cells, the overall system capacity improves in the coverage area of a single macrocell. In [68], resource allocation algorithms for LTE HetNets with eICIC were proposed to deal with two important issues: (i) determining the amount of radio resources that a macrocell should allocate to small cells, and (ii) determining the association rule for pairing user terminals and small cells. The aforementioned interference cancellation techniques focus on centralized multicell radio resource management (RRM).

In principle, multicell RRM in LTE networks can be implemented by centralized or decentralized approaches [69, 70]. Centralized multicell RRM requires an additional network element to perform joint optimization among multiple cells, such as eICIC [71]. Alternatively, without an additional central RRM entity, decentralized multicell RRM can achieve the same performance as centralized multicell RRM by exchanging messages among cells and by letting each cell make its local decision based on information collected from its neighboring cells. In 5G cellular systems, decentralized multicell RRM techniques are being actively studied, especially for green transmission technologies which aim at balancing energy efficiency and spectrum efficiency. In Chapter 5, decentralized radio resource management for dense HetNets will be discussed, including the impact of power on/off of small cells on network power consumption and throughput, and the interference-aware channel-segregation-based channel assignment.

To address the interference mitigation problem in wireless networks, CoMP transmission, also known as network MIMO, has been proposed. Network MIMO can be classified into three types: joint transmission, coordinated scheduling, and coordinated beamforming. Joint transmission network MIMO processes both the CSI and the data of neighboring base stations, but coordinated scheduling and coordinated beamforming network MIMO only need to share CSI. High-speed and reliable backhaul connections are important for network MIMO in exchanging the CSI and transmitting the data between the cooperating base stations. All these base station cooperation techniques aim to reduce the intercell interference. With the help of collaborative base stations, the signal quality at the cell boundary can be enhanced significantly, since all potential interfering transmitting sources transmit in different directions or even become helpers. Furthermore, network MIMO has also been applied to coordinate a macrocell and small cells [72]. Chapter 11 will discuss CoMP transmission, interference alignment [73], and compute-and-forward for receiver cooperation [74] in detail.

Chapter 12 will introduce a wireless caching technique called physical layer (PHY) caching [75, 76]. In PHY caching, when several neighboring base stations have the same content required by different users, there are various benefits. In particular, base stations can use cache-induced opportunistic CoMP to serve the users. When global CSI at the transmitter (CSIT) is available, the interference between different users can be mitigated and there is a spatial multiplexing gain. On the other hand, when global CSIT is not available, PHY caching can still improve reliability by exploiting the opportunistic cooperative spatial diversity. The basic concepts of PHY caching, its design challenges and solutions, and a cache content placement algorithm will be presented in Chapter 12.

1.12   Energy Harvesting Communications

Recently, there has been an upsurge in research interest in green communication owing to environmental concerns. It has been reported that cellular networks world-wide consume approximately 60 billion kWh of energy per year. In particular, 80% of the electricity used in cellular networks is consumed by wireless communication base stations, emitting over a hundred million tons of carbon dioxide (CO_2) per year [77]. If no further action is taken to reduce this energy consumption, it is expected that these figures will double by the year 2020. Besides, the escalating energy consumption of wireless communication systems will become a financial burden on service providers if the energy cost per unit of transmitted information cannot be reduced. As a result, energy-efficient communication design has become an important requirement for 5G systems.

In some practical situations, energy harvesting provides a viable solution, since energy-limited wireless communication devices can harvest energy from renewable energy sources such as solar, wind, and geothermal energy. Besides, harvesting energy from "free of cost" natural resources can possibly reduce the energy costs of service providers substantially in the long term. Therefore, energy harvesting communications are expected not only to be energy-efficient but also to be self-sustaining and

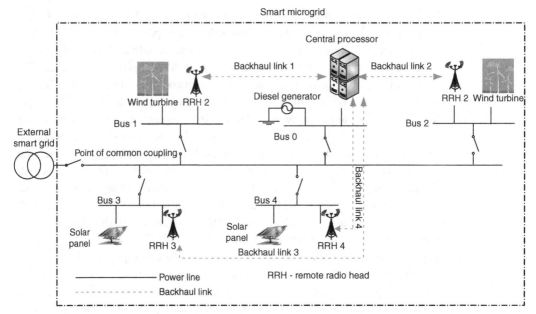

Figure 1.4 An example of energy trading and cooperative transmission in a hybrid energy harvesting communication system.

cost-effective. The integration of energy harvesters into wireless communication devices poses many interesting new challenges on system architecture and resource allocation algorithm design. In practice, renewable energy is perpetual but intermittent, introducing randomness into the energy availability at wireless communication devices. If a wireless transceiver is powered solely by a renewable energy harvester, we may not be able to maintain stable operation and to guarantee a certain QoS to end users. As a result, advanced signal-processing techniques have been proposed to realize energy harvesting communications.

Hybrid energy sources, which use different energy sources in a complementary manner, facilitate energy exchange and cooperation between transmitters via a smart grid [78]. Figure 1.4 shows an example of such a system in a C-RAN. In the example network considered, several nearby RRHs, equipped with hybrid energy sources, are connected via power lines to form a smart microgrid which enables two-way energy trading and cooperative transmission. With two-way energy trading, RRHs with a renewable energy surplus can either store the excess harvested energy or sell it to the grid to make a profit, while RRHs with a renewable energy deficit can buy additional energy from the grid or other RRHs to maintain reliable operation and communication. Data backhaul connections between RRHs enable the possibility of CoMP transmission. If power line communication is feasible, the data backhaul links can be combined with the power line connection. In fact, the CoMP system architecture inherently offers spatial diversity for combating path loss and shadowing. This important feature can reduce the energy consumption of the system further and improve the efficiency of energy harvesting communication systems. However, transferring the data of all users from the central processor to all RRHs, as is required for full cooperation, may be infeasible when the capacity of the backhaul links is limited by power line communications. Besides, conventional cooperation schemes for wireless communications enabling energy saving utilize energy in a conservative manner. These schemes cannot be directly applied to communication systems with hybrid energy sources and energy trading where renewable energy must be fully exploited. In Chapter 13, a cost-aware cooperative communication scheme will be discussed which exploits renewable energy sources.

1.13   Visible Light Communication

Wireless communication systems operating in RF bands have been deployed worldwide. However, the existing RF frequency spectrum is limited and is already congested. Hence, it is expected that the supply of RF frequency spectrum will gradually be overtaken by the demand for data traffic. Therefore, a complementary communication technology is needed to assist RF-based communication. The available bandwidth of visible light is $10,000$ times larger than that of the microwave RF band [79]. As a result, the use of visible light communication (VLC) provides a viable solution to alleviating the spectrum shortage in the RF band [80]. VLC exploits solid-state lighting such as light-emitting diodes (LEDs) for data communication. In particular, a data stream can be modulated and transmitted by varying the intensity of luminaires. In other words, an LED can serve as a dual-purpose device, providing lighting and communication simultaneously. In fact, the visible light spectrum is unregulated. Therefore, it can be used freely for communication purposes, which significantly reduces the costs for wireless communication service operators.

There are several benefits to implementing VLC. First, VLC systems facilitate simple co-channel interference management and offer a high level of communication security. Indoors, the signals from VLC systems are well confined by the walls of rooms, which prevents interference leakage to other VLC communication systems in other rooms. Besides, potential eavesdroppers cannot intercept VLC signals outside an apartment or room. Second, VLC is energy-efficient and expected to be popular. Traditional indoor lighting or illumination is ubiquitous in the modern world and accounts for approximately 11% of the total electricity consumption in the USA [81]. Owing to environmental concerns, governments around the world are advocating the replacement of energy-inefficient halogen lamps and incandescent lamps with energy-efficient LEDs. This trend is paving the way for the development of ubiquitous VLC systems as LED illumination begins to spread widely. More importantly, the energy used for communication in VLC is essentially free, as lighting is required anyway.

VLC not only has the potential to offer higher data rates but also has various important applications in the IoT, intelligent transportation systems, indoor positioning, and entertainment [82]. For the realization of truly mobile communication systems, a complete networking solution is required, which leads to the concept of light-fidelity (LiFi), analogous to Wi-Fi. LiFi is a subset of VLC that exhibits high-speed, bidirectional, fully networked communications. For instance, in order to achieve communication with multiple VLC receivers in an indoor environment, a cellular network structure composed of several small optical attocells has been proposed [83]. The goal is to provide seamless coverage and high spectral efficiency to multiple users simultaneously. In practice, an optical attocell network can be realized by installing multiple LED access points in the ceiling of a room. However, co-channel interference increases rapidly when multiple optical base stations are densely deployed in optical attocells in the same indoor environment. Thus, it is necessary to study the impact of co-channel interference on the system data rate and outage probability. A thorough discussion of optical attocells in the context of 5G will be provided in Chapter 14.

Acknowledgments

This work was supported by the Natural Sciences and Engineering Council of Canada (NSERC). Derrick Wing Kwan Ng is supported by the Australian Research Council's Discovery Early Career Researcher Award (DECRA) DE170100137. We would like to thank Hamed Shah-Mansouri and Yong Zhou from the University of British Columbia for their editorial help.

References

[1] Cisco, "Cisco visual networking index: Global mobile data traffic forecast update, 2015–2020 white paper," Feb. 2016. Available at www.cisco.com/c/en/us/solutions/colla teral/service-provider/visual-networking-index-vni/mobile-white-paper-c11-520862.html.

[2] J. Andrews, S. Buzzi, W. Choi, S. Hanly, A. Lozano, A. Soong, and J. Zhang, "What will 5G be?" *IEEE J. Sel. Areas Commun.*, vol. 32, no. 6, pp. 1065–1082, Jun. 2014.

[3] F. Boccardi, R. Heath, A. Lozano, T. Marzetta, and P. Popovski, "Five disruptive technology directions for 5G," *IEEE Commun. Mag.*, vol. 52, no. 2, pp. 74–80, Feb. 2014.

[4] N. Bhushan, J. Li, D. Malladi, and R. Gilmore, "Network densification: The dominant theme for wireless evolution into 5G," *IEEE Commun. Mag.*, vol. 52, no. 2, pp. 82–89, Feb. 2014.

[5] A. Osseiran, F. Boccardi, V. Braun, K. Kusume, P. Marsch, M. Maternia, O. Queseth, M. Schellmann, H. Schotten, H. Taoka, H. Tullberg, M. A. Uusitalo, B. Timus, and M. Fallgren, "Scenarios for 5G mobile and wireless communications: The vision of the METIS project," *IEEE Commun. Mag.*, vol. 52, no. 5, pp. 26–35, May 2014.

[6] Ericsson, "5G use cases," 2015. Available at www.ericsson.com/res/docs/2015/5g-use-cases.pdf.

[7] 5G Infrastructure Association, "The 5G infrastructure public private partnership: The next generation of communication networks and services," Feb. 2015. Available at https://5g-ppp.eu/wp-content/uploads/2015/02/5G-Vision-Brochure-v1.pdf.

[8] China Mobile, "C-RAN: The road towards green RAN," Tech. Rep., version 3.0, Dec. 2013. Available at http://labs.chinamobile.com/cran/wp-content/uploads/2014/06/20140613-C-RAN-WP-3.0.pdf.

[9] A. Checko, H. L. Christiansen, Y. Yan, L. Scolari, G. Kardaras, M. S. Berger, and L. Dittmann, "Cloud RAN for mobile networks – A technology overview," *IEEE Commun. Surv. Tutor.*, vol. 17, no. 1, pp. 405–426, Mar. 2015.

[10] M. Peng, Y. Li, Z. Zhao, and C. Wang, "System architecture and key technologies for 5G heterogeneous cloud radio access networks," *IEEE Netw.*, vol. 29, no. 2, pp. 6–14, Mar. 2015.

[11] S. H. Park, O. Simeone, O. Sahin, and S. Shamai Shitz, "Fronthaul compression for cloud radio access networks: Signal processing advances inspired by network information theory," *IEEE Signal Process. Mag.*, vol. 31, no. 6, pp. 69–79, Nov. 2014.

[12] G. Wang, Q. Liu, R. He, F. Gao, and C. Tellambura, "Acquisition of channel state information in heterogeneous cloud radio access networks: Challenges and research directions," *IEEE Wireless Commun.*, vol. 22, no. 3, pp. 100–107, Jun. 2015.

[13] F. Bonomi, R. Milito, J. Zhu, and S. Addepalli, "Fog computing and its role in the Internet of things," in *Proc. of MCC Workshop on Mobile Cloud Computing*, Aug. 2012.

[14] A. Botta, W. De Donato, V. Persico, and A. Pescape, "On the integration of cloud computing and Internet of things," in *Proc. of International Conf. on Future Internet of Things and Cloud (FiCloud)*, Aug. 2014.

[15] Cisco, "Fog computing and the Internet of things: Extend the cloud to where the things are," 2015. Available at www.cisco.com/c/dam/en_us/solutions/trends/iot/docs/computing-over view.pdf.

[16] T. Cover and J. Thomas, *Elements of Information Theory*, Wiley, 2006.

[17] Y. Saito, Y. Kishiyama, A. Benjebbour, T. Nakamura, A. Li, and K. Higuchi, "Non-orthogonal multiple access (NOMA) for cellular future radio access," in *Proc. IEEE Vehicular Technology Conf.*, Jun. 2013.

[18] Z. Ding, Z. Yang, P. Fan, and H. V. Poor, "On the performance of non-orthogonal multiple access in 5G systems with randomly deployed users," *IEEE Signal Process. Lett.*, vol. 21, no. 12, pp. 1501–1505, Dec. 2014.

[19] 3rd Generation Partnership Project (3GPP), "Study on downlink multiuser superposition transmission (MUST) for LTE," TR 36.859 V0.1.0, Apr. 2015.

[20] L. Dai, B. Wang, Y. Yuan, S. Han, C.-L. I, and Z. Wang, "Non-orthogonal multiple access for 5G: Solutions, challenges, opportunities, and future research trends," *IEEE Commun. Mag.*, vol. 53, no. 9, pp. 74–81, Sep. 2015.

[21] Y. Tao, L. Liu, S. Liu, and Z. Zhang, "A survey: Several technologies of non-orthogonal transmission for 5G," *China Commun.*, vol. 12, no. 10, pp. 1–15, Oct. 2015.

[22] H. Nikopour and H. Baligh, "Sparse code multiple access," in *Proc. of IEEE Conf. on Personal Indoor and Mobile Radio Commun. (PIMRC)*, Sep. 2013.

[23] Z. Ding, F. Adachi, and H. V. Poor, "The application of MIMO to non-orthogonal multiple access," *IEEE Trans. Wireless Commun.*, vol. 15, no. 1, pp. 537–552, Jan. 2016.

[24] Z. Ding, R. Schober, and H. V. Poor, "A general MIMO framework for NOMA downlink and uplink transmissions based on signal alignment," *IEEE Trans. Wireless Commun.*, vol. 15, no. 6, pp. 4438–4454, Jun. 2016.

[25] G. Fettweis, M. Krondorf, and S. Bittner, "GFDM – Generalized frequency division multiplexing," in *Proc. of IEEE Vehicular Technology Conf. (VTC-Spring)*, Apr. 2009.

[26] T. L. Marzetta, "Noncooperative cellular wireless with unlimited number of base station antennas," *IEEE Trans. Wireless Commun.*, vol. 9, no. 11, pp. 3590–3600, Nov. 2010.

[27] G. J. Foschini and M. J. Gans, "On limits of wireless communications in a fading environment when using multiple antennas," *Wireless Personal Commun.*, vol. 6, no. 3, pp. 311–335, Mar. 1998.

[28] E. Telatar, "Capacity of multi-antenna Gaussian channels," *Eur. Trans. Telecommun.*, vol. 10, no. 6, pp. 585–595, Nov.–Dec. 1999.

[29] G. Caire and S. Shamai, "On the achievable throughput of a multi-antenna Gaussian broadcast channel," *IEEE Trans. Inf. Theory*, vol. 49, no. 7, pp. 1691–1706, Jul. 2003.

[30] S. Vishwanath, N. Jindal, and A. Goldsmith, "Duality, achievable rates, and sum-rate capacity of Gaussian MIMO broadcast channels," *IEEE Trans. Inf. Theory*, vol. 49, no. 10, pp. 2658–2668, Oct. 2003.

[31] D. Gesbert, M. Kountouris, R. Heath, C. Chae and T. Salzer, "Shifting the MIMO paradigm," *IEEE Signal Process. Mag.*, vol. 24, no. 5, pp. 36–46, Sep. 2007.

[32] S. Verdú, *Multiuser Detection*, Cambridge University Press, 1998.

[33] C. Peel, B. M. Hochwald, and A. L. Swindlehurst, "A vector-perturbation technique for near-capacity multiantenna multiuser communication – Part I: Channel inversion and regularization," *IEEE Trans. Commun.*, vol. 53, no. 1, pp. 195–202, Jan. 2005.

[34] T. Yoo and A. Goldsmith, "On the optimality of multiantenna broadcast scheduling using zero-forcing beamforming," *IEEE J. Sel. Areas Commun.*, vol. 24, no. 3, pp. 528–541, Mar. 2006.

[35] H. Ngo and E. Larsson, "EVD-based channel estimation in multicell multiuser MIMO systems with very large antenna arrays," in *Proc. of International ITG Workshop on Smart Antennas*, Mar. 2012.

[36] R. R. Müller, L. Cottatellucci, and M. Vehkaperä, "Blind pilot decontamination," *IEEE J. Sel. Signal Process.*, vol. 8, no. 5, pp. 773–786, Oct. 2014.

[37] H. Q. Ngo, E. G. Larsson, and T. L. Marzetta, "Energy and spectral efficiency of very large multiuser MIMO systems," *IEEE Trans. Commun.*, vol. 61, no. 4, pp. 1436–1449, Apr. 2013.

[38] J. Zhu, R. Schober, and V. Bhargava, "Secure transmission in multicell massive MIMO systems," *IEEE Trans. Wireless Commun.*, vol. 13, no. 9, pp. 4766–4781, Sep. 2014.

[39] Y. Wu, R. Schober, D. W. K. Ng, E. Lo, C. Xiao, and G. Caire, "Secure massive MIMO transmission with an active eavesdropper," *IEEE Trans. Inf. Theory*, vol. 62, no. 7, pp. 3880–3900, Jul. 2016.

[40] S. Zarei, W. Gerstacker, J. Aulin, and R. Schober, "I/Q imbalance aware widely-linear receiver for uplink multi-cell massive MIMO systems: Design and sum rate analysis," *IEEE Trans. Wireless Commun.*, vol. 15, no. 5, pp. 3393–3408, May 2016.

[41] A. El Hajj and Z. Dawy, "Dynamic joint switching point configuration and resource allocation in TDD-OFDMA wireless networks," in *Proc. of IEEE Global Communications Conf. (Globecom)*, Dec. 2011.

[42] M. Chen and A. Yener, "Power allocation for multi-access two-way relaying," in *Proc. of IEEE International Conf. on Communications (ICC)*, Jun. 2009.

[43] 3GPP, "Full duplex configuration of Un and Uu subframes for type I relay," 3GPP TSG RAN WG1 R1-100139, Tech. Rep., Jan. 2010.

[44] 3GPP, "Text proposal on inband full duplex relay for TR 36.814," 3GPP TSG RAN WG1 R1-101659, Tech. Rep., Feb. 2010.

[45] D. Kim, H. Lee, and D. Hong, "A survey of in-band full-duplex transmission: From the perspective of PHY and MAC layers," *IEEE Commun. Surv. Tutor.*, vol. 17, no. 4, pp. 2017–2046, Fourth Quarter 2015.

[46] H. Ju, E. Oh, and D. Hong, "Catching resource-devouring worms in next-generation wireless relay systems: Two-way relay and full-duplex relay," *IEEE Commun. Mag.*, vol. 47, no. 9, pp. 58–65, Sep. 2009.

[47] T. Riihonen, S. Werner, and R. Wichman, "Optimized gain control for single-frequency relaying with loop interference," *IEEE Trans. Wireless Commun.*, vol. 8, no. 6, pp. 2801–2806, Jun. 2009.

[48] B. Day, A. Margetts, D. Bliss, and P. Schniter, "Full-duplex bidirectional MIMO: Achievable rates under limited dynamic range," *IEEE Trans. Signal Process.*, vol. 60, no. 7, pp. 3702–3713, Jul. 2012.

[49] J. I. Choi, M. Jain, K. Srinivasan, P. Levis, and S. Katti, "Achieving single channel, full duplex wireless communication," in *Proc. of ACM MobiCom*, Sep. 2010.

[50] M. A. Khojastepour, K. Sundaresan, S. Rangarajan, X. Zhang, and S. Barghi, "The case for antenna cancellation for scalable full duplex wireless communications," in *Proc. of ACM Workshop on Hot Topics in Networks (HotNets)*, Nov. 2011.

[51] M. Duarte and A. Sabharwal, "Full-duplex wireless communications using off-the-shelf radios: Feasibility and first results," in *Proc. of Asilomar Conf. on Signals, Systems and Computers*, Nov. 2010.

[52] C. Hansen, "WiGiG: Multi-gigabit wireless communications in the 60 GHz band," *IEEE Wireless Commun.*, vol. 18, no. 6, pp. 6–7, Dec. 2011.

[53] H. Mehrpouyan, R. Khanzadi, M. Matthaiou, A. Sayeed, R. Schober, and Y. Hua, "Improving bandwidth efficiency in E-band communication systems," *IEEE Commun. Mag.*, vol. 52, no. 3, pp. 121–128, Mar. 2014.

[54] T. Rappaport, S. Sun, R. Mayzus, H. Zhao, Y, Azar, K, Wang, G. Wong, J. Schulz, M. Samimi, and F. Gutierrez, "Millimeter wave mobile communications for 5G cellular: It will work!," *IEEE Access*, vol. 1, pp. 335–349, May 2013.

[55] T. Rappaport, F. Gutierrez, E. Ben-Dor, J. Murdock, Y. Qiao, and J. Tamir, "Broadband millimeter-wave propagation measurements and models using adaptive-beam antennas for outdoor urban cellular communications," *IEEE Trans. Antennas Propag.*, vol. 61, no. 4, pp. 1850–1859, Jun. 2013.

[56] A. Alkhateeb, O. El Ayach, G. Leus, and R. Heath, "Channel estimation and hybrid precoding for millimeter wave cellular systems," *IEEE J. Sel. Topics Signal Process.*, vol. 8, no. 5, pp. 831–846, Sep. 2014.

[57] H. Shokri-Ghadikolaei, C. Fischione, G. Fodor, P. Popovski, and M. Zorzi, "Millimeter wave cellular networks: A MAC layer perspective," *IEEE Trans. Commun.*, vol. 63, no. 10, pp. 3437–3458, Oct. 2015.

[58] A. Adhikary, E. Safadi, M. Samimi, R. Wang, G. Caire, T. Rappaport, and A. Molisch, "Joint spatial division and multiplexing for mm-wave channels," *IEEE J. Sel. Areas Commun.*, vol. 32, no. 6, pp. 1239–1255, Jun. 2014.

[59] L. Swindlehurst, E. Ayanoglu, P. Heydari, and F. Capolino, "Millimeter-wave massive MIMO: The next wireless revolution?" *IEEE Commun. Mag.*, vol. 52, no. 9, pp. 56–62, Sep. 2014.

[60] C. Dehos, J. Gonzalez, A. De Domenico, D. Ktenas, and L. Dussopt, "Millimeter-wave access and backhauling: The solution to the exponential data traffic increase in 5G mobile communications systems?" *IEEE Commun. Mag.*, vol. 52, no. 9, pp. 88–95, Sep. 2014.

[61] S. Hur, T. Kim, D. Love, J. Krogmeier, T. Thomas, and A. Ghosh, "Millimeter wave beamforming for wireless backhaul and access in small cell networks," *IEEE Trans. Commun.*, vol. 61, no. 10, pp. 4391–4403, Oct. 2013.

[62] Qualcomm, "LTE-U/LAA, MuLTEfire and Wi-Fi: Making best use of unlicensed spectrum," Sep. 2015. Available at www.qualcomm.com/documents/making-best-use-unlicensed-spectrum-presentation.

[63] 3GPP RP-140808, "Review of regulatory requirements for unlicensed spectrum," Jun. 2014.

[64] 3GPP, "LTE in unlicensed spectrum," Jun. 2014. Available at www.3gpp.org/news-events/3gpp-news/1603-lte_in_unlicensed.

[65] S. Sen, C. Joe-Wong, S. Ha, and M. Chiang, *Smart Data Pricing*, Wiley, 2014.

[66] A. Al-Fuqaha, M. Guizani, M. Mohammadi, M. Aledhari, and M. Ayyash, "Internet of things: A survey on enabling technologies, protocols, and applications," *IEEE Commun. Surv. Tutor.*, vol. 17, no. 4, pp. 2347–2376, Fourth Quarter 2015.

[67] M. R. Palattella, M. Dohler, A. Grieco, G. Rizzo, J. Torsner, T. Engel, and L. Ladid, "Internet of things in the 5G era: Enablers, architectures and business models," *IEEE J. Sel. Areas Commun.*, vol. 34, no. 3, pp. 510–527, March 2016.

[68] S. Deb, P. Monogioudis, J. Miernik, and J. P. Seymour, "Algorithms for enhanced inter-cell interference coordination (eICIC) in LTE HetNets," *IEEE/ACM Trans. Networking*, vol. 22, no. 1, pp. 137–150, Feb. 2014.

[69] Nokia Solutions and Networks, "Multi-cell radio resource management: Centralized or decentralized?" NSN White Paper, Jan. 2014.

[70] D. Wübben, P. Rost, J. Bartelt, M. Lalam, V. Savin, M. Gorgoglione, A. Dekorsy, and G. Fettweis, "Benefits and impact of cloud computing on 5G signal processing," *IEEE Signal Process. Mag.*, vol. 31, no. 6, pp. 35–44, Nov. 2014.

[71] 3GPP, "Requirements for support of radio resource management," Tech. Rep. TS.36.133, Mar. 2016.

[72] V. Junginckel, K. Manolakis, W. Zirwas, B. Panzner, V. Braun, M. Lossow, M. Sternad, R. Apelfrojd, and T. Svensson, "The role of small cells, coordinated multipoint, and massive MIMO in 5G," *IEEE Commun. Mag.*, vol. 52, no. 5, pp. 44–51, May 2014.

[73] S. A. Jafer, *Interference Alignment: A New Look at Signal Dimensions in a Communications Network*, NOW Publishers, 2011.

[74] B. Nazer and M. Gastpar, "Compute-and-forward: Harnessing interference through structured codes," *IEEE Trans. Inf. Theory*, vol. 57, no. 10, pp. 6463–6486, Oct. 2011.

[75] A. Liu and V. K. N. Lau, "Exploiting base station caching in MIMO cellular networks: Opportunistic cooperation for video streaming," *IEEE Trans. Signal Process.*, vol. 63, no. 1, pp. 57–69, Jan. 2015.

[76] W. Han, A. Liu, and V. Lau, "Degrees of freedom in cached MIMO relay networks," *IEEE Trans. Signal Process.*, vol. 63, no. 15, pp. 3986–3997, Aug. 2015.

[77] G. P. Fettweis and E. Zimmermann, "ICT energy consumption – Trends and challenges," in *Proc. of International Symposium on Wireless Personal Multimedia Communications*, Sep. 2008.

[78] J. Xu and R. Zhang, "CoMP meets smart grid: A new communication and energy cooperation paradigm," *IEEE Trans. Veh. Technol.*, vol. 64, no. 6, pp. 2476–2488, Jun. 2015.

[79] S. Wu, H. Wang, and C. H. Youn, "Visible light communications for 5G wireless networking systems: From fixed to mobile communications," *IEEE Netw.*, vol. 28, no. 6, pp. 41–45, Nov. 2014.

[80] A. Jovicic, J. Li, and T. Richardson, "Visible light communication: Opportunities, challenges and the path to market," *IEEE Commun. Mag.*, vol. 51, no. 12, pp. 26–32, Dec. 2013.

[81] U.S. Energy Information Administration, "How much electricity is used for lighting in the United States?" 2015. Available at www.mouser.com/pdfdocs/Powercast-Overview-2011-01-25.pdf.

[82] Disney Research, "Visible light communication," 2015. Available at www.disneyresearch.com/project/visible-light-communication/.

[83] C. Chen, D. Tsonev, and H. Haas, "Joint transmission in indoor visible light communication downlink cellular networks," in *Proc. of IEEE Globecom Workshops (GC Wkshps)*, Dec. 2013.

Part I

Communication Network Architectures for 5G Systems

2 Cloud Radio Access Networks for 5G Systems

Chih-Lin I, Jinri Huang, Xueyan Huang, Rongwei Ren, and Yami Chen

According to a report from Cisco [1], global mobile data traffic will continue to grow rapidly from 2015 to 2020. Meanwhile, the fifth generation (5G) is required to enhance the telecommunications infrastructure and provide new information services to support vertical applications in a variety of industrial areas, such as agriculture, medicine, finance, transportation, manufacturing, and education. Therefore, 5G requires innovative solutions to meet new demands from both the mobile Internet and the Internet of Things (IoT) in terms of user-experienced data rate improvement, latency reduction, connection density and area capacity density enhancement, mobility enhancement, and spectral efficiency and energy efficiency improvements.

According to the International Telecommunication Union (ITU), the current 5G scenarios can be divided into three categories: enhanced mobile broadband (eMBB), massive machine-type communications (mMTC), and ultra-reliable low-latency communications (URLLC). Hotspots (indoor/outdoor), wide-area coverage, and high speed are typical use cases. Performance measures of human-centric communications such as the ultimate user experience are primary targets in the eMBB scenario. Use cases of mMTC include the monitoring and automation of buildings and infrastructure, smart agriculture, logistics, tracking, and fleet management. A high connection density, low complexity and cost, and long battery life are essential objectives in the mMTC scenario. There are many representative use cases related to URLLC, such as remote machinery and intelligent transportation systems. Low latency and high reliability are key points that need to be taken into account in the design of the radio technology in order to solve the problem of the specific requirements of URLLC scenarios.

2.1 Rethinking the Fundamentals for 5G Systems

The 5G network is anticipated to be soft, green, and superfast [2]. To meet the critical requirements for various scenarios, it is simply not enough for 5G to evolve from current fourth generation (4G) systems. Rather, it requires a revolutionary path. In [2–4], it was proposed to rethink the fundamentals from seven perspectives, such as architectures, protocols, and functions, to revolutionarily redesign future 5G networks, including:

1. Rethinking Shannon, which is to take a green metric such as the energy efficiency as a key performance indicator of wireless systems.
2. Rethinking Ring and Young, which is to break the boundary of conventional cells. As we move toward the timeline of 2020 with the introduction of heterogeneous networks (HetNets) and ultra-dense networks (UDNs), multiple layers of radio networks have come into being. The traditional homogeneous cell-centric design of mobile networks does not match the traffic variations and diverse radio environments. The principle of "no more cells" (NMC) or the "user-centric cell" (UCC) has therefore been proposed, departing from cell-based coverage, resource management, and signal processing.
3. Rethinking signaling and control, which is to make the network become context-aware and service-customized. In the 5G era, the user and traffic characteristics will be even more diversified and differentiated, and the resource-contending environment will be more complex. Therefore, mobile networks must be capable of providing different network functions such as mobility management, and security control, with customized signaling control based on differentiated user and traffic characteristics. Networks need to serve the various requirements with high efficiency.
4. Rethinking the antenna, which is to make the base station invisible and to drive toward novel low-cost antenna implementation. Targeting significant capacity enhancement by 2020, the 5G network is expected to be ultra-densed, with massive antennas deployed in either a distributed or a centralized manner. However, accommodating a few hundred antennas and transceiver chains all on one structure in a traditional cell site appears to be nearly impossible. It is proposed to fundamentally change the future appearance of cellular networks: to make base stations invisible, by configuring active antenna arrays in a flexible manner on the walls of city buildings and town houses. Meanwhile, novel antenna implementations need to be designed to achieve this goal.
5. Rethinking the spectrum and air interface, which is to enable the wireless signal to "dress for the occasion." To provide a high data rate with the capability of all-spectrum access, the 5G air interface must provide a flexible configuration according to the diverse service requirements. The traditional "one size fits all" air interface paradigm therefore needs to undergo a fundamental change. A software-defined air interface (SDAI) will meet the diverse demands of 5G by reconfiguring itself from an energy-efficiency–spectrum-efficiency (EE-SE) co-optimized set of combinations of the physical layer building blocks, including the frame structure, duplex modes, waveforms, multiple access schemes, modulation and coding schemes, and spatial-processing schemes.
6. Rethinking the fronthaul, which in essence is to redesign the fronthaul interface to make the fronthaul data become user traffic dependent and antenna independent, while still supporting cell coordination algorithms. In this way, the fronthaul transport network could be of much higher efficiency and scalability compared with traditional fronthaul solutions and, therefore, provide better support for key 5G technologies such as the cloud radio access network (C-RAN) and massive multiple-input multiple-output (MIMO).

7. Rethinking the protocol stack to manage user equipment (UE) and cells as two separate resource entities, to enhance multicell cooperation and cross-cell experience, and to better support dense network deployment in 5G.

2.2   User-Centric Networks

To date, the radio access network (RAN) architecture is uniform toward all users in terms of the network elements involved, the network signaling and control, and network protocols. However, as mentioned, in the 5G era, not only will services be more diversified, but the network access points and topologies will also be more complex. In order to achieve the 5G vision, the RAN architecture must be redesigned to accommodate the above-mentioned seven "rethinking points." The answer lies in user-centric networks (UCNs). The essence of the UCN is that it is user-centric rather than cell-centric, with the following key properties:

- The network should support uniform access and seamless mobility across conventional cells on demand from users to achieve the target user-experienced data rate.
- The network functions should be flexibly distributed over different radio network entities to efficiently support diverse services and to effectively organize HetNet entities. Different network entities jointly provide a complete set of network functions on service demand. For example, a centralized network entity provides the common high-layer network services, while distributed network access points serve the users with multiple links. As with URLLC services, all network functions can still be pushed to the distributed access points.
- The UCN should be aware of the user services and provide specific optimal network decisions to differentiate different network services and optimize service experience. It should make smart decisions about local cached traffic, local traffic distribution, and service-specific RAN optimizations.
- To achieve low-cost and high-efficiency network operation, IT technologies such as software-defined networks (SDNs), network functions virtualization (NFV), and big data analytics can be borrowed so that the network can be automatically optimized with continually collected network knowledge.

The concept of the UCN reveals the design principles for 5G. When it comes to the realization, it has been suggested that centralized, collaborative, cloud-based, and clean RAN, or, in brief, cloud RAN (C-RAN) [5, 6], is one of the key enablers.

2.3   C-RAN Basics

The concept of C-RAN was first proposed by China Mobile in a white paper in 2009. The basic idea of C-RAN starts from centralization, which is to aggregate different baseband units (BBUs) that in a traditional deployment are geographically separated into the same location. Once centralized in a central room, it should be possible to make different BBUs communicate with each other in a more timely way by connecting them with high-speed switch networks, therefore allowing implementation of cooperative algorithms to improve system performance.

Simple centralization does not support resource sharing among different physical BBUs. With virtualization, this is about to change. The ultimate goal of C-RAN is to realize the feature called resource cloudification, which is to pool the baseband processing resources so that these resources can be managed and dynamically allocated on demand.

Virtualization is the key to realizing resource cloudification. The virtualization of RANs aims to architect RANs with evolving standard IT virtualization technology to consolidate traditional network equipment types into industry-standard high-volume servers, switches, and storage, which would be located in wireless sites or central offices. This involves the implementation of parts of RAN functions in software that can run on a range of industry-standard server hardware and can be moved to, or instantiated in, various locations in the network as required, without the need for installation of new equipment.

2.3.1   C-RAN Challenges Toward 5G

C-RAN has been studied in the literature for a long time and it is well known that the major challenges of C-RAN lie in two aspects: the fronthaul (FH) issue and the virtualization for cloudification.

An FH link is a link between a BBU and an RRU. Typical FH interfaces include the Common Public Radio Interface (CPRI) and the Open Base Station Architecture Initiative (OBSAI). Since the CPRI is the most widely used FH protocol in the industry, we will use the CPRI in this chapter, unless otherwise mentioned, to represent the FH to describe the various issues. In C-RAN with centralization, fibers are used to connect the BBU pool with the remote RRUs. The larger the centralization scale, the more the fibers are needed. In other words, centralization may consume a large amount of fiber resources, which is unaffordable to most operators given the scarcity of fiber. The FH issue has been widely studied and several schemes have been proposed, including various compression techniques, wavelength division multiplexing (WDM), optical transport network (OTN), and microwave transmission. Readers can find more information in [5, 6]. Although WDM-based solutions are feasible for 4G systems, when it comes to 5G with its diverse services and requirements, there needs to be a new solution. In the following section, the Next Generation Fronthaul Interface (NGFI) will be introduced.

The other challenge of C-RAN lies in how to realize the cloudification feature. It is widely believed that one of the keys to this goal is virtualization technology, which has been pervasive as a key cloud computing technology in data centers in the IT industry for many years. However, the use of virtualization in telecom networks is far more complicated owing to the unique features of wireless communications, especially the baseband processing in the RAN. Carrier-grade telecom functions usually have extremely strict requirements for real-time processing. For time division duplex (TDD) LTE systems, it is required that an acknowledgment (ACK/NACK) must be produced and sent back to the UE or eNodeB (eNB) within 3 ms after a frame is received [6]. Traditional data center virtualization technology cannot meet this requirement. Therefore, the virtualization technology and commercial off-the-shelf (COTS) platforms need to be optimized and even customized in various respects, from the I/O interface, hypervisor, and operating systems to management systems, in order to be competent for real-time, computation-intensive baseband processing.

2.4   Next Generation Fronthaul Interface (NGFI): The FH Solution for 5G C-RAN

Several solutions including CPRI compression and WDM transport technologies have been proposed to address the C-RAN FH issue. In essence, the idea of all these solutions is to "accommodate" the FH without changing the FH interfaces themselves. It should be realized that the root cause of the FH challenge lies exactly in the FH interfaces themselves. Take CPRI as an example. The CPRI specification has defined several classes of line rates. For a time division (TD) LTE carrier with a frequency of 20 MHz and eight antennas, the CPRI rate could be as high as 9.8 Gb/s [7]. Moreover, the rate is constant regardless of the dynamically changing mobile traffic, which leads to low transmission efficiency. In addition, existing CPRI interfaces have other shortcomings such as low scalability and flexibility, which impedes C-RAN large-scale deployment. Therefore, the authors of [4, 8, 9] proposed to redefine the CPRI and brought forward a new concept, called the Next Generation Fronthaul Interface (NGFI). NGFI possesses the following desirable features [4, 8]:

- Its data rate will be traffic dependent and therefore support statistical multiplexing.
- The mapping between BBU and RRH will be one-to-many and flexible.
- It will be independent of the number of antennas.
- It will be packet-based, i.e., the FH data can be packetized and transported via packet-switched networks.

The key to achieving NGFI is to repartition the function layout between the BBUs and the RRUs. Traditionally, all the baseband functions, including the physical layer (PHY), medium access control (MAC), and packet data convergence protocol (PDCP), are processed on the BBU side, while the RRU deals mainly with the radio-related functions. The signal transmitted in CPRI is a high-bandwidth I/Q sampling signal. From the perspective of effective information, any data can be transported between the baseband protocol stacks (e.g., between MAC and PHY). The basic idea of function splitting is to move partial baseband functions to the RRU to reduce the bandwidth without losing any information.

There have been some related studies in the literature on this topic. To achieve NGFI in general, the function splitting must decouple the bandwidth from the antennas, which can be achieved by moving antenna-related functions (downlink antenna mapping, fast Fourier transform (FFT), channel estimation, equalization, etc.) to the RRH. It has been shown that the FH bandwidth of an LTE carrier may decrease to the order of 100 Mb/s no matter how many antennas are used [10]. In addition, it has been suggested that the UE processing functions should be decoupled from the cell processing functions. In this way, the FH bandwidth will be lowered and, more importantly, it will become load-dependent. The load-dependent feature gives an opportunity to exploit the statistical multiplexing gain when it comes to FH transport network design for C-RAN deployment. Thanks to statistical multiplexing, the bandwidth needed for transport of a number of FH links in C-RAN can be reduced greatly, consequently decreasing the cost.

Support for collaborative technologies is another key factor in the design of function splitting. Coordinated multipoint (CoMP) has been viewed as one of the key technologies in 4G and 5G to mitigate interference. CoMP can be divided into two classes: MAC layer coordination and physical layer coordination. For example, collaborative scheduling (CS) is one of the MAC layer coordination mechanisms. Joint reception (JR) and joint transmission (JT) are physical layer coordination technologies. In [11] it was found that the performance gain from JR/JT decreases significantly as the number of antennas increases. Moreover, in [8] it was found from field trial data that MAC-level collaborative technologies can bring comparable performance gains with lower complexity, easier implementation, and fewer constraints. Based on these observations, it was suggested that the function splitting for NGFI does not have to support PHY layer coordination technology. It is enough to achieve a considerable performance gain by supporting MAC layer coordination technologies.

Function splitting is just the first step for NGFI. When it comes to FH networks in the context of C-RAN, there is a radical change compared with the original WDM and other existing FH solutions. Thanks to the packet-based features, it is expected that packet-switching networks will be used to transport the NGFI packets. This is when Ethernet can come into play. Thanks to its ubiquity, low cost, high flexibility, and scalability, it has been proposed that Ethernet should be adopted as the NGFI FH solution. There are several benefits. First, an Ethernet interface is the most common interface on standard IT servers, and the use of Ethernet makes C-RAN virtualization easier and cheaper. Second, Ethernet can make full use of the dynamic nature of NGFI to realize statistical multiplexing. Third, flexible routing capabilities could also be used to realize multiple paths between BBU pools and the RRH [8].

The main challenges for the use of Ethernet as an FH solution lie in the high timing and synchronization requirements imposed by the NGFI interface. Although the exact NGFI has not so far been specified, it is possible that NGFI may keep some requirements of CPRI, such as synchronization requirements. The allowable radio frequency error for a CPRI link is 2 ppb, and the timing alignment error must not exceed 65 ns in order to support MIMO and transmission diversity [12]. In order to meet the timing requirements, both the BBU and the RRUs must be perfectly synchronized, which therefore requires a very accurate clock distribution mechanism. Potential solutions may include any combination of the Global Positioning System (GPS), IEEE 1588, and synchronous Ethernet (Sync-E). Finally, the transport protocols on top of Ethernet such as multiprotocol label switching (MPLS) and the packet transport network (PTN) that establish transport paths for the FH traffic need to be defined.

2.4.1   Proof-of-Concept Development of NGFI

In order to provide a scalable platform for the industry for exploring innovative functional designs and assessment of the issues, so that these emerging network technologies for 5G standards will flourish, an NGFI proof-of-concept (PoC) design platform, called the CPRI gateway, based on a standard Xilinx evaluation board, was developed and is introduced in this section.

2.4.1.1   Design Principles

The CPRI gateway is designed to work with the existing network topologies and have the flexibility to address future evolution of these topologies. As a first step, it makes sense to retain the CPRI links to existing RRUs in the installed base and aggregate a number of those radio heads into a single Ethernet link to the base station with the CPRI gateway in the middle.

The CPRI gateway, in the downlink case, receives packets from the base station, with each packet containing data for one or more antennas connected to the CPRI links. The gateway parses the packet, determines which antenna the I/Q data is destined for, and fills a buffer on the CPRI side in the correct position for the CPRI frame map in use on the CPRI link. In the uplink direction, the I/Q samples are assembled into per-antenna Ethernet frames, and, as each frame is completed to a given length, it is posted immediately for transmission back to the base station for further processing. In such a configuration, the timing of the downlink frame egress in the base station has to be advanced to the point that any given I/Q sample is ready in the gateway for egress to the radio head at the correct point in the radio frame timing. The gateway is designed to have some buffering inside to allow for some uncertainty in packet delay for those data packets across the Ethernet interface.

In addition to reformatting the I/Q samples in Ethernet frames into CPRI payloads and vice versa, the gateway also needs to recreate the 10 ms radio strobe and clock frequency reference for the CPRI masters in the gateway. In the initial phase of the design, this was accomplished using external inputs to the evaluation boards.

2.4.1.2   System Description

The block diagram in Figure 2.1 shows one configuration of the current CPRI gateway, as implemented on a Xilinx ZC706 evaluation board. Throughout the design, AXI4-Stream interfaces are used to interconnect the data paths, and AXI4-Lite interfaces are used for control via the ARM processor. The key functional blocks of the platform are listed below.

- *Packet parser and demultiplexer*. This block is the first block that the downlink Ethernet packet stream hits after the MAC/PHY pair. This block inspects each frame for a specific header that denotes it as an I/Q sample packet. If the packet matches the header, it is sent on to the rest of the data path. Otherwise, it is forwarded to the ARM processor for further control plane processing.
- *Ethernet to CPRI (E2C) packet processor*. This block contains individual data buffers, one for each CPRI interface. Each header is inspected further to determine

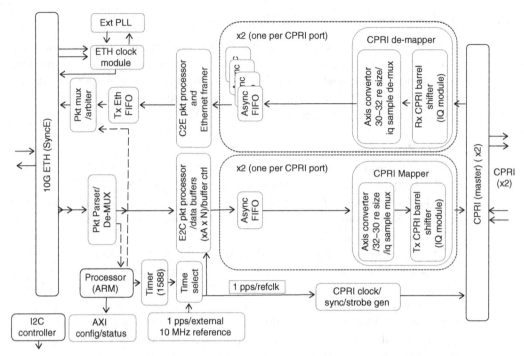

Figure 2.1 The function blocks of the CPRI gateway design for use as an NGFI platform.

which antenna the packet belongs to and the timing of the data with respect to the radio timing. It then writes data to the buffer such that the buffer is in the correct sample order for the CPRI interface. When the frame-timing circuit in the gateway determines that the CPRI frame needs to be sent out, it is moved across the asynchronous first-in-first-out (FIFO) buffer to the CPRI mapper.

- *CPRI mapper.* The samples coming from the Ethernet-to-CPRI Packet processor are 15+15 bits wide. The CPRI mapper takes the 15 bit samples and packs them into the continuous 32 bit-wide data bus that the CPRI core requires for egress.
- *CPRI demapper.* This performs the opposite function to the mapper, in that it unpacks a continuous 32 bit bus into 15 bit I/Q samples. The CPRI demapper then writes each antenna's samples into individual asynchronous FIFO buffers to move them back into the Ethernet clock domain and to the CPRI-to-Ethernet (C2E) packet processor.
- *CPRI-to-Ethernet packet processor.* This collects each antenna's samples into per-antenna Ethernet packets. Once the Ethernet packet is completed, it is queued in a FIFO buffer to be sent out on the Ethernet link to the server.
- *Packet multiplexer.* This is a simple priority arbiter to merge Ethernet streams from the I/Q data path and the ARM processor. The data path has priority to avoid buffer overflows in the packet processor.
- *Strobe generation.* This block uses time and frequency references from either on-board or off-board sources to create the 10 ms radio frame boundary for the CPRI core. In addition, it provides control signals on the transmit data path to trigger

transfers of data down the pipeline so that the asynchrous FIFO and data buffers do not run empty in normal operation.

2.4.1.3   The Potential Capability of the Platform

With the data path translation from CPRI to Ethernet and from Ethernet to CPRI completed, the current work on the CPRI gateway design is focused on two separate areas: frequency and time synchronization, and layer-1 physical layer offload. For frequency and time synchronization, technologies such as synchronous Ethernet can be used to provide frequency synchronization to the gateway from the base station, and IEEE 1588 (the Precision Time Protocol) can be used to provide time synchronization to the gateway. There are then a number of techniques that can be used to take these sources and regenerate the combined frequency and time reference that the CPRI master port provides to the remote radio head in a traditional CPRI-based system. The CPRI gateway platform can be used to experiment with these techniques in a lab environment, and also to evaluate the quality of time and frequency synchronization needed to maintain the quality of operation required at the RF level to maintain network access performance in the field.

For layer-1 physical offload, there are elements of the layer-1 LTE stack that are effectively associated with the antenna infrastructure rather than with the user traffic. As the remainder of the baseband processing becomes more virtualized in the cloud, it makes sense to move some of the layer-1 processing to the CPRI gateway and out of the base station. Initial work will move the FFT/IFFT blocks into the gateway to gain some efficiency increase from transporting OFDM symbols rather than I/Q samples over the Ethernet link. Further on, we can extend the integration up the LTE physical stack toward the boundary with the layer-2 processing. However, moving the processing out of the base station will create challenges in itself, such as concerns about the Hybrid-ARQ retransmission mechanism. At that time the platform could be used to evaluate the trade-offs in the potential split points of the stack. The split can be static or dynamic depending on the underlying silicon technology.

2.5   Proof-of-Concept Verification of Virtualized C-RAN

The challenges of virtualization implementation in C-RAN were described in the previous section. Since virtualization is mainly an implementation issue, in the work presented this section, a PoC was developed to verify the applicability of virtualized C-RAN.

In the PoC, a common server platform provided wireless services by optimizing the system itself with key technology adaptations such as the Data Plane Development Kit (DPDK). Whether a common platform was able to offer a qualified wireless protocol stack communication was validated in this test, and some relative performance indexes are measured.

This prototype simulated a full-duplex wireless communication path from the UE to Evolved Packet Core (EPC). In order to support wireless communication services,

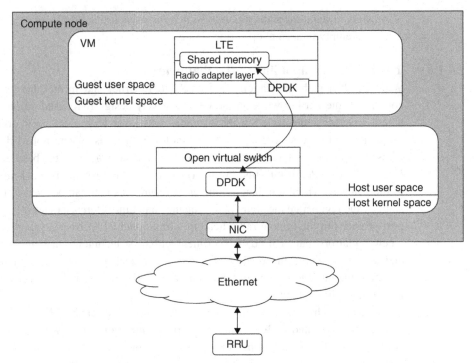

Figure 2.2 Transmission route of virtualized C-RAN prototype.

the performance of the common platform had to be improved, including increasing the service transmission speed, the memory size, and the reading/writing speed. The management system was based on an OpenStack cloud computing platform, and the LTE protocol stack services were implemented as a virtualized network function (VNF), which was running in one virtual machine (VM) or multiple VMs on the platform. For data transmission, we can take the uplink as an example. In the uplink LTE protocol stack packets are transmitted from the RRU to the platform. The packets enter the common platform through the network interface card (NIC). With the implementation of DPDK, packets were sent to the host OS user space directly. After that, the packets were forwarded by the OVS and entered specific VM(s) with the LTE protocol stack running within them. Finally, the packets from the UE were processed by the LTE protocol stack. The detailed transmission route is shown in Figure 2.2.

In this system, DPDK was used to accelerate data transmission. A server with a 3 GHz frequency and 16 cores was used as the common server. Other equipment information is listed below:

- Server: HP DL380 Gen8 with two sockets, eight cores per socket;
- CPU: Intel® Xeon®, E-2690 v2, 3.00 GHz;
- NIC: Intel 82599;
- memory: 64 GB (compute node).

Based on the physical framework described above, a Wind River real-time system was built first. The Wind River system performed optimization of a KVM system to improve the real-time performance in order to meet the wireless communication requirements.

The software running on the system is listed below:

- Wind River Titanium server based on Wind River Linux OVP 6 (Open Virtualization Platform);
- Enhanced OpenStack kilo;
- Wind River advanced open switch solutions.

The introduction of wireless network functions virtualization can support one-to-many or many-to-many relationships between the common processing platform and the RRU. To achieve data transmitting as in current communication networks, two types of connection streams needed to be established in this prototype. One was a baseband data stream and the other a control and management data stream. In the current prototype design, the first was designed as an I/Q data stream that carried data information. The second was designed as a control signaling stream called *Count*. The *Count* stream contained clock information that indicated the order of hyperframes. In the current system, the interval between hyperframes was set to 66.67 μs by the LTE protocol stack. Each hyperframe contained eight data packets and the size of each packet was 1024 bytes. Memory space was opened within the VM to store the Ethernet packet payload from the RRU. Through reading the memory, the LTE protocol stack could extract data for further processing.

2.5.1   Data Packets

In this prototype, Ethernet was used to connect the common platform and the RRU, and therefore the wireless data was encapsulated in the form of Ethernet packets. As synchronization of high accuracy is required by the LTE stack, an extra header was added to the packet to realize timing alignment. In this prototype, the header had to include the MAC header, IP header, UDP header, and radio over Ethernet (RoE) header.

The RoE header indicated the sequence of sending of Ethernet data to the LTE stack. It contained an eight-byte header and a payload. Key parameters in the header included:

- flow_id: the identification of a flow to distinguish different flows.
- length: to indicate the length of the RoE payload.
- ordering info: to indicate the order of packets. After the Ethernet header was disassembled, this parameter was used to offer data in the correct order for the LTE protocol stack. The memory arrangement process was simplified by this frame structure.

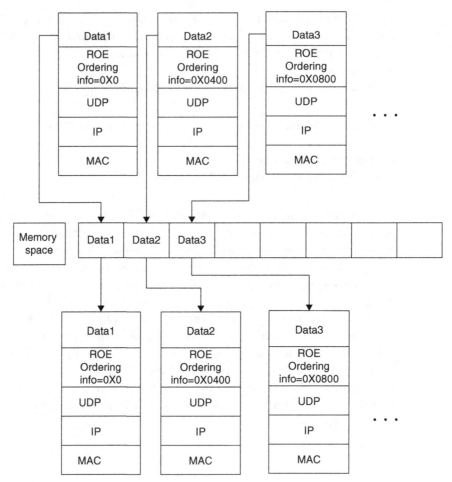

Figure 2.3 IQ packet processing of a hyperframe.

2.5.2   Test Procedure

In the test, data packets were simulated with a high-precision Ethernet tester. Packets traveled from the RRU to the virtualization platform. The tester and platform formed a test loop. The I/Q and *Count* packets were simulated by the Ethernet tester and entered a 10G network interface. After passing through a self-loop within the VM, they were exported from another 10G port back to the Ethernet tester. As shown in Figure 2.2, packet processing in the platform stopped at the shared memory without VM processing.

Figure 2.3 shows the processing of the I/Q packets of a hyperframe on the common platform. After an I/Q packet enters the VM, the DPDK running in the VM disassembles the packet. After the ordering information is read, the payload is placed in the right slots in memory, which simulates the uplink path of the wireless communication. For the downlink, DPDK reads data from memory and encapsulates them as Ethernet packets. The packets are sent by the virtualization platform back to the Ethernet tester.

Table 2.1 Test results in the case of a one-antenna RRU.

Type	Average latency (μs)	Average jitter (μs)	Packet loss rate (%)
Count	17.44	7.49	0
Data	20.588	6.98	0

Table 2.2 Test results in the case of a four-antenna RRU.

Type	Average latency (μs)	Average jitter (μs)	Packet loss rate (%)
Count	16	5	0.2
Data	29	0.8	0.2

Table 2.3 Test results in the case of an eight-antenna RRU.

Type	Average latency (μs)	Average jitter (μs)	Packet loss rate (%)
Count	18.956	5.3	0.3
Data	42.76	0.085	0.3

2.5.3   Test Results

Based on the scenario described above, the tester generated two data streams, and transmitted them to and received them from the virtualization platform in a loop. The packet loss rate, latency, and jitter were measured as output and recorded. According to the requirements of the wireless telecommunication application, the transmission rate of *Count* (the clock stream) was set to 123.596 Mbit/s. In other words, 150 Ethernet packets were sent every 10 ms. The I/Q packet size was set to 1240 bytes. With various bandwidth conditions, the parameters of average latency, jitter, and loss ratio were measured. The results are shown in Tables 2.1, 2.2, and 2.3.

Compared with the traditional L2-forward behavior on the common platform, the prototype platform had an increased latency due to the wireless services running on it. The first reason is that the wireless service threads and virtualization platform system threads were running simultaneously, contending for resources and therefore resulting in additional latency. Second, disassembling of Ethernet packets also increased the latency. On a traditional common platform, only a MAC header is removed when L2-forward packets are transmitted, and other higher-layer headers are not of concern. However, with wireless services running on the platform, the wireless data needs to be encapsulated with additional headers, including the MAC header, IP header, UDP header, and RoE header. The encapsulation and de-encapsulation of the headers lead to a latency increase. Third, memory copying increased the overall latency as well.

In our test, it was observed that the emulation of LTE services by the tester could operate normally for a long time without any break. Therefore, the test initially demonstrated our platform's capability to support wireless services with tolerable transmission latency and jitter. The test has laid a foundation and provided a guideline for further research. It is also worth pointing out that future commercial systems may

require even lower latency and jitter. The performance of the current platform still needs
to be improved to meet its goals.

2.6   Rethinking the Protocol Stack for C-RAN

In traditional base stations, the system software architecture is designed based on
vendors' traditional proprietary platforms, consisting of digital signal processors
(DSPs), application-specific integrated circuits (ASICs), etc., to meet the cell-centric
purpose. For C-RAN in 5G, systems will operate based on a COTS platform consisting
of standardized IT servers, switches, storage, and so on. All the resources will
be in the cloud and allocated on demand according to user needs. Thanks to the
differences between COTS platforms and traditional DSP-constituted platforms, and,
more importantly, the difference in the design principles between cell-centric and
user-centric, the whole software system architecture in C-RAN needs to be reconsidered
to exploit the cloud computing features and capabilities of COTS platforms as much as
possible.

2.6.1   Motivation

Although the framework of the traditional LTE protocol stack architecture is clear,
the signaling interaction in the protocol stack architecture is complex. The traditional
LTE protocol stack architecture is unable to support a high-density 5G network, a
massive number of users, and various kinds of businesses using 5G. We need to rethink
the protocol stack architecture in the 5G era. The protocol stack architecture should
be "user-centric," and provide a flexible air interface and reduce the frequency of
radio resource control (RRC) signaling transmission. Meanwhile, the protocol stack
architecture should take full advantage of the cloud, with its enormous computing
capability. Considering the high density of users and cells, with big data, the protocol
stack architecture needs to implement an optimized configuration of air interface
resources, for example frequency domain resources, time domain resources, and radio
resources.

2.6.2   Multilevel Centralized and Distributed Protocol Stack

In traditional LTE/LTE-A, the basic element of a communication network is the "cell,"
which manages the radio resources and the users connected to it. In the traditional LTE
protocol, which is shown in Figure 2.4, the UE context can only be established based
on a specific cell. In the case of carrier aggregation (CA), the UE context is established
based on the primary cell (Pcell) rather than the secondary cell (Scell), which only
provides channels for data transmission/receiving. In wireless systems, the interaction
between cells, for example handover, always leads to a change in the user context, and
therefore the rate of the change will impact the system performance to some extent. For
example, in the procedure for cell handover, the signaling interaction between cells is

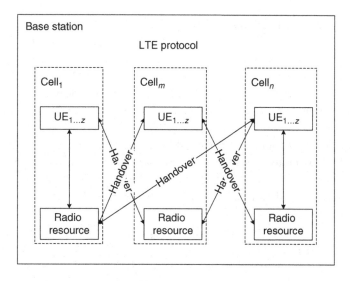

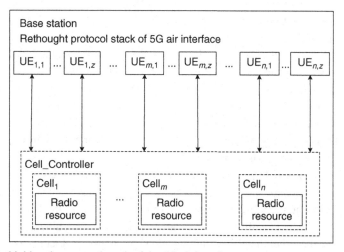

Figure 2.4 Rethinking the protocol stack of the 5G air interface.

slow, and the duration is on scale of several seconds or even minutes. In the case of intercell interference coordination (ICIC), the signaling interaction is semistatic.

In 5G networks, the UCN is introduced to solve the problem of the explosive growth of data traffic and the increasing density of deployed base stations. The signaling interaction in the UCN gets in the way of control plane/user plane (CP/UP) decoupling. In addition, several technologies such as full duplex and hybrid PHY have been introduced to enhance the capabilities of the air interface. One of the key requirements for the implementation of these technologies is that the interaction among the cells should occur in a real-time way, typically on the level of the transmission time interval (TTI), which is difficult for the existing cell-centric protocol stack framework. The signaling interaction is simply too slow to fulfill the 5G requirements with the traditional architecture.

Motivated by this, it was proposed in [2] that it is necessary to rethink the protocol stack of the air interface for the requirements of 5G. The traditional network, which has been characterized as cell-centric, has been proved to be a simple and practical method of radio resource management [13]. The protocol stack of the 5G air interface should inherit the advantages of traditional networks, be rethought to meet the requirements of 5G networks, and coexist with the traditional network protocol stack.

In this section, we introduce the concept of the Multilevel Centralized and Distributed (MCD) protocol stack for the 5G air interface, in which the UE makes decisions independently. In our proposal, the "UE" and the "cell" are both basic elements of communication networks, and cells become the dedicated radio resource management elements.

The difference between the rethought protocol stack and the traditional LTE protocol stack is shown in Figure 2.4. In the traditional LTE protocol stack, all the signaling and context of the UE treat the connected cell as the only key label; for example, the scope of the Cell Radio Network Temporary Identifier (C-RNTI) for each cell is 0–65535 [14]. The Data Resource Bearer (DRB) and Signaling Resource Bearer (SRB) are both allocated and managed in the scope of one cell, as well as the process of mapping to the E-UTRAN Radio Access Bearer (E-RAB). With CA, the UE can use the resources of more than one cell. However, as a supplementary technology to the LTE protocol stack, CA cannot assist the LTE network in solving the problem in 5G. In summary, with "cell" as the key label, the LTE protocol stack simplifies the process of radio resource management for the network [15]. On the other hand, such a protocol stack increases the complexity of management and leads to a long delay in UE mobility, and it is hard to meet the needs of 5G with it.

In the MCD protocol stack, "UE" becomes a basic element as well as "cell." On the one hand, the UE is responsible for the management of all the information itself, including the UE context, the mapping process from the DRB and SRB to the E-RAB, the channel quality, the dedicated radio resources allocated to the UE, and so on. On the other hand, as another element of the protocol, the cell manages all radio resources that are not allocated to any users. As shown in Figure 2.4, the Cell_Controller module manages cells, which allocate their own radio resources based on the allocation result of the Cell_Controller. According to the specific requirements of each UE, a Cell_Controller allocates corresponding cells to it with the required radio resources to fulfill the demands of the UE. The radio resources which are allocated to the UE become specific attributes of the UE to manage radio resources. The UE returns radio resources to a cell when the transmission process ends. Compared with the traditional protocol stack, the MCD protocol stack for resource allocation is in the form of changes to the UE attributes, which work in the same way as UE context modification. Such operation avoids changes to the DRB and logical channel when the radio resources change. From the macroscopic perspective, when UE mobility leads to radio resource changes, handover is replaced by modification of the UE radio resource attribute via faster radio resource deployment of the air interface. With the MCD protocol stack, it is

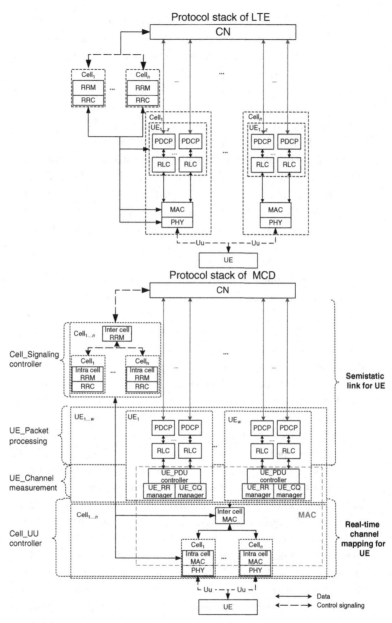

Figure 2.5 MCD protocol stack.

expected that the radio resource process could be carried out in a much faster way than in traditional systems.

As shown in Figure 2.5, the functional blocks in the protocol stack can be divided into two categories: those to maintain the semistatic link for the UE, and those with real-time channel mapping for the UE.

The semistatic link for the UE focuses on logical channels, the DRB/SRB and E-RAB links, and the links indicating the specific type of Service (ToS) of the UE. To implement unbundling the ToS with a specific PHY mode in Hybrid PHY, these links work in a semistatic way. When the protocol runs in a C-RAN network, these links work only with the modes of establishment, reconfiguration, and deletion. When a UE moves to the coverage of another C-RAN, the handover of UE context is realized by redefining handover signaling. In this way, a semistatic pattern is used to realize the zero-handover, which is critical for scenarios with high-density cells.

For the real-time channel mapping, the mapping of the logical channels to the transport channels and of the transport channels to the physical channels is done in real time [16]. The UE provides parameters to the MAC layer such as the quality of the channel, the buffer occupied, the request for PHY, and the characteristics of the allocated radio resources. Based on the available radio resources of all the cells and the parameters received, the MAC layer configures a cell and its radio resources as attributes of the UE. In that way, the logical channel matches the appropriate cell first, and then maps the transport channel to the physical channel in that cell. With real-time modification of the air interface rather than handover, the UE can receive data from different cells.

In the traditional LTE protocol stack architecture, a handover process is necessary when the UE moves across cells. However, in the 5G MAC protocol stack architecture, this real-time mapping replaces the handover process. Furthermore, the handover of a semistatic link for the UE module takes place only when the UE moves to another C-RAN. In addition, flexible control of the air interface could be achieved by decreasing the frequency and complexity of handover, decreasing the signal interaction delay, and increasing the reliability at the same time.

The MCD protocol stack architecture with bundling/unbundling of "UE" and "cell" keeps the advantages of the traditional LTE protocol stack for radio resource management but with increased flexibility and stability. The mapping between UE and cell is completed by fast MAC mapping, which ensures the flexibility of the UE. On the other hand, it requires little change in the link in the upper-level protocol, which keeps the stability of the cell.

When it comes to implementation, a COTS platform could be a good place for the MCD protocol stack. COTS platforms have been viewed as important platforms for realizing future 5G networks because of their strong computational capabilities and their dynamically adaptive hardware management capabilities. The MCD protocol stack redesigns the architecture of the protocol stack from the point of view of the centralized and distributed control plane and user plane. With MCD, traditional resource management functions which are realized in the RRC could be divided further into two parts. Some functions remain in the RRC, while others go to the MAC layer. It is proposed that the RRC manages radio resources with low latency requirements, and the MAC layer manages radio resources with high latency requirements at the UE level.

MCD uses hierarchical centralized wireless access control and implements fast/slow resource scheduling, coordination, and handover across access points. It implements real-time collaboration of air interfaces across cells at the same time. The centralized control plane and user plane can both make full use of the computing capability of the cloud platform. The distributed control plane dynamically adjusts according to the load on the network, which makes full use of the adaptation abilities of the hardware resources of the cloud platform. The distributed user plane can rapidly initiate different I/O when massive data are transmitted between the layers of the protocol stack. The MCD protocol stack divides the PHY and the intracell MAC from the intercell MAC and the upper-layer protocol, and the functions of the PHY and intracell MAC could be implemented in accelerators.

2.7   Conclusion

Future 5G networks should be user-centric rather than cell-centric as traditional, to fulfill diverse scenarios, applications, and user demands. To this end, C-RAN has been viewed as the key enabler. In this chapter, C-RAN, with its major advantages, key challenges, and potential solutions has been presented.

Fronthaul has been a critical issue for C-RAN centralization. When it comes to 5G, the problem will become more obvious. To address this challenge, a new FH interface called the Next Generation Fronthaul Interface has been introduced. NGFI aims to address the shortcomings of traditional FH interfaces, such as low transmission efficiency and poor scalability, and ultimately aims to facilitate large-scale deployment of C-RAN as well as to support other 5G technologies. The key to realizing NGFI lies in the redesign of the BBU–RRH function split and packetization of FH data. An NGFI platform prototype has been developed which allows us to experiment with key NGFI-related technologies, including frequency and time synchronization, and layer-1 PHY offload in the future.

One of the ultimate goals of C-RAN is to realize resource cloudification with implemented virtualization technology. As a pioneer on this front, we have developed a prototype of optimized virtualization platform with enhanced real-time performance. We have demonstrated the platform's capability to support wireless services with tolerable transmission latency and jitter. Furthermore, it lays down a foundation and provides a guideline for further research.

Protocol stack restructuring is a new but important topic in the context of C-RAN to support 5G. In this chapter, the Multilevel Centralized and Distributed (MCD) protocol stack was proposed. With MCD, both "UE" and "cell" are basic elements of communication networks, and cells become the dedicated radio resource management elements. MCD requires a protocol stack repartition and it is proposed that the functional blocks in the protocol stack could be divided into two categories: those to maintain the semistatic link for the UE and those with real-time channel mapping for the UE. MCD is expected to enhance multicell cooperation and cross-cell experience, and to better support dense network deployment in 5G.

Acknowledgments

We would like to express our sincere gratitude to our partners, including Xilinx for the development of the CPRI gateway for the NGFI platform and Intel and Wind River for the development of the virtualized RAN PoC. In addition, we thank the experts Dr. Perminder Tumber and Dr. Gareth Edwards from Xilinx for their contributions to the chapter. We also owe special thanks to all C-RAN team members in China Mobile for helpful discussions and valuable comments. This work was partly supported by the National High-Tech R&D Program of China (Grants No. 2014AA01A703 and No. 2014AA01A704).

References

[1] Cisco Systems, "Cisco visual networking index: Global mobile data traffic forecast update, 2015–2020," Feb. 2016. Available at www.cisco.com/c/en/us/solutions/collateral/service-provider/visual-networking-index-vni/mobile-white-paper-c11-520862.html.

[2] C. I, S. Han, Z. Xu, S. Qi, and Z. Pan, "5G: Rethink mobile communications for 2020+," *Philos. Trans. R. Soc. A: Math., Phys. and Eng. Sci.*, vol. 374, no. 2062, 20140432, Jan. 2016.

[3] C. I, C. Rowell, S. Han, Z. Xu, G. Li, and Z. Pan, "Toward green and soft: A 5G perspective," *IEEE Commun. Mag.*, vol. 52, no. 2, pp. 66–73, Feb. 2014.

[4] C. I, Y. Yuan, J. Huang, S. Ma, R. Duan, and C. Cui, "Rethink fronthaul for soft RAN," *IEEE Commun. Mag.*, vol. 53, no. 9, pp. 82–88, Sep. 2015.

[5] China Mobile Research Institute, "C-RAN: The road towards green RAN," White paper, version 3.0, Dec. 2013. Available at http://labs.chinamobile.com/cran/wp-content/uploads/2014/06/20140613-C-RAN-WP-3.0.pdf.

[6] C. I, J. Huang, R. Duan, C. Cui, J. Jiang, and L. Li, "Recent progress on C-RAN centralization and cloudification," *IEEE Access*, vol. 2, pp. 1030–1039, Sep. 2014.

[7] Ericsson, Huawei Technologies, NEC, Alcatel Lucent and Nokia Siemens Networks, "Common Public Radio Interface (CPRI) specification (v6.0)," Tech. Rep., Aug. 2013. Available at www.cpri.info/downloads/CPRI_v_6_0_2013-08-30.pdf.

[8] China Mobile Research Institute, "White paper of next generation fronthaul interface," White paper, Version 1.0. Available at http://labs.chinamobile.com/cran/wp-content/uploads/WhitePaperofNextGenerationFronthaulInterface.PDF 2015.

[9] C. I, J. Huang, Y. Yuan, S. Ma, and R. Duan, "NGFI, the xHaul," in *Proc. of IEEE Globecom Workshops (GC Wkshps)*, Dec. 2015.

[10] D. Wubben, P. Rost, J. Bartelt, M. Lalam, and V. Savin, "Benefits and impact of cloud computing on 5G signal processing: Flexible centralization through cloud-RAN," *IEEE Signal Process. Mag.*, vol. 31, no. 6, pp. 35–44, Nov. 2014.

[11] A. Davydov, G. Morozov, I. Bolotin, and A. Papathanassiou, "Evaluation of joint transmission CoMP in C-RAN based LTE-A HetNets with large coordination areas," in *Proc. of IEEE Globecom Workshops (GC Wkshps)*, Dec. 2013.

[12] ETSI, "LTE; Evolved Universal Terrestrial Radio Access (E-UTRA); Base station (BS) radio transmission and reception (release 11)," ETSI TS 136 104, Version 11.4.0, 2013.

Available at www.etsi.org/deliver/etsi_ts/136100_136199/136104/11.04.00_60/ts_136104 v110400p.pdf.

[13] ETSI, "E-UTRAN S1 general aspects and principles," ETSI TS 36.410, Version 12.1.0, Dec. 2014.

[14] ETSI, "E-UTRAN S1 application protocol," ETSI TS 36.413, Version 12.6.0, Jun. 2015.

[15] ETSI, "E-UTRAN overall description," ETSI TS 36.300, Version 12.6.0, Jun. 2015.

[16] ETSI, "E-UTRAN physical channels and modulation," ETSI TS 36.211, Version 12.6.0, Jun. 2015.

3 Fronthaul-Aware Design for Cloud Radio Access Networks

Liang Liu, Wei Yu, and Osvaldo Simeone

3.1 Introduction

The cloud radio access network (C-RAN) is an emerging paradigm for fifth generation (5G) wireless cellular networks, according to which the traditional physical layer base station (BS) transmission and reception infrastructure is virtualized using cloud computing techniques. The virtualization of wireless access also enables centralized control and management of wireless access points, which provides significant benefit from a transmission spectral efficiency perspective. In the current third generation (3G) and fourth generation (4G) cellular networks, each user, also called the user equipment or UE, is served solely by its own BS. This traditional single-cell processing paradigm, shown in Figure 3.1, suffers from considerable intercell interference, especially for cell edge users. In the C-RAN paradigm shown in Figure 3.2, as the BSs are coordinated centrally from the cloud, they can potentially transmit and receive radio signals to/from users jointly, thereby creating the possibility for interference cancellation, which can significantly improve the overall network throughput.

In a C-RAN architecture, the traditional BSs essentially become remote radio heads (RRHs) that serve to relay information between the mobile users and the central processor (CP) in the cloud. Baseband processing, together with its associated decoding/encoding complexities, is implemented in the cloud rather than taking place locally at each BS as in traditional 3G/4G networks. As the RRHs in C-RAN require only rudimentary wireless access capabilities, they are much more cost-effective to deploy, therefore allowing the C-RAN architecture to be more easily scaled geographically, leading to denser deployment of remote antennas and the ability of the network to support many more users. Furthermore, as baseband units (BBUs) are now implemented centrally at the CP, the C-RAN architecture allows the pooling of computational resources across the entire network, leading to better utilization of the computational units and higher energy efficiency for the network.

Because of both the distributed nature of wireless antenna placement and the centralized nature of cloud computing resources in C-RAN, the communication links between the RRHs and the CP are of central importance in C-RAN design. These links are often referred to as *fronthaul* links, as they connect the radio front end with the BBUs implemented in the cloud (in contrast to the *backhaul* links between the traditional

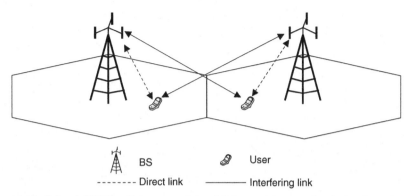

Figure 3.1 Traditional 3G/4G cellular network: each BS serves its associated users in each cell; cell edge users suffer from severe interference.

3G/4G BSs and the backbone network). The fronthaul links are typically implemented with fiber optics, but they can also be implemented as wireless links, especially for pico- and femto-BSs in heterogeneous networks (HetNets), where self-backhauling is increasingly desirable.

The capacity and the latency performance of the fronthaul links have significant impact on the design of C-RAN. For example, the current standardized Common Public Radio Interface (CPRI), which defines the communication protocol between the RRH and the BBU, specifies fronthaul rates ranging from hundreds of Mbps to tens of Gbps. When multiple RRHs are aggregated, the deluge of data required to be transported between the RRHs and the cloud can easily overwhelm the physical limitations of practical fronthaul implementations. Furthermore, as the C-RAN architecture now allows the BBUs to be physically located much further away from the RRHs, the ensuing latency would have a significant impact on the overall delay performance of the network.

This chapter aims to illustrate how the physical and data link layer design of a C-RAN system can adapt to the capacity and latency limitations of the fronthaul links. The first part of the chapter provides an information-theoretical evaluation of the achievable rates of C-RAN with the impact of finite-capacity fronthaul links taken into account. Toward this end, a practical user–RRH clustering strategy and various fronthaul techniques for implementing uplink and downlink beamforming in C-RAN are considered. The achievable rates subject to the fronthaul capacity constraints are evaluated. In the second part of the chapter, the effect of fronthaul latency on the throughput and efficiency of the data link layer is discussed. Toward this end, a novel design of a hybrid automatic repeat request (hybrid ARQ or HARQ) protocol is proposed to circumvent the additional delay caused by the fronthaul transmission.

3.2 Fronthaul-Aware Cooperative Transmission and Reception

A key benefit of the C-RAN architecture as compared with traditional single-cell processing is that it enables cooperative transmission and reception across multiple

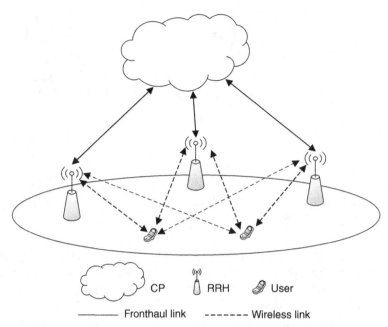

Figure 3.2 C-RAN: the RRHs serve the users under the coordination of the central processor (CP) via finite-capacity fronthaul links; intercell interference can be effectively mitigated.

RRHs via *beamforming*. This section illustrates beamforming design techniques for both uplink and downlink C-RAN and characterizes the theoretical achievable rates for users of a C-RAN deployment as functions of the fronthaul capacity constraints.

Consider a typical C-RAN deployment as depicted in Figure 3.2, where a cluster of RRHs, each equipped with multiple antennas, cooperatively serve multiple single-antenna users under the coordination of the CP via finite-capacity fronthaul links. The fronthaul links are modeled as finite-capacity noiseless digital links with some fixed capacity for each link. We remark that although analog optical modulation using the radio-over-fiber technique is also an alternative, the digital fronthaul model is adopted here, because it is easier to implement in practice.

To illustrate the benefits of cooperation in C-RAN, the following network model is adopted in this section. Assuming a network with N RRHs each equipped with M antennas, each user is associated with its strongest RRH. From its associated users, each RRH schedules $K < M$ users at each time slot for service. Furthermore, to achieve a cooperation gain, each scheduled user is jointly served by a cooperative cluster of RRHs. Specifically, in the uplink, each RRH forwards its received signal to the CP over its fronthaul link. The CP then decodes each user's message based on the signals received from the RRHs in the cooperative cluster of the user. In the downlink, with coordination from the CP, the transmit signal of each RRH is designed as a function of the messages of users whose cooperation cluster includes the RRH. This enables joint transmission across the cluster of RRHs to each user. Note that the size of the cooperation cluster depends on the ability to acquire the channel state information (CSI) for the channels between the RRHs and the users. For simplicity, the cluster size is assumed to be fixed in this section.

This section assumes a *user-centric* clustering strategy, in which each user is always placed at the center of its cooperation cluster, but the clusters for different users may overlap. As compared with disjoint clustering (which partitions the entire network into disjoint sets of cooperating RRHs), user-centric clustering has the advantage of completely eliminating cluster edges, hence resulting in better fairness in rate distribution across the users [1].

The remainder of this section describes a particular zero-forcing (ZF) beamforming strategy across the user-centric clusters using various fronthaul techniques for both uplink and downlink C-RAN. Notationally, let $\mathcal{K}$ denote the set of scheduled users in a particular time slot, and let Θ_k denote the cooperative cluster of RRHs for user $k \in \mathcal{K}$, with $D_k = |\Theta_k|$ being the cluster size. As each RRH in Θ_k schedules K active users, a sensible strategy is to zero-force all the interference due to the signals of the scheduled users in the cluster. We use Ω_k to denote the set of users scheduled by all the RRHs in Θ_k (i.e., the cluster for user k), so that $|\Omega_k| = KD_k$. Finally, from the RRHs' perspective, it is convenient to define the set of all users whose signals are zero-forced by RRH n by Φ_n, i.e., $\Phi_n = \{k : n \in \Theta_k, k = 1, \cdots, K\}$. Throughout this section, the wireless channels between the RRHs and the users are assumed to be quasi-static flat-fading channels over a fixed bandwidth of B Hz. The fronthaul capacity between each RRH and the CP is assumed to be C bits per second (bps).

3.2.1 Uplink

In the uplink C-RAN, at each time slot each RRH's observed signal is a superposition of the signals sent from all the scheduled users in the set $\mathcal{K}$. Specifically, let x_k^{ul} denote the transmit signal of user k, and $\mathbf{y}_n^{\text{ul}} \in \mathbb{C}^{M \times 1}$ denote the received signal at RRH n, then

$$\mathbf{y}_n^{\text{ul}} = \sum_{k \in \mathcal{K}} \mathbf{h}_{n,k}^{\text{ul}} x_k^{\text{ul}} + \mathbf{z}_n^{\text{ul}}, \quad \forall n, \tag{3.1}$$

where $\mathbf{h}_{n,k}^{\text{ul}} \in \mathbb{C}^{M \times 1}$ is the uplink channel from user k to RRH n, and $\mathbf{z}_n^{\text{ul}} \in \mathbb{C}^{M \times 1} \sim \mathcal{CN}(\mathbf{0}, \sigma_{\text{ul}}^2 \mathbf{I})$ denotes additive white Gaussian noise (AWGN).

Observe that the received signal of each RRH contains useful information even for the users that are not associated with it. However, the single-cell processing mechanism of the current 3G/4G cellular networks cannot take advantage of this information, as it restricts decoding to being done locally at each BS for its own associated users. To make the best utilization of the received signals in the entire network, the RRHs in C-RAN need to relay their observed signals to the CP over the fronthaul links, so that the messages of each user can be decoded by the CP based on the observations of all the RRHs serving that user.

If the fronthaul links have infinite capacities, each RRH can convey perfectly its observed signal to the CP, and the full joint-decoding gain can be achieved. In practice, however, the fronthaul links have finite capacities, thus each RRH can only convey an approximate version of its received signal. An interesting but essential question arises: What is the appropriate way for the RRHs to preserve as much information as possible when relaying their observations to the CP, while satisfying the finite-capacity constraints of the fronthaul links?

The basic strategy is that the RRHs should compress their received signal. Below, we describe two uplink fronthaul compression techniques. When the C-RAN system is fully loaded, i.e., when the spatial dimensions at the RRHs are fully occupied by all the users, a *compress–forward* strategy [2, 3] works very well. When the C-RAN is lightly loaded, i.e., when there are excess spatial dimensions compared with the number of active users, it may be advantageous to employ a *beamform–compress–forward* strategy [3, 4].

3.2.1.1   Compress-Forward Strategy

In the compress–forward strategy, the RRHs first downconvert their received RF signals to baseband signals, which are analog in nature, and then compress the baseband signals and send the corresponding compression indices, which are represented by digital codewords, to the CP. After receiving the compression indices, the CP first decompresses these quantized signals in order to recover a distorted version of the received signals across all the RRHs, and then decodes the user messages based on the entire set of decompressed signals.

Intuitively, the compression resolution is determined by the available fronthaul capacity, i.e., a more stringent fronthaul capacity constraint would imply "coarser compression," which in rate–distortion theory is reflected by a larger quantization noise. We note that the optimal compression in the C-RAN setting would involve vector quantization across the antennas and Wyner–Ziv compression across the RRHs, which are techniques capable of taking advantage of the fact that the received signals across the antennas and across the RRHs are correlated. But, for simplicity, the model in the rest of this section assumes scalar quantization modeled by independent additive Gaussian quantization noise.

With independent compression across RRHs and scalar quantization across antennas at each RRH, we can now describe the compress–forward strategy in uplink C-RAN as follows. For simplicity, we assume that all users transmit with an identical power, denoted by p_u, so that the transmit signal of user k is expressed as $x_k^{ul} = \sqrt{p_u} s_k^{ul}$, where $s_k^{ul} \sim \mathcal{CN}(0, 1)$ denotes the message for user k, chosen from a Gaussian codebook. With the channel model as described in (3.1), the discrete-time baseband signal received by RRH n is given by

$$y_n^{ul} = \sum_{k \in \mathcal{K}} \sqrt{p_u} h_{n,k}^{ul} s_k^{ul} + z_n^{ul}, \quad \forall n. \tag{3.2}$$

The scalar quantization process at the mth antenna of the nth RRH is modeled as a Gaussian test channel with the uncompressed signal as the input and the compressed signal as the output, i.e.,

$$\tilde{y}_{n,m}^{ul} = y_{n,m}^{ul} + e_{n,m}^{ul} = \sum_{k \in \mathcal{K}} \sqrt{p_u} h_{n,m,k}^{ul} s_k^{ul} + z_{n,m}^{ul} + e_{n,m}^{ul}, \quad \forall n, m, \tag{3.3}$$

where $h_{n,m,k}^{\mathrm{ul}}$ is the channel from user k to the mth antenna of RRH n, and $z_{n,m}^{\mathrm{ul}}$ is the Gaussian noise at the mth antenna of RRH n; $e_{n,m}^{\mathrm{ul}} \sim \mathcal{CN}(0, q_{n,m}^{\mathrm{ul}})$ denotes the quantization noise in the compression of $y_{n,m}^{\mathrm{ul}}$, and $q_{n,m}^{\mathrm{ul}}$ denotes its variance. Note that since independent compression is employed across the RRHs and scalar quantization is employed at each RRH, the quantization noises $e_{n,m}^{\mathrm{ul}}$ are independent over the RRHs and the antennas.

With the above Gaussian test channel model, the design of the compression codebook is equivalent to setting the variances of the compression noise. To achieve higher compression resolution, which leads to higher achievable rates in the uplink C-RAN, the quantization noise must be made as small as possible at each RRH. However, the minimum amount of quantization noise is also limited by the fronthaul capacity, as given by rate–distortion theory. In practice, assuming a gap Γ_{q} from the rate–distortion limit, the fronthaul capacity in bps required to transmit $\tilde{y}_{n,m}^{\mathrm{ul}}$ to the CP can be expressed as

$$C_{n,m}^{\mathrm{ul}} = B \log_2 \left(\frac{\Gamma_{\mathrm{q}} \left(\sum_{k \in \mathcal{K}} p_{\mathrm{u}} |h_{n,m,k}^{\mathrm{ul}}|^2 + \sigma_{\mathrm{ul}}^2 \right) + q_{n,m}^{\mathrm{ul}}}{q_{n,m}^{\mathrm{ul}}} \right), \quad \forall m, n. \tag{3.4}$$

For simplicity, the total fronthaul capacity of RRH n is assumed to be equally allocated to its M antennas, i.e., $C_{n,m}^{\mathrm{ul}} = C/M$, $\forall m$. From (3.4), the variance of the quantization noise for the compression $y_{n,m}^{\mathrm{ul}}$ is then given by

$$q_{n,m}^{\mathrm{ul}} = \frac{\Gamma_{\mathrm{q}} \left(\sum_{k \in \mathcal{K}} p_{\mathrm{u}} |h_{n,m,k}^{\mathrm{ul}}|^2 + \sigma_{\mathrm{ul}}^2 \right)}{2^{C/BM} - 1}, \quad \forall n, m. \tag{3.5}$$

This allows us to derive the achievable rate of each user as follows. The CP decodes the message of user k based on the signals sent from its serving RRHs in the cooperation cluster, i.e., the set Θ_k. We denote the received signal across Θ_k by $\tilde{\mathbf{y}}^{\mathrm{ul},k} = [\cdots, \tilde{y}_{n,1}^{\mathrm{ul}}, \cdots, \tilde{y}_{n,M}^{\mathrm{ul}}, \cdots]_{n \in \Theta_k}^{\mathrm{T}} \in \mathbb{C}^{MD_k \times 1}$, $\forall k$. For convenience, we define $\mathbf{g}_{k,i}^{\mathrm{ul}} = [\cdots, (\mathbf{h}_{n,i}^{\mathrm{ul}})^{\mathrm{T}}, \cdots]_{n \in \Theta_k}^{\mathrm{T}} \in \mathbb{C}^{MD_k \times 1}$ as the collective channel vector from user i to the RRHs in Θ_k. Then,

$$\tilde{\mathbf{y}}^{\mathrm{ul},k} = \underbrace{\sqrt{p_{\mathrm{u}}} \mathbf{g}_{k,k}^{\mathrm{ul}} s_k^{\mathrm{ul}}}_{\text{desired signal}} + \underbrace{\sum_{i \neq k, i \in \Omega_k} \sqrt{p_{\mathrm{u}}} \mathbf{g}_{k,i}^{\mathrm{ul}} s_i^{\mathrm{ul}}}_{\text{intracluster interference}} + \underbrace{\sum_{j \notin \Omega_k} \sqrt{p_{\mathrm{u}}} \mathbf{g}_{k,j}^{\mathrm{ul}} s_j^{\mathrm{ul}}}_{\text{intercluster interference}} + \tilde{\mathbf{z}}_k^{\mathrm{ul}} + \tilde{\mathbf{e}}_k^{\mathrm{ul}}, \quad \forall k, \tag{3.6}$$

where $\tilde{\mathbf{z}}_k^{\mathrm{ul}}$ and $\tilde{\mathbf{e}}_k^{\mathrm{ul}}$ are the collective AWGN and quantization noises across the RRHs in Θ_k, with covariances $\mathbf{S}_{\tilde{z},k}^{\mathrm{ul}} = \mathbb{E}[\tilde{\mathbf{z}}_k^{\mathrm{ul}}(\tilde{\mathbf{z}}_k^{\mathrm{ul}})^{\mathrm{H}}] = \sigma_{\mathrm{ul}}^2 \mathbf{I}$ and $\mathbf{S}_{\tilde{e},k}^{\mathrm{ul}} = \mathbb{E}[\tilde{\mathbf{e}}_k^{\mathrm{ul}}(\tilde{\mathbf{e}}_k^{\mathrm{ul}})^{\mathrm{H}}] = \mathrm{diag}([\cdots, q_{n,1}^{\mathrm{ul}}, \cdots, q_{n,M}^{\mathrm{ul}}, \cdots]_{n \in \Theta_k}^{\mathrm{T}})$, respectively.

The CP applies a linear beamformer $\mathbf{w}_k^{\text{ul}} \in \mathbb{C}^{MD_k \times 1}$ with unit norm to $\tilde{\mathbf{y}}^{\text{ul},k}$ for decoding s_k^{ul}:

$$\hat{s}_k^{\text{ul}} = \sqrt{p_{\text{u}}}(\mathbf{w}_k^{\text{ul}})^{\text{H}}\mathbf{g}_{k,k}^{\text{ul}}s_k^{\text{ul}} + \sum_{i \neq k, i \in \Omega_k} \sqrt{p_{\text{u}}}(\mathbf{w}_k^{\text{ul}})^{\text{H}}\mathbf{g}_{k,i}^{\text{ul}}s_i^{\text{ul}}$$

$$+ \sum_{j \notin \Omega_k} \sqrt{p_{\text{u}}}(\mathbf{w}_k^{\text{ul}})^{\text{H}}\mathbf{g}_{k,j}^{\text{ul}}s_j^{\text{ul}} + (\mathbf{w}_k^{\text{ul}})^{\text{H}}\bar{\mathbf{z}}_k^{\text{ul}} + (\mathbf{w}_k^{\text{ul}})^{\text{H}}\bar{\mathbf{e}}_k^{\text{ul}}, \quad \forall k. \tag{3.7}$$

For the practical choice of $\mathbf{w}_k^{\text{ul}}$, this section considers the ZF beamforming technique, where the intracluster interference due to users in Ω_k is completely eliminated, i.e., $(\mathbf{w}_k^{\text{ul}})^{\text{T}}\mathbf{g}_{k,i}^{\text{ul}} = 0$, $\forall i \in \Omega_k$ and $i \neq k$. To achieve this goal, the following ZF beamforming vectors can be utilized:

$$\mathbf{w}_k^{\text{ul}} = \frac{(\mathbf{I} - \mathbf{G}_{-k}^{\text{ul}}(\mathbf{G}_{-k}^{\text{ul}})^{\dagger})\mathbf{g}_{k,k}^{\text{ul}}}{\|(\mathbf{I} - \mathbf{G}_{-k}^{\text{ul}}(\mathbf{G}_{-k}^{\text{ul}})^{\dagger})\mathbf{g}_{k,k}^{\text{ul}}\|_2}, \quad \forall k, \tag{3.8}$$

where $\mathbf{G}_{-k}^{\text{ul}} = [\cdots, \mathbf{g}_{k,i}^{\text{ul}}, \cdots]_{i \neq k, i \in \Omega_k} \in \mathbb{C}^{MD_k \times (KD_k - 1)}$ denotes the collection of channels from the intracluster users (excluding user k) to the RRHs serving user k, and $(\mathbf{G}_{-k}^{\text{ul}})^{\dagger}$ denotes the pseudo-inverse of $\mathbf{G}_{-k}^{\text{ul}}$.

The achievable rate of user k under the above compress–forward scheme can now be characterized as

$$r_k^{\text{ul,CF}} = B\log_2\left(1 + \frac{p_{\text{u}}\left|(\mathbf{w}_k^{\text{ul}})^{\text{H}}\mathbf{g}_{k,k}^{\text{ul}}\right|^2}{\left(\sum_{j \neq \Omega_k} p_{\text{u}}\left|(\mathbf{w}_k^{\text{ul}})^{\text{H}}\mathbf{g}_{k,j}^{\text{ul}}\right|^2 + \sigma_{\text{ul}}^2 + (\mathbf{w}_k^{\text{ul}})^{\text{H}}\mathbf{S}_{\tilde{\mathbf{e}},k}^{\text{ul}}\mathbf{w}_k^{\text{ul}}\right)\Gamma_{\text{s}}}\right), \tag{3.9}$$

where Γ_{s} is the signal-to-interference-plus-noise ratio (SINR) gap due to the coding and modulation scheme used in practice.

3.2.1.2   Beamform–Compress–Forward Strategy

The compress–forward strategy discussed above compresses the received signals at each antenna independently; the fronthaul capacity is shared across all antennas. This compression strategy can be shown to be close to optimum when the system is fully loaded (i.e., it schedules as many users as there are antennas) and operates at high signal-to-quantization-noise ratio (SQNR) with equal quantization noise level (rather than equal allocation of quantization bits) applied across the antennas [3]. In general, however, it may not be the most efficient use of the limited fronthaul, especially for systems with many more antennas at the RRHs than the number of scheduled users. This is an increasingly possible scenario with the emerging massive multiple-input multiple-output (MIMO) technology. To address this issue, in the following we propose a beamform–compress–forward scheme, where each RRH first performs beamforming of its received signals across its antennas, followed by compression in a reduced-dimensional space.

Specifically, the beamforming operation applied by RRH n to its received signal $\mathbf{y}_n^{\text{ul}}$ given in (3.2) can be modeled as

$$\hat{\mathbf{y}}_n^{\text{ul}} = \mathbf{V}_n^{\text{ul}} \mathbf{y}_n^{\text{ul}} = \sum_{k \in \mathcal{K}} \sqrt{p_u} \mathbf{V}_n^{\text{ul}} \mathbf{h}_{n,k}^{\text{ul}} s_k^{\text{ul}} + \mathbf{V}_n^{\text{ul}} \mathbf{z}_n^{\text{ul}}, \quad \forall n, \tag{3.10}$$

where $\mathbf{V}_n^{\text{ul}} = [\mathbf{v}_{n,1}^{\text{ul}}, \cdots, \mathbf{v}_{n,L_n}^{\text{ul}}]^{\text{T}} \in \mathbb{C}^{L_n \times M}$ denotes the beamforming matrix at RRH n, with $L_n \leq M$ denoting the (reduced) dimension of the output signal $\hat{\mathbf{y}}_n^{\text{ul}}$ after beamforming. The beamformers $\mathbf{V}_n^{\text{ul}}$ at the RRHs essentially transform the effective channel between the users and the RRHs from $\mathbf{h}_{n,k}^{\text{ul}} \in \mathbb{C}^{M \times 1}$ to $\mathbf{V}_n^{\text{ul}} \mathbf{h}_{n,k}^{\text{ul}} \in \mathbb{C}^{L_n \times 1}$, $\forall k, n$, while transforming the effective noise at the RRHs from $\mathbf{z}_n^{\text{ul}} \sim \mathcal{CN}(\mathbf{0}, \sigma_{\text{ul}}^2 \mathbf{I})$ to $\mathbf{V}_n^{\text{ul}} \mathbf{z}_n^{\text{ul}} \sim \mathcal{CN}(\mathbf{0}, \sigma_{\text{ul}}^2 \mathbf{V}_n^{\text{ul}} (\mathbf{V}_n^{\text{ul}})^{\text{H}})$, $\forall n$. As a result, the aforementioned compress–forward strategy can be applied to the new uplink C-RAN channel model given in (3.10), and the achievable rate of user k under the beamform–compress–forward scheme can be derived similarly to (3.9) with the new effective channels and noises. The remaining question is how to determine the beamforming vectors at each RRH to maximize the achievable rates.

The above question can be answered through an optimization framework involving quantization noise covariance matrices across the RRHs [3], but such an optimization is complex and it assumes vector quantization across the antennas. Below, we offer a heuristic approach of first determining L_n, the dimension of the beamforming matrix at each RRH n, and then finding a beamformer through identifying the principal component of the received signal. The proposed heuristic is to set

$$L_n = \min(|\Phi_n|, M) \tag{3.11}$$

at each RRH n. Recall that Φ_n is the set of users served by RRH n. Thus, for a lightly loaded C-RAN system in which the number of users being served is less than the number of antennas, each RRH n compacts its received signal $\mathbf{y}_n^{\text{ul}}$ into $|\Phi_n|$ dimensions to preserve the useful information for its served users.

Next, to extract the L_n most informative dimensions, we perform a singular-value decomposition of the covariance matrix of the received signal of RRH n as $\mathbf{S}_{\text{y},n}^{\text{ul}} = \mathbf{U}_n \Lambda_n \mathbf{U}_n^{\text{H}}$, where

$$\mathbf{S}_{\text{y},n}^{\text{ul}} = \mathbb{E}[\mathbf{y}_n^{\text{ul}} (\mathbf{y}_n^{\text{ul}})^{\text{H}}] = \sum_{k \in \mathcal{K}} p_u \mathbf{h}_{n,k}^{\text{ul}} (\mathbf{h}_{n,k}^{\text{ul}})^{\text{H}} + \sigma_{\text{ul}}^2 \mathbf{I}, \quad \forall n. \tag{3.12}$$

To access information in the strongest subspace associated with the largest eigenvalues, the beamforming matrix adopted by RRH n is chosen as a collection of the first L_n dominant eigenvectors of $\mathbf{S}_{\text{y},n}^{\text{ul}}$ in $\mathbf{U}_n$, i.e.,

$$\mathbf{V}_n^{\text{ul}} = [\mathbf{u}_{n,1}, \cdots, \mathbf{u}_{n,L_n}]^{\text{H}}, \quad \forall n. \tag{3.13}$$

As mentioned above, the achievable rates of the beamform–compress–forward scheme can now be obtained in the same manner as for the compress–forward scheme, but with the new effective channels $\mathbf{V}_n^{\text{ul}} \mathbf{h}_{n,k}^{\text{ul}}$ and effective noises $\mathbf{V}_n^{\text{ul}} \mathbf{z}_n^{\text{ul}}$.

Table 3.1 System parameters of the numerical example.

Channel bandwidth	20 MHz
Distance between cell sites	0.8 km
Number of RRHs per cell	3
Number of antennas per RRH (M)	12
Number of scheduled users per RRH	2
User transmit power (p_u)	23 dBm
Antenna gain	15 dBi
Path loss model	$140.7 + 36.7 \log_{10}(d)$ dB
Log-normal shadowing	8 dB
Rayleigh small-scale fading	0 dB
SINR gap (Γ_s)	6 dB
Rate–distortion gap (Γ_q)	4.3 dB
AWGN power spectrum density	-169 dBm/Hz

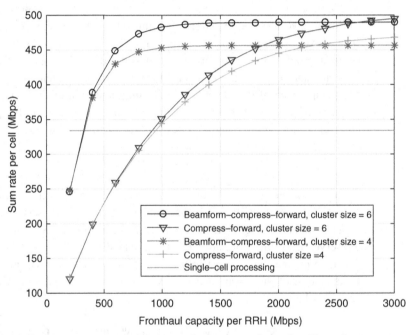

Figure 3.3 Performance comparison between the compress–forward and beamform–compress–forward strategies in the uplink C-RAN.

3.2.1.3   Performance Evaluation

To compare the performance of the beamform–compress–forward strategy and the compress–forward strategy, we present a numerical example of a 19-cell wrapped-around cellular network simulation topology with the parameters listed in Table 3.1. Each cell was sectorized using three RRHs. The per-cell sum rate as a function of the per-RRH fronthaul capacities with different cluster sizes is shown in Figure 3.3.

In this example, each RRH is equipped with 12 antennas but serves only two users in each time slot. Thus, the system is not fully loaded; beamform–compress–forward is expected to show an advantage compared with compress–forward. Indeed, at moderate fronthaul capacity, a sum-rate gain of 20–50% is observed, owing to the fact that beamform–compress provides better utilization of the fronthaul. The difference between the two strategies diminishes for large fronthaul capacities, because the quantization noise is no longer the limiting factor.

It is of interest to note that the C-RAN system significantly outperforms the single-cell-processing baseline. Furthermore, the amount of fronthaul capacity required in order to reap the full benefit of uplink C-RAN is about six times the access rate using beamform–compress–forward, making the practical implementation of C-RAN feasible.

3.2.2   Downlink

In the downlink C-RAN, each user's observed signal is a superposition of the signals sent from all the RRHs. Specifically, let $x_n^{\text{dl}} \in \mathbb{C}^{M \times 1}$ denote the transmit signal of RRH n, and y_n^{dl} denote the received signal at RRH n. Then, the received signal at user k can be modeled as

$$y_k^{\text{dl}} = \sum_{n=1}^{N} (\mathbf{h}_{k,n}^{\text{dl}})^{\text{H}} \mathbf{x}_n^{\text{dl}} + z_k^{\text{dl}}, \quad \forall k, \tag{3.14}$$

where $\mathbf{h}_{k,n}^{\text{dl}} \in \mathbb{C}^{M \times 1}$ is the downlink channel from RRH n to user k, and $z_k^{\text{dl}} \sim \mathcal{CN}(0, \sigma_{\text{dl}}^2)$ denotes the AWGN at user k.

In the current 3G/4G cellular network, each scheduled user is served by one BS and sees interference from all neighboring BSs. The benefit of the C-RAN architecture arises from the ability of multiple RRHs to cooperatively serve users, thereby minimizing the effect of undesired interference. As the messages intended for different users in C-RAN all originate from the CP, the CP can relay useful information about the user messages to the RRHs via the fronthaul links, thus allowing the RRHs to perform network-wide beamforming in order to achieve cooperative transmission.

If the fronthaul links have infinite capacity, the CP can convey the data of all users in the C-RAN perfectly to each RRH, thus achieving full cooperation. With finite-capacity fronthaul links, however, the CP can only send a limited amount of information to each RRH. As a result, a key task for the CP is to convey information about the user messages to the RRHs in the most succinct form in order to enable as much interference cancellation as possible.

One possibility is to use a *compression-based* strategy [5], which can be thought of as the dual operation of the compress–forward strategy used in the uplink. The idea is to preform the cooperative beamformed signals to be transmitted by the RRHs at the CP. The analog signals to be transmitted by the antennas are then compressed and sent digitally to the corresponding RRHs via the fronthaul links for cooperative transmission. As a simpler alternative, the CP may opt to share the user messages directly with the RRHs via fronthaul links, leading to a *data-sharing* strategy [6, 7]. With the user messages at hand, the RRHs can beamform their transmit signals on their own and then transmit them to the users. In the following, we quantify the achievable rates and the fronthaul requirements for the downlink C-RAN using the compression-based and data-sharing strategies.

3.2.2.1 Compression–Forward Strategy

Under the compression-based strategy, the transmit signal of each RRH n, i.e., $\mathbf{x}_n^{\mathrm{dl}}$, is a compressed version of the beamformed signal at the CP, denoted by $\tilde{\mathbf{x}}_n^{\mathrm{dl}}$. Specifically, the CP first forms the beamformed transmit signal for each RRH n as follows:

$$\tilde{\mathbf{x}}_n^{\mathrm{dl}} = \sum_{k \in \Phi_n} \mathbf{w}_{n,k}^{\mathrm{dl}} \sqrt{p_b} s_k^{\mathrm{dl}}, \quad \forall n, \tag{3.15}$$

where $s_k^{\mathrm{dl}} \sim \mathcal{CN}(0,1)$ denotes the message intended for user k, chosen from the Gaussian codebook, $\mathbf{w}_{n,k}^{\mathrm{dl}} = [w_{n,1,k}^{\mathrm{dl}}, \cdots, w_{n,M,k}^{\mathrm{dl}}]^{\mathrm{T}} \in \mathbb{C}^{M \times 1}$ denotes the beamforming vector at RRH n for user k so that the overall unit-norm beamformer for user k across all of its serving RRHs is $\mathbf{w}_k^{\mathrm{dl}} = [\cdots, (\mathbf{w}_{n,k}^{\mathrm{dl}})^{\mathrm{T}}, \cdots]_{n \in \Theta_k}^{\mathrm{T}}$, and p_b is the power of the beamformers, assumed to be identical across all the $\mathbf{w}_k^{\mathrm{dl}}$'s.

As in the case of the uplink, we assume here that the CP applies scalar quantization to compress each component of the beamformed signals independently (although we note here that multivariate compression across the RRHs is also possible [5], although with much higher complexity). The compression process is modeled as a Gaussian test channel with independent additive Gaussian quantization noise. As a result, the transmit signal from the mth antenna of RRH n is given by

$$x_{n,m}^{\mathrm{dl}} = \tilde{x}_{n,m}^{\mathrm{dl}} + e_{n,m}^{\mathrm{dl}} = \sum_{k \in \Phi_n} w_{n,m,k}^{\mathrm{dl}} \sqrt{p_b} s_k^{\mathrm{dl}} + e_{n,m}^{\mathrm{dl}}, \quad \forall n, m, \tag{3.16}$$

where $\tilde{x}_{n,m}^{\mathrm{dl}}$ denotes the mth component of $\tilde{\mathbf{x}}_n^{\mathrm{dl}}$, $e_{n,m}^{\mathrm{dl}} \sim \mathcal{CN}(0, q_{n,m}^{\mathrm{dl}})$ denotes the quantization noise for quantizing $\tilde{x}_{n,m}^{\mathrm{dl}}$, and $q_{n,m}^{\mathrm{dl}}$ denotes the variance of the quantization noise. Note that the $x_{n,m}^{\mathrm{dl}}$ are transmitted from the CP to the corresponding RRHs as quantization indices in digital format. By rate–distortion theory, the fronthaul capacity in terms of bps required for sending $x_{n,m}^{\mathrm{dl}}$ is given by

$$C_{n,m}^{\mathrm{dl}} = B \log_2 \left(\frac{\Gamma_q \sum_{k \in \Phi_n} p_b |w_{n,m,k}^{\mathrm{dl}}|^2 + q_{n,m}^{\mathrm{dl}}}{q_{n,m}^{\mathrm{dl}}} \right), \quad \forall n, m. \tag{3.17}$$

If, for simplicity, we assume further that the fronthaul capacity is evenly allocated to all the antennas of each RRH, i.e., $C_{n,m}^{\mathrm{dl}} = C/M$, $\forall n, m$, the quantization noise resulting from compressing the transmit signal of the mth antenna of RRH n is given by

$$q_{n,m}^{\mathrm{dl}} = \frac{\Gamma_q \sum_{k \in \Phi_n} p_b |w_{n,m,k}^{\mathrm{dl}}|^2}{2^{C/MB} - 1}, \quad \forall n, m. \tag{3.18}$$

Next, we discuss the choice of the per-beam transmit power p_b so that the transmit power constraint is satisfied. For convenience, we assume a sum-power constraint for all the RRHs, denoted by P_R. According to (3.16) and (3.18), the total transmit power across all the RRHs is given by

$$p^{dl} = \sum_{n=1}^{N} \mathbb{E} \|\mathbf{x}_n^{dl}\|^2 = \frac{2^{C/MB} - 1 + \Gamma_q}{2^{C/MB} - 1} \cdot \sum_{k \in \mathcal{K}} p_b. \tag{3.19}$$

By setting the total transmit power to be equal to the sum-power constraint, i.e., $p^{dl} = P_R$, the per-beam transmit power p_b is given by

$$p_b = \frac{(2^{C/MB} - 1)P_R}{(2^{C/MB} - 1 + \Gamma_q)|\mathcal{K}|}. \tag{3.20}$$

Lastly, we quantify the achievable rates of the downlink C-RAN under the above compression-based strategy. For convenience, we define $\mathbf{g}_{k,i}^{dl} = [\cdots, (\mathbf{h}_{k,n}^{dl})^{T}, \cdots]_{n \in \Theta_i}^{T} \in \mathbb{C}^{MD_k \times 1}$ as the collective channel from user i's serving set of RRHs (i.e., Θ_i) to user k. Then, the received signal at user k given in (3.14) can be expressed as

$$y_k^{dl} = \underbrace{(\mathbf{g}_{k,k}^{dl})^{H}\mathbf{w}_k^{dl}\sqrt{p_b}s_k^{dl}}_{\text{desired signal}} + \underbrace{\sum_{i \neq k, i \in \Omega_k} (\mathbf{g}_{k,i}^{dl})^{H}\mathbf{w}_i^{dl}\sqrt{p_b}s_i^{dl}}_{\text{intracluster interference}}$$

$$+ \underbrace{\sum_{j \notin \Omega_k}(\mathbf{g}_{k,j}^{dl})^{H}\mathbf{w}_j^{dl}\sqrt{p_b}s_j^{dl}}_{\text{intercluster interference}} + z_k^{dl} + \sum_{n=1}^{N}(\mathbf{h}_{k,n}^{dl})^{H}\mathbf{e}_n^{dl}, \quad \forall k, \tag{3.21}$$

where $\mathbf{e}_n^{dl} = [e_{n,1}^{dl}, \cdots, e_{n,M}^{dl}]^{T}$ denotes the collective quantization noise for compressing the signal at RRH n across its M antennas. Since scalar quantization is applied, the covariance matrix of $\mathbf{e}_n^{dl}$ is diagonal, i.e., $\mathbf{S}_{e,n}^{dl} = \mathbb{E}[\mathbf{e}_n^{dl}(\mathbf{e}_n^{dl})^{H}] = \mathrm{diag}(q_{n,1}^{dl}, \cdots, q_{n,M}^{dl})$.

As in the case of the uplink, we apply ZF beamforming so that the downlink transmit beamforming vectors $\mathbf{w}_k^{dl}$ are designed to completely cancel the intracluster interference term in (3.21), i.e., $(\mathbf{g}_{i,k}^{dl})^{H}\mathbf{w}_k^{dl} = 0$, $\forall k \in \Omega_i$. Specifically, we define $\mathbf{G}_{-k}^{dl} = [\cdots, \mathbf{g}_{i,k}^{dl}, \cdots]_{i \neq k, k \in \Omega_i}$ as the collection of channel vectors from all the RRHs serving user k to its intracluster users (excluding user k). Similarly to the uplink ZF beamforming design given in (3.8), the downlink beamforming vectors to zero-force the intracluster interference can be obtained as follows:

$$\mathbf{w}_k^{dl} = \frac{(\mathbf{I} - \mathbf{G}_{-k}^{dl}(\mathbf{G}_{-k}^{dl})^{\dagger})\mathbf{g}_{k,k}^{dl}}{\|(\mathbf{I} - \mathbf{G}_{-k}^{dl}(\mathbf{G}_{-k}^{dl})^{\dagger})\mathbf{g}_{k,k}^{dl}\|_2}, \quad \forall k, \tag{3.22}$$

where $(\mathbf{G}_{-k}^{dl})^{\dagger}$ denotes the pseudo-inverse of $\mathbf{G}_{-k}^{dl}$. The achievable rate of user k with the compression-based transmission strategy is then given by

$$r_k^{dl,CP} = B\log_2\left(1 + \frac{p_b\left|(\mathbf{g}_{k,k}^{dl})^{H}\mathbf{w}_k^{dl}\right|^2}{\left(\sum_{j \neq \Omega_k} p_b\left|(\mathbf{g}_{k,j}^{dl})^{H}\mathbf{w}_j^{dl}\right|^2 + \sigma_{ul}^2 + \sum_{n=1}^{N}(\mathbf{h}_{k,n}^{dl})^{H}\mathbf{S}_{e,n}^{dl}\mathbf{h}_{n,k}^{dl}\right)\Gamma_s}\right). \tag{3.23}$$

3.2.2.2   Data-Sharing Strategy

An alternative to the compression strategy for the downlink C-RAN is that instead of sending a compressed version of the beamformed signals, the CP can directly share user messages with the RRHs, which then perform beamforming locally and transmit the beamformed signals cooperatively to the users.

Specifically, the message of each user k, i.e., s_k^{dl}, is sent from the CP to all of the RRHs serving that user, i.e., Θ_k, via the fronthaul links. In this case, the transmit signal of RRH n is given by

$$\mathbf{x}_n^{dl} = \sum_{k \in \Phi_n} \mathbf{w}_{n,k}^{dl} \sqrt{p_b} s_k^{dl}, \quad \forall n. \tag{3.24}$$

Observe that there is no quantization noise term in the above. As a consequence, if the ZF beamforming design given in (3.22) is applied for all users, the achievable rate of user k under the data-sharing strategy can be obtained similarly to what was done for (3.23), but without the quantization noise, i.e.,

$$r_k^{dl,DS} = B \log_2 \left(1 + \frac{p_b \left| (\mathbf{g}_{k,k}^{dl})^H \mathbf{w}_k^{dl} \right|^2}{\left(\sum_{j \neq \Omega_k} p_b \left| (\mathbf{g}_{k,j}^{dl})^H \mathbf{w}_j^{dl} \right|^2 + \sigma_{ul}^2 \right) \Gamma_s} \right), \quad \forall k, \tag{3.25}$$

where the per-beam transmit power is now simply $p_b = P_R/|\mathcal{K}|$.

It is worth noting that although the data-sharing strategy does not suffer from quantization noise, the cluster size is severely limited by the finite-capacity fronthaul links. Specifically, if user k is served by RRH n, i.e., $k \in \Phi_n$, then s_k needs to be sent to RRH n at a rate of $r_k^{dl,DS}$ bps. As a result, given a clustering strategy defined by Φ_n, the fronthaul capacity required for each RRH n is the sum of all the rates for the users served by it, i.e.,

$$C_n^{dl} = \sum_{k \in \Phi_n} r_k^{dl,DS}, \quad \forall n. \tag{3.26}$$

It is thus essential to design the cluster size carefully under the data-sharing strategy such that the fronthaul traffic does not exceed the fronthaul capacity at each link, i.e., $C_n^{dl} \leq C, \forall n$. For example, compressed sensing techniques can be used to choose the serving cluster for each user in an intelligent fashion [6].

3.2.2.3   Performance Evaluation

A per-cell sum-rate comparison between the compression and data-sharing strategies in downlink C-RAN under user-centric clustering is shown in Figure 3.4 for various per-RRH fronthaul capacities. The network setup was similar to that of the uplink as given in Table 3.1, with three RRHs per cell, except that the number of antennas at each RRH was set to $M = 4$ and the average transmit power of each RRH was set to 43 dBm. The user-centric cluster size for the compression strategy was fixed, while the cluster size for the data-sharing strategy ranged from 1 to 10. Observe that as in the case of the uplink, C-RAN brings a considerable performance gain as compared with

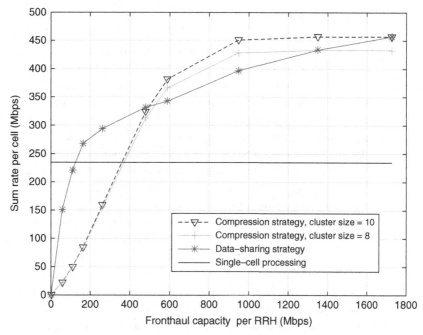

Figure 3.4 Performance comparison between compression-based and data-sharing strategies in the downlink C-RAN under user-centric clustering.

the single-cell baseline. Cooperative transmission is able to almost double the sum rate at a fronthaul capacity cost of about six times the access rate.

Observe also that at low fronthaul capacity, data-sharing outperforms compression, while at high fronthaul capacity, compression outperforms data-sharing. The reason is that in the data-sharing strategy, the message of each user is repeatedly transmitted over different fronthaul links to its serving RRHs; this is not the most efficient use of the fronthaul when the cluster size is large.

We remark that this section assumes a single-hop C-RAN with direct links between the CP and RRHs. If the fronthaul network consists of edge routers and network processors with multiple hops, routing strategies can also play a significant role. In particular, as the data-sharing strategy amounts to multicasting of user messages to multiple RRHs, network coding techniques can be applied to improve the efficiency of the fronthaul network [8].

3.3   Fronthaul-Aware Data Link and Physical Layers

So far, we have discussed the impact of fronthaul capacity limitations on the spectral efficiency of C-RAN. This section addresses latency, which is an equally important system objective that affects the performance of a 5G deployment. An important case in point is the HARQ protocol, which runs across the data link and physical layers, and has the role of guaranteeing reliable communication over fading channels. HARQ accomplishes this goal via the transmission of additional information about data frames that have been previously transmitted but not correctly received and acknowledged by the receiver. Given that baseband processing and HARQ retransmission decisions are implemented at the BBU, delays in communication between the RRH and the BBU due to fronthaul transmission entail an increased latency between successive retransmission attempts. This may disrupt the operation of existing HARQ protocols or cause excessive delays in latency-sensitive applications. As an example, in Long Term Evolution (LTE), a latency larger than 3 ms in the uplink is treated as a system outage in the data link layer [9].

Fronthaul latency can be partly mitigated by deploying shorter, dedicated rather than multihop, fronthaul links between each RRH and the BBU, so as to reduce the transit time between the RRH and BBU. Furthermore, one can enhance the computing power at the BBU, so as to limit the time required to process signals at the BBU for fronthaul transmission and reception. To provide some reference values, the times for two-way fronthaul transmission, excluding processing times, for single-hop fronthaul links may be of the order of 0.5 ms, while processing times at the BBU and UE can amount to a few milliseconds [10].

This section discusses solutions for scenarios in which the performance limitations incurred owing to fronthaul latency constraints cannot be satisfactorily dealt with by means of the approaches outlined above within the standard C-RAN architecture. Specifically, we consider the potential advantages that could be gained by leveraging *alternative functional splits* between the BBU and the RRHs, whereby the RRHs implement some control functionalities of the HARQ protocol.

As discussed, the performance degradation of HARQ protocols in C-RAN is to be ascribed to the need to transfer baseband signals, as well as retransmission request (NAK) or positive acknowledgment (ACK) messages, between the RRH and the BBU, given that both the control and the data plane functionalities are implemented solely in the BBU. With the aim of reducing latency, the class of solutions investigated in this section allows for *HARQ control functions to be carried out at the RRH*.

In equipping the RRH with sufficient intelligence to perform some baseband functions as well as control decisions, the solutions considered deviate from the standard C-RAN architecture and are in line with the *alternative functional splits* being investigated in the literature and in industry (see [11, 12]). In defining such alternative splits, emphasis will be given here to solutions that require RRHs with reduced complexity as compared with conventional base stations. This choice excludes, for instance, approaches that require full data decoding at the RRHs in the uplink or data encoding and precoding in the downlink.

The functional splits considered can be interpreted as implementing an instance of the *separation of control and data planes* that is currently being advocated for next-generation wireless network architectures in a variety of guises (see [13] for a review). In particular, in the solutions at hand, the control functions associated with the HARQ protocol are carried out at the network edge, namely at the RRHs, while functionalities related to the data plane are still performed remotely at the BBU as in a conventional C-RAN system. The key advantages of this architecture are that (i) retransmission control is not subject to fronthaul latency constraints; (ii) the complexity

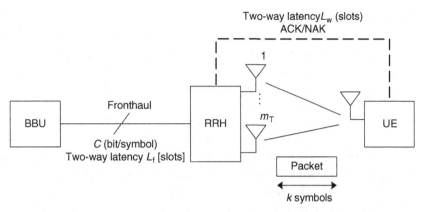

Figure 3.5 System model considered in Section 3.3 (with $m_R = 1$ receive antennas).

of the RRHs can be kept in check given that data plane processing is performed at the BBU; and (iii) joint baseband processing of data plane information at the BBU yields spectral-efficiency gains in the physical layer, as discussed in the previous section.

In the rest of this section, we describe HARQ protocols for both the uplink and the downlink based on the separation of control and data planes discussed. In order to simplify the discussion, instead of treating a C-RAN system with cooperative transmission and reception, this section focuses on the throughput and latency of a simplified distributed RAN (D-RAN) system in which each RRH serves its own set of users. Here, we use the term "D-RAN" to refer to an intermediate architecture between a standard cellular system and a C-RAN in which the BBU of each base station is hosted at a remote site (see, e.g., [11, 14, 15]). We note that "D-RAN" is also used to indicate the conventional cellular architecture in 5G documents. In a D-RAN, unlike a C-RAN, the BBUs of different RRHs are physically distinct. Note that joint baseband decoding is generally not feasible in a D-RAN, since this would require the exchange of baseband signals among BBUs, rather than user plane data as allowed by the X2 interface that may connect BBUs (see, e.g., [15]). The focus on D-RAN allows us to concentrate on setups in which any given UE is assigned to a single RRH–BBU pair and to distill the essence of the effect of latency on the throughput performance.

3.3.1 Uplink

In this subsection, we consider the uplink and describe a solution, first proposed in [16, 17], whereby HARQ control functions are carried out at the RRH. The scheme works as follows: rather than waiting for an ACK or NAK message to be received from the BBU, the RRH assigned to a UE estimates the uplink channel based on the received signal, and preemptively makes the control decision to send a NAK message if the signal-to-noise ratio (SNR) is found to be below a threshold and an ACK message otherwise. Importantly, the RRH does not perform data decoding, and hence its complexity is kept significantly lower than that of a conventional base station. We now give details of the system model and an analysis of its performance.

3.3.1.1   System Model

We concentrate on a D-RAN system in which a UE transmits on a dedicated spectral resource to an RRH, as illustrated in Figure 3.5. The RRH is connected by means of a dedicated fronthaul link to a BBU. The BBU performs decoding, while the RRH is assumed to have limited baseband processing functionalities that allow resource demapping and estimation of CSI. Note that different UEs are assumed to be served with distinct time–frequency resources, as is done for instance in LTE, and hence we limit our attention to the performance of a given UE.

Each packet transmitted by the UE contains k encoded complex symbols and is transmitted within a coherent time–frequency interval of the channel, which is referred to as a *slot*. The transmission rate of the first transmission of an information message is defined to be r bits per symbol, so that kr is the number of information bits in the information message.

Each transmitted packet is acknowledged via the transmission of a feedback message by the RRH to the UE. We assume that these feedback messages are correctly decoded by the UE. The same information message may be transmitted in up to n_{max} successive slots using a HARQ protocol. Here, we adopt the incremental redundancy (IR) protocol, which operates across the physical and data link layers and is implemented in standards such as LTE [9]. With HARQ-IR, the UE transmits new encoded symbols at each transmission attempt and the BBU performs decoding based on all of the received packets. Furthermore, we assume a backlogged UE that always has packets to transmit, and a selective repeat retransmission policy in which only frames from unsuccessfully received information messages are retransmitted.

In order to capture the impact of the fronthaul latency, we assume that a two-way delay of L_f slots is incurred in transmission between the RRH and the BBU. Furthermore, we assume that the round-trip transmission delay between the UE and RRH, including the decoding delay at the UE, is L_w slots. As discussed, these two-way latencies may amount to a few milliseconds, and typically encompass multiple transmission intervals, for example multiple transmission time intervals (TTIs) in LTE. The fronthaul capacity is denoted by C and is measured in bits per symbol of the wireless channel or, equivalently, bits per second per Hz with respect to the wireless bandwidth.

The UE is equipped with m_T transmitting antennas, while m_R receiving antennas are available at the RRH. The received signal in the nth slot can be expressed as

$$\mathbf{y}_n = \sqrt{\frac{s}{m_T}}\mathbf{H}_n\mathbf{x}_n + \mathbf{z}_n, \tag{3.27}$$

where s measures the average SNR per receive antenna; $\mathbf{x}_n \in \mathbb{C}^{m_T \times 1}$ represents the symbols sent by the transmit antennas of the UE during a given channel use, whose average power is normalized such that $E[||\mathbf{x}_n||^2] = 1$; $\mathbf{H}_n \in \mathbb{C}^{m_R \times m_T}$ is the channel matrix, which is assumed to have independent identically distributed (i.i.d.) $\mathcal{CN}(0,1)$ entries (Rayleigh fading); and $\mathbf{z}_n \in \mathbb{C}^{m_R \times 1}$ is an i.i.d. Gaussian noise vector with $\mathcal{CN}(0,1)$ entries. The channel matrix $\mathbf{H}_n$ changes independently in each slot n. Moreover, it is assumed to be known to the RRH and to the BBU. We assume the use of Gaussian codebooks with an equal power allocation across the transmit antennas, although the analysis could be extended to arbitrary power allocation and antenna selection schemes.

The main performance metrics of interest are as follows:

- *Throughput T*. The throughput measures the average rate, in bits per symbol, at which information can be successfully delivered from the UE to the BBU.
- *Probability P_s of success*. The metric P_s measures the probability of a successful transmission within a given HARQ session, which is the event that, in one of the n_{max} allowed transmission attempts, the information message is decoded correctly at the BBU.
- *Average latency D*. The average latency D measures the average number N of transmission attempts per information message.

A few remarks are in place. First, the three metrics are interdependent. In particular, based on standard renewal theory arguments, the throughput can be calculated as [18]

$$T = \frac{rP_s}{\mathrm{E}[N]}, \tag{3.28}$$

where we recall that r is the transmission rate, and the random variable N denotes the number of transmission attempts for a given information message. As will be discussed, the average latency D is an increasing function of $\mathrm{E}[N]$. Therefore, given r, any two of the metrics defined above determine the third. Secondly, errors in the HARQ sessions, which occur with probability $1 - P_s$, are typically dealt with by higher layers, as is done by the radio link control (RLC) layer in LTE [16]. Finally, as will be seen, the proposed schemes aim at trading a decrease in the throughput T for a reduction in the average latency D.

3.3.1.2 Conventional D-RAN

In a conventional D-RAN system, all processing and retransmission decisions are made at the BBU. Therefore, each transmission requires a two-way latency of $L_w + L_f$ slots, owing to the need to communicate in both directions on the wireless channel and on the fronthaul link. Assuming, as mentioned, a backlogged UE, the throughput can be written as in (3.28), where the average number of transmissions can be computed as

$$\mathrm{E}[N] = \sum_{n=1}^{n_{max}-1} nP(\mathrm{ACK}_n) + n_{max}P(\mathrm{NAK}_{n_{max}-1}). \tag{3.29}$$

We denote by ACK_n the event that an ACK message is sent to the UE after exactly n transmission attempts, hence terminating the retransmission process. Note that, in a conventional D-RAN implementation, this implies that the BBU decodes the message successfully after exactly n transmission attempts. We also denote by NAK_n the event in which a NAK message is sent to the UE for all transmission attempts up to and including the nth one. In a conventional D-RAN implementation, this implies that the BBU does not decode the message successfully in up to and including the nth transmission attempt. We observe that, by definition, we have the relationship

$P(\text{ACK}_n) = P(\text{NAK}_{n-1}) - P(\text{NAK}_n)$. Furthermore, for a conventional D-RAN system, the probability of a successful transmission is given by

$$P_s = 1 - P(\text{NAK}_{n_{\max}}). \tag{3.30}$$

In order to evaluate the probabilities of error, we adopt the finite-block-length Gaussian approximation proposed in [19], based on [20]. Accordingly, the probability $P_e(r, k, \mathbf{H})$ of a decoding error for a transmission at rate r in a slot of k channel uses when the channel matrix is $\mathbf{H}$ can be approximated as

$$P_e(r, k, \mathbf{H}) = Q\left(\frac{C(\mathbf{H}) - r}{\sqrt{V(\mathbf{H})/k}}\right), \tag{3.31}$$

where we have defined

$$C(\mathbf{H}) = \sum_{j=1}^{m_{\text{RT}}} \log_2\left(1 + \frac{s\lambda_j}{m_{\text{T}}}\right) \text{ and } V(\mathbf{H}) = \left(m_{\text{RT}} - \sum_{j=1}^{m_{\text{RT}}} \frac{1}{\left(1 + s\lambda_j/m_{\text{T}}\right)^2}\right) \log_2^2 e, \tag{3.32}$$

where $m_{\text{RT}} = \min(m_{\text{R}}, m_{\text{T}})$, $\{\lambda_j\}_{j=1,...,m_{\text{RT}}}$ are the eigenvalues of the matrix $\mathbf{H}^{\text{H}}\mathbf{H}$, and $Q(\cdot)$ is the Gaussian complementary cumulative distribution function.

With this approximation, since with HARQ-IR a set of n transmission attempts for a given information message can be treated as a transmission over n parallel channels (see, e.g., [18]), the error probability at the nth transmission can be computed as $E[P_e(r, k, \mathcal{H}_n)]$, where $\mathcal{H}_n = \text{diag}([\mathbf{H}_1, \ldots, \mathbf{H}_n])$ and the expectation is taken with respect to the channel distribution [19]. Moreover, the probability of a decoding error up to and including the nth transmission can be bounded from above by the probability of error *at* the n transmission as

$$P(\text{NAK}_n) \leq E[P_e(r, k, \mathcal{H}_n)]. \tag{3.33}$$

Using (3.33) in (3.28) and (3.30), we obtain lower bounds on the throughput and the probability of success, respectively (within the approximation (3.32) for the probability of error). For the average latency, given the discussion above, we can calculate

$$D = E[N](L_w + L_f). \tag{3.34}$$

An upper bound on D can be computed using (3.29) and (3.33).

3.3.1.3   Edge-Based Retransmission

A low-latency edge-based HARQ control scheme was first proposed in [16, 17]. This approach assumes an RRH–BBU functional split where each RRH can perform synchronization and resource demapping, so as to be able to perform CSI estimation of the channel $\mathbf{H}_n$ at each transmission attempt n. The RRH is also assumed to have obtained the modulation and coding scheme (MCS) used for data transmission from the BBU, which selects the MCS during scheduling. The MCS information amounts here to the rate r and the packet length k. Based on this information, the RRH can compute the probability of error for decoding at the BBU. This could be done, for example, by using an analytical approximation, such as $P_e(r, k, \{\mathbf{H}_i\}_{i \leq n})$ in (3.31), or a precomputed lookup table. Note that this probability depends on all channel matrices $[\mathbf{H}_1, \cdots, \mathbf{H}_n]$ corresponding to prior and current transmission attempts. In the following, we will assume that the approximation $P_e(r, k, \{\mathbf{H}_i\}_{i \leq n})$ is used by the RRH, although the discussion applies more generally.

The gist of the approach is to allow the RRH to make preemptive decisions regarding the feedback of ACK/NAK messages to the UE without waiting for the L_f slots required for two-way communication on the fronthaul link. This is done as follows: if the decoding-error probability $P_e(r, k, \{\mathbf{H}_i\}_{i \leq n})$ is smaller than a given threshold P_{th}, the RRH sends an ACK message to the UE, predicting a positive decoding event at the BBU; otherwise, a NAK message is transmitted, that is, the following rule is used by the RRH:

$$P_e(r, k, \{\mathbf{H}_i\}_{i \leq n}) \underset{\text{NAK}}{\overset{\text{ACK}}{\lessgtr}} P_{th}. \qquad (3.35)$$

While reducing the average latency as a result of the implementation of control decisions at the RRH, the edge-based HARQ scheme discussed above introduces a possible mismatch between the RRH's decisions and the actual outcome of the decoding at the BBU. In particular, there are two types of error. In the first type of error, the transmitted packet is not decodable at the BBU, but an ACK message is sent by the RRH. As seen, this type of mismatch needs to be dealt with by higher layers. In the second type of error, the received packet is decodable at the BBU, but a NAK message is sent by the RRH. In this case, the UE performs an unnecessary HARQ retransmission, unless the maximum number n_{max} of transmission attempts has already been reached. These errors generally cause a reduction in the throughput and probability of success, in return for which the edge-based scheme discussed here promises significant gains in terms of delays. To see this, we note that the average latency until a packet is acknowledged, positively or negatively, to the UE is given by

$$D = \mathrm{E}[N]L_w, \qquad (3.36)$$

since no use of the fronthaul link is required in order to complete the HARQ process. This value may be significantly smaller than that in (3.34), depending on the relative values of L_w and L_f.

With regard to the optimization of the threshold P_{th} in (3.35), one needs to strike a balance between the probability of success P_s, which would call for a smaller P_{th} and hence more retransmissions, and the throughput T, which may generally be improved by a larger P_{th}, resulting in the transmission of new information.

The throughput and probability of success can be computed in a manner similar to that for the conventional D-RAN implementation. The main caveat is the definition of a successful event: a transmission is considered as successful here if an ACK message is sent to the UE within one of the n_{max} allowed transmission attempts *and* if the BBU can correctly decode the packet. Hence, by the law of total probability, the probability of success P_s can be written as

$$P_s = \sum_{n=1}^{n_{\max}} P(S_n | \text{ACK}_n) P(\text{ACK}_n), \qquad (3.37)$$

where S_n is the event that the BBU can successfully decode the packet at the nth transmission, while the event ACK_n is defined as above.

The probabilities needed to compute the throughput (3.28) and the probability of success (3.37) can be obtained using the Gaussian approximation discussed above, as follows. The probability of a NAK message being sent up to the n transmission attempt is

$$P(\text{NAK}_n) = P(P_e(r, k, \mathcal{H}_n) > P_{\text{th}}). \qquad (3.38)$$

Note that this is due to the monotonicity of the probability $P_e(r, k, \mathcal{H}_n)$ as a function of each eigenvalue, so that the probability $P_e(r, k, \mathcal{H}_n)$ is no larger than $P_e(r, k, \mathcal{H}_{n-1})$. In a similar manner, we can also calculate

$$P(S_n | \text{ACK}_n) = 1 - \text{E}[P_e(r, k, \mathcal{H}_n) | \mathcal{A}(P_{\text{th}})], \qquad (3.39)$$

where we have defined the event $\mathcal{A}(P_{\text{th}}) = \{\{P_e(r, k, \mathcal{H}_{n-1}) > P_{\text{th}}\} \bigcap \{P_e(r, k, \mathcal{H}_n) \le P_{\text{th}}\}\}$.

3.3.1.4   Numerical Example

We now corroborate the analysis presented in the previous sections by providing insights into a performance comparison of conventional D-RAN and edge-based-scheme systems via a numerical example. Figure 3.6 shows the throughput loss of the edge-based scheme as compared with the conventional D-RAN implementation, as a function of the block length k, for two rates $r = 1$ bit/symbol and $r = 3$ bit/symbol. We have set $s = 4$ dB and $n_{\max} = 10$ and focused on a single-antenna link, i.e., $m_T = m_R = 1$. For every value of k, the threshold P_{th} was optimized to maximize the throughput T

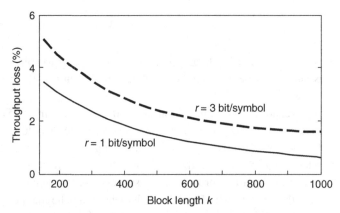

Figure 3.6 Throughput loss of the edge-based scheme with respect to the standard D-RAN implementation versus a block-length-k system ($s = 4$ dB, $n_{\max} = 10$, $m_T = 1$, $m_R = 1$, $P_s > 0.99$ for $r = 1$ bit/symbol and $r = 3$ bit/symbol).

under the constraint that the probability of success satisfies the requirement $P_s > 0.99$. This is typically assumed to be acceptable in existing systems (see, e.g., [9]).

It can be seen that, as the block length increases, the throughput loss with local feedback decreases significantly. In this regime, the latency reduction afforded by edge-based control comes at a minor cost in terms of throughput loss. This reflects a fundamental insight: the performance loss with local feedback is due to the fact that local decisions are taken by the RRH based only on channel state information, without reference to the specific channel noise realization that affects the received packet. Therefore, as the block length k increases, and hence the errors due to atypical channel noise realizations become less likely, the local decisions tend to be consistent with the actual decoding outcomes at the BBU. In other words, as the block length k grows larger, it becomes easier for the RRH to predict the decoding outcome at the BBU: in the Shannon regime of infinite k, successful or unsuccessful decoding depends deterministically on whether the rate r is above or below the capacity.

3.3.2 Downlink

In this section, we consider a low-latency HARQ protocol in which control is carried out at the RRH for the downlink of a D-RAN system. Similarly to the case of the uplink, the key idea is that of enabling the RRH to make low-latency retransmission control decisions, while still retaining all baseband encoding capabilities at the BBU so as to reduce the complexity of the RRH. According to the protocol, at the first transmission attempt, the RRH stores the transmitted baseband signal, which is encoded by the BBU and received by the RRH on the fronthaul link. If a NAK is received, the RRH then retransmits the stored baseband signal without further baseband processing and without assistance from the BBU. We observe that, unlike the uplink mechanism discussed above, here no CSI estimation is needed at the RRH, but the RRH still needs to be equipped with sufficient baseband capabilities to enable the detection of ACK/NAK messages. Furthermore, the RRH is assumed to have enough memory to store previously transmitted baseband signals.

3.3.2.1 System Model

We consider downlink communication in the same D-RAN system as studied above and shown in Figure 3.5, consisting of an RRH that is connected to a BBU through a fronthaul link. We focus on the performance of a given single-antenna UE, i.e., $m_T = 1$, which is allocated dedicated spectral resources for downlink transmission. The RRH transmits a packet of length k symbols in each slot, and the UE sends an ACK or NAK message depending on the decoding outcome, which is assumed to be correctly decoded at the RRH or BBU. The key parameters r, L_w, and L_f are defined as for the uplink (see Figure 3.5).

The received signal at the UE in a time slot t can be written as

$$y_t = \mathbf{h}_t^\dagger \mathbf{x}_t + \mathbf{z}_t, \tag{3.40}$$

where $\mathbf{h}_t$ is the $m_R \times 1$ channel vector, with m_R being the number of transmit antennas of the RRH; $\mathbf{x}_t$ is the $m_R \times 1$ signal transmitted by the RRH, with power constraint $E[\|\mathbf{x}_t\|^2] = P$; and z_t is complex Gaussian noise with unitary power, i.e., distributed as $\mathcal{CN}(0,1)$. Note that, unlike the case for the uplink, we find it convenient here to express the received signal as a function of the slot index t rather than of the retransmission attempt index n. In particular, this allows us to keep track of the channel variations, which, as elaborated on further in the rest of this subsection, play a key role in the downlink. To this end, we assume that the channel vector process $\mathbf{h}_t$ is correlated across two successive slots according to a stationary autoregressive model of order one, namely

$$\mathbf{h}_t = \rho \mathbf{h}_{t-1} + \mathbf{v}_t \qquad (3.41)$$

for $t \in \{\ldots, -2, -1, 0, 1, 2, \ldots\}$, with a correlation coefficient $\rho \in [0,1)$, where $\mathbf{v}_t$ has independent entries distributed as $\mathcal{CN}(0, 1 - \rho^2)$. Full CSI is assumed at the UE.

3.3.2.2   Conventional D-RAN

In a conventional D-RAN implementation, as shown in Figure 3.7, the RRH delivers the ACK/NAK feedback message, as well as updated CSI information, from the UE to the BBU, and the BBU carries out the encoding of the data plane information of a previously transmitted packet when a NAK is received, or of a new packet when an ACK is obtained. Therefore, a round trip delay of L_w slots on the wireless channel and a two-way fronthaul latency of L_f slots elapses between the transmission of a downlink packet and the time in which that packet may be retransmitted to the UE.

To elaborate, as for the uplink, we assume that the BBU implements the IR protocol. Accordingly, at any time t, the RRH sends a new part of an encoded information frame that was last transmitted at slot $t - (L_w + L_f)$ if a NAK is received at the RRH at time $t - L_f$; otherwise, it transmits a packet from a new information frame.

In either case, the information-bearing symbol $s_t \sim \mathcal{CN}(0,1)$ to be transmitted at slot t is linearly precoded at the BBU by means of an $m_R \times 1$ beamforming vector $\mathbf{w}_t$, which is matched to the last channel realization that is available from the cloud, namely $\mathbf{h}_{t-L_f}$, that is, $\mathbf{w}_t = \mathbf{h}_{t-L_f}/\|\mathbf{h}_{t-L_f}\|$. Then, the precoded signal $\tilde{\mathbf{x}}_t$ at the BBU is given as

$$\mathbf{x}_t = \sqrt{P}\frac{\mathbf{h}_{t-L_f}}{\|\mathbf{h}_{t-L_f}\|}s_t. \qquad (3.42)$$

3.3.2.3   Edge-Based Retransmission

To potentially alleviate the performance limitations (in terms of delay) of the standard D-RAN system described above due to two-way fronthaul latency, we now consider a solution based on the implementation of the HARQ control function at the RRH. Accordingly, we allow for low-latency retransmissions by the RRH, under the working assumption that the RRH can store the previously transmitted baseband signals as well as decode the ACK/NAK feedback messages on the uplink.

As illustrated in Figure 3.8, if a NAK is fed back by the UE regarding a packet previously sent at time $t - L_w$, the RRH autonomously retransmits the previously transmitted packet at time t without waiting for a newly encoded packet from the BBU.

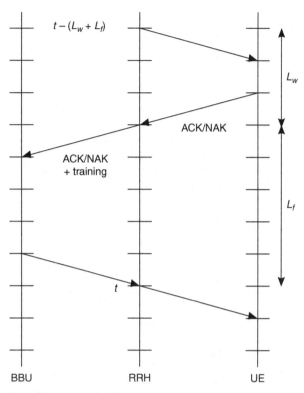

Figure 3.7 Illustration of the conventional implementation of downlink HARQ in D-RAN.

If an ACK is instead fed back by the UE, the RRH asks for an encoded packet associated with a new information frame from the BBU. As a result, for the first transmission of a packet, a signal (3.42) is transmitted, and, for each retransmission, the same signal is retransmitted by the RRH. Note that this strategy cannot adapt to channel variations and is implicitly based on HARQ Type I or Chase combining [21].

3.3.2.4   Numerical Results and Discussion

In this section, we compare the throughput and latency performance of the conventional D-RAN and edge-based retransmission. Throughout this section, we assume that the number of transmit antennas at the RRH is $m_R = 4$, the length of a packet in each slot is $k = 100$ symbols, the transmit power P is 10 dB, the maximum number of transmission attempts is $n_{max} = 10$, and the two-way latencies are equal to $L_f = L_w = 2$ slots. We also assume a fronthaul capacity C of $C = 3$ bit/symbol.

We first plot the throughput as a function of the correlation coefficient ρ, which defines the time variability of the channel, in Figure 3.9 for $r = 2$ and $r = 3$ bit/symbol. The throughput loss of the edge-based scheme, which is caused by the lack of adaptation to the varying channel conditions in the retransmission attempts and by the simpler HARQ protocol, depends on the correlation coefficient ρ and the transmission rate R. Specifically, for a lower transmission rate requiring with high probability no more than one retransmission, such as 2 bit/symbol, the throughput loss is minor. However,

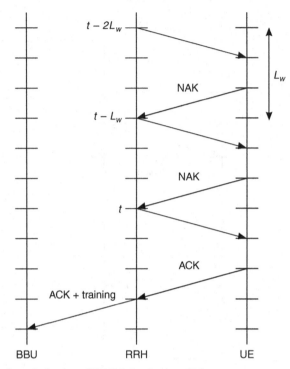

Figure 3.8 Illustration of edge-based HARQ for the downlink.

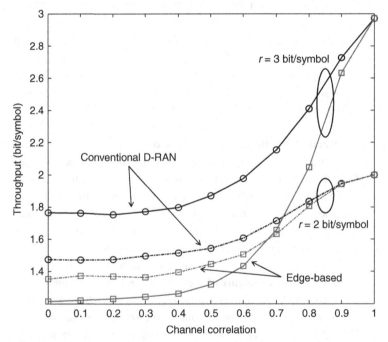

Figure 3.9 Throughput T versus correlation coefficient ρ for $r = 2$ and $r = 3$ bit/symbol ($m_T = 4$, $k = 100$, $P = 10$ dB, $C = 3$ bit/symbol, $L_w = L_f = 2$ slots, and $n_{max} = 10$).

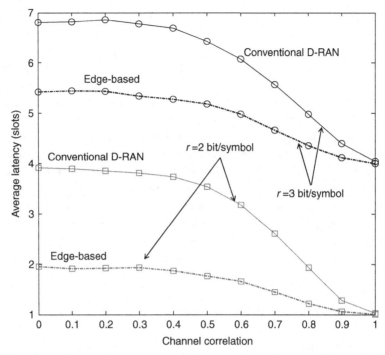

Figure 3.10 Latency D versus the correlation coefficient ρ for $r=2$ and $r=3$ bit/symbol ($m_T = 4$, $k = 100$, $P = 10$ dB, $C = 3$ bit/symbol, $L_w = L_f = 2$ slots, and $n_{max} = 10$).

for a larger transmission rate, which calls for more retransmissions, the loss may be substantial unless the correlation coefficient ρ is large enough. For instance, with $\rho = 0.8$, which corresponds to a speed of 60 km/h and a carrier frequency of 2.6 GHz for a slot duration of 1 ms according to Clarke's standard model, the loss is 2% for $r = 2$ bit/symbol and 18% for $r = 3$ bit/symbol.

The implementation of edge-based retransmission is justified if the throughput loss discussed above is deemed to be acceptable when compared with the given reduction in latency. This was investigated as shown in Figure 3.10, which shows the latency as a function of the correlation coefficient ρ for the same parameters as above. It can be seen that the reduction in latency can be very significant, particularly for sufficiently small rates and/or for slowly varying channels, i.e., for large enough ρ. As examples, for $\rho = 0.8$ as considered above, the latency can be reduced by 3.2 slots at the cost of a throughput reduction of 0.05 bit/symbol if $r = 2$ bit/symbol, while the latency reduction is 3 slots at the cost of a throughput loss of 0.35 bit/symbol if $r = 3$ bit/symbol.

3.4   Conclusion

The cloud radio access network architecture enables a significant increase in area spectral efficiency for 5G wireless cellular networks by allowing dense, distributed deployment of remote antennas together with a centralized capability for joint baseband processing across cooperative antenna clusters. The advent of such an architecture also necessitates rethinking of both the physical layer and the data link layer operations. This chapter has presented some of the challenges and opportunities in C-RAN design. The first part of the chapter illustrated that in the physical layer, coherent transmission and reception of user signals across multiple remote radio heads can provide a significant rate gain, but the design of the cooperative communication strategies needs to be adapted according to the capacity constraints of the fronthaul. We utilized a compression strategy as the fundamental technique for both uplink and downlink, and quantified the effect of limited fronthaul capacity on the overall spectral efficiency for a zero-forcing-based user-centric cooperative communication strategy. The second part of this chapter has made a further contribution by pointing out that the latency in a C-RAN architecture can be managed and controlled in an intelligent way by rethinking the design of the HARQ protocol for a multihop network. Together, these techniques point to ways that could make radio access via cloud computing a reality.

Acknowledgments

Liang Liu and Wei Yu would like to acknowledge the support of the Natural Sciences and Engineering Research Council (NSERC) of Canada. Osvaldo Simeone would like to thank Shahrouz Khalili (NJIT), Wonju Lee, and Joonhyuk Kang (KAIST) for their collaboration on the material covered in Section 3.3. The work of Osvaldo Simeone was partially supported by the US NSF through grant 1525629.

References

[1] C. Zhu and W. Yu, "Stochastic analysis of user-centric network MIMO," in *Proc. of IEEE International Workshop on Signal Processing Advances in Wireless Communications (SPAWC)*, Jul. 2016.
[2] Y. Zhou and W. Yu, "Optimized backhaul compression for uplink cloud radio access network," *IEEE J. Sel. Areas Commun.*, vol. 32, no. 6, pp. 1295–1307, Jun. 2014.
[3] Y. Zhou and W. Yu, "Fronthaul compression and transmit beamforming optimization for multi-antenna uplink C-RAN," *IEEE Trans. Signal Process.*, vol. 64, no. 16, pp. 4138–4151, Aug. 2016. Available at http://arxiv.org/abs/1604.05001v1.
[4] L. Liu and R. Zhang, "Optimized uplink transmission in multi-antenna C-RAN with spatial compression and forward," *IEEE Trans. Signal Process.*, vol. 63, no. 19, pp. 5083–5095, Oct. 2015.
[5] S. H. Park, O. Simeone, O. Sahin, and S. Shamai, "Joint precoding and multivariate backhaul compression for the downlink of cloud radio access networks," *IEEE Trans. Signal Process.*, vol. 61, no. 22, pp. 5646–5658, Nov. 2013.
[6] B. Dai and W. Yu, "Sparse beamforming and user-centric clustering for downlink cloud radio access network," *IEEE Access*, vol. 2, pp. 1326–1339, 2014.

[7] P. Patil, B. Dai, and W. Yu, "Performance comparison of data-sharing and compression strategies for cloud radio-access networks," in *Proc. of European Signal Processing Conf. (EUSIPCO)*, Sep. 2015.

[8] L. Liu and W. Yu, "Joint sparse beamforming and network coding for downlink multi-hop cloud radio access networks," in *Proc. of IEEE Global Communications Conf. (Globecom)*, Dec. 2016.

[9] E. Dahlman, S. Parkvall, J. Skold, and P. Bemin, *3G Evolution: HSPA and LTE for Mobile Broadband*, Academic Press, 2nd edn., 2008.

[10] NGMN Alliance, "Further study on critical C-RAN technologies," Tech. Rep., Mar. 2015.

[11] A. Checko, H. Christiansen, Y. Yan, L. Scolari, G. Kardaras, M. Berger, and L. Dittmann, "Cloud RAN for mobile networks – A technology overview," *IEEE Commun. Surv. Tutor.*, vol. 17, no. 1, pp. 405–426, Mar. 2015.

[12] O. Simeone, A. Maeder, M. Peng, O. Sahin, and W. Yu, "Cloud radio access network: Virtualizing wireless access for dense heterogeneous systems," *J. Commun. Netw.*, vol. 18, no. 2, Apr. 2016.

[13] A. Mohamed, O. Onireti, M. Imran, A. Imran, and R. Tafazolli, "Control-data separation architecture for cellular radio access networks: A survey and outlook," *IEEE Commun. Surv. Tutor.*, vol. 18, no. 1, pp. 446–465, Jan. 2016.

[14] China Mobile, "C-RAN: The road towards green RAN," White Paper, Oct. 2011.

[15] M. Nahas, A. Saadani, J. Charles, and Z. El-Bazzal, "Base stations evolution: Toward 4G technology," in *Proc. of International Conf. on Telecommunications (ICT)*, Apr. 2012.

[16] U. Dotsch, M. Doll, H.-P. Mayer, F. Schaich, J. Segel, and P. Sehier, "Quantitative analysis of split base station processing and determination of advantageous architectures for LTE," *Bell Labs Tech. J.*, vol. 18, no. 1, pp. 105–128, Jun. 2013.

[17] P. Rost and A. Prasad, "Opportunistic hybrid ARQ-enabler of centralized-RAN over nonideal backhaul," *IEEE Wireless Commun. Lett.*, vol. 3, no. 5, pp. 481–484, Oct. 2014.

[18] G. Caire and D. Tuninetti, "The throughput of hybrid-ARQ protocols for the Gaussian collision channel," *IEEE Trans. Inform. Theory*, vol. 47, no. 5, pp. 1971–1988, Jul. 2001.

[19] W. Yang, G. Durisi, T. Koch, and Y. Polyanskiy, "Quasi-static MIMO fading channels at finite blocklength," Apr. 2014. Available at http://arxiv.org/pdf/1311.2012v2.pdf.

[20] Y. Polyanskiy, H. Poor, and S. Verdú, "Channel coding rate in the finite blocklength regime," *IEEE Trans. Inf. Theory*, vol. 56, no. 5, pp. 2307–2359, May 2010.

[21] S. Khalili and O. Simeone, "Uplink HARQ for distributed and cloud RAN via separation of control and data planes," Dec. 2015. Available at http://arxiv.org/pdf/1508.06570v3.pdf.

4 Mobile Edge Computing

Ben Liang

4.1 Introduction

The ongoing development of the fifth generation (5G) wireless technologies is taking place in a unique landscape of recent advancement in information processing, marked by the emerging prevalence of cloud-based computing and smart mobile devices. These two technologies complement each other by design, with cloud servers providing the engine for computing and smart mobile devices naturally serving as human interfaces and untethered sensory inputs. Together, they are transforming a wide array of important applications such as telecommunications, industrial production, education, e-commerce, mobile healthcare, and environmental monitoring. We are entering a world where computation is ubiquitously accessible on local devices, global servers, and processors everywhere in between. Future wireless networks will provide communication infrastructure support for this ubiquitous computing paradigm, but at the same time they can also utilize the new-found computing power to drastically improve communication efficiency, expand service variety, shorten service delay, and reduce operational expenses.

The previous generations of wireless networks are passive systems. Residing near the edge of the Internet, they serve only as communication access pathways for mobile devices to reach the Internet core and the public switched telephone network (PSTN). Improvements to these wireless networks have focused on the communication hardware and software, such as advanced electronics and signal processing in the transmitters and receivers. Even for 5G, substantial research effort has been devoted to densification techniques, such as small cells, device-to-device (D2D) communications, and massive multiple-input multiple-output (MIMO). The successes of this communication-only wireless evolution reflect the classical view of an information age centered on information consumption through the Internet.

Yet, in many emerging applications, communication and computation are no longer separated, but interactive and unified. For example, in an augmented-reality application, which might be displayed on smart eyeglasses, the user's mobile device continuously records its current view, computes its own location, and streams the combined information to the cloud server, while the cloud server performs pattern recognition and information retrieval and sends back to the mobile device contextual augmentation labels, to be seamlessly displayed overlaying the actual scenery. As it can be seen from this example, there is a high level of interactivity between the communication and computing functions, and a low tolerance for the total delay due to information transmission and information processing. A fitting analog of this may be found in biological systems, where the computation performed by neurons and the communication among them are inseparable. Indeed, we are moving toward an info-computation age characterized by tight coupling between communication and computation. With an explosion of available data and the consequent need for enormous data-processing capabilities in the emerging big data movement and the Internet of Things (IoT) environment, both communication and computation will be paramount in future wireless applications and services.

Therefore, 5G and future wireless systems are expected to transition away from the model of passive information conduits into active providers and creators of info-computation resources in an integrated communication–computation paradigm. To this end, one major characteristic of future-generation wireless systems will be seamless integration between hardware and software. For example, the 5G Infrastructure Public Private Partnership (5G-PPP) has acknowledged the central role of software and recognized several key software-driven components in 5G standardization, including software-defined networking (SDN), network functions virtualization (NFV), and mobile edge computing (MEC) [1]. Through these technologies, it is envisioned that network functions will be provided over multiple points of presence, especially near the edge of the Internet, in order to achieve the 5G targets for performance, scalability, and agility.

In this chapter, we focus on MEC in 5G systems and beyond. Here, the term *mobile edge* refers to the radio access network (RAN) side of the Internet. It signifies the position of mobile devices with respect to the core computing servers residing in cloud centers. This is in contrast to the network-centric view, where all equipment attached to the Internet, including mobile devices and cloud computing servers, is considered as the edge. We will first describe the MEC functionalities and architecture, then present some examples of use cases, and finally discuss some relevant research challenges.

4.2   Mobile Edge Computing

The concept of MEC is built on recent advances in mobile cloud computing (MCC). In MCC, cloud computing servers produce shared pools of always-on computing resources (e.g., processors, software, and storage), while mobile devices consume these external resources through RANs and the Internet. Cloud computing resources can be rapidly provisioned without the usually prohibitive capital and management costs incurred by users of traditional, self-managed computing servers. MCC is aimed at making computation a ubiquitously accessible utility, similar to but more agile than physical utilities such as electricity and water.

The computing resources in MCC may be centralized or distributed. In the conventional centralized form of MCC, computing resources are provided to mobile users from large remote cloud centers such as the Amazon Elastic Compute Cloud and

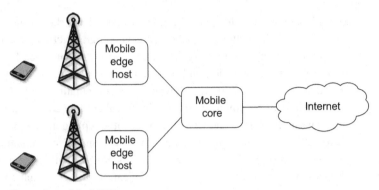

Figure 4.1 Illustration of the MEC system.

Microsoft Azure. These cloud centers provide virtually unlimited computation capacity to augment the processors in mobile devices. However, the communication between mobile users and remote cloud centers is often over a long distance, adding to the latency in cloud computation. Therefore, alternate forms of MCC have been proposed, where computing resources may be accessed in a distributed manner, from smaller local servers such as computing-augmented base stations and Wi-Fi access points or from nearby mobile devices with excess computing capacity. The latter two scenarios of MCC are sometimes named *micro cloud centers* [22], *cloudlets* [35], or *fog computing* [4]. They supplement centralized cloud centers by offering lower access delay and more local awareness.

MEC, as defined by the European Telecommunications Standards Institute (ETSI) [17], refers to a distributed MCC system where computing resources are installed within the RANs, close to the mobile-device end of the Internet. An illustration of the MEC system is given in Figure 4.1, where the mobile edge hosts are computing equipment installed at or near base stations. Unlike centralized cloud servers or peer-to-peer mobile devices, MEC is managed locally by the network operator. The generic computing resources within the mobile edge hosts are virtualized and are exposed via application program interfaces (APIs), so that they are accessible by both user and operator applications.

The mobile edge hosts provide local virtual machines (VMs) to serve the computation needs of mobile devices, often with much lower latency than that of remote cloud centers. They also serve some functions of the traditional mobile core, such as user content caching and traffic monitoring, as well as new functions such as local information aggregation and user location services. Thus, the MEC system may be viewed as a natural outcome of the evolution of mobile base stations from passively serving purely communication functions to becoming an integral part of the new communication–computation paradigm. It is both a midway stop for mobile access to information and computation in the Internet, and a cross-layer bridge that promotes more efficient integration between mobile devices and the mobile core, facilitating the operations of both.

4.3   Reference Architecture

A simplified illustration of the MEC reference architecture proposed by ETSI is given in Figure 4.2. The MEC system resides between the user equipment (UE) and the mobile core networks. It consists of management and functional blocks at the mobile edge host level and the mobile edge system level.

At the mobile edge host level, *mobile edge applications* run VMs supported by the *virtualization infrastructure* within the mobile edge host. They provide services such as computational job execution, radio network information, bandwidth management, and UE location information. The *mobile edge platform* hosts mobile edge services. It interacts with mobile edge applications so that they can advertise, discover, offer, and consume mobile edge services. The *mobile edge platform manager* provides element management functions to the mobile edge platform and administers application essentials such as the life cycle, service requirements, operational rules, domain name system (DNS) configuration, and security.

At the mobile edge system level, the *mobile edge orchestrator* serves the central role of coordinating the UEs, the mobile edge hosts, and the network operator. It records accounting and topological information about the mobile edge hosts deployed, the available resources, and the available mobile edge services. It interfaces with the virtualization infrastructure and maintains the authentication and validation of application packages before their on-boarding. It is also in charge of triggering application instantiation, termination, and relocation, based on its choice of appropriate mobile edge hosts to satisfy the application's requirements and constraints. MEC-capable *UE applications* run within the UE and interact with the mobile edge system to request the on-boarding, instantiation, termination, and relocation of mobile edge applications.

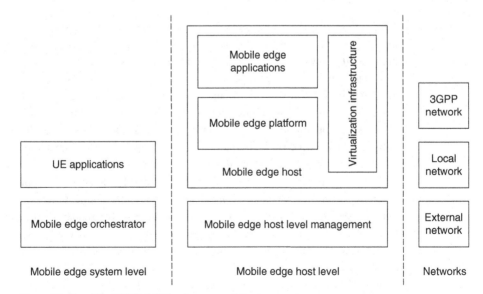

Figure 4.2 Simplified ETSI MEC reference architecture.

An important service provided by the mobile edge orchestrator to an MEC-capable UE application is the timely migration of mobile edge applications between mobile edge hosts, to support UE handoff between different network attachment points.

4.4   Benefits and Application Scenarios

Because of its proximity to mobile devices, the MEC system offers low-latency and high-bandwidth mobile access to both information and computation resources. At the same time, by locally absorbing some of the communication traffic and computational functions of the mobile core, it reduces the resource demand on the mobile backhaul. Furthermore, because of its unique location within the RAN, it is also capable of monitoring and reporting the local network conditions to the mobile core, which promotes awareness of the communication environment at the edge for improved operational efficiency.

Developing the MEC system will benefit a wide range of concerned parties in various application scenarios. MEC serves end users directly by placing information and computation resources in close proximity to them. For network operators, MEC also serves important roles in improving wireless system performance and reducing the cost of operation. For hardware and software developers, the availability of mobile edge platforms and virtualization infrastructure promotes the creation of new applications and consumer products. The following are some example use cases suggested by ETSI for the users and operators of MEC [16].

4.4.1   User-Oriented Use Cases

- *Application computation offloading.* Computationally intensive jobs can be processed by the mobile edge host instead of by mobile devices. Examples of such jobs include high-speed browsing, 3D rendering, video analysis, sensor data processing, and language translation. They tend to require many CPU cycles and are major sources of drainage of mobile on-board batteries. With the option to cheaply offload heavy computation to nearby mobile edge hosts for accelerated processing, the mobile device's computational capability and energy consumption will no longer be the bottleneck for delivering rich applications. This is particularly appealing in the IoT environment, where the mobile devices are likely to consist mostly of small sensors and other equipment with minuscule processors and limited energy supply.
- *Gaming, virtual reality, and augmented reality.* These applications all require low latency. On the one hand, the user could implement the rendering pipeline on the mobile device itself, but the heavy computational requirements of jobs such as physical simulation and artificial intelligence might overwhelm the limited processing capability of the mobile device. On the other hand, offloading these jobs to a remote cloud server might incur too much latency. Instead, with MEC, part of the computational load can be offloaded to some mobile edge application running

on a mobile edge host. Thus, MEC provides an appropriate balance between computational power and proximity.

- *Edge video orchestration.* During a local event with a large audience, such as sports game or a concert, a huge number of mobile devices simultaneously access videos of the same event. These videos are often rich in content and include multiple streams, since they may include both real-time and on-demand versions from multiple camera angles. However, they are all locally generated. Instead of these videos being sent back and forth through the mobile core and the Internet to and from a video content server, they can be processed and delivered by the mobile edge hosts. In addition to relieving the traffic demand on the backhaul, edge video orchestration also provides a more convenient framework to control the service quality, for example through better matching between the hardware capabilities of mobile devices and scalable video coding.

- *Vehicle-to-infrastructure communication.* Vehicle-to-infrastructure communication services, such those between cars and roadside units, are important to the safety and efficiency of transportation systems. MEC can enable the application of 5G wireless to serve such communication functionalities, as it can more readily provides global coverage than can dedicated short-range communications (DSRC). Applications that provide these functionalities can be loaded into mobile edge hosts, which are installed inside RANs along the roadway. Owing to their proximity to vehicles and roadside equipment, hazard warnings and other latency-sensitive messages can reach targeted vehicles within extremely stringent timescale requirements.

4.4.2   Operator-Oriented Use Cases

- *Local content caching at the mobile edge.* The prevalence of widely shared large files on social media, such as high-definition videos, stresses the backhaul network capacity. However, this viral content tends to be consumed by many users in the same geographical area and in the same time period. Therefore, such content can be cached locally at the mobile edge hosts to drastically reduce the traffic demand on the backhaul. Furthermore, with a mobile edge application that proactively moves the cached content between mobile edge hosts in anticipation of the user movement, the service quality can be improved. Similar benefits extend also to broadcast videos, such as mobile TV, where the same video stream is viewed by many mobile users near a mobile edge host.

- *Data aggregation and analytics.* Many services provided by the operator or by third-party vendors, such as security monitoring and massive sensor information processing, depend on a large amount of data collected from sensors and mobile devices. Often it is inefficient to send all of the raw data collected from each sensor and mobile device to back-end servers. In particular, the massive influx of IoT devices may overwhelm the core network. Instead, a distributed mobile edge application running on multiple mobile edge hosts can process the data first to extract the metadata of interest, before forwarding the metadata to back-end servers.

- *Mobile media streaming with bandwidth feedback.* The use of the Hypertext Transfer Protocol (HTTP) over the Transmission Control Protocol (TCP) is ubiquitous in media streaming. Yet, TCP is highly inefficient when applied over conventional wireless networks. TCP is designed to view packet loss as a signal of network congestion, but in wireless networks, packet loss is commonplace and is often due to drastic fluctuations in the radio channel conditions. Thus, the streaming performance can suffer because of miscalculation of the available bandwidth by TCP's congestion control algorithm. With MEC, a radio analytics application running on the mobile edge host can monitor the available wireless bandwidth and forward it to an MEC-capable back-end video server. Such information can then be used by TCP at the video server to adjust its sending rate properly.
- *Mobile backhaul optimization.* Conventional wireless networks lack coordination between the RANs and the backhaul, which prevents effective utilization of backhaul capacity shared by multiple RANs. With MEC, the traffic and performance at the RAN level can be monitored and processed by mobile edge applications. Such localized real-time information, along with other pertinent information such as RAN scheduling and user application profiles, can then be made available to the backhaul network. Thus, the backhaul can be optimized through techniques such as application traffic shaping, traffic routing, and capacity provisioning.
- *Location-based services.* MEC provides more effective location-based services in two ways. First, it allows user location tracking using advanced analytical techniques beyond measuring the received signal strength. Second, mobile edge applications can provide more appropriate location-based service recommendations. A mobile edge application can take into account its knowledge of the context of the user's location, such as a shopping mall or museum, as well as the user's behavior pattern, to make recommendations. It may also utilize advanced machine learning techniques and interface with big data analysis at back-end servers to further improve the accuracy and usefulness of its recommendations.

4.5   Research Challenges

MEC is a complex system that creates a new framework to support tight integration between wireless networking and cloud computing. On the networking side, it addresses all layers of the network protocol stack. On the cloud computing side, it involves all three cloud service modes: Infrastructure as a Service (IaaS), Platform as a Service (PaaS), and Software as a Service (SaaS). Because of the synergistic nature of MEC based on both communication and computation, there are wide-ranging challenges in the related research and development.

4.5.1   Computation Offloading

Offloading computation-intensive and resource-hungry applications to resource-rich servers can augment the capabilities of mobile devices. The reduction in energy consumption by mobile devices that can be achieved through computation offloading has been demonstrated in experimental tests and analytical studies [26]. However, energy is only one of many important factors in mobile resource optimization [11, 13]. In MEC, there are multiple mobile devices sharing the same mobile edge host, so its limited capacity in terms of communication and computation is a major concern. The usage cost of the back-end cloud servers and the MEC system should also be considered.

For each job, a decision needs to be made about whether to execute it locally on the mobile device, send it to the mobile edge host, or forward it further to remote cloud servers. Even assuming that the processing times and costs of each job at different locations are known in advance, this leads to an integer programming problem that is often extremely difficult to solve [15]. Moreover, the processing time and costs are likely to be unknown before a job has been executed, owing to the lack of exact information about the number of operations required for the job, as well as the randomness in each processor's available computing cycles [6, 38]. New analytical tools are needed to handle such uncertainty in order to design MEC systems with both average and worst-case performance guarantees.

Yet another challenge is the placement constraints imposed on each job. With the coexistence of multiple mobile operating systems and the vast heterogeneity in hardware and software requirements, not all mobile edge hosts are able to execute every job. This patchwork of placement constraints and user utility preferences will substantially complicate the optimization of computation offloading [37].

4.5.2   Communication Access to Computational Resources

The benefits of computation offloading cannot be enjoyed without efficient access to the mobile edge hosts and the remote cloud servers. In conventional grid computing and cloud computing over wired links, a common assumption is that the communication pathway between the user and the cloud center is unimpeded [38], so that much research effort has been focused instead on ensuring wired connectivity between grid computers or between cloud nodes. However, one of the main vulnerabilities of MCC is the instability of the wireless access links between mobile devices and cloud computing resources [27]. The access links may be interrupted without notice, leading to random service outages. Therefore, the joint allocation of communication resources and computational resources is paramount for the successful deployment of MEC.

There are three important communication concerns in MEC: wireless access while offloading to the mobile edge host; backhaul access while offloading to a remote cloud server; and the communication among mobile devices, mobile edge hosts, and remote cloud servers when they collaboratively execute multiple jobs. Both the wireless and the backhaul access links have limited capacity and must be properly shared among multiple mobile devices, in a similar way to that in which the computing resources of the mobile edge host are shared [10, 52]. Therefore, a joint communication–computation resource-sharing framework is required in MEC. Furthermore, such resource-sharing decisions are strongly influenced by the requirement for data exchange between jobs that are executed at different locations. This data requirement induces a dependency relation among jobs, where the execution of one job must wait for the completion of some other jobs and their output data. Thus, the jobs are equivalent to vertices in a directed graph, with the edges representing dependencies and their weights modeling data communication costs, such as the number of bits, the price per bit, or the transmission delay [42, 44]. Hence, joint communication–computation resource sharing in MEC engenders a complicated graph partitioning problem on a directed graph where the edge weights also need to be suitably optimized.

The design of computation offloading may be challenged further by the need to satisfy job execution deadlines, for example when the common application supported by these jobs is delay sensitive. The problem of minimizing the makespan of executing multiple jobs in parallel, even in the simple case where there is no requirement for communication among the jobs, is NP-hard [5]. Therefore, the general case of scheduling jobs with dependencies and execution deadlines is among the hardest problems in MEC.

4.5.3   Multi-resource Scheduling

The fairness and efficiency of resource allocation are fundamental problems in the field of distributed computing systems. MEC requires the allocation of multiple communication and computation resources. Within each mobile edge host, the communication bandwidth, processor cycles, memory bandwidth, and on-board storage are shared among multiple users and applications. Scheduling jobs that require multiple resources is a challenging problem owing to the combinatorial nature of allocating the blocks of different resources. Known results from the classical flow-shop problem suggest optimal scheduling is often intractable when three or more resources are considered [33].

The multi-resource fairness problem has been studied in the field of operations research [50], and seminal work in the context of networking and computing appeared in [20]. Most existing work has focused on either a single machine or an isolated group of servers. In contrast, MEC has a unique scheduling environment with multiple access networks, mobile edge hosts, and remote servers. New multi-resource scheduling designs for MEC must account for this complexity. Furthermore, multi-resource scheduling necessarily leads to a trade-off between fairness among jobs and system efficiency [25, 46], which does not exist in traditional single-resource scheduling with work conservation. This is further exacerbated by the heterogeneous computation and communication capabilities of local and remote equipment [18, 48]. It is particularly challenging to distribute jobs across multiple heterogeneous mobile edge hosts, since these hosts are shared by other users and applications with uncertain workloads, and the quality of access to these hosts is also random owing to wireless channel fluctuations, spectrum sharing, and user mobility.

4.5.4   Mobility Management

One main characteristic of MEC is device mobility. When a mobile user is handed off from one base station to another, its association with all of its active applications running on the mobile edge hosts and the remote servers must remain intact. Furthermore, when the user moves too far away from its serving mobile edge host, the VMs created on that host for that user must migrate to a more suitable new host. Thus, mobility management in MEC involves both communication handoff and computation handoff. VM migration is costly in general [12], and in the case of MEC, it is further complicated by the heterogeneity of the device software and mobile edge host capabilities, and the stringent delay constraints of some applications. From the operator's point of view, device mobility brings substantial challenges to a wide range of functionalities such as content caching, data analytics, and backhaul optimization.

For optimal system design, it will be necessary to accurately model the impact of mobile handoff. Furthermore, because of the high cost of handoff in MEC, it may be prudent to reduce the handoff frequency of a mobile device as it moves through the system. This can be achieved by disregarding opportunities for stronger connections with nearer base stations while the device's weaker connection with its present base station remains useful [3, 40], and by delaying the migration of VMs even when the device has connected to a new base station [43]. The former will be facilitated by the high density of base stations in 5G systems and beyond, with a trade-off between handoff frequency and data rate. The latter will be supported by virtual network connections, with a trade-off between handoff frequency and application latency. Thus, one of the main challenges in MEC is in optimally balancing these trade-offs in a vastly complex system with many mobile devices and complicated connections among the mobile edge hosts.

4.5.5   Resource Allocation and Pricing

MEC offers unprecedented opportunities for innovative applications and services. These applications and services have diverse resource requirement profiles. Some have real-time demands with strict timeliness requirements, while others can tolerate some delay, and yet others may benefit from reserving resources for future use. Any resource allocation scheme will need to balance these diverse needs of different applications [9]. In particular, separate pools of resources may need to be reserved for steady long-term usage contracts and for the unpredictable arrival of urgent demands [45, 51]. Moreover, the mobile edge hosts serve both user application jobs and MEC service jobs, and the allocation of communication and computation resources between these two types of jobs will reflect the relative importance that the operator places on them. This adds another dimension of difficulty to the resource allocation problem.

Resource allocation is also tightly coupled to resource pricing. The mobile operator may charge users a price for offloading user applications to mobile edge hosts. This may follow the pricing schemes used by current cloud computing service providers, for example charging higher amounts for on-demand VMs and giving discounts for

VMs that are reserved ahead of time. However, joint optimization of resource pricing and resource allocation remains an open problem for general cloud computing systems. Furthermore, since an offloaded user application in MEC may be forwarded further to remote cloud servers, which can be paid for directly by the users themselves or be brokered by the mobile operator [39, 47], the problem is substantially more complicated in the case of MEC.

4.5.6   Network Functions Virtualization

As a core enabling technology, NFV is employed by mobile edge hosts to create a wide variety of network appliances, such as routers, packet gateways, and Internet Protocol (IP) multimedia subsystems, using generic hardware. By separating software from hardware, NFV allows dynamic provisioning of services and flexible deployment of network functions. However, the performance of current NFV implementations often falls well below that of dedicated hardware network equipment [24]. This is a particularly acute issue for small cells, where the base stations and their associated mobile edge hosts need to have a light footprint for flexible installation. Therefore, one difficult challenge in MEC is to improve the performance of virtualized services inside moderately endowed mobile edge hosts [29].

Furthermore, because of the performance limitations of NFV, it is essential to make judicious decisions about whether to keep particular network functions within the mobile core, or to virtualize them and move them to the mobile edge hosts. This decision must make a trade-off between hardware equipment performance and NFV service flexibility [31]. An additional dimension to this issue is service latency, as the proximity of services to mobile devices is of paramount importance to the overall characteristics of MEC. Finally, since the mobile edge hosts are shared by user applications and NFV appliances, the dynamics of user communication traffic and computation offloading demand also have a high impact on the decisions about NFV.

4.5.7   Security and Privacy

MEC is a complex amalgamation of diverse technologies, including wireless networking, distributed computing, and the virtualization of networking equipment and computing servers. This opens up multiple fronts for malicious attacks, and a wide range of security measures are needed to thwart them [36, 41]. In particular, owing to the user-friendly location and limited size of mobile edge hosts, mobile edge hosts cannot enjoy the physical protection afforded to large data centers. Furthermore, even if each component technology is individually secured, because of the complicated interactions and dependencies among them, there is no guarantee that the entire MEC system is secure.

Meanwhile, the MEC objective of ubiquitous and low-latency availability of information and computation resources, and the consequent requirement for a high density and diversity of connections everywhere, makes it difficult to set up a global secure perimeter. Vulnerabilities in a single mobile edge host can provide the means to launch an attack vector against the whole system. Furthermore, the existence of many VMs, spread across multiple mobile edge hosts, increases the chances of multiple compromised VMs coordinating large-scale attacks such as distributed denial-of-service (DDoS) attacks. These security risks did not exist in previous generations of wireless networks, so there are few studies in the literature on counter measures.

Furthermore, the ever-present conflict between system security and service agility is amplified in MEC. On the one hand, MEC is aimed at efficient and responsive services, so that the resource overheads for security, such as authentication, access control, and intrusion detection, need to be minimal. On the other hand, the unique MEC architecture, with its amalgamation of numerous heterogeneous components, requires strong security protection through complex multidimensional strategies. Adding to this challenge, the software nature of the NFV-based implementation of security protocols can lead to a severe performance bottleneck in MEC [32]. We require efficient security strategies that match the unique characteristics of MEC, for example distributed authentication services implemented in mobile edge hosts [49].

Privacy is an important component of the security of the overall system. In MEC, with increased hardware and software connections between mobile devices and the network operator, there is a higher risk of information leakage. MEC applications such as computation offloading, content caching, and augmented reality bring the frontline of privacy out of the relatively safe core mobile network. At the same time, they require a large amount of user data and user interaction, which adds to the difficulty of privacy protection. Whether it is information gathering by the operator or privacy breaches by malicious attackers, in MEC there are more opportunities for security mis-steps. Networking and computing security functionalities must be installed in mobile edge hosts to balance the needs for both privacy protection and service performance, for example by establishing trusted local proxies for anonymous access to information and computation [21], and replacing deep packet inspection with locally aware machine learning techniques for traffic classification and anomaly detection [30].

4.5.8   Integration with Emerging Technologies

MEC joins other advanced technologies in the 5G ecosystem, such as D2D communication [14] and cloud RAN (C-RAN) [8]. It is also expected to coexist with other emerging information and computation technologies such as information-centric networking (ICN) [2], intelligent vehicular systems [19], and hybrid private–public cloud computing [23]. The successful deployment of MEC will depend on its integration with these new technologies.

Although the traffic of D2D communication does not pass through base stations, the mobile edge hosts can serve important coordination functions. With their unique abilities for monitoring and processing near the D2D nodes, and their interconnectedness through the backhaul, the mobile edge hosts are well positioned to assist with traffic scheduling and interference management in D2D communication. The main concern here is scalability as the D2D communication group increases in size, particularly in the

IoT environment. The integration between MEC and C-RAN is also challenging, since C-RAN promotes lightweight remote radio heads. The mobile edge hosts may reside at the remote radio heads or the baseband units of the C-RAN system. In either case, the corresponding MEC platform and virtualization designs will need to account for the unique characteristics of C-RAN.

The mobile edge hosts provide convenient hardware and software resources for ICN, allowing content caching directly within the RANs. However, the current MEC framework is described under the assumption of a traditional TCP/IP network. It needs to evolve as the role of ICN becomes more prominent in the future Internet [34]. For intelligent vehicular systems, the ETSI example use case of MEC given earlier covers only vehicle-to-infrastructure communication, while in vehicle-to-vehicle communication based on either D2D links or DSRC, the mobile edge hosts will serve only coordination functionalities. Proper operation of such a vehicular system requires seamless integration of these two communication modes [28]. Finally, in hybrid cloud computing, an enterprise maintains local private cloud centers at the same time as it employs remote public cloud services [7]. Application scheduling and resource allocation in MEC will depend on whether the mobile edge hosts or the VMs residing within them have access to the private cloud.

4.6   Conclusion

MEC is a natural outcome of the emerging convergence between communication and computation. It requires multidimensional study of a complex amalgamation of diverse subjects in communication systems, distributed computing, software engineering, and system optimization. Its standardization alongside 5G and future wireless technologies is poised to bring drastic changes to how wireless systems are designed, operated, and utilized. In this early stage of development of MEC, there remain many challenging open problems and ample opportunities for future innovation.

References

[1] 5G Infrastructure Public Private Partnership, "5G vision: The 5G infrastructure public private partnership: The next generation of communication networks and services," Feb. 2015.

[2] B. Ahlgren, C. Dannewitz, C. Imbrenda, D. Kutscher, and B. Ohlman, "A survey of information-centric networking," *IEEE Commun. Mag.*, vol. 50, no. 7, pp. 26–36, Jul. 2012.

[3] W. Bao and B. Liang, "Stochastic geometric analysis of user mobility in heterogeneous wireless networks," *IEEE J. Sel. Areas Commun.*, vol. 33, no. 10, pp. 2212–2225, Oct. 2015.

[4] F. Bonomi, R. Milito, J. Zhu, and S. Addepalli, "Fog computing and its role in the Internet of Things," in *Proc. of ACM SIGCOMM Workshop on Mobile Cloud Computing*, Helsinki, Finland, Aug. 2012.

[5] D. G. Cattrysse and L. N. Van Wassenhove, "A survey of algorithms for the generalized assignment problem," *Eur. J. Oper. Res.*, vol. 60, no. 3, pp. 260–272, 1992.

[6] J. P. Champati and B. Liang, "Semi-online task partitioning and communication between local and remote processors," in *Proc. of IEEE International Conf. on Cloud Networking (CLOUDNET)*, Niagara Falls, Canada, Oct. 2015.

[7] J. P. Champati and B. Liang, "One-restart algorithm for scheduling and offloading in a hybrid cloud," in *Proc. of IEEE/ACM International Symposium on Quality of Service (IWQoS)*, Portland, OR, Jun. 2015.

[8] A. Checko, H. L. Christiansen, Y. Yan, L. Scolari, G. Kardaras, M. S. Berger, and L. Dittmann, "Cloud RAN for mobile networks – A technology overview," *IEEE Commun. Surv. Tutor.*, vol. 17, no. 1, pp. 405–426, First Quarter 2015.

[9] Y. Chen, A. Das, W. Qin, A. Sivasubramaniam, Q. Wang, and N. Gautam, "Managing server energy and operational costs in hosting centers," in *Proc. of ACM International Conf. on Measurement and Modeling of Computer Systems (SIGMETRICS)*, Banff, Canada, Jun. 2005.

[10] M.-H. Chen, M. Dong, and B. Liang, "Joint offloading decision and resource allocation for mobile cloud with computing access point," in *Proc. of IEEE International Conf. on Acoustics, Speech and Signal Processing (ICASSP)*, Shanghai, China, Mar. 2016.

[11] B. Chun, S. Ihm, P. Maniatis, M. Naik, and A. Patti, "CloneCloud: Elastic execution between mobile device and cloud," in *Proc. of the European Conf. on Computer Systems (EuroSys)*, Salzburg, Austria, Apr. 2011.

[12] C. Clark, K. Fraser, S. Hand, J. G. Hansen, E. Jul, C. Limpach, I. Pratt, and A. Warfield, "Live migration of virtual machines," in *Proc. of USENIX Symposium on Networked Systems Design and Implementation (NSDI)*, Berkeley, CA, May 2005.

[13] E. Cuervo, A. Balasubramanian, D. Cho, A. Wolman, S. Saroiu, R. Chandra, and P. Bahl, "MAUI: Making smartphones last longer with code offload," in *Proc. of ACM International Conf. on Mobile Systems, Applications, and Services (MobiSys)*, San Francisco, CA, Jun. 2010.

[14] K. Doppler, M. Rinne, C. Wijting, C. B. Ribeiro, and K. Hugl, "Device-to-device communication as an underlay to LTE-Advanced networks," *IEEE Commun. Mag.*, vol. 47, no. 12, pp. 42–49, Dec. 2009.

[15] M. Drozdowski, *Scheduling for Parallel Processing*, Springer, 2009.

[16] ETSI, "Mobile edge computing (MEC); Technical requirements," ETSI GS MEC 002 V1.1.1 (2016-03), Mar. 2016.

[17] ETSI, "Mobile edge computing (MEC); Framework and reference architecture," ETSI GS MEC 003 V1.1.1 (2016-03), Mar. 2016.

[18] E. Friedman, A. Ghodsi, and C.-A. Psomas, "Strategyproof allocation of discrete jobs on multiple machines," in *Proc. of ACM Conf. on Economics and Computation (EC)*, Palo Alto, CA, Jun. 2014.

[19] M. Gerla and L. Kleinrock, "Vehicular networks and the future of the mobile Internet," *Comput. Netw.*, vol. 55, no. 2, pp. 457–469, Feb. 2011.

[20] A. Ghodsi, M. Zaharia, B. Hindman, A. Konwinski, S. Shenker, and I. Stoica, "Dominant resource fairness: Fair allocation of multiple resource types," in *Proc. of USENIX Symposium on Networked Systems Design and Implementation (NSDI)*, Boston, MA, Mar. 2011.

[21] G. Ghinita, P. Kalnis, and S. Skiadopoulos, "PRIVE: Anonymous location-based queries in distributed mobile systems," in *Proc. of ACM International Conf. on World Wide Web (WWW)*, Banff, Canada, May 2007.

[22] A. Greenberg, J. Hamilton, D. A. Maltz, and P. Patel, "The cost of a cloud: Research problems in data center networks," *ACM SIGCOMM Computer Commun. Rev.*, vol. 39, no. 1, pp. 68–73, Dec. 2008.

[23] T. Guo, U. Sharma, P. Shenoy, T. Wood, and S. Sahu, "Cost-aware cloud bursting for enterprise applications," *ACM Trans. Internet Technol.*, vol. 13, no. 3, 10:1–10:24, May 2014.

[24] B. Han, V. Gopalakrishnan, L. Ji, and S. Lee, "Network function virtualization: Challenges and opportunities for innovations," *IEEE Commun. Mag.*, vol. 53, no. 2, pp. 90–97, Feb. 2015.

[25] C. Joe-Wong, S. Sen, T. Lan, and M. Chiang, "Multi-resource allocation: Fairness–efficiency tradeoffs in a unifying framework," in *Proc. of IEEE International Conference on Computer Communications (INFOCOM)*, Orlando, FL, Mar. 2012.

[26] K. Kumar and Y.-H. Lu, "Cloud computing for mobile users: Can offloading computation save energy?" *IEEE Computer*, vol. 43, no. 4, pp. 51–56, Apr. 2010.

[27] K. Kumar, J. Liu, Y.-H. Lu, and B. Bhargava, "A survey of computation offloading for mobile systems," *Mobile Netw. Appl.*, vol. 18, no. 1, pp. 129–140, Feb. 2013.

[28] E. Lee, E. K. Lee, M. Gerla, and S. Y. Oh, "Vehicular cloud networking: Architecture and design principles," *IEEE Commun. Mag.*, vol. 52, no. 2, pp. 148–155, Feb. 2014.

[29] A. Manzalini, R. Minerva, F. Callegati, W. Cerroni, and A. Campi, "Clouds of virtual machines in edge networks," *IEEE Commun. Mag.*, vol. 51, no. 7, pp. 63–70, Jul. 2013.

[30] T. T. T. Nguyen and G. Armitage, "A survey of techniques for Internet traffic classification using machine learning," *IEEE Commun. Surv. Tutor.*, vol. 10, no. 4, pp. 56–76, Fourth Quarter 2008.

[31] D. Oppenheimer, B. Chun, D. Patterson, A. C. Snoeren, and A. Vahdat, "Service placement in a shared wide-area platform," in *Proc. of USENIX Annual Technical Conf.*, Boston, MA, Jun. 2006.

[32] G. Pek, L. Buttyan, and B. Bencsath, "A survey of security issues in hardware virtualization," *ACM Comput. Surv.*, vol. 45, no. 3, pp. 40:1–40:34, Jul. 2013.

[33] M. L. Pinedo, *Scheduling: Theory, Algorithms, and Systems*, Springer, 2012.

[34] R. Ravindran, X. Liu, A. Chakraborti, X. Zhang, and G. Wang, "Towards software defined ICN based edge-cloud services," in *Proc. of IEEE International Conf. on Cloud Networking (CloudNet)*, San Francisco, CA, Nov. 2013.

[35] M. Satyanarayanan, P. Bahl, R. Caceres, and N. Davies, "The case for VM-based cloudlets in mobile computing," *IEEE Pervasive Comput.*, vol. 8, no. 4, pp. 14–23, Oct.–Dec. 2009.

[36] M. Satyanarayanan, G. Lewis, E. Morris, S. Simanta, J. Boleng, and K. Ha, "The role of cloudlets in hostile environments," *IEEE Pervasive Comput.*, vol. 12, no. 4, pp. 40–49, Oct.–Dec. 2013.

[37] B. Sharma, V. Chudnovsky, J. L. Hellerstein, R. Rifaat, and C. R. Das, "Modeling and synthesizing task placement constraints in Google compute clusters," in *Proc. of the ACM Symposium on Cloud Computing (SoCC)*, Cascais, Portugal, Oct. 2011.

[38] D. B. Shmoys, J. Wein, and D. P. Williamson, "Scheduling parallel machines on-line," *SIAM J. Comput.*, vol. 24, no. 6, pp. 1313–1331, Dec. 1995.

[39] Y. Song, M. Zafer, and K.-W. Lee, "Optimal bidding in spot instance market," in *Proc. of IEEE International Conf. on Computer Communications (INFOCOM)*, Orlando, FL, Mar. 2012.

[40] E. Stevens-Navarro, Y. Lin, and V. W. S. Wong, "An MDP-based vertical handoff decision algorithm for heterogeneous wireless networks," *IEEE Trans. Veh. Technol.*, vol. 57, no. 2, pp. 1243–1254, Mar. 2008.

[41] I. Stojmenovic, S. Wen, X. Huang, and H. Luan, "An overview of fog computing and its security issues," *Concurr. Comput.: Pract. Exp.*, vol. 28, no. 10, pp. 2991–3005, Jul. 2016.

[42] S. Sundar and B. Liang, "Communication augmented latest possible scheduling for cloud computing with delay constraint and task dependency," in *Proc. of IEEE INFOCOM Workshop on Green and Sustainable Networking and Computing (GSNC)*, San Francisco, CA, Apr. 2016.

[43] R. Urgaonkara, S. Wang, T. He, M. Zafer, K. Chan, and K. K. Leung, "Dynamic service migration and workload scheduling in edge-clouds," *Perform. Eval.*, vol. 91, pp. 205–228, Sep. 2015.

[44] C. Wang and Z. Li, "Parametric analysis for adaptive computation offloading," in *Proc. of ACM SIGPLAN Conf. on Programming Language Design and Implementation (PLDI)*, Washington, DC, Jun. 2004.

[45] W. Wang, B. Li, and B. Liang, "Towards optimal capacity segmentation with hybrid cloud pricing," in *Proc. of IEEE International Conf. on Distributed Computing Systems (ICDCS)*, Macau, China, Jun. 2012.

[46] W. Wang, C. Feng, B. Li, and B. Liang, "On the fairness–efficiency tradeoff for packet processing with multiple resources," in *Proc. of ACM SIGCOMM International Conf. on Emerging Networking Experiments and Technologies (CoNEXT)*, Sydney, Australia, Dec. 2014.

[47] W. Wang, D. Niu, B. Liang, and B. Li, "Dynamic cloud resource reservation via IaaS cloud brokerage," *IEEE Trans. Parallel Distrib. Syst.*, vol. 26, no. 6, pp. 1580–1593, Jun. 2015.

[48] W. Wang, B. Liang, and B. Li, "Multi-resource fair allocation in heterogeneous cloud computing systems," *IEEE Trans. Parallel Distrib. Syst.*, vol. 26, no. 10, pp. 2822–2835, Oct. 2015.

[49] R. Yahalom, B. Klein, and T. Beth, "Trust relationships in secure systems – A distributed authentication perspective," in *Proc. of IEEE Computer Society Symposium on Research in Security and Privacy*, Oakland, CA, May 1993.

[50] H. P. Young, *Equity: In Theory and Practice*, Princeton University Press, 1994.

[51] Q. Zhang, Q. Zhu, and R. Boutaba, "Dynamic resource allocation for spot markets in cloud computing environments," in *Proc. of IEEE International Conf. on Utility and Cloud Computing (UCC)*, Victoria, Australia, Dec. 2011.

[52] B. Zhou, A. V. Dastjerdi, R. N. Calheiros, S. N. Srirama, and R. Buyya, "A context sensitive offloading scheme for mobile cloud computing service," in *Proc. of IEEE International Conf. on Cloud Computing (CLOUD)*, New York, Jun. 2015.

5 Decentralized Radio Resource Management for Dense Heterogeneous Wireless Networks

Abolfazl Mehbodniya and Fumiyuki Adachi

5.1 Introduction

The number of devices connected to the wireless infrastructure is increasing significantly owing to recent developments such as the concept of the Internet of Things (IoT). As a result, the load on wireless networks is increasing, as well as the energy consumption [1]. Therefore, we need to develop energy-efficient mechanisms for resource allocation in wireless networks [1, 2]. The use of a heterogeneous network (HetNet), i.e., a set of small-cell base stations (BSs) overlaid by macrocell base stations (MBSs), can improve the energy efficiency (EE) [3] of wireless systems, mainly because it brings the transmitters and receivers closer together, thereby combating the path loss effect. However, owing to the proximity of BSs, a co-channel interference (CCI) problem arises which may degrade the overall system performance to a great extent [4]. As a result, CCI management is a crucial task in the next-generation dense HetNets.

In the existing literature, there are several studies regarding EE in HetNets, such as BS placement, load balancing, power control, and dynamic BS sleep–wake mechanisms [4–6]. All these studies provide good solutions to improving EE. However, they are all based on centralized control approaches and need to collect network information in order to make a unified decision. In [7], a distributed energy-efficient BS ON/OFF switching algorithm based on game theory was proposed. This algorithm considers a utility function combined with the total power consumption and the load on the BSs. Later, by evaluating this utility function, the BSs independently choose a predefined transmission power level. This algorithm can provide improvements in terms of system EE and the overall load reduction compared with conventional approaches in a distributed manner.

Another major issue in wireless communications is the scarcity of wireless channels, i.e., channels in the frequency, time, and space dimensions. Hence, the same channel needs to be reused among BSs, which consequently limits the network capacity owing to the introduced CCI. Dynamic channel assignment (DCA) is an effective technique for reusing the same channel in wireless communications. This technique has been abundantly studied in the literature [8–11]. Previously, we proposed a technique called interference-aware channel-segregation based DCA (IACS-DCA) [12], in which each BS measures the CCI of different channels on a periodic basis and calculates the average

CCI powers. Later, the BS updates a channel priority table in increasing order of average CCI power level. Then, the channel with the lowest average CCI is chosen for data communication in the next phase. It has been shown that the proposed IACS-DCA technique can form a channel reuse pattern with low CCI which adapts to the BS distribution [13, 14].

In this chapter, we introduce two novel channel segregation (CS) techniques based on the IACS-DCA approach for the dense user equipment (UE) scenarios. Both techniques are combined with a learning-based game-theoretic algorithm for the ON/OFF switching of the BSs similar to that presented in [7]. The first technique, i.e., the UE-based CS algorithm (UECSA) [15], considers the CCI environment experienced at the UE. The CCI experienced at the UE varies over time and the channel allocation needs to cope with this changing environment. These variations are mainly caused by power control, UE location updates, and the BS ON/OFF switching patterns. In UECSA, each UE informs its corresponding BS about its channel priority information and the BSs select the best channel by analyzing such information. The second proposed technique, i.e., the BS-based CS algorithm (BSCSA), adopts a similar procedure. However, it is performed at the BS. As a result, no signaling overhead is necessary for reporting decisions from the UEs to the BSs. Simulation results show that combining the aforementioned ON/OFF switching algorithm with the two proposed techniques not only improves the EE but also achieves a significant gain in terms of spectral efficiency (SE). Moreover, UECSA outperforms BSCSA at the cost of increased signaling overhead. Another notable characteristic of the proposed scheme is that all UECSA, BSCSA, and sleep mode algorithms are executed in a fully distributed manner at each BS, without the need for information exchange between BSs.

The remainder of this chapter is organized as follows. In Section 5.2, a system model is presented along with expressions for the power, signal-to-interference-plus-noise ratio (SINR), load, and cost. Section 5.3 discusses the various steps of our two proposed CS algorithms. Section 5.4 provides simulation results and an evaluation of our algorithm, and, finally, Section 5.5 concludes the chapter.

5.2   System Model

The mathematical notation and formulation used in this chapter are similar to those used in [15]. We focus on downlink transmission in a HetNet. Figure 5.1 shows the system model, which consists of a set of BSs, $S = \{1,\ldots,s,\ldots,S\}$. We assume only one MBS, $s = 1$, which encompasses a variable number of small BSs (SBSs), $S = \{2,\ldots,s,\ldots,S\}$. A total of X users, $\mathbb{X} = \{1,\ldots,x,\ldots,X\}$, are available in the system.

The power consumption of the sth BS, BS(s), at time t is given by

$$P_s^{\text{All}}(t) = \frac{P_s(t)}{\eta\alpha(1-\alpha_{\text{feed}})} + P_s^{\text{Back}} + P_s^{\text{Idle}}, \tag{5.1}$$

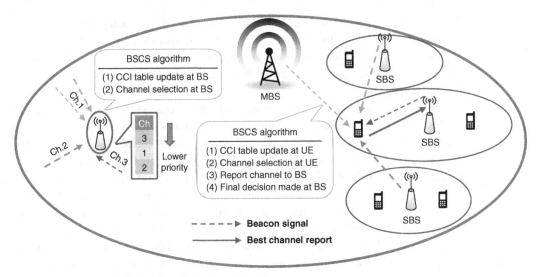

Figure 5.1 Illustration of the system model.

with

$$\alpha = (1 - \alpha_{\mathrm{DC}})(1 - \alpha_{\mathrm{main}})(1 - \alpha_{\mathrm{cool}}) \tag{5.2}$$

and

$$P_s^{\mathrm{Idle}} = \frac{P_{\mathrm{radio}} + P_{\mathrm{base}}}{\alpha}, \tag{5.3}$$

where $P_s(t)$ is the transmission power, η is the efficiency of the power amplifier, P_s^{Back} is the power consumption in the backhaul, and P_s^{Idle} is the power consumption in OFF mode. The variables α_{feed}, α_{DC}, α_{main}, and α_{cool} represent the loss fractions of the feeder, AC–DC conversion, mains supply, and cooler system, respectively. P_{radio} and P_{base} are the power consumptions of the radio frequency and baseband units, respectively.

We assume a total of C available frequency channels, and $\mathfrak{C} = \{1, \ldots, c, \ldots, C\}$. Each channel uses orthogonal frequency division multiple access (OFDMA) with N_c subcarriers. A frequency-selective block Rayleigh fading channel is considered which is composed of Q distinct paths. The impulse response of the propagation channel at time t is modeled according to

$$h(\tau;t) = \sum_{q=0}^{Q-1} h_q(t)\delta(\tau - \tau_q), \tag{5.4}$$

where $h_q(t)$ denotes the time-varying complex-valued path gain, with $E[\sum_{q=0}^{Q-1} |h_q(t)|^2] = 1$ ($E[.]$ denotes the ensemble average operation), and τ_q denotes the time delay of the qth path.

5.2.1   SINR Expression

The downlink CCI experienced at the xth UE connected to the sth BS, UE_s^x, comes from the co-channel BSs using the same $c(s)$th channel and is given by

$$
I_{\mathrm{UE}_s^x}(t) = \sum_{\substack{s' \in \mathrm{BSG}(c(s)) \\ s' \neq s}} P_{s'}(t)\, \xi_{\mathrm{UE}_s^x,\mathrm{BS}(s')} \sum_{k' \in N_s^x} \left| H_{\mathrm{UE}_s^x,\mathrm{BS}(s')}(t;k',c(s)) \right|^2,
\tag{5.5}
$$

where $\mathrm{BSG}(c(s))$ denotes the BS group which uses the same $c(s)$th channel, $P_{s'}(t)$ is the transmit power of $\mathrm{BS}(s')$, $\xi_{\mathrm{UE}_s^x,\mathrm{BS}(s')}$ is the path loss between UE_s^x, $\mathrm{BS}(s')$, $H_{\mathrm{UE}_s^x,\mathrm{BS}(s')}(t;k,c(s))$ is obtained from the Fourier transform of the impulse response of the channel between UE_s^x and $\mathrm{BS}(s')$ at time t, and N_s^x is the set of subcarriers assigned by $\mathrm{BS}(s)$ to UE_s^x. Here, we do not consider any optimal subcarrier allocation for the OFDMA system. Subcarriers are divided equally and sequentially between the UEs connected to each BS and we assume $N_c \gg X_s$, where X_s is the total number of UEs connected to $\mathrm{BS}(s)$. The downlink instantaneous SINR $\lambda_{\mathrm{UE}_s^x}$ experienced at UE_s^x's antenna is given by

$$
\lambda_{\mathrm{UE}_s^x}(t) = \frac{P_{s'}(t)\xi_{\mathrm{UE}_s^x,\mathrm{BS}(s')} \sum_{k' \in N_s^x} \left| H_{\mathrm{UE}_s^x,\mathrm{BS}(s)}(t;k',c(s)) \right|^2}{I_{\mathrm{UE}_s^x}(t) + N_s^x N_0},
\tag{5.6}
$$

where N_0 is the noise power.

5.2.2   Load and Cost Function Expressions

We assume that the UEs have different quality-of-service (QoS) requirements, i.e., they have different packet arrival rates and mean packet sizes. We define the instantaneous load density of each BS as the sum of the loads of all individual UEs connected to it according to

$$
v_s(t) = \sum_{x=1}^{X_s} \frac{\gamma_s^x}{\mu_s^x \omega' \log_2(1 + \lambda_{\mathrm{UE}_s^x}(t))},
\tag{5.7}
$$

where γ_s^x and μ_s^x are the packet arrival rate and mean packet size, respectively, for the xth UE connected to the sth BS, $\omega' = \omega/C$ is the bandwidth of each BS, and ω is the total system bandwidth. The load of each BS is inversely related to the throughput for the UEs that are being served. The average value of the load, $\hat{v}_s(t)$, is calculated as

$$
\hat{v}_s(t) = \hat{v}_s(t-1) + l(t)\left(v_s(t-1) - \hat{v}_s(t-1)\right),
\tag{5.8}
$$

where $l(t)$ is the learning rate. In order to ensure system stability, $l(t)$ is chosen such that the load averaging is sufficiently slower than the UE association process.

The cost function which simultaneously captures the load and the energy consumption for the sth BS is defined by

$$
\Psi_s(t) = \frac{\sum_{x=1}^{X_s} \omega' \log_2(1 + \lambda_{\mathrm{UE}_s^x}(t))}{P_s^{\mathrm{All}}(t)},
\tag{5.9}
$$

Table 5.1 Transmission power levels.

Strategy identification number (i)	Transmission power level ($\zeta_s(t)$)
1	0
2	1/3
3	2/3
4	1

where P_s^{Max} is the maximum allowed transmission power for the BSs in the system, and φ_s and ψ_s are the weighting parameters which define the impact of energy and load, respectively.

5.3   Joint BSCSA/UECSA ON/OFF Switching Scheme

Our goal is to design a fully distributed solution which can minimize the cost function in (5.9) for all BSs. The proposed algorithm is summarized in Algorithm 5.1.

5.3.1   Strategy Selection and Beacon Transmission

In the first step of the algorithm, the BSs decide their action, i.e., their ON/OFF mode. A noncooperative game is used to design the decision process. In this game, the players are the BSs, the strategies are the different power levels chosen by the BSs, and the utility function of each BS is defined according to $u_s(t) = -\Psi_s(t)$. Here, we use the variable i to refer to each strategy, and $a_s(t)$ captures the strategy selected by the sth BS at time t. Each BS selects its action $a_s(t)$ at time t based on a probability distribution $p_{s,i}(t-1)$ associated with each action, i.e., $p_{s,i}(t-1)$ is a mixed strategy. Assuming f is a probabilistic mapping function, $a_s(t)$ is given by $a_s(t) = f\left(p_{s,i}(t-1)\right)$, where $a_s(t) \in i = \{1,\ldots,4\}$. Each strategy i introduces a transmission power level, which is defined in terms of the transmission power level $\zeta_s(t)$ and, at time t, is given by

$$P_s(t) = \zeta_s(t)P_s^{\mathrm{Max}}. \tag{5.10}$$

Table 5.1 shows the selected transmission power level for different strategies. Note that the MBS can select only two strategies, $i = 1$ and $i = 4$, whereas the SBSs can select all four available strategies. After strategy selection, each BS periodically broadcasts a beacon signal on the selected channel along with its load estimate.

5.3.2   UE Association

If the UE belongs to the set of BSs that have recently slept, $\mathcal{W}$, or if it belongs to the set of UEs which have dropped owing to overload, $\mathcal{O}$, or if the UE has newly joined the network, then it must be assigned to a new BS. In order to connect to a new BS, the UE receives load estimates for all BSs through the beacon signal and chooses the BS to which it wants to connect by evaluating an association function. This association

---

Algorithm 5.1 Joint BSCSA/UECSA ON/OFF switching algorithm.

1: **Input:** $\mathcal{C}(s,t)$, $\hat{u}_{s,i}(t)$, $\hat{r}_{s,i}(t)$, $p_{s,i}(t)$ **Output:** $a_s(t-1)$, $\mathcal{A}(x,t+1)$
2: **Initialization:** $\mathcal{S}=\{1,\ldots,S\}$; $\mathcal{X}=\{1,\ldots,X_s\}$; $\mathfrak{C}=\{1,\ldots,C\}$; $\Phi_c^s \in [0,X_s]$
3: **while do**
4:      $t \leftarrow t+1$,
5:      **for** $\forall s \in \mathcal{S}$ **do**
6:          Find $a_s(t) = f\left(p_{s,i}(t-1)\right)$,
7:      **end for**
8:      Beacon signal transmission and load advertising, $\hat{\upsilon}_s(t)$ from (5.8)
9:      **for** $\forall x \in \mathcal{X}$ **do**
10:         **if** $(x \in \mathcal{W}) \vee (x \in \mathcal{O})$ **then**
11:             Find $\mathcal{A}(x,t)$ from (5.11)
12:         **end if**
13:     **end for**
14:     **if** UECSA is chosen **then**
15:         **for** $\forall s \in \mathcal{S}$ **do**
16:             **for** $\forall x \in \mathcal{X}$ **do**
17:                 **for** $\forall c \in \mathfrak{C}$ **do**
18:                     CCI measurement, $I_{\mathrm{UE}_s^x}(t;c)$ (5.5) and avg. $\bar{I}_{\mathrm{UE}_s^x}(t;c)$ from (5.12)
19:                 **end for**
20:                 UE table update (Table 5.2) and channel selection, $\mathcal{C}(\mathrm{UE}_s^x,t)$ (5.13)
21:             **end for**
22:             **for** $\forall x \in \mathcal{X}$ **do**
23:                 **for** $\forall c \in \mathfrak{C}$ **do**
24:                     **if** $c = \mathcal{C}(\mathrm{UE}_s^x,t)$ **then**
25:                         $\Phi_c^s \leftarrow \Phi_c^s + 1$,
26:                     **end if**
27:                 **end for**
28:             **end for**
29:             BS selected channel $\mathbb{C}(s,t) = \arg\max_{c \in \mathfrak{C}} \Phi_c^s$
30:         **end for**
31:     **else if** BSCSA is chosen **then**
32:         **for** $\forall s \in \mathcal{S}$ **do**
33:             **for** $\forall c \in \mathfrak{C}$ **do**
34:                 CCI measurement, $I_{\mathrm{BS}(s)}(t;c)$ (5.14) and avg. $\bar{I}_{\mathrm{BS}(s)}(t;c)$ (5.15)
35:             **end for**
36:             BS table update (Table 5.2) and channel selection, $\mathcal{C}(s,t)$, (5.16)
37:         **end for**
38:     **end if**
39:     Update instantaneous values, $\upsilon_s(t)$, $\Psi_s(t)$, (5.7), (5.9)
40:     Average, $\hat{u}_{s,i}(t)$, $\hat{r}_{s,i}(t)$, $p_{s,i}(t)$, (5.17), (5.18), (5.19)
41: **end while**

---

function is based on two metrics, i.e., the received signal power and the load condition of each BS. The reason for choosing two metrics is to ensure a minimum required QoS for the UE and at the same time prevent overloading of the BSs. The UE's association criterion is determined according to

$$A(x,t) = \underset{s \in S}{\arg\max} \left\{ \hat{\upsilon}_s(t)^{-\varepsilon} P_s(t) \, \xi_{\text{UE}_s^x, \text{BS}(s')} \sum_{k=1}^{N_c} \left| H_{\text{UE}^x, \text{BS}(s)}(t;k,c(s)) \right|^2 \right\}, \tag{5.11}$$

where ε is a coefficient which indicates the impact of each BS's traffic load.

5.3.3   Proposed Channel Segregation Algorithms

5.3.3.1   UECSA

We use first-order filtering to compute the average CCI power received at each UE. The average CCI power, $\bar{I}_{\text{UE}_s^x}(t;c)$, computed for the xth UE connected to the sth BS on the cth channel at time t is given by

$$\bar{I}_{\text{UE}_s^x}(t;c) = (1-\beta)I_{\text{UE}_s^x}(t;c) + \beta\bar{I}_{\text{UE}_s^x}(t-1;c), \tag{5.12}$$

where $0 < \beta < 1$ denotes the forgetting factor. Using the average CCI powers on all available channels, the CCI table is updated for all available channels ($c = 1, \dots, C$). Later, each UE selects the channel that has the lowest CCI power according to

$$C(\text{UE}_s^x, t) = \underset{c \in \mathcal{C}}{\arg\min} \, \bar{I}_{\text{UE}_s^x}(t;c), \tag{5.13}$$

which is used until the next CCI table-updating time $1+t$. The averaging interval of the first-order filtering is given by $1/(1-\beta)$. If too small a value of β is used, the averaging is not sufficient and the measured average CCI power varies like the instantaneous CCI power. Therefore, the channel reuse pattern varies at every CCI table-updating time. Hence, $\beta \approx 1$ is recommended [13]; $\beta = 0.95$ was used in our computer simulation.

After all UEs have reported their first-priority channel to their corresponding BS, the BS updates its lookup table accordingly, as shown in Figure 5.2. Based on this table, the BS chooses the channel which is reported by most UEs, $\mathbb{C}(s,t)$.

5.3.3.2   BSCSA

In channel selection from the point of view of the BS, all the processes, i.e., the interference measurements, averaging, and forming the lookup table, are performed at the BS. As a result, the computational complexity and control signaling are less. However, there may be some performance degradation, as the knowledge about the interference is not diverse.

The instantaneous CCI power $I_{\text{BS}(s)}(t;c)$ measured at BS(s) on the cth channel ($c = 0, \dots, C-1$) at time t is represented by

$$I_{\text{BS}(s)}(t;c) = \frac{1}{N_c} \sum_{\substack{s' \in \text{BSG}(c(s)) \\ s' \neq s}} P_{s'}(t) \, \xi_{\text{BS}(s), \text{BS}(s')} \sum_{k=0}^{N_c-1} \left| H_{\text{BS}(s), \text{BS}(s')}(t;k,c) \right|^2, \tag{5.14}$$

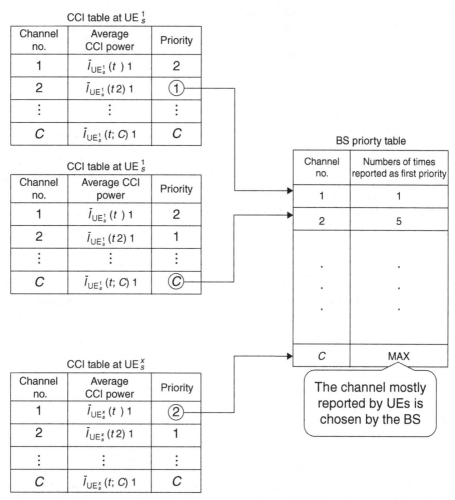

Figure 5.2 Mechanism for updating priority tables [15]. Reproduced with permission from the IEEE.

where $\xi_{\mathrm{BS}(s),\mathrm{BS}(s')}$ is the normalized distance between BS(s) and BS(s'), $H_{\mathrm{BS}(s),\mathrm{BS}(s')}(t;k,c)$ is obtained from the Fourier transform of the impulse response of the channel between BS(s) and BS(s') at time t, and $E[|H_{\mathrm{BS}(s),\mathrm{BS}(s')}(t;k,c)|^2]=1$. We assume that the beacon signal contains the load information about the BS as well. A more precise method would be to measure the interference at the UE, as we are studying the downlink transmission. However, intuitively, we can deduce that interference measurement at the BS can still serve as a good approximation, given that the coverage of the SBSs is much smaller than that of the MBS.

The average CCI power $\bar{I}_{\mathrm{BS}(s)}(t;c)$ computed for the sth BS on the cth channel at time t is given by

$$\bar{I}_{\mathrm{BS}(s)}(t;c) = (1-\beta)I_{\mathrm{BS}(s)}(t;c) + \beta\bar{I}_{\mathrm{BS}(s)}(t-1;c), \qquad (5.15)$$

Table 5.2 CCI table at BS(s).

Channel number	Average CCI power	Priority
1	$\bar{I}_{\mathrm{BS}(s)}(t;1)$	2
2	$\bar{I}_{\mathrm{BS}(s)}(t;2)$	1
...	$\bar{I}_{\mathrm{BS}(s)}(t;3)$	...
C	$\bar{I}_{\mathrm{BS}(s)}(t;4)$	n

where β denotes the forgetting factor. Using the average CCI powers on all available channels, the CCI table is updated for all available channels ($c = 0,\ldots,C-1$) (Table 5.2). The channel having the lowest average CCI power is selected as

$$C(s,t) = \arg\min_{c \in \mathcal{C}} \bar{I}_{\mathrm{BS}(s)}(t;c), \tag{5.16}$$

which is used until the next CCI table-updating time $t+1$. The averaging interval of the first-order filtering is given by $1/(1-\beta)$, and $\beta = 0.95$ was used in our computer simulation.

5.3.4 Mixed-Strategy Update

The utility estimation $\hat{u}_{s,i}(t+1)$, regret $\hat{r}_{s,i}(t+1)$, and probability distribution $p_{s,i}(t+1)$ of the ith strategy for the sth BS at time $t+1$ are given by

$$\hat{u}_{s,i}(t+1) = \hat{u}_{s,i}(t) + c_b(t)\mathbf{1}(t)\left(u_s(t) - \hat{u}_{s,i}(t)\right),$$
$$\hat{r}_{s,i}(t+1) = \hat{r}_{s,i}(t) + d_s(t+1)\left(\hat{u}_{s,i}(t) - u_s(t) - \hat{r}_{s,i}(t)\right), \tag{5.17}$$
$$p_{s,i}(t+1) = p_{s,i}(t) + e_s(t+1)\left(G_{s,i}(\hat{r}_{s,i}(t)) - p_{s,i}(t)\right),$$

with

$$\mathbf{1}(t) = \begin{cases} 1 & \text{if } a_s(t+1) = a_s(t), \\ 0 & \text{otherwise,} \end{cases} \tag{5.18}$$

and

$$G_{s,i}(\hat{r}_{s,i}(t)) = \frac{\exp\left(\sigma_s\hat{r}_{s,i}(t)\right)}{\sum_{i' \in A_s}\exp\left(\sigma_s\hat{r}_{s,i'}(t)\right)}, \tag{5.19}$$

where $G_{s,i}(\hat{r}_{s,i}(t))$ is the Boltzmann–Gibbs (BG) distribution, which is used to encourage actions with lower regret and discourage actions with higher regret. In (5.19), σ_s is the temperature parameter. For further information about the BG distribution and the role it plays in this game to allow it to reach equilibrium see [16]. The variables $c_b(t)$, $d_b(t)$, and $e_b(t)$ are learning rates which decay in inverse proportion to time and must satisfy the following conditions:

Table 5.3 Simulation parameters.

Parameter	Value
Network	
Noise variance, N_0	-174 dBm/Hz
Arrival rate, γ_s^x/μ_s^x	1.8 Mbps
Number of channels, C	2
MBS	
Max. trans. power, P_s^{Max}	46 dBm
Min. MBS–SBS distance (m)	75
Min. MBS–UE distance (m)	35
SBS	
Max. trans. power, P_s^{Max}	30 dBm
Min. SBS–SBS distance (m)	40
Min. SBS–UE distance (m)	10
Path loss (d, distance between UE and BS) (units: dB)	
UE–MBS, $\xi_{\mathrm{UE,MBS}}$	$15.3 + 37.6\log_{10}(d)$
UE–SBS, $\xi_{\mathrm{UE,SBS}}$	$27.9 + 37.6\log_{10}(d)$
Learning parameters	
Boltzmann temperature, σ_s	10
Weighting parameters, φ_s, ψ_s	10, 5

$$\lim_{t\to\infty}\sum_{m=1}^{t} c_s(m) = +\infty, \quad \lim_{t\to\infty}\sum_{m=1}^{t} d_s(m) = +\infty,$$

$$\lim_{t\to\infty}\sum_{m=1}^{t} e_s(m) = +\infty, \quad \lim_{t\to\infty}\sum_{m=1}^{t} c_s^2(m) < +\infty,$$

$$\lim_{t\to\infty}\sum_{m=1}^{t} d_s^2(m) < +\infty, \quad \lim_{t\to\infty}\sum_{m=1}^{t} e_s^2(m) < +\infty,$$

$$\lim_{t\to\infty}\frac{d_s(t)}{c_s(t)} = 0, \qquad \lim_{t\to\infty}\frac{e_s(t)}{d_s(t)} = 0. \tag{5.20}$$

5.4 Computer Simulation

We simulated dense UE/BS scenarios, assuming the system model in Figure 5.1, in which an MBS is collocated with several SBSs. The simulation parameters are summarized in Table 5.3. Two benchmarks were considered for comparison with our proposed algorithm. The first benchmark was a baseline approach, in which all BSs are always on. The second benchmark introduced the ON/OFF switching algorithm proposed in [7] without any CS scheme, i.e., the same bandwidth is shared among all the BSs, or $C = 1$.

Figure 5.3 plots the average UE throughput versus the number of UEs for $S = 11$ BSs and $C = 2$. The proposed CS algorithms, i.e., BSCSA ON/OFF switching and UECSA

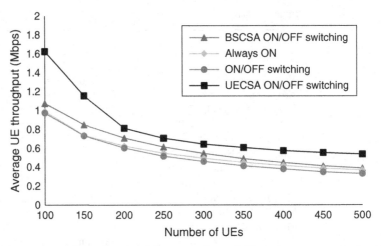

Figure 5.3 Average UE throughput for different numbers of UEs in the system, with $S = 11$ BSs and $C = 2$.

ON/OFF switching, are compared with the two benchmarks. We observe that both of the proposed CS algorithms outperform the benchmarks. However, UECSA ON/OFF switching has the best performance in all cases. Specifically, for lower numbers of UEs ($X < 200$), UECSA ON/OFF switching has the highest performance gap, with a maximum improvement of 30% over BSCSA ON/OFF switching, while for higher numbers of UEs ($X > 200$), the improvement is on average about 15%. The reason is that, indirectly, UECSA employs a wider knowledge of the CCI in the environment through reports from all of the UEs connected to the BSs. This is mostly beneficial in terms of combating the shadowing effect, as the shadowing map may vary widely, and distributed sensing and processing of the CCI by UEs that are spread throughout the environment helps increase the information about the sample space. However, this comes at the cost of increased complexity and signaling overhead. The performance of BSCSA ON/OFF switching is limited because it utilizes knowledge of the CCI only in the location of the BS, but it still outperforms the ON/OFF switching algorithm [7] by about 10%. However, its performance almost converges to the always-ON benchmark for higher numbers of UEs in the system.

Figure 5.4 plots the average percentage of dropped UEs versus the number of UEs for $S = 11$ BSs and $C = 2$. In fact, this graph shows the average percentage of UEs which are dropped owing to BS overload, i.e., $v_s(t) > 1$ in (5.7), in the system. As expected, UECSA ON/OFF switching outperforms the other schemes. Moreover, BSCSA ON/OFF switching outperforms the two benchmarks, i.e., always ON and ON/OFF switching, at lower UE densities. However, their performance converges as the UE density increases.

Figure 5.5 plots the average energy consumption per BS versus the number of UEs in the system for $S = 11$ BSs and $C = 2$. The two proposed CS algorithms have considerably better performance in terms of the energy consumption of the BSs compared with the always ON algorithm (by about 20%). UECSA ON/OFF switching

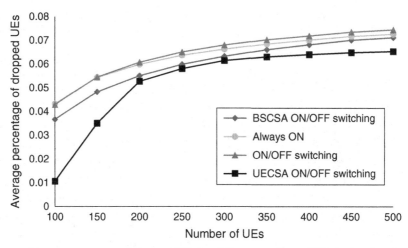

Figure 5.4 Average percentage of dropped UEs for different numbers of UEs in the system, with $S = 11$ BSs and $C = 2$.

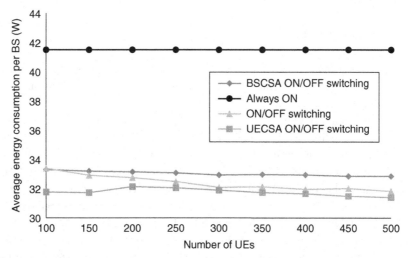

Figure 5.5 Average energy consumption per BS for different numbers of UEs in the system, with $S = 11$ BSs and $C = 2$.

has almost identical performance to the ON/OFF switching algorithm, while the performance of BSCSA ON/OFF switching is slightly lower (by about 4%) than that of the ON/OFF switching algorithm.

Finally, Figure 5.6 illustrates the average energy consumption per BS versus the number of SBSs for $X = 60$ UEs and $C = 2$. As expected, UECSA ON/OFF switching outperforms the other three algorithms. However, its performance converges to that of the UECSA ON/OFF switching and ON/OFF switching algorithms for higher SBS densities.

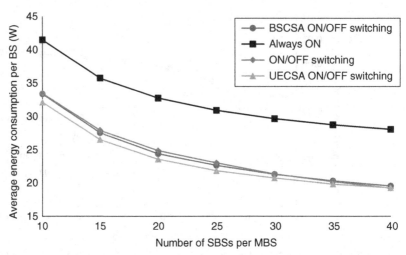

Figure 5.6 Average energy consumption per BS for different number of SBSs and $X = 60$ UEs and $C = 2$.

5.5   Conclusion

In this chapter, we have proposed two channel segregation algorithms, i.e., UECSA and BSCSA, and combined them with an ON/OFF switching algorithm. A noncooperative game-theoretic approach was used to design the solution to the BS ON/OFF switching problem. The UECSA ON/OFF switching algorithm prioritizes the available channels in the UEs based on their level of interference and chooses the one with the least average received interference. The chosen channel is reported to the BS and then the BS chooses the channel which was most reported. The BSCSA ON/OFF switching algorithm performs channel selection and interference measurement only at the location of the BS. As a result, it has a slightly lower performance than the UECSA ON/OFF switching algorithm. However, it has less computational complexity and does not need any signaling exchange between the BS and its associated UEs. Both algorithms are fully distributed and the BSs do not need to exchange any information for their ON/OFF switching or channel segregation decisions. A dense UE/BS scenario was simulated. The UECSA ON/OFF switching algorithm has the best performance compared with the other three schemes, for example a maximum of 30% and 20% improvement in terms of UE throughput and average energy consumption per BS, respectively, when compared with the conventional always ON algorithm.

Acknowledgments

The research results presented in this chapter were achieved as part of the JUNO Project #1680301 (2014.2 2017.2), "Towards Energy-Efficient Hyper-Dense Wireless

Networks with Trillions of Devices," a commissioned research project of NICT, Japan. The authors would like to thank Ms. Atefeh Hajijamali for her assistance in preparing part of the material of this chapter.

References

[1] D. Feng, C. Jiang, G. Lim, L. Cimini, G. Feng, and G. Li, "A survey of energy-efficient wireless communications," *IEEE Commun. Surv. Tutor.*, vol. 15, no. 1, pp. 167–178, Jan. 2013.

[2] H. Zhang, X. Chu, W. Ma, W. Zheng, and X. Wen, "Resource allocation with interference mitigation in OFDMA femtocells for co-channel deployment," *EURASIP J. Wireless Commun. Netw.*, vol. 2012, p. 289, Sep. 2012.

[3] S. Navaratnarajah, A. Saeed, M. Dianati, and M. Imran, "Energy efficiency in heterogeneous wireless access networks," *IEEE Wireless Commun.*, vol. 20, no. 5, pp. 37–43, Oct. 2013.

[4] K. Son, H. Kim, Y. Yi, and B. Krishnamachari, "Base station operation and user association mechanisms for energy–delay tradeoffs in green cellular networks," *IEEE J. Sel. Areas Commun.*, vol. 29, no. 8, pp. 1525–1536, Sep. 2011.

[5] M. Arshad, A. Vastberg, and T. Edler, "Energy efficiency improvement through pico base stations for a green field operator," in *Proc. of IEEE Wireless Communications and Networking Conf. (WCNC)*, Apr. 2012.

[6] S. Zhou, A. Goldsmith, and Z. Niu, "On optimal relay placement and sleep control to improve energy efficiency in cellular networks," in *Proc. of IEEE International Conf. on Communications (ICC)*, Jun. 2011.

[7] S. Samarakoon, M. Bennis, W. Saad, and M. Latva-aho, "Opportunistic sleep mode strategies in wireless small cell networks," in *Proc. of IEEE International Conf. on Communications (ICC)*, Jun. 2014.

[8] D. Goodman, S. Grandhi, and R. Vijayan, "Distributed dynamic channel assignment schemes," in *Proc. of IEEE Vehicular Technology Conf. (VTC)*, May 1993.

[9] G. Cao and M. Singhal, "Distributed fault-tolerant channel allocation for cellular networks," *IEEE J. Sel. Areas Commun.*, vol. 18, no. 7, pp. 1326–1337, Jul. 2000.

[10] H. Luo and N. Shankaranarayanan, "A distributed dynamic channel allocation technique for throughput improvement in a dense WLAN environment," in *Proc. of IEEE International Conf. on Acoustics, Speech, and Signal Processing (ICASSP)*, May 2004.

[11] Y. Furuya and Y. Akaiwa, "Channel segregation, a distributed adaptive channel assignment scheme for mobile communication systems," *IEICE Trans. Commun.*, vol. 74, no. 6, pp. 1531–1537, Jun. 1991.

[12] R. Matsukawa, T. Obara, and F. Adachi, "A dynamic channel assignment scheme for distributed antenna networks," in *Proc. of IEEE Vehicular Technology Conf. (VTC)*, May 2012.

[13] Y. Matsumura, S. Kumagai, T. Obara, T. Yamamoto, and F. Adachi, "Channel segregation based dynamic channel assignment for WLAN," in *Proc. of IEEE International Conf. on Communication Systems (ICCS)*, Nov. 2012.

[14] Y. Matsumura, K. Temma, R. Sugai, T. Obara, T. Yamamoto, and F. Adachi, "Interference-aware channel segregation based dynamic channel assignment for wireless networks," *IEICE Trans. Commun.*, vol. 98, no. 5, pp. 854–860, May 2015.

[15] A. Mehbodniya, K. Temma, R. Sugai, W. Saad, I. Guvenc, and F. Adachi, "Energy-efficient dynamic spectrum access in wireless heterogeneous networks," in *Proc. of IEEE International Conf. on Communication Workshop (ICCW)*, Jun. 2015.

[16] M. Bennis, S. Perlaza, and M. Debbah, "Learning coarse correlated equilibria in two-tier wireless networks," in *Proc. of IEEE International Conf. on Communications (ICC)*, Jun. 2012.

Part II

Physical Layer Communication Techniques

6 Non-Orthogonal Multiple Access (NOMA) for 5G Systems

Wei Liang, Zhiguo Ding, and H. Vincent Poor

Radio access technologies for cellular communications are characterized by multiple-access schemes whose purpose is to serve multiple users with limited bandwidth resources. Typical examples of multiple access schemes are frequency division multiple access (FDMA), time division multiple access (TDMA), code division multiple access (CDMA), and orthogonal frequency division multiple access (OFDMA). In the history of mobile communications, multiple access schemes have been widely investigated in various types of cellular networks, from the first generation (1G) to the fourth generation (4G). Specifically, the FDMA scheme was used in the 1G system, the TDMA scheme was employed in the second generation (2G) systems, CDMA was the dominant multiple access scheme in the third generation (3G) systems, and the OFDMA scheme has been widely used in 4G systems. In many conventional multiple access schemes, such as TDMA and OFDMA, different users are allocated to orthogonal resources in either the time or the frequency domain in order to alleviate interuser interference. However, the spectral efficiency of these orthogonal multiple access (OMA) schemes is low, since bandwidth resources occupied by users with poor channel conditions cannot be shared by others. As a result, these OMA schemes are not sufficient to handle the explosive growth in data traffic in the mobile Internet of the fifth generation (5G) networks. On the other hand, the chip rates of a CDMA system need to be much higher than the information data rates, which means that the use of CDMA in 5G is also potentially problematic owing to the ultra-high data rates expected in 5G systems. Consequently, in 5G networks, new, sophisticated multiple access technologies are needed to support massive connectivity for a very large number of mobile users and/or Internet of Things (IoT) devices with diverse quality-of-service (QoS) requirements [1]. Multiple access in 5G mobile networks is an emerging and challenging research topic, since it needs to provide massive connectivity, large system throughput and small latency simultaneously [2, 3]. Among the promising candidates for 5G multiple access, NOMA has received considerable attention in particular [4–7]. In contrast to conventional OMA, NOMA can accommodate user fairness via non-orthogonal resource allocation and, therefore, is also expected to increase the system throughput. Additionally, in comparison with OMA schemes, user fairness and system spectral efficiency can be improved significantly by employing NOMA [5] since users will be able to transmit simultaneously at the same frequency and in the same time slot.

6.1   Introduction

The aim of this chapter is to focus on NOMA, which is a promising technology that can address the key challenges of 5G communications. NOMA allows multiple users to share the same time, frequency, and spreading-code resources via power allocation [4]. In NOMA, each user can exploit the entire bandwidth for the entire communication time, which can potentially increase users' data rates and reduce latency [8]. In other words, NOMA uses the power domain for multiple access, where different users are served at different power levels. Because of its spectral efficiency, NOMA was recently included in the Long Term Evolution-Advanced (LTE-A) standard by the Third Generation Partnership Project (3GPP), where it was termed multiuser superposition transmission (MUST) [9]. Although MUST is a two-user special case of NOMA, the concept of NOMA is scalable. Specifically, non-orthogonal resource allocation indicates that the number of supported users or devices is not strictly limited by the amount of available orthogonal resources. Therefore, NOMA can accommodate significantly more users than OMA by using non-orthogonal resource allocation [1]. The key features of NOMA are as follows:

- *Utilization of the power domain for user multiplexing.* In NOMA, the power domain, which has not been used for multiple access in conventional cellular networks, is used for user multiplexing. Let us take a scenario with single-antenna nodes as an example. Users are multiplexed in the power domain using superposition coding at the transmitter side, and demultiplexing is done by employing successive interference cancellation (SIC) at the receivers [4]. In particular, users whose messages are at high power levels are capable of decoding their own messages by treating other users' information as noise. On the other hand, users whose messages are at lower power levels can employ the SIC technique to decode their own messages by first removing the other users' information in a successive manner.
- *Utilization of different channel conditions.* In contrast to conventional OMA schemes, the diverse channel conditions among users are utilized in the NOMA scheme. Again, let us take a scenario with single-antenna nodes as an example. The power allocation coefficients, which are the combination coefficients for superposition coding, are determined by the users' channel conditions. Specifically, a message for a user with a strong channel gain is allocated less transmission power and a message for a user with poor channel conditions is allocated more power. This is a win–win scenario since both users, with strong and weak channel gains, may satisfy their needs. The superiority of NOMA can be explained as follows [10]. Suppose that we have two users, namely user m and user n. The channel conditions of these users are dependent on their connections to the base station (BS). The channel gain between the BS and user m is denoted by $|h_m|$. When $|h_m|$ is small, the data rate supported by user m's channel is small as well. Allocating a dedicated bandwidth resource to user m is less spectrally efficient, since the bandwidth allocated to this user cannot be fully utilized. In the NOMA system, user n also has access to the resource allocated to user m. If $|h_n| - |h_m|$ is small, i.e., user n's channel quality is similar to user m's, the benefit of

Table 6.1 Notation adopted in this chapter.

Notation	Description
α_m	Power allocation
γ	Transmit signal-to-noise ratio
M	Total number of mobile users in the SISO-NOMA scheme
P	Probability
U	Number of PUs in the CR-NOMA scheme
Q	Number of CUs in the CR-NOMA scheme
u	PU index
q	CU index
N_t^B	Number of antennas in the BS
N_r^U	Number of antennas for each user
L	Number of clusters in the MIMO-NOMA scheme
l	Cluster index in the MIMO-NOMA scheme
K	Number of users in each cluster in the MIMO-NOMA scheme
k	User index in a cluster
$\mathbf{P}$	Precoding matrix
$\mathbf{w}_{l,k}$	Detection vector of the kth user in the lth cluster

using NOMA is limited. But if $|h_n| \gg |h_m|$, user n can use the bandwidth resource much more efficiently than user m. Therefore, a larger value of $|h_n| - |h_m|$ will result in a higher system throughput compared with that for conventional OMA.

In 2013, Saito *et al.* presented the NOMA concept for cellular future radio access (FRA) [4]. In the same year, Saito *et al.* [11] also proposed a downlink NOMA scheme, which has been considered for further enhancement of Long Term Evolution (LTE) as well as for FRA. In 2014, Ding *et al.* [12] investigated the NOMA scheme in a cellular downlink scenario in which the mobile users are randomly deployed. A NOMA scheme based on random linear network coding for multicast services was proposed by Park and Cho in 2015 [13], in order to improve packet success probability. Cooperative communication in conjunction with a NOMA scheme was characterized by Ding *et al.* [6], where a cooperative NOMA scheme was proposed that exploited the fact that users with better channel conditions have prior information about a message for another user. A NOMA scheme for coordinated direct and relay transmission was proposed by Kim and Lee [14] for the purpose of improving the diversity order for users with poor channel conditions. Ding *et al.* [15] proposed a cognitive radio inspired NOMA network in 2015, in order to improve the total sum rate of both users in a pair and also the individual rates of each user. In 2016, the concept of NOMA was applied in multiple research areas, such as multiple-input multiple-output (MIMO) systems [16], high-rate visible light communication (VLC) downlink networks [8], simultaneous wireless information and power transfer (SWIPT) [17], and security of the physical layer [18].

The notation used in the following sections is listed in Table 6.1.

6.2   NOMA in Single-Input Single-Output (SISO) Systems

In this section, we focus on the NOMA scheme in SISO scenarios. First we consider a single NOMA scheme with fixed power allocation, termed F-NOMA, to be described in Section 6.2.1. The F-NOMA scheme can offer a larger sum rate compared with that of conventional multiple access (MA) scheme, and the performance gain of F-NOMA over OMA scheme can be increased further by selecting users whose channel conditions are very distinctive [15]. In practice, it may not be realistic to ask all of the users in the system to perform NOMA jointly. A promising alternative is to build a hybrid multiple access system, in which NOMA is combined with conventional OMA. More specifically, the users in the system can be divided into multiple groups, where NOMA is implemented within each group and different groups are allocated orthogonal bandwidth resources. Therefore, the performance of this hybrid multiple access scheme depends on which users are grouped together, and the effects of this user pairing in F-NOMA will be investigated in Section 6.2.2.

Moreover, NOMA can be viewed as a special case of a cognitive radio (CR) system, as mentioned in [16, 19]. A CR-inspired NOMA (CR-NOMA) scheme will be introduced in Section 6.2.3. In CR-NOMA, the key idea is to opportunistically serve the users with strong channel conditions provided that the QoS requirements of the users with poor channel conditions are guaranteed. In particular, the transmit power allocated to a user with a strong channel condition is constrained by its partner's target signal-to-interference-plus-noise ratio (SINR). The implementation of the CR-NOMA scheme with fixed QoS requirements, as well as dynamic QoS requirements for the users with poor channel conditions, will be discussed. Additionally, the application of matching theory to user pairing in CR-NOMA networks, which reduces the system complexity and achieves a high system throughput, will be introduced at the end of this section.

6.2.1   The Basics of NOMA

In this section, we consider a cellular downlink communication scenario with one BS and M mobile users. The Rayleigh-fading channel gain between the BS and user m is denoted by h_m. Without loss of generality, we assume that the users' channels have been ordered as $|h_1|^2 \leq |h_2|^2 \cdots \leq |h_M|^2$. To clearly explain the concept of NOMA, a special case with two users is illustrated in Figure 6.1.

In Figure 6.1, we consider two users, namely user m and user n, $1 \leq m < n \leq M$, that are paired to perform NOMA. By using the F-NOMA concept, the BS allocates a fixed amount of transmit power to each user. The power allocation coefficients of these two users are denoted by α_m and α_n, respectively. Additionally, these power coefficients are not functions of instantaneous channel realizations, and we assume that $\alpha_m + \alpha_n = 1$.

According to the principle of NOMA, the users' power allocation coefficients are given by $\alpha_m \geq \alpha_n$. Additionally, we assume that all users have the same transmit signal-to-noise ratio (SNR). By assuming an additive white Gaussian noise (AWGN)

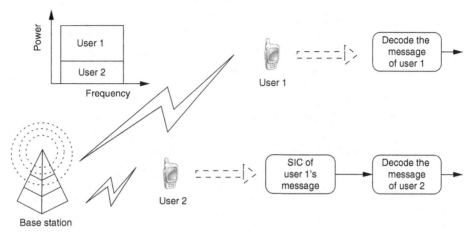

Figure 6.1 Illustration of the basic NOMA scheme with a SIC receiver.

channel and user n's message as Gaussian noise, the achievable rate of user m is given by

$$R_m = \log_2 \left(1 + \frac{|h_m|^2 |\alpha_m|^2}{|h_m|^2 |\alpha_n|^2 + 1/\gamma}\right). \tag{6.1}$$

where γ is the transmit SNR. After SIC is carried out, the achievable rate of user n is expressed as follows:

$$R_n = \log_2 \left(1 + \gamma |\alpha_n|^2 |h_n|^2\right). \tag{6.2}$$

Note that R_n is achievable if user n can detect a message for user m.

Unlike the case for NOMA, the data rate of user i, $i \in \{m,n\}$, in a conventional OMA scheme such as TDMA is given by

$$\bar{R}_i = \frac{1}{2} \log_2 \left(1 + \gamma |h_i|^2\right), \tag{6.3}$$

where the term $\frac{1}{2}$ is the multiplexing loss in the OMA system. Different choices of users have different effects on the performance gain of NOMA. Therefore, the impact of user pairing on the sum rate and the individual user rates achieved in F-NOMA will be investigated in the following section.

6.2.2 Impact of User Pairing on NOMA

In this section, we will investigate the impact of user pairing on the performance gain of NOMA over conventional MA schemes.

6.2.2.1 Impact of User Pairing on the Sum Rate

First, we focus on how user pairing affects the probability that NOMA achieves a lower sum rate than conventional MA schemes, which can be expressed as

$$P(R_m + R_n < \bar{R}_m + \bar{R}_n), \tag{6.4}$$

where R_m and R_n are the achievable rates of user m and user n in the F-NOMA scheme as described in (6.1) and (6.2). $\bar{R}_i$ is user i's achievable data rate, $i \in \{m,n\}$, in a conventional OMA scheme as shown in (6.3). Theorem 6.1 provides a simple closed-form high-SNR approximation for the probability in (6.4).

Theorem 6.1 *Suppose that user m and user n are paired to perform NOMA, where the channel conditions of the two users are sorted such that $|h_m|^2 \le |h_n|^2$. The probability that F-NOMA achieves a lower sum rate than conventional OMA can be approximated at high SNR [15] as follows:*

$$P(R_m + R_n < \bar{R}_m + \bar{R}_n) \approx \frac{1}{\gamma^n}\left(\frac{\psi_3 \psi_2^n}{n} - \psi_1 \psi\right),\tag{6.5}$$

where

$$\psi = \sum_{i=0}^{n-1-m}\binom{n-1-m}{i}\frac{(-1)^i}{m+i}\int_{\psi_4}^{\psi_2} y^{n-1-m-i}\left(y^{m+i} - \left[\frac{\psi_2 - y}{(1+y)}\right]^{m+i}\right)dy,$$

$$\psi_1 = \frac{M!}{(m-1)!(n-1-m)!(M-n)!}, \psi_2 = \frac{1-2\alpha_n^2}{\alpha_n^4},$$

$$\psi_3 = \frac{M!}{(n-1)!(M-n)!}$$

and

$$\psi_4 = \sqrt{1 + \frac{1-2\alpha_n^2}{\alpha_n^4}} - 1.$$

Theorem 6.1 indicates that $P(R_m + R_n < \bar{R}_m + \bar{R}_n)$ quickly goes to zero with increasing transmit SNR. Also, it is certain that the rate performance of F-NOMA is superior to that of conventional OMA, particularly at higher SNR. Furthermore, the decay rate of the probability $P(R_m + R_n < \bar{R}_m + \bar{R}_n)$ is approximately $1/\gamma^n$, i.e., the quality of user n's channel determines the decay rate of this probability, or equivalently, the diversity order. A proof of Theorem 6.1 is given in [15].

The above probability measures the probability of the event that NOMA outperforms OMA, and another important criterion for performance evaluation is how large this performance gap is. Assume that there is a target performance gain, namely R_{target}. The probability of measuring the difference between the sum rate of F-NOMA and that of OMA can be defined as follows:

$$P(R_m + R_n - \bar{R}_m - \bar{R}_n < R_{\text{target}}).\tag{6.6}$$

When the target rate $R_{\text{target}} \ne 0$, the probability in (6.6) can be expressed asymptotically as $P\left(|h_n|^2/|h_m|^2 < 2^{2R_{\text{target}}}\right)$ [15]. Therefore, the gap between the sum rates achieved by F-NOMA and conventional OMA is determined by how different the two users' channel conditions are. A larger difference between the two users' channel conditions results in a larger performance gap.

6.2.2.2   Impact of User Pairing on Individual Data Rates

We will focus on the individual data rates achieved by the F-NOMA scheme in comparison with those for the conventional OMA scheme. The probability that the F-NOMA scheme achieves a higher rate compared with OMA for user m is given by

$$P(R_m > \bar{R}_m) = P\left(\left(1 + \frac{|h_m|^2 \alpha_m^2}{|h_m|^2 \alpha_n^2 + \frac{1}{\gamma}}\right)^2 > (1 + \gamma |h_m|^2)\right). \tag{6.7}$$

By referring to [15], (6.7) can be approximated at high SNR as follows:

$$P(R_m > \bar{R}_m) \approx \frac{M!}{(m-1)!(M-m)!} \frac{(1 - 2\alpha_n^2)^m}{m \gamma^m \alpha_n^{4m}}, \tag{6.8}$$

which shows that $P(R_m > \bar{R}_m)$ decays at a rate of $1/\gamma^m$.

Additionally, the probability that user n experiences better performance with the F-NOMA system than with the OMA scheme is given by

$$P(R_n > \bar{R}_n) = P\left(\log_2\left(1 + \gamma \alpha_n^2 |h_n|^2\right) > \frac{1}{2}\log_2(1 + \gamma |h_n|^2)\right),$$

and a high-SNR approximation is as follows:

$$P(R_n > \bar{R}_n) \approx 1 - \frac{M!}{(n-1)!(M-n)!} \frac{(1 - 2\alpha_n^2)^n}{n \gamma^n \alpha_n^{4n}}. \tag{6.9}$$

As seen from (6.8) and (6.9), these users have totally different experiences in NOMA systems. Particularly, a user with a strong channel condition is more willing to use NOMA since $P(R_n > \bar{R}_n) \to 1$, which is not the case for a user with a poor channel condition. Accordingly, it is preferable to pair two users whose channel conditions are significantly different, since (6.8) and (6.9) imply that m should be as small as possible and n should be as large as possible. Moreover, pairing users with different channel conditions not only improves the sum rate of the two users, but also has the potential to improve each individual user's rate.

In Figure 6.2, the probability that F-NOMA achieves a lower sum rate than OMA, $P(R_m + R_n < \bar{R}_m + \bar{R}_n)$, is shown as a function of SNR. The power allocation factors of user m and user n are given by $\alpha_m^2 = \frac{4}{5}$ and $\alpha_n^2 = \frac{1}{5}$. As seen from Figure 6.2(a), this probability approaches zero as the SNR increases without bound. The simulation results in Figure 6.2 also demonstrate the accuracy of the high-SNR approximation results given by Theorem 6.1. Specifically, when the value of SNR is sufficiently large, the curves achieved by the computer simulations match perfectly with those of the high-SNR approximation. It can be observed from Figure 6.2 that increasing the index of user n, i.e., scheduling a user with a better channel condition, makes the probability decrease at a faster rate. This observation is consistent with the high-SNR approximation results in Theorem 6.1 since the slopes of the outage probability curves are based on the diversity gain achieved by the NOMA transmission scheme. In Figure 6.2(b), the impact of user pairing on the individual rates is studied. Again, the accuracy of the approximation results presented in (6.8) and (6.9) is confirmed by the simulation results. Additionally, Figure 6.2(b) also demonstrates that users with strong

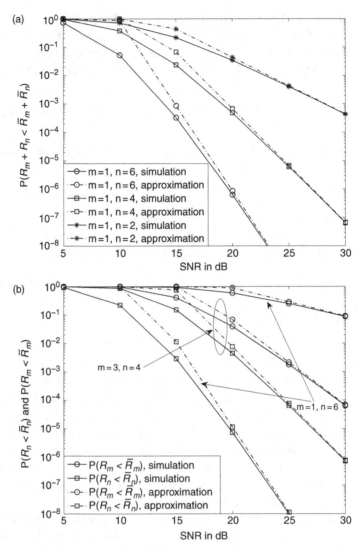

Figure 6.2 The probability that F-NOMA achieves a lower sum rate than conventional MA, when the total number of users is $M = 6$. The analytical results are based on Theorem 6.1. (a) Sum rate of both users; (b) individual rates of each user.

channel conditions prefer to participate in NOMA, since they will experience larger individual data rates compared with those for OMA.

6.2.3   Cognitive Radio Inspired NOMA

NOMA can be interpreted as a special case of a CR system [4, 19], in which a user with a strong channel condition, viewed as a cognitive user (CU), accesses the spectrum occupied by a user with a poor channel condition, viewed as a primary user (PU). Following the concept of the CR network, a variation of NOMA, termed CR-NOMA, can be designed. In the CR-NOMA scheme, the CUs transmit simultaneously with the PUs under the QoS constraints of the PUs. The interference imposed on the CUs can be canceled since the channel conditions of the CUs are strong. On the other hand, the PUs may treat the interference from the CUs as noise. Along the lines of [15], each PU is paired with one particular CU for the purpose of performing NOMA. These two users should have very distinct channel conditions, as mentioned in Section 6.2. We focus on a downlink CR-NOMA scenario in this section, consisting of a unique BS, U PUs, and Q CUs. We assume that the users in each pair can transmit in the same spectrum and that different pairs are served using different orthogonal bandwidth resources in order to avoid any interpair interference. Then a CU can be admitted to a PU's channel under the condition that the CU does not cause too much performance degradation to the PU.

According to the NOMA protocol described in Section 6.2.1, the BS allocates different power levels to the uth PU and the qth CU which are in the same pair. Referring to (6.1) and (6.2), we assume further that the power allocation coefficients allocated to the uth PU and qth CU, α_u and α_q, satisfy the following constraint:

$$\frac{|h_u|^2\alpha_u^2}{|h_u|^2\alpha_q^2 + 1/\gamma} \geq I_{\text{target}}, \tag{6.10}$$

where the targeted SINR at the uth PU is defined as I_{target} and is a constant. Thus, the maximum transmit power that can be allocated to the qth CU is given by

$$\alpha_q^2 = \max\left\{0, \frac{|h_u|^2 - I_{\text{target}}/\gamma}{|h_u|^2(1 + I_{\text{target}})}\right\}, \tag{6.11}$$

where $\alpha_q = 0$ if $|h_u|^2 < I_{\text{target}}/\gamma$ [15]. Note that the choice of α_q in (6.11) is a function of the channel coefficient h_u, which is different from the constant choice of α_q used by F-NOMA in Section 6.2.2.

Since the uth PU's QoS can be guaranteed owing to (6.10), we need only to study the performance experienced by the qth CU. Therefore, the outage performance of the qth CU is defined as

$$P_o^q \triangleq P\left(\log_2(1 + \alpha_q^2\gamma|h_q|^2) < l_{\text{target}}\right), \tag{6.12}$$

where P_o^q denotes the outage probability of the qth CU achieved in the CR-NOMA scheme. As detailed in [15], the diversity order achieved by the CR-NOMA transmission protocol can be obtained from Theorem 6.2 below.

Theorem 6.2 *Suppose that the transmit power allocated to qth CU satisfies the predetermined SINR threshold I_{target} shown in (6.11). Then the diversity order achieved by CR-NOMA is given by [15]*

$$\lim_{\gamma \to \infty} -\frac{\log P_o^q}{\log \gamma} = u.$$

Theorem 6.2 demonstrates the interesting phenomenon that the diversity order of the CR-NOMA scheme experienced by qth CU is determined by how good uth PU's channel quality is. The reason is that qth CU is permitted to access this channel, occupied by uth PU, only if uth PU's QoS requirement is satisfied. More specifically, if

uth PU's channel condition is poor and its target SINR is high, the BS may allocate all the power to uth PU in this pair, and then the qth CU has no chance of being served.

As described in Section 6.2.2, the F-NOMA scheme achieves a diversity gain of n for user n. In F-NOMA, user n is the one with the good channel condition, and can be viewed as the CU in our CR-NOMA scheme. Thus the diversity order achieved by the CR-NOMA scheme can be much smaller than that achieved by the F-NOMA scheme, particularly if $q \gg u$. This performance difference is due to the imposed power constraint, as shown in (6.11).

It is important to point out that CR-NOMA can strictly guarantee uth PU's QoS requirement, and achieve better fairness compared with the F-NOMA scheme. More specifically, the use of CR-NOMA can ensure that a diversity order of u is achievable for qth CU, and that admitting qth CU to the same channel as uth PU does not cause too much performance degradation to uth PU.

If it does not share spectrum with qth CU, uth PU occupies the bandwidth resources solely. The achievable rate for OMA of the PU in this case is defined as follows:

$$\tilde{R}_u^{\mathrm{PU}} = \log_2 \left(1 + \gamma |h_u|^2 \right). \tag{6.13}$$

The sum rate gap between CR-NOMA and OMA is illustrated in the following equation. When $|h_q|^2 \geq |h_u|^2$, the use of CR-NOMA always achieves a larger sum rate since

$$R_u^{\mathrm{PU}} + R_q^{\mathrm{CU}} - \tilde{R}_u^{\mathrm{PU}} \tag{6.14}$$

$$= \log_2 \left(1 + \frac{|h_u|^2 \alpha_u^2}{|h_u|^2 \alpha_q^2 + 1/\gamma} \right) + \log_2 \left(1 + \gamma \alpha_q^2 |h_q|^2 \right) - \log_2 \left(1 + \gamma |h_u|^2 \right)$$

$$= \log_2 \frac{1 + \gamma \alpha_q^2 |h_q|^2}{1 + \gamma \alpha_q^2 |h_u|^2} \geq 0.$$

Based on (6.14), the CR-NOMA scheme achieves superior performance compared with the OMA scheme, since the key idea of CR-NOMA is to serve a user with a strong channel condition without causing too much performance degradation to a user with a weak channel condition.

6.2.3.1   Matching Theory in CR-NOMA Scheme

There are two aspects to designing a CR-NOMA scheme. One is to consider the power allocation between the PUs and CUs, and another is to decide how to rationally pair the PUs and CUs. It is desirable to achieve a balance between user fairness and total sum-rate maximization. In order to realize this balanced trade-off in a low-complexity manner, we can apply matching theory to CR-NOMA [20, 21]. More specifically, we can construct small game units among the PUs and the CUs, in which a PU with a poorer channel condition is matched to a CU that has a good channel condition by considering an appropriate power allocation for the sake of improving the total sum rate. We model our CR-NOMA scheme as a two-sided one-to-one matching market problem [22] as illustrated in Example 6.1, which is a distributed approach to finding the suitable power allocations for the PUs and CUs. We aim at increasing the sum rate of all the matched pairs without degrading the PUs' performance. Specifically, we consider a matching problem designed for multiple PUs and CUs, where the PUs and CUs are appropriately paired to ensure that both of their individual rate requirements are satisfied.

In particular, we assume that each PU will participate in NOMA transmission if it can experience a larger data rate than it can with the OMA scheme, i.e.,

$$R_{u,\text{req}}^{\text{PU}} \geq \frac{1}{2}\log_2(1 + \gamma |h_u|^2) . \tag{6.15}$$

Additionally, the minimum rate requirement of a CU is assumed to be a constant, namely $R_{q,\text{req}}^{\text{CU}}$. Moreover, the PUs negotiate with the CUs regarding the amount of power made available for them, as well as concerning their agreed-upon power allocation coefficients. The CUs then agree to the PUs' offers if both the PUs' and the CUs' minimum rate requirements are satisfied. Otherwise, the CUs reject the offers from the PUs. To illustrate this method, we consider the following example.

Example 6.1 Suppose that there are U PUs and Q CUs in the same cell, with a unique BS. Before any offer is made to the CUs, each PU constructs a preferred list of CUs, namely $PULIST_u$, which lists the CUs that can satisfy the PU's rate requirement as defined in (6.15). Similarly, each CU also has its preferred PU list, called $CULIST_q$, containing PUs that satisfy the minimum rate requirement of the CU, i.e., if the CU transmits in the spectral band occupied by the preferred PUs then its achievable transmission rate is higher than its minimum rate requirement.

We first initialize the value of the power allocation factor of each PU to $\alpha_{u,q} = \alpha_{\text{init}}$, and the power allocated to each CU is then $1 - \alpha_{u,q}$. Then we set the value of the power step counter to τ. We construct $PULIST_u$ and $CULIST_q$, for the uth PU and the qth CU, respectively, by using the initialized value α_{init}.

An offer of $\alpha_{u,q}$ is made by uth PU to the top qth CU in its preference list $PULIST_u$, since the top candidate in its preference list is the best choice for this user. Then qth CU grants a power allocation factor $1 - \alpha_{u,q}$ to itself.

When qth CU receives an offer, it has two options, namely either to reject the offer, if uth PU is not in qth CU's $CULIST_q$, or to accept it, if uth PU is in $CULIST_q$, which leads to a matching pair made up of uth PU and qth CU. However, if two or more PUs send a request to the same CU, a conflict will happen [23]. In order to resolve this conflict, the CU matches only to the PU that can provide the greatest revenue. Specifically, if the intended qth CU has already been matched to another PU, PU_{cur}, and PU_{cur} fails to provide lower power to qth CU, i.e., we have

$$1 - \alpha_{u_{\text{cur}},q} < 1 - \alpha_{u,q} , \tag{6.16}$$

then qth CU will discard its current matching in favor of the new matching. Moreover, the rejected PU_{cur} updates the value of the power allocation factor by setting it to

$$\alpha_{u_{\text{cur}},q}^* = \alpha_{u_{\text{cur}},q} - \tau , \tag{6.17}$$

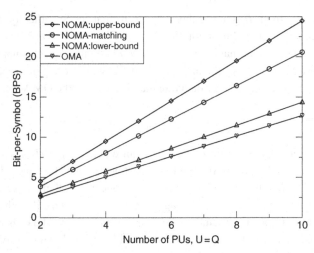

Figure 6.3 Performance in terms of average throughput versus number of PUs obtained by considering matching theory in downlink CR-NOMA communications.

and then this failed PU reconstructs its preference list based on $\alpha^*_{u,q_{cur}}$ and repeats the matching process.

This algorithm aims to find a specific value of power to be allocated that can be accepted by both uth PU and qth CU. The algorithm terminates when each PU has found its appropriate match, provided that all users' rate requirements or targeted SINRs are satisfied.

It can be observed from Figure 6.3 that our CR-NOMA schemes achieve a higher sum rate than that for the conventional OMA scheme with the same number of PUs when the matching theory technique is applied. Particularly, the CR-NOMA scheme with matching theory applied as described in Example 6.1 consistently attains a higher rate than that in the lower bound curve, which is obtained by randomly pairing the PUs and CUs. The gap between the two NOMA schemes can be enlarged further by increasing the number of PUs. In addition, the CR-NOMA scheme with application of matching theory provides a distributed way to pair the PUs and CUs appropriately. As shown in Figure 6.3, the rate performance obtained by employing matching theory is reasonably close to that in the upper-bound curve, which was obtained by applying an exhaustive search to achieve user pairing, and this result is obtained with less complexity.

6.3   NOMA in MIMO Systems

In this section, we focus on the application of MIMO to NOMA downlink communication systems. The concept of MIMO-NOMA has been validated by using a systematic implementation [24–28], which demonstrated that the use of MIMO in NOMA networks can outperform conventional MIMO-OMA. For example, MIMO-NOMA with random beamforming was proposed in [24], in which a BS equipped with multiple antennas randomly forms transmitting beams, and the users falling into these beams are served in an opportunistic manner. The application of MIMO technology to NOMA is important, since MIMO provides more degrees of freedom for further performance improvement. In [27], a multiple-input single-output (MISO) scenario was considered, in which the BS has multiple antennas and the users are each equipped with a single antenna. In [28], a BS with multiple antennas used the NOMA approach to serve two multiple-antenna users simultaneously, and algorithms were proposed for solving the optimization problem of maximizing the system throughput. In this section, we focus on one particular type of MIMO-NOMA approach, which decomposes a MIMO-NOMA downlink into multiple separated SISO-NOMA downlink channels. It is worth pointing out that this decomposition approach is also applicable to NOMA uplink communication scenarios, as well as other types of 5G multiple access techniques, such as sparse code multiple access (SCMA) and low-density spreading (LDS).

6.3.1 System Model for MIMO-NOMA Schemes

In this section, we consider a MIMO-NOMA downlink communication scenario in which a BS is communicating with multiple users. The BS is equipped with N_t^B antennas and each mobile user is equipped with N_r^U antennas. Note that in order to maximize the multiplexing gain, the BS needs to send at least N_t^B distinct messages to the users.

6.3.1.1 The Network Topology Considered

In the network topology that we have designed, the users are assumed to be uniformly deployed on a disk, denoted by $\mathcal{D}$. The radius of this disk is denoted by r, and the BS is located at the center of $\mathcal{D}$. Many existing studies of NOMA have demonstrated that in order to reduce the load on the system, it is ideal to pair two users whose channel conditions are very distinctive [4, 15]. Following this insight, we divide the disk $\mathcal{D}$ into two parts. One is a smaller disk, namely $\mathcal{D}_1$, with a radius r_1 ($r_1 < r$), and with the BS located at its center. The other is a ring, denoted by $\mathcal{D}_2$, which is constructed from $\mathcal{D}$ by removing $\mathcal{D}_1$. Suppose that there are L pairs of users deployed in $\mathcal{D}$. In the lth pair, user 1 is randomly located in $\mathcal{D}_1$, and is paired with user 2, randomly located in $\mathcal{D}_2$. We consider the use of a composite channel model with both quasi-static Rayleigh fading and large scale path loss. In particular, we define the channel matrix from the BS to user k in the lth cluster as

$$\mathbf{H}_{l,k} = \frac{\mathbf{G}_k}{\sqrt{L(d_k)}}, \tag{6.18}$$

where $\mathbf{G}_k$ is an $N_r^U \times N_t^B$ matrix consisting of the Rayleigh-fading channel gains. $L(d_k)$ is the path loss, which can be computed as

$$L(d_k) = \begin{cases} d_k^a & \text{if } d_k > r_0, \\ r_0^a & \text{otherwise,} \end{cases}$$

where d_k denotes the distance from the BS to the kth mobile user in the lth cluster, and a is the path loss exponent.

6.3.1.2   A General Framework for MIMO-NOMA Transmission

In order to simplify the notation, we define $L \triangleq N_t^B$, where l is the cluster (pair) index; namely, $l \in L$. The signals transmitted from the BS are given by

$$\mathbf{x} = \mathbf{P}\tilde{\mathbf{s}} , \tag{6.19}$$

where the length-L vector $\tilde{\mathbf{s}}$ is defined as

$$\tilde{\mathbf{s}} = \begin{bmatrix} \alpha_{1,1}s_{1,1} + \alpha_{1,2}s_{1,2} \\ \vdots \\ \alpha_{L,1}s_{L,1} + \alpha_{L,2}s_{L,2} \end{bmatrix} \triangleq \begin{bmatrix} \tilde{s}_1 \\ \vdots \\ \tilde{s}_L \end{bmatrix} , \tag{6.20}$$

and where $s_{l,1}$ denotes the information-bearing signal to be transmitted to the first user in the lth ($l \in L$) cluster. Similarly, $s_{l,2}$ stands for the information transmitted to the second user in the lth cluster. The factor $\alpha_{l,k}$ denotes the NOMA power allocation coefficient of user k in the lth cluster. We will introduce two approaches to designing the $N_t^B \times N_t^B$ precoding matrix $\mathbf{P}$ in Sections 6.3.2 and 6.3.3.

Without loss of generality, we focus on the users in the first cluster. The observation of user 2 in the first cluster is given by

$$\mathbf{y}_{1,2} = \mathbf{H}_{1,2}\mathbf{P}\tilde{\mathbf{s}} + \mathbf{n}_{1,2} , \tag{6.21}$$

where $\mathbf{H}_{1,2}$ is the $N_r^U \times N_t^B$ Rayleigh-fading channel matrix from the BS to the second user in the first cluster, and $\mathbf{n}$ defines the additive Gaussian noise vector. $\mathbf{w}_{l,k}$ is the detection vector used by the kth user in the lth cluster. After applying the detection vector for the second user in the first cluster, the signal model in (6.21) can be rewritten as follows:

$$\mathbf{w}_{1,2}^H\mathbf{y}_{1,2} = \mathbf{w}_{1,2}^H\mathbf{H}_{1,2}\mathbf{P}\tilde{\mathbf{s}} + \mathbf{w}_{1,2}^H\mathbf{n}_{1,2} \tag{6.22}$$

$$= \mathbf{w}_{1,2}^H\mathbf{H}_{1,2}\mathbf{p}_1 \left(\alpha_{1,1}s_{1,1} + \alpha_{1,2}s_{1,2} \right) + \sum_{l=2}^{L} \mathbf{w}_{1,2}^H\mathbf{H}_{1,2}\mathbf{p}_l\tilde{s}_l + \mathbf{w}_{1,2}^H\mathbf{n}_{1,2},$$

where $\mathbf{p}_l$ denotes the lth column of $\mathbf{P}$. Because user 1 is located in $\mathcal{D}_1$ and user 2 is located in $\mathcal{D}_2$, the power allocation coefficients of these two users can be compared as follows:

$$\alpha_{1,1} \leq \alpha_{1,2} ,$$

by following the principle of NOMA.

According to the above assumptions, the SINR for user 2 in the first cluster is given by

$$\text{SINR}_{1,2} = \frac{|\mathbf{w}_{1,2}^H\mathbf{H}_{1,2}\mathbf{p}_1|^2\alpha_{1,2}^2}{|\mathbf{w}_{1,2}^H\mathbf{H}_{1,2}\mathbf{p}_1|^2\alpha_{1,1}^2 + \sum_{l=2}^{L}|\mathbf{w}_{1,2}^H\mathbf{H}_{1,2}\mathbf{p}_l|^2 + |\mathbf{w}_{1,2}|^2(1/\gamma)}, \tag{6.23}$$

where again γ denotes the transmit SNR.

Therefore, user 2 needs to decode the messages from the users with poorer channel conditions first,* before detecting its own information. Moreover, user 1 in the first cluster needs to decode user 2's message. If the decoding is successful, user 1 will decode its own message with the following SINR:

$$\text{SINR}_{1,1}^1 = \frac{|\mathbf{w}_{1,1}^H \mathbf{H}_{1,1} \mathbf{p}_1|^2 \alpha_{1,1}^2}{\sum_{l=2}^{L} |\mathbf{w}_{1,1}^H \mathbf{H}_{1,1} \mathbf{p}_l|^2 + |\mathbf{w}_{1,1}|^2 1/\gamma}. \tag{6.24}$$

The design of the precoding and detection matrices when there is limited channel state information at the transmitter (CSIT) will be discussed in Section 6.3.2.

6.3.2   Design of Precoding and Detection Matrices with Limited CSIT

For the technique introduced in this section, it is assumed that the mobile users have more antennas than the BS, namely $N_r^U \geq N_t^B$, which can be justified as follows. It is envisioned that small cells will be ultra-densely deployed in 5G networks, in which low-cost, low-power small-cell BSs will be used [29]. It is very likely that such a low-cost small-cell BS will have the same number of antennas as a user handset, or even fewer, particularly given the rapidly growing capabilities of smartphones and tablets. Cloud radio access networks (C-RANs) provide another example of a 5G application of the proposed scheme, in which users are served by a small number of low-cost remote radio heads (RRHs) in order to reduce the fronthaul overhead. The proposed scheme may also be applicable to aspects of the IoT, such as smart homes, in which the capability of a home BS is similar to that of a laptop and other digital devices.† In this section, we consider a general model, in which the design of the precoding and detection can be applied for any number of users. To completely remove intercluster interference among L clusters, the precoding and detection matrices need to satisfy the following constraints:

$$\mathbf{w}_{l,k}^H \mathbf{H}_{l,k} \mathbf{p}_i = 0. \tag{6.25}$$

As described in Section 6.3.1, k here is the index of a user in a cluster and l is the index of the cluster. For all cases, we have $i \neq l$. In order to reduce the system overhead caused by acquiring channel state information (CSI) at the BS, it is assumed that the BS does not have global CSI. Therefore, the choice of $\mathbf{P}$ can be expressed as follows:

$$\mathbf{P} = \mathbf{I},$$

where $\mathbf{I}$ is the $N_t^B \times N_t^B$ identity matrix. The above choice means that the BS broadcasts the users' messages without manipulating them. The advantage of this choice is that

---

* With careful design of the precoding and detection matrices, intercluster interference can be canceled completely, as described in Section 6.3.2. Therefore, the MIMO-NOMA system can be degraded into simple SISO-NOMA channels, for which the use of SIC can yield optimal performance. This performance is identical to that obtained by nonlinear approaches, such as nonlinear SIC and dirty paper coding.
† In the case where the BS has more antennas than the users have, different approaches need to be used to implement MIMO-NOMA. For example, one possible way is to allocate different beamforming vectors to different users individually, and the precoding matrices at the BS can be optimized by taking user fairness into consideration.

it avoids asking the users to feed all of their CSI back to the BS, which significantly reduces system overhead. More specifically, however, the BS still needs to know the order of the users' effective channel gains in order to implement NOMA. Since each user needs to feed back only a scalar effective channel gain, this imposes much less demanding requirements compared with the case where all the users' channel matrices need to be known at the BS.

With this choice of $\mathbf{P}$, the constraints in (6.25) can be derived as

$$\mathbf{w}_{l,k}^H \mathbf{h}_{n_t,lk} = 0, \tag{6.26}$$

where $\mathbf{h}_{n_t,lk}$ is the n_tth column of $\mathbf{H}_{l,k}$, and $n_t \in N_t^B$. Therefore, for the kth user in the lth cluster, the constraints can be rewritten as

$$\mathbf{w}_{l,k}^H \underbrace{\begin{bmatrix} \mathbf{h}_{1,lk} & \cdots & \mathbf{h}_{n_t-1,lk} & \mathbf{h}_{n_t+1,lk} & \cdots & \mathbf{h}_{N_t^B,lk} \end{bmatrix}}_{\tilde{\mathbf{H}}_{l,k}} = 0. \tag{6.27}$$

Note that the dimension of $\tilde{\mathbf{H}}_{l,k}$ is $N_r^U \times (N_t^B - 1)$ since it is a submatrix of $\mathbf{H}_{l,k}$ formed by removing one column. As a result, $\mathbf{w}_{l,k}$ can be obtained from the null space of $\tilde{\mathbf{H}}_{l,k}$, which is

$$\mathbf{w}_{l,k} = \mathbf{U}_{l,k} \mathbf{z}_{l,k}, \tag{6.28}$$

where $\mathbf{U}_{l,k}$ contains all the left singular vectors of $\tilde{\mathbf{H}}_{l,k}$ corresponding to zero singular values, and $\mathbf{z}_{l,k}$ is an $(N_r^U - N_t^B + 1) \times 1$ normalized vector to be optimized later. In order to ensure the existence of $\mathbf{w}_{l,k}$, $N_r^U \geq N_t^B$ is assumed. It is worth pointing out that this MIMO-NOMA scheme can also be employed in cases in which users have different numbers of antennas, as long as the total number of users' antennas is greater than or equal to that for the BS, in order to ensure the existence of a solution of (6.27).

By using the above precoding and detection matrices, the SINR for the last user K in the first cluster is given by

$$\text{SINR}_{1,K} = \frac{|\mathbf{w}_{1,K}^H \mathbf{h}_{1,1K}|^2 \alpha_{1,K}^2}{\sum_{q=1}^{K-1} |\mathbf{w}_{1,K}^H \mathbf{h}_{1,1K}|^2 \alpha_{1,q}^2 + |\mathbf{w}_{1,K}|^2 (1/\gamma)}, \tag{6.29}$$

where intercluster interference has been removed. At user k, $1 < k < K$, the messages $s_{1,q}$, $K \geq q \geq (k+1)$, will be detected with the following SINR:

$$\text{SINR}_{1,k}^k = \frac{|\mathbf{w}_{1,k}^H \mathbf{h}_{1,1k}|^2 \alpha_{1,k}^2}{\sum_{q=1}^{K-1} |\mathbf{w}_{1,k}^H \mathbf{h}_{1,1k}|^2 \alpha_{1,q}^2 + |\mathbf{w}_{1,k}|^2 (1/\gamma)}. \tag{6.30}$$

If the decoding is successful, $s_{1,k}$ will be removed from the kth user's observation, and SIC will be carried out until the kth user's own message is decoded with this SINR, namely $\text{SINR}_{1,k}^k$.

The first user in the first cluster decodes the other users' messages with $\text{SINR}_{1,1}^k$, $k \geq q \geq 2$. If the decoding is successful, this user decodes its own message with the following SINR:

$$
\text{SINR}_{1,1}^1 = \gamma \frac{|\mathbf{w}_{1,1}^H \mathbf{h}_{1,11}|^2 \alpha_{1,1}^2}{|\mathbf{w}_{1,1}|^2}. \tag{6.31}
$$

As can be seen from the above SINR expressions, $\mathbf{z}_{l,k}$ determines the SINRs through $|\mathbf{w}_{l,k}^H \mathbf{h}_{l,lk}|^2$. Therefore, one possible choice of $\mathbf{z}_{l,k}$ can be made by using maximal ratio combining (MRC). In particular, the choice of $\mathbf{z}_{l,k}$ based on MRC is given by

$$
\mathbf{z}_{l,k} = \frac{\mathbf{U}_{l,k}^H \mathbf{h}_{l,lk}}{|\mathbf{U}_{l,k}^H \mathbf{h}_{l,lk}|}. \tag{6.32}
$$

Theorem 6.3 provides an exact expression for the outage probability achieved by MIMO-NOMA, and a high-SNR approximation to it.

Theorem 6.3 *Assume that the users in one cluster are ordered as follows:*

$$
\alpha_{1,1} \leq \cdots \leq \alpha_{1,K}. \tag{6.33}
$$

When MIMO-NOMA is applied, the outage probability experienced by the kth ordered user in the lth cluster, using a high-SNR approximation for the case without path loss, is given by [16]

$$
P_{l,k}^{\text{o}} \approx \frac{K! \left[\left((\epsilon_{l,k}^*)^{N_{\mathrm{r}}^U - N_{\mathrm{t}}^B + 1} / (N_{\mathrm{r}}^U - N_{\mathrm{t}}^B + 1)! \right) \right]^{K-k+1}}{(K-k)!(k-1)!(K-k+1)}, \tag{6.34}
$$

where

$$
\epsilon_{l,k}^* = \max \left\{ \frac{2^{R_{l,k}} - 1}{\gamma \left(\alpha_{l,k}^2 - (2^{R_{l,k}} - 1) \sum_{k=1}^{K-1} \alpha_{l,k}^2 \right)}, \cdots, \frac{2^{R_{l,k}} - 1}{\gamma \left(\alpha_{l,k}^2 - (2^{R_{l,k}} - 1) \sum_{l=1}^{K-1} \alpha_{l,k}^2 \right)} \right\},
$$

for $2 \leq k \leq K$.

Recall that the achievable rate for user k in the lth cluster is $\log(1 + \text{SINR}_{l,k})$ for $1 \leq k \leq K$, since $\log(1 + \text{SINR}_{l,q}^k) > \log(1 + \text{SINR}_{l,p}^k)$ for $1 \leq q < p \leq k$, where $\text{SINR}_{l,k}^k \triangleq \text{SINR}_{l,k}$. It is straightforward to show that $\lim_{\gamma \to \infty} (\log(1 + \text{SINR}_{l,k})/\log \gamma) = 0$ for $2 \leq k \leq K$, and that $\lim_{\gamma \to \infty} (\log(1 + \text{SINR}_{l,1})/\log \gamma) = 1$, since only the first user, which is the one with the strongest CSI, can remove all the interference from the other users completely. Therefore, the maximum multiplexing gain for the best user of each cluster is one, and a multiplexing gain of zero is achieved by the remaining users.

Theorem 6.4 *The MIMO-NOMA scheme achieves a diversity gain of $(N_{\mathrm{r}}^U - N_{\mathrm{t}}^B + 1)(K - k + 1)$, which is same as that in the conventional MIMO-OMA scheme [16].*

Specifically, this diversity gain is achieved by allowing all K users in the same cluster to share the same bandwidth resources, and this yields better spectral efficiency.

6.3.3   Design of Precoding and Detection Matrices with Perfect CSIT

In this section, the global CSI is assumed to be perfectly known at the users and the BS. Again we assume that these two users are user 1 and user 2, in the lth cluster. Additionally, user 1 is randomly located in $\mathcal{D}_1$ defined in Section 6.3.1 and is paired with user 2, which is randomly located in $\mathcal{D}_2$. In this section, we consider the scenario in which the number of antennas at each user is greater than half the number of antennas at the BS, such that $N_r^U > N_t^B/2$, in order to implement the concept of signal alignment. To make it possible to remove intercluster interference, the following constraint should ideally be applied:

$$\begin{bmatrix} \mathbf{w}_{l,1}^H \mathbf{G}_{l,1} \\ \mathbf{w}_{l,2}^H \mathbf{G}_{l,2} \end{bmatrix} \mathbf{p}_i = \mathbf{0}_{2\times 1}, \tag{6.35}$$

where $\mathbf{G}_{l,1}$ denotes the $N_r^U \times N_t^B$ matrix whose elements represent the Rayleigh-fading channel gains as described in (6.18). Then, we assume that $i \neq 1$. $\mathbf{0}_{n_t \times n_r}$ is an $n_t \times n_r$ all-zero matrix, where $n_t \in N_t^B$ and $n_r \in N_r^U$. Without loss of generality, we focus on $\mathbf{p}_1$, which needs to satisfy the following constraint:

$$\begin{bmatrix} \mathbf{G}_2^H \mathbf{w}_2 & \mathbf{G}_{2'}^H \mathbf{w}_{2'} & \cdots & \mathbf{G}_{N_t^B}^H \mathbf{w}_{N_t^B} & \mathbf{G}_{N_t^{B'}}^H \mathbf{w}_{N_t^{B'}} \end{bmatrix}^H \mathbf{p}_1 = \mathbf{0}_{2(N_t^B-1)\times 1}. \tag{6.36}$$

Note that the dimension of the matrix in (6.36) is $2(N_t^B - 1) \times N_t^B$. Therefore, there is no solution for such a vector $\mathbf{p}_i$ satisfying (6.36). In order to ensure the existence of $\mathbf{p}_i$, one straightforward approach is to serve fewer user pairs, which means that the number of user pairs should be less than $(L/2 + 1)$. In this way, the overall system throughput will be reduced.

By applying the concept of interference alignment, we find that the user's detection vector needs to satisfy the following constraint [30, 31]:

$$\mathbf{w}_{l,1}^H \mathbf{G}_{l,1} = \mathbf{w}_{l,2}^H \mathbf{G}_{l,2} \tag{6.37}$$

or, equivalently,

$$\begin{bmatrix} \mathbf{G}_{l,1}^H & -\mathbf{G}_{l,2}^H \end{bmatrix} \begin{bmatrix} \mathbf{w}_{l,1} \\ \mathbf{w}_{l,2} \end{bmatrix} = \mathbf{0}_{N_t^B \times 1}. \tag{6.38}$$

We define a $\mathbf{U}_{l,1}$ as a $2N_r^U \times (2N_r^U - N_t^B)$ matrix containing $(2N_r^U - N_t^B)$ right singular vectors of $\begin{bmatrix} \mathbf{G}_{l,1}^H & -\mathbf{G}_{l,2}^H \end{bmatrix}$ corresponding to its zero singular values. The detection vectors at the users can then be expressed as follows:

$$\begin{bmatrix} \mathbf{w}_{l,1} \\ \mathbf{w}_{l,2} \end{bmatrix} = \mathbf{U}_{l,1} \mathbf{x}_{l,1}, \tag{6.39}$$

where $\mathbf{x}_{l,1}$ is a $(2N_r^U - N_t^B) \times 1$ vector. We assume that $\mathbf{x}_{l,1}$ is normalized to 2, i.e., $|\mathbf{x}|^2 = 2$, in order to facilitate the performance analysis. It is straightforward to show that the choice of the detection vectors in (6.39) satisfies $\begin{bmatrix} \mathbf{G}_{l,1}^H & -\mathbf{G}_{l,2}^H \end{bmatrix} \mathbf{U}_{l,1} \mathbf{x}_{l,1} = \mathbf{0}_{N_t^B \times 1}$.

The effect of signal alignment based on (6.37) is to project the channels of the two users in the same cluster onto the same direction. We define $\mathbf{g}_{l,1} \triangleq \mathbf{G}_{l,1}^H \mathbf{w}_{l,1}$, where $\mathbf{g}_{l,1}$ is the effective channel vector shared by these two users, user 1 and user 2. As a result of the projection, the number of rows in the matrix in (6.36) can be reduced significantly. In particular, the constraint on $\mathbf{p}_i$ in (6.36) can be rewritten as follows:

$$\begin{bmatrix} \mathbf{g}_1 & \cdots & \mathbf{g}_{i-1} & \mathbf{g}_{i+1} & \cdots & \mathbf{g}_{N_t^B} \end{bmatrix}^H \mathbf{p}_i = \mathbf{0}_{(N_t^B-1)\times 1} \ . \tag{6.40}$$

Note that $\begin{bmatrix} \mathbf{g}_1 & \cdots & \mathbf{g}_{i-1} & \mathbf{g}_{i+1} & \cdots & \mathbf{g}_{N_t^B} \end{bmatrix}^H$ is an $(N_t^B - 1) \times N_t^B$ matrix, which means that $\mathbf{p}_i$ exists.

We define $\mathbf{G} \triangleq \begin{bmatrix} \mathbf{g}_1 & \cdots & \mathbf{g}_{N_t^B} \end{bmatrix}^H$. Then, the zero-forcing-based precoding matrix at the BS can be designed as follows:

$$\mathbf{P} = \mathbf{G}^{-H}\mathbf{D} , \tag{6.41}$$

where $\mathbf{D}$ is a diagonal matrix to ensure power normalization at the BS, i.e., $\mathbf{D}^2 = \mathrm{diag}\{1/(\mathbf{G}^{-1}\mathbf{G}^{-H})_{1,1}, \cdots, 1/(\mathbf{G}^{-1}\mathbf{G}^{-H})_{N_t^B, N_t^B}\}$. As a result, the transmission power at the BS is constrained as follows:

$$\mathrm{tr}\{\mathbf{PP}^H\}\gamma = \mathrm{tr}\{\mathbf{G}^{-H}\mathbf{DD}^H\mathbf{G}^{-1}\}\gamma \tag{6.42}$$
$$= \mathrm{tr}\{\mathbf{G}^{-1}\mathbf{G}^{-H}\mathbf{D}^2\}\gamma = N_t^B\gamma \ .$$

Based on (6.37) and (6.41), the signal model at user 1 is given by

$$\mathbf{w}_{l,1}^H\mathbf{y}_{l,1} = \frac{\mathbf{g}_{l,1}^H}{\sqrt{L(d_{l,1})}}\mathbf{p}_{l,1}(\alpha_{l,1}s_{l,1} + \alpha_{l,2}s_{l,2}) \tag{6.43}$$

$$+ \sum_{i\neq 1}\frac{\mathbf{g}_{l,1}^H}{\sqrt{L(d_{l,1})}}\mathbf{p}_i(\alpha_i s_i + \alpha_{i'}s_{i'}) + \mathbf{w}_{l,1}^H(\mathbf{n}_{l,1})$$

$$= \frac{(\alpha_{l,1}s_{l,1} + \alpha_{l,2}s_{l,2})}{\sqrt{(L(d_{l,1}))(\mathbf{G}^{-1}\mathbf{G}^{-H})_{1,1}}} + \mathbf{w}_{l,1}^H(\mathbf{n}_{l,1}) \ .$$

For notational simplicity, we define $y_{l,1} = \mathbf{w}_{l,1}^H\mathbf{y}_{l,1}$,

$$h_{l,1} = \frac{1}{\sqrt{L(d_{l,1})(\mathbf{G}^{-1}\mathbf{G}^{-H})_{1,1}}},$$

and $n_{l,1} = \mathbf{w}_{l,1}^H\mathbf{n}_{l,1}$. Therefore, the use of the signal-alignment-based precoding and detection matrices can decompose the multiuser MIMO-NOMA channels into L pairs of SISO-NOMA channels. More specifically, within each pair, user 1 and user 2 receive the following scalar observations:

$$y_{l,1} = h_{l,1}(\alpha_{l,1}s_{l,1} + \alpha_{l,2}s_{l,2}) + n_{l,1}, \tag{6.44}$$

and

$$y_{l,2} = h_{l,2}(\alpha_{l,1}s_{l,1} + \alpha_{l,2}s_{l,2}) + n_{l,2} , \tag{6.45}$$

where $y_{l,2}$ and $n_{l,2}$ are defined similarly to $y_{l,1}$ and $n_{l,1}$. Additionally,

$$h_{l,2} = \frac{1}{\sqrt{L(d_{l,2})(\mathbf{G}^{-1}\mathbf{G}^{-H})_{1,1}}},$$

and it is important to point out that $h_{l,1}$ and $h_{l,2}$ share the same small fading gain at different distances.

Note that user 1 and user 2 are selected from $\mathcal{D}_1$ and $\mathcal{D}_2$, respectively, and that their distances are ordered as $d_{l,1} < d_{l,2}$. Therefore the two users in one pair are ordered without any ambiguity, which simplifies the design of the power allocation coefficients, namely $\alpha_{l,1} \leq \alpha_{l,2}$, by obeying the NOMA principle. User 2 decodes its message with the following SINR:

$$\mathrm{SINR}_{l,2} = \frac{\gamma |h_{l,2}|^2 \alpha_{l,2}^2}{\gamma |h_{l,2}|^2 \alpha_{l,1}^2 + |\mathbf{w}_{l,2}|^2} . \tag{6.46}$$

User 1 carries out SIC by first removing the message from user 2 with the SINR $\mathrm{SINR}_{1,2} = \gamma |h_{l,1}|^2 \alpha_{l,2}^2 / (\gamma |h_{l,1}|^2 \alpha_{l,1}^2 + |\mathbf{w}_{l,1}|^2)$, and then decodes its own message with the following SINR:

$$\mathrm{SINR}_{l,1} = \frac{\gamma |h_{l,1}|^2 \alpha_{l,1}^2}{|\mathbf{w}_{l,1}|^2} . \tag{6.47}$$

In Figure 6.4, three benchmark schemes are considered in order to better illustrate the performance gain of the proposed framework. The design for the two schemes without precoding can be found in [16], where each user carries out antenna selection. The path loss exponent was set at $a = 4$. The sizes of $\mathcal{D}_1$ and $\mathcal{D}_2$ were set by choosing $r = 10$ m, $r_1 = 5$ m, and $r_0 = 1$ m. In Figure 6.4(a), the downlink outage sum rate $R_{l,2}(1 - P_{l,2}) + R_{l,2}(1 - P_{l,2})$ is shown as a function of the SNR γ, and the outage probability is shown in Figure 6.4(b). As can be seen from Figure 6.4(a), the two NOMA schemes achieve larger outage sum rates compared wih the two OMA schemes, which demonstrates the superior spectral efficiency achieved by NOMA. In Figure 6.4(b), the two precoding schemes can achieve better outage performance than the two schemes without precoding, owing to the use of the spatial degrees of freedom at the BS. By comparing the performance of the SISO-NOMA scheme with that of the NOMA scheme proposed in [16], we observe that their outage sum-rate performance is similar, but SISO-NOMA offers better outage performance, particularly at high SNR. In terms of individual rates, SISO-NOMA can outperform the OMA scheme with precoding for $P_{l,1}$, but suffers a small loss for $P_{l,2}$. This is consistent with the finding in [15] that the user with poor channel conditions will suffer some performance loss due to co-channel interference from its partner.

6.4   Summary and Future Directions

In this chapter, we have reviewed developments in NOMA for 5G systems. We have discussed the basic concept of NOMA in a SISO scenario in Section 6.2, and studied the impact of user pairing on NOMA with fixed power allocation. In general, F-NOMA can offer a larger sum rate compared with OMA, and the performance gain of F-NOMA over conventional OMA can be increased further by selecting users whose channel

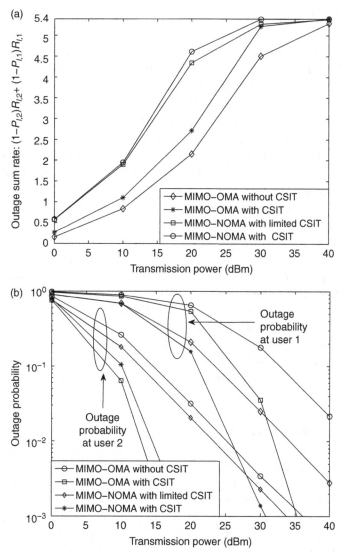

Figure 6.4 Comparison of performance three benchmark schemes for downlink MIMO-NOMA transmissions. (a) outage rate; (b) outage probability.

conditions are more distinctive. A CR-inspired NOMA scheme was introduced in Section 6.2.3. In particular, we saw that the principles of the NOMA scheme can be applied to the concept of a CR network, in which the user with the poorer channel can be treated as the PU and the user with the better channel can be considered as the CU. In CR-NOMA, the channel quality of a user with a poor channel condition is critical, since the transmit power allocated to the other user is constrained following the concept of a CR network. Additionally, a distributed approach to dynamic user pairing/grouping by applying matching theory was discussed in Section 6.2.3, where the power allocation coefficients change dynamically, according to the rate requirements of both the PUs and the CUs'. In Section 6.3, we studied the application of MIMO to NOMA systems. The key idea of the MIMO-NOMA scheme is to decompose MIMO-NOMA into a simple SISO-NOMA scheme, for which the use of SIC can yield optimal performance. Two new designs of precoding and detection matrices for MIMO-NOMA were proposed in Sections 6.3.2 and 6.3.3. Additionally, closed-form analytical results were developed which demonstrate that NOMA can outperform conventional OMA, even with simple choices of power allocation coefficients.

As we have discussed, NOMA can be employed to improve the system throughput as well as to enable the very large number of connections in a 5G network. NOMA can also be used in massive MIMO transmission in order to further improve the spectral efficiency. Specifically, it is preferable to serve as many users as possible to reduce user latency and improve user fairness. The main challenge in considering the use of the NOMA technique in massive MIMO scenarios is to specify a suitable assumption about the CSI. The assumption of perfect CSI in massive MIMO-NOMA corresponds in practice to the use of excessive bandwidth resources. Additionally, if the CSI is not required at the transmitter, then a larger number of antennas is needed at the receiver than at the transmitter. In the scheme presented in [32], the mobile users do not need to feed their channel matrices back to the BS. In particular, a massive MIMO-NOMA scheme can be decomposed into multiple separated SISO-NOMA channels, where each channel adopts the basic NOMA schemes described in Section 6.2. Moreover, when users are located far away in a downlink multicell NOMA system, the cell edge users will experience increased interference from neighboring cells. Therefore, another challenge in this NOMA system is to mitigate the intercell interference among user pairs. To mitigate intercell interference, power adaptation, BS cooperation, and other spatial modulations can be considered. Additionally, NOMA can also be applied together with various wireless networking paradigms, such as cooperative communication systems, CR networks, and physical layer security systems. However, this does not mean that conventional OMA could be replaced universally by the NOMA technique in future 5G networks. Conventional OMA is a better choice when the cell is small and also the number of mobile users is small. Specifically, when the channel conditions of the users are bad and/or the users are concentrated, SCMA is a suitable choice because of its shaping gain. In general, coexistence of SCMA and NOMA is desired in 5G systems for the sake of fulfilling the diverse requirements of different user demands and applications.

References

[1] L. Dai, B. Wang, Y. Yuan, S. Han, C. L. I, and Z. Wang, "Non-orthogonal multiple access for 5G: Solutions, challenges, opportunities, and future research trends," *IEEE Commun. Mag.*, vol. 53, no. 9, pp. 74–81, Sep. 2015.

[2] Q. C. Li, H. Niu, A. T. Papathanassiou, and G. Wu, "5G network capacity: Key elements and technologies," *IEEE Veh. Technol. Mag.*, vol. 9, no. 1, pp. 71–78, Mar. 2014.

[3] Huawei Technologies, "5G: A technology vision," White paper, Nov. 2013.

[4] Y. Saito, Y. Kishiyama, A. Benjebbour, T. Nakamura, A. Li, and K. Higuchi, "Non-orthogonal multiple access (NOMA) for cellular future radio access," in *Proc. of IEEE Vehicular Technology Conf. (VTC Spring)*, Jun. 2013.

[5] J. H. Choi, "Non-orthogonal multiple access in downlink coordinated two-point systems," *IEEE Commun. Lett.*, vol. 12, no. 2, pp. 313–316, Feb. 2014.

[6] Z. Ding, M. Peng, and H. V. Poor, "Cooperative non-orthogonal multiple access in 5G systems," *IEEE Commun. Lett.*, vol. 19, no. 8, pp. 1462–1465, Aug. 2015.

[7] M. Imari, P. Xiao, M. A. Imran, and R. Tafazolli, "Uplink non-orthogonal multiple access for 5G wireless networks," in *Proc. of International Symposium on Wireless Communications Systems (ISWCS)*, Aug. 2014.

[8] H. Marshoud, V. M. Kapinas, G. K. Karagiannidis, and S. Muhaidat, "Non-orthogonal multiple access for visible light communications," *IEEE Photon. Technol. Lett.*, vol. 28, no. 1, pp. 51–54, Jan. 2016.

[9] 3GPP, TSG RAN WG1 Meeting 82 R1-154999, May 2015.

[10] A. Benjebbour, Y. Saito, Y. Kishiyama, A. Li, A. Harada, and T. Nakamura, "Concept and practical considerations of non-orthogonal multiple access (NOMA) for future radio access," in *Proc. of International Symposium on Intelligent Signal Processing and Communications Systems (ISPACS)*, Nov. 2013.

[11] Y. Saito, A. Benjebbour, Y. Kishiyama, and T. Nakamura, "System-level performance evaluation of downlink non-orthogonal multiple access (NOMA)," in *Proc. of IEEE International Symposium on Personal Indoor and Mobile Radio Communications (PIMRC)*, Sep. 2013.

[12] Z. Ding, Z. Yang, P. Fan, and H. V. Poor, "On the performance of non-orthogonal multiple access in 5G systems with randomly deployed users," *IEEE Signal Process. Lett.*, vol. 21, no. 12, pp. 1501–1505, Dec. 2014.

[13] S. Park and D. H. Cho, "Random linear network coding based on non-orthogonal multiple access in wireless networks," *IEEE Commun. Lett.*, vol. 18, no. 7, pp. 1273–1276, Jul. 2015.

[14] J. B. Kim and I. H. Lee, "Non-orthogonal multiple access in coordinated direct and relay transmission," *IEEE Commun. Lett.*, vol. 19, no. 11, pp. 2037–2040, Nov. 2015.

[15] Z. Ding, P. Z. Fan, and H. V. Poor, "Impact of user pairing on 5G non-orthogonal multiple access downlink transmissions," *IEEE Trans. Veh. Technol.*, vol. 65, no. 8, pp. 6010–6023, Aug. 2016.

[16] Z. Ding, F. Adachi, and H. V. Poor, "The application of MIMO to non-orthogonal multiple access," *IEEE Trans. Wireless Commun.*, vol. 15, no. 1, pp. 537–552, Jan. 2016.

[17] Y. W. Liu, G. Z. Ding, M. Elkashlan, and H. V. Poor, "Cooperative non-orthogonal multiple access with simultaneous wireless information and power transfer," *IEEE J. Sel. Areas Commun.*, vol. 34, no. 4, pp. 938–953, Mar. 2016.

[18] Y. Zhang, H. M. Wang, Q. Yang, and Z. Ding, "Secrecy sum rate maximization in non-orthogonal multiple access," *IEEE Commun. Lett.*, vol. 20, no. 5, pp. 930–933, Mar. 2016.

[19] A. Goldsmith, S. A. Jafar, I. Maric, and S. Srinivasa, "Breaking spectrum gridlock with cognitive radios: An information theoretic perspective," *Proc. IEEE*, vol. 97, no. 5, pp. 894–914, May 2009.

[20] D. Gusfield and R. W. Irving, *The Stable Marriage Problem: Structure and Algorithms*, MIT Press, 1989.

[21] A. Roth and M. Sotomanyor, *Two Sided Matching: A Study in Game-Theoretic Modeling and Analysis*, 1st edn, Cambridge University Press, 1989.

[22] S. Bayat, R. H. Y. Louie, Z. Han, B. Vucetic, and Y. H. Li, "Physical-layer security in distributed wireless networks using matching theory," *IEEE Trans. Inf. Forens. Security*, vol. 8, no. 5, pp. 717–732, May 2013.

[23] Y. Xiao, K. C. Chen, C. Yuen, Z. Han, and L. A. DaSilva, "A Bayesian overlapping coalition formation game for device-to-device spectrum sharing in cellular networks," *IEEE Trans. Wireless Commun.*, vol. 14, no. 7, pp. 4034–4051, Jul. 2015.

[24] Y. Hayashi, Y. Kishiyama, and K. Higuchi, "Investigations on power allocation among beams in non-orthogonal access with random beamforming and intra-beam (SIC) for cellular (MIMO) downlink," in *Proc. of IEEE Vehicular Technology Conf. (VTC Fall)*, Sep. 2013.

[25] Y. Lan, A. Benjebboiu, X. Chen, A. Li, and H. Jiang, "Considerations on downlink non-orthogonal multiple access (NOMA) combined with closed-loop SU-MIMO," in *Proc. IEEE International Conf. on Signal Processing and Communication Systems (ICSPCS)*, Dec. 2014.

[26] X. Chen, A. Benjebbour, Y. Lan, A. Li, and H. Jiang, "Impact of rank optimization on downlink non-orthogonal multiple access (NOMA) with SU-MIMO," in *Proc. of IEEE International Conf. on Communications Systems (ICCS)*, Nov. 2014.

[27] J. Choi, "Minimum power multicast beamforming with superposition coding for multiresolution broadcast and application to NOMA systems," *IEEE Trans. Commun.*, vol. 63, no. 3, pp. 791–800, Mar. 2015.

[28] Q. Sun, S. Han, C. L. I, and Z. Pan,"On the ergodic capacity of MIMO NOMA systems," *IEEE Wireless Commun. Lett.*, vol. 4, no. 4, pp. 405–408, Aug. 2015.

[29] X. Ge, H. Cheng, M. Guizani, and T. Han, "5G wireless backhaul networks: Challenges and research advances," *IEEE Netw.*, vol. 28, no. 6, pp. 6–11, Nov. 2014.

[30] N. Lee and J. B. Lim, "A novel signaling for communication on MIMO Y channel: Signal space alignment for network coding," in *Proc. of IEEE International Symposium on Information Theory (ISIT)*, Jul. 2009.

[31] Z. Ding, T. Wang, M. Peng, and W. Wang, "On the design of network coding for multiple two-way relaying channels," *IEEE Trans. Wireless Commun.*, vol. 10, no. 6, pp. 1820–1832, Jun. 2011.

[32] Z. Ding and H. V. Poor, "Design of massive-MIMO-NOMA with limited feedback," *IEEE Signal Process. Lett.*, vol. 23, no. 5, pp. 629–633, May 2016.

7 Flexible Physical Layer Design

Maximilian Matthé, Martin Danneberg, Dan Zhang, and Gerhard Fettweis

7.1 Introduction

The role of software in mobile communication systems has increased over time. For the upcoming fifth generation (5G) mobile networks, the concept of software-defined networking (SDN) can ease network management by enabling anything as a service. Software-defined radio (SDR) enables radio virtualization, where several radio components are implemented in software. Cognitive radio (CR) goes one step further by using a software-based decision cycle to self-adapt the SDR parameters and consequently optimize the use of communication resources. This proposal leads to the possibility of having real-time communication functionalities at virtual machines in cloud computing data centers, instead of deploying specialized hardware. Network functions virtualization (NFV) claims to provide cloud-based virtualization of network functionalities. The perspective is that all these software-based concepts should converge while 5G networks are being designed. A new breakthrough will be achieved when all these software paradigms are applied to the physical layer (PHY), where its functionalities are defined and controlled by software as well.

A flexible PHY design is particularly beneficial considering the diverse applications proposed for 5G [1]. In fact, these applications typically have conflicting design objectives and face extreme requirements. Broadband communication will play an important role in, for instance, offering video streaming services with high resolution for TV and supporting high-density multimedia such as 4K and 3D videos in smartphones. Data rates up to 10 gigabits per second (Gbps) are therefore being targeted in 5G. The Tactile Internet [2] enables one to control virtual or real objects via wireless links with haptic feedback. This implies that the end-to-end latency constraint in 5G must be dropped by at least one order of magnitude compared with current fourth generation (4G) technologies. The Internet of Things (IoT) is aimed at connecting a massive amount of devices. Wireless sensor networks need to provide in-service monitoring at low cost and with a long battery life. Smart vehicles improve safety and actively avoid accidents by exchanging their driving status, such as position, breaking, acceleration, and speed, with surrounding vehicles and infrastructure via challenging doubly dispersive channels. Overall, this demands asynchronous multiple access, ultra-low latency, and ultra-high reliability. As the final frontier for mobile communication, areas with low population densities await the day that fifth generation (5G) systems can provide a sustainable and economically feasible wireless regional-area network (WRAN) mode of operation to deliver the service [3].

The 4G PHY is based on a waveform termed orthogonal frequency division multiplexing (OFDM). It has been intensively studied in recent decades [4–6] because of its discrete Fourier transform (DFT) based energy-efficient implementation and one tap frequency domain equalization to tackle multipath fading channels. Furthermore, its orthogonality allows for easy integration with multiple-input multiple-output (MIMO) techniques. Nevertheless, considering the challenging requirements of 5G, OFDM has limitations and thus requires enhancement. Its high out-of-band (OOB) emission, due to rectangular pulse shaping in the time domain, results in severe interference among asynchronous users. This is a key limitation in broadband communication, IoT, and wireless regional area network (WRAN) scenarios. In order to maintain orthogonality, the time misalignment of the OFDM signals coming from several users must lie within the cyclic prefix (CP) and cyclic suffix (CS). In other words, the demand for fine synchronization in orthogonal frequency division multiplexing (OFDM) not only consumes a considerable amount of time and frequency resources, but also increases the computational energy and processing latency at the receiver. Evidently, this characteristic of OFDM will become problematic for Tactile Internet and IoT applications. We also note that the length of the CP of an OFDM symbol is typically defined by the maximum channel delay length, which is independent of the frame design. For short-length packets or channels with a very long channel impulse response (CIR), the CP overhead can become too high, significantly degrading the spectral efficiency. Last but not least, the high peak-to-average power ratio (PAPR) of OFDM requires expensive radio frequency (RF) amplifiers, otherwise the OOB emission is increased owing to nonlinearities in the front end. To overcome these limitations, 5G system design resorts to several alternative waveforms. Major 5G waveform candidates include filtered OFDM (F-OFDM) [7], universal-filtered OFDM (UF-OFDM) [8], filter bank multicarrier (FBMC) waveforms [9], and generalized frequency division multiplexing (GFDM) [10].

Both F-OFDM and UF-OFDM, as simple variants of OFDM, aim at mostly keeping the orthogonality of OFDM, while the OOB emission outside the subbands allocated to each user is improved by means of linear filtering. The key difference between them is that F-OFDM keeps the CP, while zero padding (ZP) is adopted in UF-OFDM. In principle, additional linear filtering introduces interference between adjacent OFDM blocks. On the one hand, the filter should be shortened to alleviate this interference. The reduction of OOB emission, on the other hand, requires a longer filter length. Therefore, filter design plays a critical role in determining the performance of F-OFDM and UF-OFDM. The high-PAPR problem remains in both filtered versions of OFDM. However, the discrete Fourier transform (DFT) spreading adopted in OFDM (also known as single-carrier frequency domain multiple access (SC-FDMA)) to reduce PAPR can be used straightforwardly in these variants.

In general, subband filtering mainly affects the spectrum of the subcarriers on the edge of the allocated subband. It is an effective solution to reducing the energy leakage outside the subband at low cost. However, in situations where the spectrum of each subcarrier matters, subband filtering can become insufficient and we must resort to pulse shaping of each subcarrier, for example using FBMC and GFDM. For instance, mobility causes Doppler spreading, affecting each subcarrier. Therefore, FBMC and GFDM are expected to provide robustness against the Doppler effect. Furthermore, both of these techniques have one additional degree of freedom in the time domain. Namely, each subcarrier can carry more than one data symbol. This additional degree of freedom allows the time and frequency spacing of each data symbol to be adapted flexibly according to the channel properties and the type of application. However, the benefits of FBMC and GFDM come at the cost of system complexity. Their non-orthogonality challenges the transceiver design. In contrast to FBMC, GFDM adopts the CP and circular rather than linear filtering. The circulant structure embedded in the signal is beneficial for developing algorithms with significantly reduced complexity. Moreover, GFDM is able to emulate OFDM with and without DFT spreading [10]. This is an important feature to support migration from 4G to 5G systems. Therefore, we focus here on GFDM and present it as a flexible air interface that possesses a sufficient design space to serve different applications with complexity affordable by today's hardware technologies.

This chapter is organized as follows. We start from a general description of GFDM in Section 7.2. Section 7.3 focuses on the hardware implementation of a GFDM modem with run-time reconfigurability. In Section 7.4, we shed some light on the receiver design for GFDM, paying particular attention to how to tackle the non-orthogonality issue with manageable complexity. Finally, a summary and outlook are presented in Section 7.5.

7.2   Generalized Frequency Division Multiplexing

As a multicarrier waveform, GFDM applies circular filtering on a subcarrier basis, with the aim of confining the subcarrier in both the time and the frequency domain. Assume that a GFDM block consists of K subcarriers and each subcarrier carries M temporally equally spaced data symbols (Figure 7.1). Denoting the time spacing between any two consecutive data symbols in a subcarrier by T_s, one GFDM block with a duration T equal to MT_s and a subcarrier spacing $1/T_s$ is given as

$$s(t) = \sum_{k=0}^{K-1}\sum_{m=0}^{M-1} d_{k,m} g(t - mT_s) e^{j\,2\pi kt/T_s}, \tag{7.1}$$

where $d_{k,m}$ represents the mth data symbol transmitted on subcarrier k^* and the circular pulse-shaping filter $g(t)$ has a period equal to T. If $s(t)$ is sampled at the rate K/T_s, the

---

* $d_{k,m}$ only takes a nonzero value if subcarrier k is active.

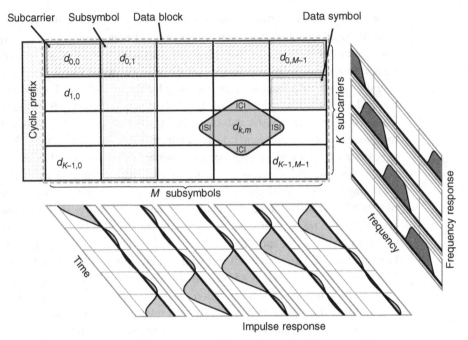

Subcarrier   Subsymbol   Data block   Data symbol

Cyclic prefix

$d_{0,0}$ $d_{0,1}$ $d_{0,M-1}$

$d_{1,0}$

ICI

ISI $d_{k,m}$ ISI

ICI

$d_{K-1,0}$ $d_{K-1,M-1}$

M subsymbols

K subcarriers

Frequency response

frequency

Time

Impulse response

Figure 7.1 Illustration of one GFDM block positioned in a time–frequency grid.

resulting $N = MK$ samples are

$$s[n] = \sum_{k=0}^{K-1}\sum_{m=0}^{M-1} d_{k,m}g[\langle n - mK \rangle_N]e^{j\,2\pi nk/K}, \tag{7.2}$$

where $g[n]$ corresponds to the nth sample of $g(t)$ and $\langle a \rangle_b$ denotes a modulo b.

In general, GFDM, as a non-orthogonal waveform, is subject to intersymbol interference (ISI) and intercarrier interference (ICI). However, by choosing $g(t)$ with good time and frequency localization, for example a raised cosine filter, the interference experienced can be limited to adjacent subcarriers and subsymbols as illustrated in Figure 7.1. This property is useful for reducing the complexity of resolving interference at the receiver side (see Section 7.4). Furthermore, the choice of filter also determines the properties of the signal at the block boundaries. OFDM adopts rectangular pulse shaping. The resulting discontinuous boundary results in high OOB emission in the frequency domain. Filtering can help by smoothing the transition of the signal to the boundary so that the OOB emission of the signal can be reduced. For more details of GFDM modulation, we refer readers to [10].

In accordance with (7.2), the design space of GFDM includes the number K of subcarriers, the number M of subsymbols, the temporal spacing T_s, and the pulse-shaping filter $g[n]$. These variables jointly characterize two key features of GFDM, namely a flexible time–frequency grid and a two-fold cyclic structure. Specifically, the former three parameters define the time–frequency grid occupied by each GFDM block. By having one subsymbol in the time direction (i.e., $M = 1$) and rectangular pulse shaping, the system becomes identical to OFDM. A conventional single-carrier system simply implies a single subcarrier, $K = 1$. If we choose $M > 1$ and $K > 1$, we can effectively treat GFDM as a combination of both systems such that their benefits, for example robustness against multipath fading and low PAPR, can be jointly exploited. Furthermore, OFDM, being a corner case of GFDM, enables backward compatibility with existing fourth generation (4G) technologies. As reported in [11], GFDM systems can reuse the Long Term Evolution (LTE) master clock and be fully aligned with the LTE resource grid. This identification suggests that the use of GFDM will ensure smooth convergence of the existing 4G networks to 5G.

Similarly to OFDM, GFDM adopts a CP to combat time-dispersive fading channels. Since all subsymbols can be protected by a single CP, it is more spectrally efficient than OFDM when rich multipath fading channels are experienced. Within the data block, the circular pulse-shaping filter brings in a second level of circularity, which can be exploited for energy-efficient signal processing in the frequency domain. Overall, the twofold cyclic structure is beneficial for spectrum sensing and synchronization (e.g., [12–14]). Another advantage of circular filtering is that tail biting gets rid of the filter tail. As such, the GFDM signal is temporally compact for achieving an efficient use of time resources.

On the basis of (7.2), additional degrees of freedom can be applied in GFDM. First, offset quadrature amplitude modulation (OQAM) can be adopted in GFDM to achieve orthogonality in the real domain [15, 16]. Second, time domain windowing and guard symbols can be used to improve further the OOB emission performance of GFDM [10]. Third, it is possible to introduce two factors v_t and v_f to scale the time spacing T_s between two consecutive subsymbols and the subcarrier spacing $1/T_s$, respectively. The use of $v_t < 1$ reflects an acceleration of the insertion of subsymbols in time, and is aligned with the concept of faster-than-Nyquist (FTN) signaling [17]. Analogously, $v_f < 1$ increases the density of subcarriers in the allocated spectrum, leading to spectrally efficient frequency division multiplexing (SEFDM) [18].

In summary, GFDM is equipped with a sufficiently large design space to efficiently use both time and frequency resources for conveying information under various conditions. However, enabling such flexibility challenges the hardware implementation. The next section therefore focuses on a field-programmable gate array (FPGA) implementation of a GFDM modem, which supports run-time reconfigurability.

7.3   Software-Defined Waveform

In this section, we describe how a GFDM modem was implemented on a National Instruments (NI) SDR (USRP-RIO 2953R), which had a built-in Kintex-7 FPGA (XC7K410T) and a radio front end equipped with two antennas. While GFDM modulation is executed on the FPGA, a control personal computer (PC) connected to the USRP-RIO via a PXIe-PCIe8371 card is used to configure the parameters, generate data streams, and capture and evaluate the outcomes.

7.3.1   Time Domain Processing

Let us start by reformulating (7.2) as

$$s[n] = \sum_{m=0}^{M-1} g[\langle n - mK \rangle_N] \cdot \left[\sum_{k=0}^{K-1} d_{k,m} e^{j\,2\pi nk/K} \right]. \qquad (7.3)$$

The sum in [·] effectively corresponds to a K-point inverse DFT $\vec{W}_K$ applied to the K data symbols transmitted in the mth time slot, i.e., $\vec{W}_K \vec{d}_m$, where the column vector $\vec{d}_m$ is a compact notation for $d_{k=0,m}, \ldots, d_{k=K-1,m}$. On this basis, the sequence of N samples is simply obtained by repetition of $\vec{W}_K \vec{d}_m$ M times. Accordingly, the N samples of $g[n]$ can be segmented into M parts, i.e., $\vec{g}_0, \ldots, \vec{g}_{M-1}$, with length K each. With a properly block-circulant shift, the two sequences are multiplied. The outcome corresponds to the mth column of the matrix given by

$$\begin{bmatrix} \vec{g}_0 \circ (\vec{W}_K \vec{d}_0) & \vec{g}_{M-1} \circ (\vec{W}_K \vec{d}_1) & \cdots & \vec{g}_1 \circ (\vec{W}_K \vec{d}_{M-1}) \\ \vec{g}_1 \circ (\vec{W}_K \vec{d}_0) & \vec{g}_0 \circ (\vec{W}_K \vec{d}_1) & \cdots & \vec{g}_2 \circ (\vec{W}_K \vec{d}_{M-1}) \\ \vdots & \vdots & \cdots & \vdots \\ \vec{g}_{M-1} \circ (\vec{W}_K \vec{d}_0) & \vec{g}_{M-2} \circ (\vec{W}_K \vec{d}_1) & \cdots & \vec{g}_0 \circ (\vec{W}_K \vec{d}_{M-1}) \end{bmatrix} \in \mathbb{C}^{N \times M},$$

$$(7.4)$$

where ∘ stands for the Hadamard product. Consequently, the GFDM sample $s[n]$ equals the sum of row n of the above matrix. Overall, the number of multiplications required for such time domain processing is $MK \log K + NM$.

It is also possible to generate $s[n]$ by means of frequency domain processing [19]. As the pulse-shaping filter is often chosen with good frequency localization, the large number of zero coefficients in the frequency response of the filter reduces the number of multiplications. However, since the time domain signal $s[n]$ is our ultimate goal, frequency domain processing requires an N-point DFT for domain conversion [19]. For multicarrier modulation, we can identify the DFT as the most computationally intensive task. Since K is M times smaller than N, the K-point DFT needed for time domain processing outperforms the N-point DFT in terms of computational energy and processing latency.

7.3.2   Implementation Architecture

Figure 7.2 visualizes the implementation of the GFDM modem by means of a multirate diagram (MRD) provided by the NI LabVIEW Communications Systems Design Suite. This software also serves as a development tool that enables quick prototyping of new algorithms. The actual implementation, presented in [20], was done in a VHDL-like clock-driven logic manner, permitting flexible implementation. For illustration purposes, the example depicted considers $K = 512$ subcarriers and $M = 5$ subsymbols, where the input signal is the outcome of coding and modulation. The architecture presented can easily be extended to larger values of K and M. The design in Figure 7.2 consists of nine functional blocks:

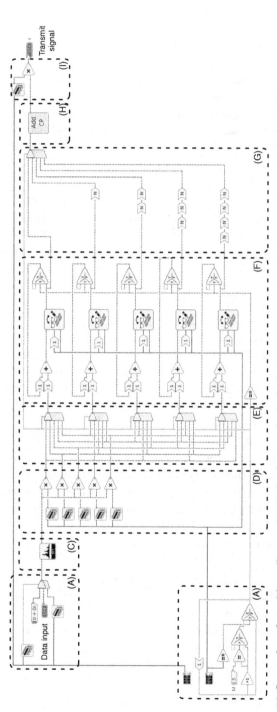

Figure 7.2 Screenshot of a multirate diagram (MRD) implementation in the LabVIEW Communication Design Suite to outline the different modulation steps.

(A) The address counter controls the processing steps conducted in different modules. It increases at every cycle and provides three address signals. The first one, k, with a range $[0, K-1]$, is associated with the subcarrier assignment, while the subsymbol assignment is reflected by the second address signal, m. Here, the assumption is that the subcarrier assignment has higher priority than the subsymbol assignment. Namely, the second address signal only increases when the first one has reached its maximum value, i.e., $K-1$. The third address signal, n, gives information about the total number of processed samples, ranging from 0 to $N-1$.

(B) The input samples to the GFDM modem are provided via three different channels, holding (1) null symbols, (2) information-bearing data symbols, and (3) training symbols. The flexible resource mapper here feeds the modem with the desired type of input samples, whereas the mapping itself is predefined at the host controller and stored in a memory. This memory is read out in an order where subcarriers are considered first, analogously to the address counter. Each entry of the mapping configures the multiplexer such that one of the three input channels is assigned to the input of the next DFT module. The resource mapper can easily be extended for additional input channels, for example a control channel. The stored mapping can be rewritten at any time by the host to support different configurations without reloading the FPGA firmware.

(C) A K-point inverse DFT is executed.

(D) M identical branches process the pulse shaping for the samples in a parallel manner. For each branch, the outcome of the DFT is multiplied by the filter coefficients of the prototype filter $g[n]$ stored in lookup tables, which can be updated on-the-fly. The pulse-shaping filter is thereby divided into M parts containing K values each. So, each set of K samples is filtered by different segments of the pulse-shaping filter (see (7.4)).

(E) The multiplexers (MUXs) are controlled by the subsymbol counter m. The sequence of input samples from the pulse-shaping filter are routed to M memory banks in such a way that the output of the MUXs has a block-circular structure.

(F) The memory banks have the purpose of accumulating the signal contributions from all M subsymbols. Each memory contains one complete subsymbol. The memories are flushed whenever a new modulation process starts. The previous values are read out during the flushing process.

(G) The delay chains followed by a MUX facilitate a serial-to-parallel conversion to ensure that the finished subsymbols in each accumulation memory are read out in the correct order. If a legacy-compatible OFDM frame is requested, then a subsymbol-wise CP is added here.

(H) A CP for the whole GFDM block is added in this step.

(I) In order to further reduce the OOB emission, an additional time domain windowing is applied here to smooth the transitions between several GFDM blocks.

The parameters K and M are adaptable during run-time. Changing K is a memory operation, meaning that only a part of the address range is in use. To obtain an efficient implementation of the DFT, K is typically changed to a power of 2. The number of subsymbols M can be changed by switching the respective blocks in (D) in Figure 7.2 on and off and reconfiguring the multiplexers. The maximum possible values for K and M are defined by the capabilities of the FPGA, particularly the number of available random access memory (RAM) blocks. The current implementation, using LabVIEW clock-driven logic, supports up to 2048 subcarriers and 16 subsymbols. Additionally, it supports run-time reconfiguration of the pulse-shaping filter, the length of the CP for each subsymbol and the whole block, and the time window applied at the end. Thus, the modulator can process not only GFDM, but also one LTE frame with 14 subsymbols for backward compatibility.

7.4   GFDM Receiver Design

Figure 7.3 depicts a generic block diagram of a digital receiver whose design principles have been presented in detail in [21, 22]. In short, to meet the requirements, each block must be specifically designed, exploiting the structure of the transmitted signal. The main functional blocks are listed as follows, and their tasks are briefly described.

- *Synchronization:* Initially, the receiver needs to acquire knowledge about at which time and at which carrier frequency a particular data packet arrives. Even though knowledge about the carrier frequency is mostly available beforehand, frequency offsets due to local-oscillator mismatches between transmitter and receiver commonly occur. Moreover, mobility results in a frequency shift (Doppler shift) that needs to be estimated and corrected at the receiver.

  Time synchronization is necessary to determine the starting point of a data packet such that the receiver knows when to start the decoding process. Timing misalignments occur in particular owing to the propagation delay of the signal from the transmitter to the receiver.

- *Channel estimation:* During transmission over the air, the wireless communication channel distorts the transmit signal. In order to retrieve the information conveyed by the signal, the receiver needs to estimate the instantaneous channel state information (CSI) such that channel impairments can be properly compensated. In particular, fading and multipath propagation are two key aspects of wireless communication channels. The transmit signal arrives at the receiver by multiple paths. Each path can introduce different propagation delays, path losses, phase rotations, and so on. When the coherence bandwidth of the channel is smaller than the bandwidth of the signal,

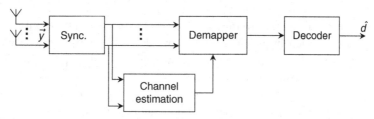

Figure 7.3 Generic block diagram of a digital receiver.

the impact of the channel on the transmit signal becomes frequency dependent. The channel also introduces additive noise. A knowledge of the noise power is useful for deriving reliability information while performing soft signal detection.

- *Signal demapping and decoding:* Once it is equipped with an estimate of the CSI, the ultimate goal of the synchronized receiver is to retrieve the transmitted information bit sequence. To this end, a demapper extracts the original information-bearing codeword from the received signal in a waveform. The outcome contains not only the sign information about each code bit, but also the reliability of that information. Under the constraints of the code, the subsequent channel decoder makes a hard decision on the information bit sequence, which is then forwarded to the upper layers as a received packet.

7.4.1   Synchronization Unit

The close relationship between GFDM and OFDM allows existing OFDM synchronization algorithms to be easily adapted to GFDM systems. Namely, estimation of the symbol timing offset (STO) and carrier frequency offset (CFO) can be achieved by cross-correlation and autocorrelation measures, as suggested in [23, 24]. However, the design of the synchronization sequences should maintain the overall desirable features of GFDM such as low OOB emission and block structure.

An important aspect of low-latency communication with sporadic traffic is the ability to perform one-shot synchronization of the system. Low latency implies a short length of the training sequence for synchronization, while sporadic traffic does not permit closed-loop synchronization with feedback from the transmitter. The one-shot synchronization algorithms initially proposed for OFDM [24–26] use a separate preamble as the synchronization signal. In adapting them for GFDM, we must avoid destroying the beneficial spectral properties of the GFDM payload block. Three ways of embedding a synchronization sequence in a GFDM signal are shown in Figure 7.4. One way is to use an isolated preamble which uses a separate time domain window to reduce its OOB emission [27]. Alternatively, the synchronization sequence can be embedded in the GFDM block by means of either a midamble [28] or a pseudo-circular preamble (PCP) [15] such that good spectral properties are straightforwardly attainable.

In contrast to using an embedded synchronization sequence, an isolated synchronization preamble in GFDM allows the synchronization sequence to be observed free of interference from the data symbols. As such, isolated preambles offer better synchronization performance. However, since the OOB emission of the preamble is controlled by the time domain windowing employed, the overall frame structure follows a double-pinching pattern, which reduces the spectral efficiency of the system owing to the additional CP overhead, as can be seen in Figure 7.4. With a midamble design, no extra overhead for CP or pinching is incurred by the system; however, the synchronization sequence is polluted by interference from adjacent data symbols, leading to a reduced synchronization performance. In the special case of the PCP,

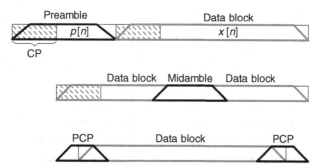

Figure 7.4 Three ways to include a preamble in the GFDM block structure for synchronization.

the synchronization sequence can also be reused as a guard interval between data blocks [15].

In order to provide robust coarse synchronization via the autocorrelation properties of the synchronization sequence, it is advisable that the sequence consists of two identical parts [24, 25]. For GFDM, this is simply achieved by forming a preamble that contains two subsymbols which transmit the same values, for example taken from a pseudonoise (PN) sequence. In general, the estimation of the CFO and STO can be achieved in three steps:

1. *Coarse estimation by means of autocorrelation.* The structure of the preamble, with two repeated parts, allows one to perform coarse synchronization based on the autocorrelation of the received signal. Specifically, the receiver window is of the same length as the preamble and the correlation between the first half and the second half of the signal within the window is calculated. As soon as the preamble is fully contained within the window, this autocorrelation metric reaches its maximum. However, since the preamble is protected by a CP, this peak extends to a plateau with the length of the CP. In order to find a good estimate of the STO, the autocorrelation metric should be integrated over time, converting the CP plateau into a peak. The position of this peak provides a coarse estimate of the STO, whereas the CFO is deduced from the phase of the autocorrelation at the peak [23, 24].

2. *Fine synchronization with cross-correlation.* After correction of the estimated CFO from the previous step, the received signal is correlated with the correct preamble signal, which is known to the receiver. If a suitable pseudo noise (PN) sequence is used, this metric contains an isolated peak at the correct STO. However, owing to the repeated structure of the synchronization signal, side peaks in the correlation occur at half of the preamble length. In order to remove them, the correlation metric is multiplied by the autocorrelation metric such that only the correct peak is preserved. By examining the phase and position of the correlation peak, we can further refine the estimates of the STO and CFO [25].

3. *Search for the first multipath component.* The algorithm presented above is suitable for frequency-flat-fading channels. When the signal is received from multiple paths, the cross-correlation metric exhibits a peak at every multipath component and the synchronization procedure presented above simply relies on the largest peak for

estimating CFO and STO. However, depending on the fading conditions, the largest peak in the correlation metric does not necessarily coincide with the first channel delay tap. Therefore, a backward search for peaks is needed. After a threshold that depends on the signal-to-noise ratio (SNR) and the fading distribution has been defined, the first peak exceeding the threshold provides the final estimate of the CFO and STO [26]. The algorithm described has been validated under conditions of severely frequency-selective [27, 28] and doubly dispersive channels [15] for GFDM, and the good spectral properties of the waveform were kept. Interested readers are referred to these references for a detailed mathematical description and performance analysis.

In addition to adapting OFDM synchronization algorithms for GFDM, the twofold circularity of the GFDM signal can be exploited to improve the synchronization performance, and is even usable for nondata-aided synchronization. In [12, 13], it was shown, in contrast to OFDM, that the cyclic autocorrelation in GFDM exhibits additional side peaks that can be used to detect the existence of a GFDM signal in CR scenarios. This property was exploited analogously in [14] to develop an efficient algorithm for estimating the CFO of the received signal. Owing to the overlapping subcarriers and the circular filtering structure of the GFDM signal, the autocorrelation of the received signal in the frequency domain exhibits an impulse train where the impulses occur at a spacing equal to the subcarrier spacing. Such an estimation algorithm has the advantage of not requiring an extra synchronization sequence but relying only on the circulant structure of the transmitted GFDM signal. By means of the autocorrelation in the time domain based on the CP, the extension of the algorithm to estimation of the STO consequently yields a robust blind synchronization algorithm for GFDM.

7.4.2   Channel Estimation Unit

In principle, by relying on some prior knowledge of the signal structure, the CSI can be extracted from the received signal. Blind channel estimation methods [29] exploit the statistical properties of the received signal, for example correlation metrics based on the CP or constellation properties. However, blind methods tend to converge slowly and hence are more suitable for static channels than for high-mobility scenarios.

Another class of channel estimation algorithms work with training sequences. The training symbols, multiplexed with the information-bearing data symbols, are transmitted over the channel. Their observation is used by the receiver to estimate the CSI. These methods can typically achieve better channel estimation quality compared with blind methods, but on the other hand sacrifice the scarce spectrum resource to transmit preknown information. Generally speaking, there exist two ways to multiplex training symbols with information-bearing data symbols. For GFDM, concentrating the training symbols into a dedicated preamble allows the CSI to be estimated without being affected by the self-introduced interference. We note that a properly designed preamble can also be reused for synchronization. However, preamble-based methods are disadvantageous in high-mobility scenarios. A rapidly changing channel requires frequent insertion of a preamble to track the time-varying CSI. The resulting overhead can be too high. For the sake of spectral efficiency, we can resort to multiplexing training symbols with the information-bearing data symbols, where the distance between adjacent training symbols should be within the coherence time of the channel. However, for non-orthogonal waveforms such as GFDM, the observation of such inserted training symbols also contains information about the unknown neighboring data symbols (cf. Figure 7.1). Hence, special care needs to be taken in designing training sequences in order to limit the impact of interference.

In the literature, a preamble-based channel estimation method in the frequency domain for GFDM which employs a least-squares criterion has been presented in [30], where the particular preamble used in the method was also reused for synchronization purposes. Alternatively, time domain estimation by means of adaptive filters [31] or weighted least-squares [32] methods can be applied in preamble-based channel estimation. In contrast, the authors of [33, 34] proposed to scatter the training symbols in the GFDM time–frequency grid. To combat the self-introduced interference, several interference cancellation techniques at the transmitter side were proposed in [33] to ensure clean pilot observations at the receiver. It should be noted that the work in [33] was based on the assumption of a nearly flat fading channel. Conversely, the channel estimator in [34], based on a linear minimum mean square error (LMMSE) criterion, treated the interference as additional Gaussian noise. It is also possible to remove interference by using the detected data symbols. This requires a feedback loop to be enabled from either the detection or the decoding unit.

7.4.3   MIMO-GFDM Detection Unit

The invention of MIMO techniques for wireless systems has led to breakthrough in increased data rates and robustness [35]. MIMO has been successfully applied in third generation (3G) and 4G cellular systems and will be an essential part of the 5G air interface. Besides the benefits of higher system capacity, however, MIMO techniques greatly increase the complexity of the algorithm at the receiver side, especially when near-optimal receivers are targeted. The situation becomes even more challenging with the application of massive MIMO techniques [36] to wireless systems. Hence, efficient algorithms for a 5G air interface that supports MIMO detection with reasonable performance and complexity have been intensively researched. A significant portion of this research is dedicated to extending OFDM alternatives to MIMO techniques, which is, depending on the waveform properties, not straightforwardly achievable.

Since GFDM is a linear modulation scheme, after synchronization, the input–output relation of a linear channel in the frequency domain can be compactly formalized as a linear system

$$\vec{y} = \mathbf{H}\vec{d} + \vec{n},\qquad(7.5)$$

where $\vec{y}$ is the received signal associated with the complete GFDM block and the noise $\vec{n}$ is often assumed to be a Gaussian random variable with zero mean and variance N_0.

The matrix $\mathbf{H}$ is the outcome of multiplying the equivalent channel matrix by the GFDM modulation matrix.

In MIMO cases, not only intercarrier interference (ICI) and intersymbol interference (ISI) but also interantenna interference (IAI) will take place. This implies that the matrix $\mathbf{H}$ is not diagonal, meaning each entry of the received vector $\overrightarrow{y}$ is a linear combination of multiple transmitted data samples followed by noise corruption. The general task of detection is then to map the observations and the channel knowledge (i.e., CSI and noise variance N_0) into a vector $\widehat{\overrightarrow{d}}$ corresponding to one possible realization of the transmitted data sequence

$$\widehat{\overrightarrow{d}} = f(\overrightarrow{y}, \mathbf{H}, N_0). \tag{7.6}$$

Without any prior knowledge of the transmitted $\overrightarrow{d}$, the optimum f is determined by the maximum likelihood detection criterion. Under the assumption of additive white Gaussian noise (AWGN), this is equivalent to a minimum-distance search problem defined by

$$\widehat{\overrightarrow{d}} = \arg\min_{\overrightarrow{d} \in \mathcal{D}} \| \overrightarrow{y} - \mathbf{H}\overrightarrow{d} \|^2, \tag{7.7}$$

where $\mathcal{D}$ denotes the feasible set of $\overrightarrow{d}$. However, as the cardinality of $\mathcal{D}$ increases exponentially with the length of the data sequence, a straightforward solution to problem (7.7) by exhaustive search is beyond the capabilities of any implementation. Feasible solutions must try to exploit the spectral structure of $\mathbf{H}$ to speed up the search for the closest point.

In an OFDM system, $\mathbf{H}$ is a block-diagonal matrix. Each block matrix corresponds to the MIMO channel transfer function obtained for one subcarrier, while its size depends on the number of transmit and receive antennas. Thus, we can decouple the large-scale linear system (7.5) into smaller systems and then conduct MIMO-OFDM detection on a subcarrier basis. For GFDM, the matrix $\mathbf{H}$ is not block diagonal, owing to the presence of inter-carrier interference (ICI) and inter-symbol interference (ISI). Figure 7.5 shows the structure of a GFDM equivalent channel matrix in the frequency domain. As can be seen, $\mathbf{H}$ has a band-diagonal structure instead. This is because the

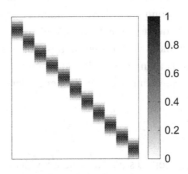

Figure 7.5 Structure of a GFDM equivalent channel matrix.

pulse-shaping filter, with good frequency localization, ensures limited ICI. However, unlike the OFDM case, the banded matrix does not allow us to reduce the system size to ease the detection problem without degrading the detection performance. As such, the MIMO-GFDM detection problem boils down to a detection problem for large-scale linear systems (e.g., [35, 36]), where the sparsity of the matrix needs to be exploited to develop low-complexity detection algorithms.

For instance, one study [37] has presented an efficient way to compute the filter coefficients of an LMMSE detector. Since the banded structure implies localized interference, a combination of successive interference cancellation with sphere decoding was proposed in [38]. This achieved significant performance gains compared with conventional OFDM systems. The localized interference also permits parallelism in the Gibbs sampling process involved in the Markov chain Monte Carlo-based MIMO-GFDM detection algorithm proposed in [39]. More advanced methods based on iterations between a demapper and a decoder can also achieve significant gains compared with noniterative solutions (e.g., [40]). However, these come with an increased amount of receiver complexity and latency, as in each iteration both the demapping and the decoding process need to be performed. Hence it is important to tailor the algorithm for the application under consideration.

7.5   Summary and Outlook

NFV is an essential paradigm in the development process of 5G networks. Our vision is to virtualize PHY functionalities. As such, the utilization rate of network resources can be maximized by means of SDN techniques. Additionally, the infrastructure can easily and seamlessly evolve along with the time-varying requirements. To enable virtual PHY, it is preferable and cost-effective to have a single PHY relying on a software-defined waveform rather than developing dedicated PHYs for each scenario. This concept leads to the SDN paradigms, with a flexible framework for services and waveforms, which at the time of standardization eliminates the need to reserve bandwidth or resource blocks for potential future upgrades. In short, it not only allows easy deployment and management of the network, but also provides centralized control over resources.

GFDM has been proposed as an enabler for flexible PHY design, aiming at diverse 5G applications. We have shown that GFDM retains necessary and sufficient degrees of freedom for waveform engineering such that the extreme requirements can be achieved by properly tuning the waveform parameters. We are also aware that 5G will not be a mere evolution of 4G. Disruptive changes are expected, but legacy waveforms cannot be forgotten. As GFDM can emulate OFDM and single carrier frequency domain multiple access (SC-FDMA), legacy systems can run in a 5G base station as a virtual 4G service.

Supporting this unprecedented flexibility requires a new design of PHY suggesting an increased implementation complexity. Fortunately, progress in hardware technologies makes this challenging objective achievable. For a proof of concept, this chapter has presented an field programmable gate array (FPGA) implementation of a flexible

GFDM modem on an NI SDR platform. For the receiver design, we have described the general concept and introduced promising algorithms. The good frequency localization of the GFDM signal was discovered to be beneficial for developing low-complexity algorithms, even though GFDM suffers from interference due to non-orthogonality. The hardware implementation of those algorithms is certainly of great interest for future work. Another aspect of future work is the joint design of a GFDM-based PHY with the medium access control (MAC) layer, exploiting its flexible time–frequency resource grid and robustness against time and frequency misalignment.

Acknowledgments

The authors thank National Instruments for unconditional technical support and for supplying software and hardware tools for prototyping. Part of the work presented in this chapter was sponsored by the Federal Ministry of Education and Research within the program "Twenty20 – Partnership for Innovation" under Contract 03ZZ0505B, "Fast Wireless." The authors also thank the Horizon 2020 project ICT-688116 "eWINE" for partially funding the work.

References

[1] G. Wunder, P. Jung, M. Kasparick, T. Wild, F. Schaich, Y. Chen, S. Brink, I. Gaspar, N. Michailow, A. Festag, L. Mendes, N. Cassiau, D. Ktenas, M. Dryjanski, S. Pietrzyk, B. Eged, P. Vago, and F. Wiedmann, "5GNOW: Non-orthogonal, asynchronous waveforms for future mobile applications," *IEEE Commun. Mag.*, vol. 52, no. 2, pp. 97–105, Feb. 2014.

[2] G. Fettweis, "The Tactile Internet: Applications and challenges," *IEEE Veh. Technol. Mag.*, vol. 9, no. 1, pp. 64–70, Mar. 2014.

[3] N. Tadayon and S. Aissa, "Modeling and analysis of cognitive radio based IEEE 802.22 wireless regional area networks," *IEEE Trans. Wireless Commun.*, vol. 12, no. 9, pp. 4363–4375, Sep. 2013.

[4] B. Hirosaki, "An orthogonally multiplexed QAM system using the discrete Fourier transform," *IEEE Trans. Commun.*, vol. 29, no. 7, pp. 982–989, Jul. 1981.

[5] G. Cariolaro and F. C. Vagliani, "An OFDM scheme with a half complexity," *IEEE J. Sel. Areas Commun.*, vol. 13, no. 9, pp. 1586–1599, Dec. 1995.

[6] E. Dahlman, S. Parkvall, and J. Skold, *4G: LTE/LTE-Advanced for Mobile Broadband*, Elsevier Academic, 2011.

[7] X. Zhang, M. Jia, L. Chen, J. Ma, and J. Qiu, "Filtered-OFDM-enabler for flexible waveform in the 5th generation cellular networks," in *Proc. of IEEE Global Communnications Conf. (GLOBECOM)*, Dec. 2015.

[8] T. Wild, F. Schaich, and Y. Chen, "5G air interface design based on universal filtered (UF-)OFDM," in *Proc. of IEEE International Conf. on Digital Signal Processing (DSP)*, Aug. 2014.

[9] M. Bellanger *et al.*, "FBMC physical layer: A primer," PHYDYAS project, Jan. 2010. Available at www.ict-phydyas.org/teamspace/internal-folder/FBMC-Primer_ 06-2010.pdf.

[10] N. Michailow, M. Matthé, I. Gaspar, A. Caldevilla, L. Mendes, A. Festag, and G. Fettweis, "Generalized frequency division multiplexing for 5th generation cellular networks," *IEEE Trans. Commun.*, vol. 62, no. 9, pp. 3045–3061, Sep. 2014.

[11] I. Gaspar, L. Mendes, M. Matthé, N. Michailow, A. Festag, and G. Fettweis, "LTE-compatible 5G PHY based on generalized frequency division multiplexing," in *Proc. of IEEE International Symposium on Wireless Communications Systems (ISWCS)*, Aug. 2014.

[12] D. Panaitopol, R. Datta, and G. Fettweis, "Cyclostationary detection of cognitive radio systems using GFDM modulation," in *Proc. of IEEE Wireless Communications and Networking Conf. (WCNC)*, Apr. 2012.

[13] R. Datta, D. Panaitopol, and G. Fettweis, "Cyclostationary detection of 5G GFDM waveform in cognitive radio transmission," in *Proc. of IEEE International Conf. on Ultra-Wide Band (ICUWB)*, Sep. 2014.

[14] T. Kadur, I. Gaspar, N. Michailow, and G. Fettweis, "Non-data aided frequency synchronization exploiting ICI in non-orthogonal systems," in *Proc. of IEEE Vehicular Technology Conf. (VTC-Fall)*, Sep. 2014.

[15] I. Gaspar, M. Matthé, N. Michailow, L. L. Mendes, D. Zhang, and G. Fettweis, "Frequency-shift offset-QAM for GFDM," *IEEE Commun. Lett.*, vol. 19, no. 8, pp. 1454–1457, Aug. 2015.

[16] M. Matthé and G. Fettweis, "Conjugate root offset-QAM for orthogonal multicarrier transmission," *EURASIP J. Adv. Signal Process.*, vol. 2016, no. 1, pp. 1–8, Apr. 2016.

[17] J. B. Anderson, F. Rusek, and V. Wall, "Faster-than-Nyquist signaling," *Proc. IEEE*, vol. 101, no. 8, pp. 1817–1830, Aug. 2013.

[18] I. Kanaras, A. Chorti, M. R. D. Rodrigues, and I. Darwazeh, "Spectrally efficient FDM signals: Bandwidth gain at the expense of receiver complexity," in *Proc. of IEEE International Conf. on Communications (ICC)*, Jun. 2009.

[19] I. Gaspar, N. Michailow, A. Navarro, E. Ohlmer, S. Krone, and G. Fettweis, "Low complexity GFDM receiver based on sparse frequency domain processing," in *Proc. of IEEE Vehicular Technology Conf. (VTC-Spring)*, Jun. 2013.

[20] M. Danneberg, N. Michailow, I. Gaspar, D. Zhang, and G. Fettweis, "Flexible GFDM implementation in FPGA with support to run-time reconfiguration," in *Proc. of IEEE Vehicular Technology Conf. (VTC-Fall)*, Sep. 2015.

[21] H. Meyr, M. Moeneclaey, and S. Fechtel, *Digital Communication Receivers: Synchronization, Channel Estimation, and Signal Processing*, John Wiley, 1997.

[22] T.-D. Chiueh and P.-Y. Tsai, *OFDM Baseband Receiver Design for Wireless Communications*, Wiley, 2008.

[23] J. J. van de Beek, M. Sandell, and P. O. Borjesson, "ML estimation of time and frequency offset in OFDM systems," *IEEE Trans. Signal Process.*, vol. 45, no. 7, pp. 1800–1805, Jul. 1997.

[24] T. M. Schmidl and D. C. Cox, "Robust frequency and timing synchronization for OFDM," *IEEE Trans. Commun.*, vol. 45, no. 12, pp. 1613–1621, Dec. 1997.

[25] H. Minn, V. K. Bhargava, and K. B. Letaief, "A robust timing and frequency synchronization for OFDM systems," *IEEE Trans. Wireless Commun.*, vol. 2, no. 4, pp. 822–839, Jul. 2003.

[26] A. B. Awoseyila, C. Kasparis, and B. G. Evans, "Improved preamble-aided timing estimation for OFDM systems," *IEEE Commun. Lett.*, vol. 12, no. 11, pp. 825–827, Nov. 2008.

[27] I. S. Gaspar, L. L. Mendes, N. Michailow, and G. Fettweis, "A synchronization technique for generalized frequency division multiplexing," *EURASIP J. Adv. Signal Process.*, vol. 2014, no. 1, pp. 1–10, Dec. 2014.

[28] I. Gaspar and G. Fettweis, "An embedded midamble synchronization approach for generalized frequency division multiplexing," in *Proc. of IEEE Global Communications Conf. (GLOBECOM)*, Dec. 2015.

[29] M. C. Necker and G. L. Stuber, "Totally blind channel estimation for OFDM on fast varying mobile radio channels," *IEEE Trans. Wireless Commun.*, vol. 3, no. 5, pp. 1514–1525, Sep. 2004.

[30] M. Danneberg, N. Michailow, I. Gaspar, M. Matthé, D. Zhang, L. L. Mendes, and G. Fettweis, "Implementation of a 2 by 2 MIMO-GFDM transceiver for robust 5G networks," in *Proc. of IEEE International Symposium on Wireless Communications Systems (ISWCS)*, Aug. 2015.

[31] A. Aghamohammadi, H. Meyr, and G. Ascheid, "Adaptive synchronization and channel parameter estimation using an extended Kalman filter," *IEEE Trans. Commun.*, vol. 37, no. 11, pp. 1212–1219, Nov. 1989.

[32] D. Kong, D. Qu, and T. Jiang, "Time domain channel estimation for OQAM-OFDM systems: Algorithms and performance bounds," *IEEE Trans. Signal Process.*, vol. 62, no. 2, pp. 322–330, Jan. 2014.

[33] U. Vilaipornsawai and M. Jia, "Scattered-pilot channel estimation for GFDM," in *Proc. of IEEE Wireless Communications and Networking Conf. (WCNC)*, Apr. 2014.

[34] S. Ehsanfar, M. Matthé, D. Zhang, and G. Fettweis, "A study of pilot-aided channel estimation in MIMO-GFDM systems," in *Proc. of IEEE Workshop on Smart Antennas (WSA)*, Mar. 2016.

[35] S. Yang and L. Hanzo, "Fifty years of MIMO detection: The road to large-scale MIMOs," *IEEE Commun. Surv. Tutor.*, vol. 17, no. 4, pp. 1941–1988, Fourth Quarter 2015.

[36] L. Lu, G. Y. Li, A. L. Swindlehurst, A. Ashikhmin, and R. Zhang, "An overview of massive MIMO: Benefits and challenges," *IEEE J. Sel. Areas Signal Process.*, vol. 8, no. 5, pp. 742–758, Oct. 2014.

[37] M. Matthé, L. L. Mendes, N. Michailow, D. Zhang, and G. Fettweis, "Widely linear estimation for space-time-coded GFDM in low-latency applications," *IEEE Trans. Commun.*, vol. 63, no. 11, pp. 4501–4509, Nov. 2015.

[38] M. Matthé, I. Gaspar, D. Zhang, and G. Fettweis, "Short paper: Near-ML detection for MIMO-GFDM," in *Proc. of IEEE Vehicular Technology Conf. (VTC-Fall)*, Sep. 2015.

[39] D. Zhang, M. Matthé, L. L. Mendes, and G. Fettweis, "A Markov chain Monte Carlo algorithm for near-optimum detection of MIMO-GFDM signals," in *Proc. of IEEE International Symposium on Personal, Indoor, and Mobile Radio Communications (PIMRC)*, Aug. 2015.

[40] D. Zhang, L. L. Mendes, M. Matthé, I. S. Gaspar, N. Michailow, and G. P. Fettweis, "Expectation propagation for near-optimum detection of MIMO-GFDM signals," *IEEE Trans. Wireless Commun.*, vol. 15, no. 2, pp. 1045–1062, Feb. 2016.

8 Distributed Massive MIMO in Cellular Networks

Michail Matthaiou and Shi Jin

8.1 Introduction

Wireless communications, together with its underlying applications, is among today's most active areas of technology development, with the demand for data rates expected to grow to unprecedented levels by 2020. Cisco's latest report predicts a monthly mobile traffic of 24.3 exabytes (2^{60} bytes) in 2019, which represents a 57% compound annual growth rate compared with 2014 [1]. The catalyst for this seminal development is 5G, the fifth generation of wireless systems, which denotes the next major phase of mobile telecommunications standards beyond the current fourth generation (4G) International Mobile Telecommunications-Advanced (IMT-Advanced) standards and promises speeds far beyond what the current 4G can offer. This represents a radically new paradigm in the field of wireless communications and promises to substantially improve the area spectral efficiency (measured in bit/s/Hz/km^2) and energy efficiency (EE) (measured in bit/J). According to [2], there are three symbiotic technologies that can support the required "data-rate boom":

1. extreme densification and offloading to serve more active nodes per unit area and Hz, also known as massive multiple-input multiple-output (MIMO);
2. increased bandwidth, primarily by moving toward and into the millimeter wave spectrum (from 30 to 300 GHz);
3. increased spectral efficiency, primarily through advances in MIMO, to support more bits/s/Hz per node.

In this chapter, we will elaborate exclusively on item 1 above, namely massive MIMO, which represents a disruptive technological paradigm and is considered by many experts as the "next big thing in wireless" [3, 4]. We will first delineate the basic principles behind the operation of massive MIMO and then review some of its applications. Finally, we will conclude this chapter by presenting some directions for future work along with open challenges in the general field of massive MIMO. Table 8.1 lists the nomenclature used in this chapter.

Table 8.1 Nomenclature adopted in this chapter.

Notation	Description
$\lvert \cdot \rvert$	Absolute norm of a complex variable
$\lVert \cdot \rVert$	Vector norm
$(\cdot)^*$	Complex conjugate
$(\cdot)^{\mathrm{H}}$	Hermitian transpose
$(\cdot)^{-1}$	Inverse of a square matrix
$(\cdot)^{\mathrm{T}}$	Transpose
$\xrightarrow{\text{a.s.}}$	Almost sure convergence
$\xrightarrow{\text{d}}$	Convergence in distribution
$\boldsymbol{A}_k$	Matrix $\boldsymbol{A}$ with the kth column removed
$[\boldsymbol{A}]_{kk}$	The (k,k)th entry of the diagonal matrix $\boldsymbol{A}$
$\mathcal{CN}(\mu,\sigma^2)$	Complex normal random variable with mean μ and variance σ^2
$\det(\boldsymbol{A})$	Determinant of the square matrix $\boldsymbol{A}$
$\mathrm{tr}(\boldsymbol{A})$	Trace of the square matrix $\boldsymbol{A}$
$\mathcal{E}\{\cdot\}$	Expectation operator
$\mathrm{var}(\cdot)$	Variance of a random variable

8.2   Massive MIMO: Basic Principles

The massive MIMO technology originates from the seminal paper of Marzetta [5], and since then has been at the forefront of wireless communications research, with numerous papers reported in the literature along with huge industrial investments. Generally speaking, in a massive MIMO topology, a number K of user terminals (UTs) communicate simultaneously with a base station (BS) over the same time–frequency resources. The BS is equipped with an unconventionally large number of antennas, M, such that $M \gg K$, which may be colocated at the same BS site, spread out on the face of a building, or distributed geographically. The massive number of BS antennas offers a plethora of staggering advantages, which can be succinctly summarized as follows [6, 7]:

- *Higher data rates.* Owing to the extremely high number of antennas, the number of spatial streams that can be transmitted increases along with the number of served terminals.
- *Improved energy efficiency.* The transmitted radio frequency (RF) energy can be more sharply focused in the space where the terminals are located and, therefore, wasted energy is minimized.
- *Relaxed hardware constraints.* Hardware accuracy constraints can be relaxed, thus allowing the deployment of lower-quality (inexpensive) components on future massive BSs, compared with today's examples. For instance, expensive ultra-linear 40 W amplifiers can be replaced by many cheap low-power devices whose combined action, only, has to meet specific tolerances.
- *Improved reliability.* If one antenna unit fails, this will not greatly affect the overall system performance, thereby providing robustness to operators and users.

• *Reduced interference*. By simple linear processing, given the high number of degrees of freedom, we can avoid transmitting in undesired directions and maintain interuser interference at very low levels.

8.2.1   Uplink/Downlink Channel Models

Let us assume that K UTs transmit simultaneously to a massive MIMO BS with M antenna elements with an average power of p_u. Then, the received signal vector $y \in \mathbb{C}^{M \times 1}$ can be expressed as

$$y = \sqrt{p_u} \sum_{K=1}^{K} g_k x_k + n = \sqrt{p_u} Gx + n, \tag{8.1}$$

where $G \in \mathbb{C}^{M \times K}$ is the uplink channel matrix and g_k is its kth column; $x = [x_1, \dots, x_K]^T$ is a vector containing the transmit symbols, which are drawn from a zero-mean Gaussian process with unit power, i.e., $\mathbb{E}\{|x_k|^2\} = 1, k = 1, \dots, K$; and n is additive white Gaussian noise (AWGN) with zero mean and unit variance. Under these conditions, p_u represents the transmit signal-to-noise ratio (SNR). Since the UTs are geographically separated, their distances from the BS will be different. Under these circumstances, the channel matrix is modeled as

$$G = HD^{1/2}, \tag{8.2}$$

where H represents the small-scale fading matrix, whose entries are assumed to be $\mathcal{CN}(0,1)$ (i.e., the Rayleigh fading model). The entries of the diagonal matrix $D \in \mathbb{C}^{K \times K}$ represent the large-scale fading coefficients, which change very slowly. Then, we can write $[D]_{kk} = \beta_k, k = 1, \dots, K$, where β_k is distributed as $\beta_k = \xi_k \left(d_g / d_k \right)^v$, and where ξ_k is a log-normal random variable with standard deviation σ_{sh}, d_g is a guard zone around the BS in which no UT is located, d_k is the distance between UT k and the BS, and v is the path loss exponent (with typical values ranging from 2 to 6) [8].

In the downlink, the BS transmits (broadcasts) signals to all K UTs in the same time–frequency resource. In the case of time division duplex (TDD) operation, it is reasonable to assume that the downlink channel matrix is reciprocal to the uplink channel matrix. Then, the received signal at the kth UT, $k = 1, \dots, K$, is given by

$$\tilde{y}_k = \sqrt{p_d} g_k^T s + \epsilon_k, \tag{8.3}$$

where $s \in \mathbb{C}^{M \times 1}$ is the transmitted signal vector, normalized such that $\mathcal{E}\{||s||^2\} = 1$, and ϵ_k is a zero-mean AWGN term at the kth UT with unit variance. Then, p_d represents the downlink SNR. By rewriting (8.3) in vector form, we get

$$\tilde{y} = \sqrt{p_d} G^T s + \epsilon, \tag{8.4}$$

where $\tilde{y} = [\tilde{y}_1, \dots, \tilde{y}_k]^T$ and $\epsilon = [\epsilon_1, \dots, \epsilon_k]^T$.

8.2.2   Favorable Propagation

A key property of massive MIMO is the concept of asymptotic *favorable propagation* [9]. This means that the channel vectors of different users become pairwise orthogonal in the limit of $M \to \infty$, thereby eliminating interuser interference. Mathematically speaking, we have that

$$\frac{1}{M} \boldsymbol{g}_k^H \boldsymbol{g}_j \to 0, \quad k \neq j, \quad k = 1, \ldots, K, \tag{8.5}$$

where $\boldsymbol{g}_k \in \mathbb{C}^{M \times 1}$ is the uplink channel vector of the kth user. The above condition leverages some fundamental results on the limits of very long random vectors, which are reviewed in the following lemma.

Lemma 8.1 *Consider two $n \times 1$ independent vectors $\boldsymbol{a} = [a_1, \ldots, a_n]^T$ and $\boldsymbol{b} = [b_1, \ldots, b_n]^T$. If the elements of $\boldsymbol{a}$ and $\boldsymbol{b}$ are independent, identically distributed (i.i.d.) random variables with $\mathbb{E}\{a_i\} = \mathbb{E}\{b_i\} = 0$, $\mathbb{E}\{|a_i|^2\} = \sigma_a^2$, and $\mathbb{E}\{|b_i|^2\} = \sigma_b^2$ for $i = 1, \ldots, n$, we have that*

$$\frac{1}{n} \boldsymbol{a}^H \boldsymbol{a} \overset{\text{a.s.}}{\to} \sigma_a^2 \text{ as } n \to \infty, \tag{8.6}$$

$$\frac{1}{n} \boldsymbol{a}^H \boldsymbol{b} \overset{\text{a.s.}}{\to} 0 \text{ as } n \to \infty, \tag{8.7}$$

and also, through the Lindeberg–Lévy theorem,

$$\frac{1}{\sqrt{n}} \boldsymbol{a}^H \boldsymbol{b} \overset{\text{d}}{\to} \mathcal{CN}(0, \sigma_a^2 \sigma_b^2) \text{ as } n \to \infty. \tag{8.8}$$

The importance of the above lemma in the massive MIMO context can be better understood by considering the instantaneous correlation matrix $\boldsymbol{G}^H \boldsymbol{G}$ (or $\boldsymbol{G}^T \boldsymbol{G}^*$) which appears in the analysis of linear detectors (or precoders) in a massive MIMO uplink (or downlink). For the channel model in (8.2), we have that

$$\frac{1}{M} \boldsymbol{G}^H \boldsymbol{G} = \frac{1}{M} \boldsymbol{D}^{1/2} \boldsymbol{H}^H \boldsymbol{H} \boldsymbol{D}^{1/2} \overset{\text{a.s.}}{\to} \boldsymbol{D} \text{ as } M \to \infty, \tag{8.9}$$

which implies that in the limit $M \to \infty$, the effects of small-scale fading are averaged out. This phenomenon is widely referred to as *channel hardening* [10] and is a critical benefit of massive MIMO systems. In the following sections, we will elaborate on the performance of linear receivers (and precoders) in a massive MIMO uplink (and downlink) and see how favorable propagation can boost the performance of massive MIMO topologies while keeping the implementation complexity to affordable levels.

8.3   Performance of Linear Receivers in a Massive MIMO Uplink

We start our analysis from the input–output relationship in (8.1) and by assuming that the BS has perfect channel state information (CSI) for the sake of simplicity. In the case of linear reception, the received signal is diagonalized by premultiplying it by a linear receiver matrix $T^H \in \mathbb{C}^{K \times M}$, such that

$$r = T^H y = \sqrt{p_u} T^H Gx + T^H n. \tag{8.10}$$

Then, the kth element of r, which can be used to extract x_k, can be expressed as follows [11]:

$$r_k = \sqrt{p_u} t_k^H Gx + t_k^H n = \underbrace{\sqrt{p_u} t_k^H g_k x_k}_{\text{desired signal}} + \underbrace{\sqrt{p_u} \sum_{j=1, k \neq j}^{K} t_k^H g_j x_j}_{\text{interuser interference}} + \underbrace{t_k^H n}_{\text{effective noise}}, \tag{8.11}$$

where t_k denotes the kth column of T. Undoubtedly, the choice of the linear receiver matrix will substantially affect the end-to-end performance. In the literature, the three prevalent linear reception schemes are the following:

1. *Maximum-ratio combining (MRC)*. This scheme seeks to maximize the power at the receiver combiner to yield the maximum SNR. This can be achieved by appropriate phase-matching of the channel vector such that $t_k^{\text{mrc}} = g_k$ [12]. In this case, the received signal-to-interference-and-noise ratio (SINR) of the kth ($k = 1, \ldots, K$) stream becomes equal to

$$\text{SINR}_k^{\text{mrc}} = \frac{p_u \|g_k\|^4}{p_u \sum_{k \neq j}^{K} |g_k^H g_j|^2 + \|g_k\|^2}. \tag{8.12}$$

In general, MRC is considered as a very viable linear reception scheme for massive MIMO topologies, as it can be implemented in a distributed manner (each component of the received signal is simply multiplied by its conjugate). Moreover, this scheme performs particularly well in the low-power regime ($p_u \to 0$) and yields nearly optimal performance when the number of antennas grows large [11, 13]. Its downside is that it neglects the effects of interuser interference, thereby delivering very poor performance in the high-power regime ($p_u \to \infty$).

2. *Zero-forcing (ZF)*. The ZF scheme seeks to eliminate interuser interference by projecting the received signal stream into the orthogonal complement subspace of the interuser interference. Mathematically speaking, the reception matrix T^{zf} reads [14]

$$T_{\text{zf}}^H = (G^H G)^{-1} G^H. \tag{8.13}$$

After some basic mathematical manipulations, it can be shown that the received SINR of the kth ($k = 1, \ldots, K$) stream becomes equal to

$$\text{SINR}_k^{\text{zf}} = \frac{p_u}{\left[(G^H G)^{-1} \right]_{kk}} = \frac{p_u \det (G^H G)}{\det (G_k^H G_k)}. \tag{8.14}$$

Compared with MRC, the ZF scheme is superior in the high-power regime, where the effects of interuser interference are more dominant. However, the ZF scheme suffers from a "noise-enhancement effect," which makes it inappropriate for operation in the low-power regime and/or when the channel matrix is ill-conditioned (i.e., when the condition number of the channel Gram matrix $G^H G$ is much larger than 1). Finally, it induces higher implementation complexity compared with MRC owing to the inverse operation in (8.13).

3. *Minimum mean squared error (MMSE).* The MMSE receiver seeks to maximize the achievable SINR by minimizing the mean squared error between r and x in (8.10):

$$T_{\mathrm{mmse}} = \arg\min_{T} \mathcal{E}\left\{\left\|T^H y - x\right\|^2\right\}. \tag{8.15}$$

The solution to this optimization problem is well known in the literature [15] and is given by

$$T^H_{\mathrm{mmse}} = \left(G^H G + \frac{1}{p_{\mathrm{u}}} I_K\right)^{-1} G. \tag{8.16}$$

Then, the output SINR of the kth stream can be shown to be equal to

$$\mathrm{SINR}_k^{\mathrm{mmse}} = g_k^H \left(\frac{1}{p_{\mathrm{u}}} I_M + \sum_{j\neq k}^{K} g_j g_j^H\right)^{-1} g_k \tag{8.17}$$

$$= \frac{1}{\left[\left(I_K + p_{\mathrm{u}} G^H G\right)^{-1}\right]_{kk}} - 1 \tag{8.18}$$

$$= \frac{\det\left(I_K + p_{\mathrm{u}} G^H G\right)}{\det\left(I_{K-1} + p_{\mathrm{u}} G_k^H G_k\right)} - 1. \tag{8.19}$$

It is worth pointing out that the MMSE receiver always yields the best performance among the three reference linear reception schemes, as it exploits the second-order statistics of the channel. Its optimality in certain scenarios with perfect decision feedback was showcased in [16] among others, while a fundamental connection between the MMSE receiver and the capacity of an optimal MIMO receiver can be found in [15].

We will now compare the performance of the three linear reception schemes as a function of the number of antennas M. In our simulations, we assumed that $K = 10$ UTs were randomly dropped in a disk of radius 1000 m and the radius of the guard zone was $d_{\mathrm{g}} = 100$ m. The standard deviation of the log-normal shadowing was set to $\sigma_{\mathrm{sh}} = 6$ dB, and the path loss exponent was $v = 4$. Our performance measure was the total (ergodic) spectral efficiency (measurable in bit/s/Hz), which is defined as follows:*

$$R^{\mathcal{U}} = \sum_{k=1}^{K} \mathcal{E}\left\{\log_2\left(1 + \mathrm{SINR}_k^{\mathcal{U}}\right)\right\}, \tag{8.20}$$

where $\mathcal{U} = \{\mathrm{mrc, zf, mmse}\}$ and the expectation is taken over small- and large-scale fading realizations. Figure 8.1 depicts the spectral efficiency of MRC, ZF, and MMSE receivers for $p_{\mathrm{u}} = 10$ dB.

The performance of the ZF and MMSE receivers is identical even for moderate numbers of antennas, while that of the MRC receiver shows a slower scaling

---

* The achievable spectral efficiency in (8.20) assumes that the total interference-plus-noise term in (8.11) is modeled as Gaussian distributed independent of the transmit symbol x_k and with variance

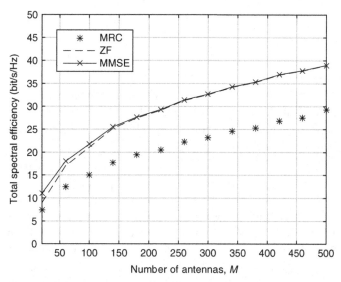

Figure 8.1 Total spectral efficiency of a massive MIMO uplink with $K = 10$ UTs, $p_u = 10$ dB, and three different linear reception schemes.

but, eventually, if will converge to the MMSE/ZF performance owing to the channel-hardening effect. This can be clearly seen by considering the power scaling strategy used in [11], where the transmit power was scaled down by the number of antennas M such that $p_u = E_u/M$, where E_u is a finite constant. In this case, it can be readily shown that the total achievable spectral efficiency of all linear reception schemes converges to the same value, given by:

$$R^{\mathcal{U}} \to \sum_{k=1}^{K} \log_2 \left(1 + E_u \beta_k\right), \quad M \to \infty. \tag{8.21}$$

This result is very important since it showcases the substantial power savings that are possible with massive MIMO topologies, while maintaining a satisfactory quality of service (QoS). It also corroborates our theoretical analysis in the previous sections and demonstrates the optimality of linear reception schemes when the number of antennas grows large. Note that very similar observations can be made for the downlink scenario as well.

8.4 Performance of Linear Precoders in a Massive MIMO Downlink

The performance of a massive MIMO downlink with perfect CSI and linear precoding at the BS can be characterized following the same line of reasoning as in Section 8.3. In particular, we refer back to (8.3) to express the preprocessed transmit vector $s \in \mathbb{C}^{M \times 1}$

$p_u \sum_{k \neq j}^{K} |t_k^H g_j|^2 + \|t_k\|^2$. This approach was first established in [17] and since then has been widely used in the MIMO literature.

as follows:

$$s = \sqrt{\lambda} \sum_{k=1}^{K} w_k z_k = \sqrt{\lambda} W z, \qquad (8.22)$$

where $W = [w_1, \ldots, w_K] \in \mathbb{C}^{M \times K}$ is the precoding vector and $z = [z_1, \ldots, z_k]^{\mathsf{T}} \in \mathbb{C}^{K \times 1}$ contains the data symbols for the K UTs. If we assume that the data symbols are all normalized to have unit power, i.e., $\mathcal{E}\{|z_k|^2\} = 1$, the constant λ satisfies the following long-term power constraint [18, 19]

$$\lambda = \frac{1}{\mathcal{E}\{\mathrm{tr}(WW^{\mathsf{H}})\}}. \qquad (8.23)$$

Then, it is straightforward to verify that $\mathcal{E}\{||s||^2\} = 1$, when the precoding matrix W maps the signals intended for the K UTs to be transmitted from the M antennas. Similarly to the uplink case, we focus on the following three linear precoding schemes: maximum-ratio transmission (MRT), ZF, and MMSE. In the first scheme, the BS aligns its transmission with the downlink channel in order to maximize the received power at the UT side, and $W^{\mathrm{mrt}} = G^*$. The ZF and MMSE precoding schemes are implemented in the same fashion as in the uplink case with an appropriate power normalization. Thus, it can be easily shown that

$$W^{\mathrm{zf}} = G^* \left(G^{\mathsf{T}} G^* \right)^{-1}, \qquad (8.24)$$

$$W^{\mathrm{mmse}} = G^* \left(G^{\mathsf{T}} G^* + \frac{K}{p_{\mathrm{d}}} I_K \right)^{-1}. \qquad (8.25)$$

8.5   Channel Estimation in Massive MIMO Systems

One of the big challenges in massive MIMO systems is the acquisition of CSI. In practice, CSI is not available at the receiver end and needs to be estimated using pilot sequences. Since the operation of massive MIMO is typically based on TDD operation, we can leverage the uplink–downlink reciprocity and make the number of pilot symbols analogous to the number of UTs K and independent of the number of BS antennas M.

It is a well-known fact that practical wireless channels experience multipath fading, and this renders their channel response varying in both time and frequency. For this reason, the estimation and payload transmission phases must fit into a time/frequency frame where the channels are approximately static. This resource frame spans T_c (s) and has a bandwidth of W_c (Hz) such that it can accommodate $\tau_c = T_c B_c$ symbols. Of the τ_c symbols, the first L symbols are reserved for uplink data training, while the remaining $\tau_c - L$ symbols are reserved for uplink and downlink data transmission.

We will now portray the different phases of channel estimation.

- *Uplink transmission.* In this phase, the K UTs transmit simultaneously orthogonal pilot sequences to the BS. Then, the BS estimates the uplink channels $\hat{g}_k, k = 1, \ldots, K$, based on the received pilot signals. The minimum duration of this phase is K symbols (or channel uses). Then, the UTs transmit their payload data to the BS (uplink transmission).

- *Downlink transmission.* Owing to channel reciprocity, knowledge of the uplink channel can be utilized by the BS to precode the downlink transmitted symbols. Another key advantage of massive MIMO is that the channel-hardening effect eliminates the need for downlink training, which can be performed either using the mean of the effective channel gain [20] or blindly at the UTs [21]. This incurs no overhead penalty in the downlink transmission.

8.5.1   Uplink Transmission

The physical modeling of the uplink transmission is done as follows. The K UTs first transmit mutually orthogonal pilot sequences of length $L \geq K$ symbols. These pilot sequences can be represented by a pilot matrix $\mathbf{\Pi} \in \mathbb{C}^{L \times K}$ with orthonormal columns, i.e., $\mathbf{\Pi}^H \mathbf{\Pi} = \mathbf{I}_K$. The average power of each pilot sequence is $p_p = L p_u$. Then, the received $M \times L$ pilot matrix at the BS is given by

$$Y_p = \sqrt{p_p} G \mathbf{\Pi}^T + N_p, \tag{8.26}$$

where N_p is the AWGN matrix, including i.i.d. $\mathcal{CN}(0,1)$ elements. The prevalent channel estimation scheme is MMSE [11, 18], which yields

$$\hat{G} = \frac{1}{\sqrt{p_p}} Y_p \mathbf{\Pi}^* \left(\frac{1}{p_p} D^{-1} + I_K \right)^{-1} \tag{8.27}$$

$$= \left(G + \frac{1}{\sqrt{p_p}} N_A \right) \left(\frac{1}{p_p} D^{-1} + I_K \right)^{-1}, \tag{8.28}$$

where $N_A = N \mathbf{\Pi}^*$, whose elements are again i.i.d. $\mathcal{CN}(0,1)$ owing to the unitary invariance of $\mathbf{\Pi}$. Let E be the channel estimation error of $\hat{G}$. Owing to the orthogonality principle used in the MMSE estimator, the channel matrix can be decomposed as follows:

$$G = \hat{G} + E, \tag{8.29}$$

where the entries of E are zero-mean Gaussian distributed random variables and the entries of the kth column have variance $\beta_k / (p_p \beta_k + 1)$. Likewise, the entries of the kth column of $\hat{G}$ are zero-mean Gaussian random variables with variance $p_p \beta_k^2 / (p_p \beta_k + 1)$. Most importantly, the matrices $\hat{G}$ and E are independent of each other.

We can now rewrite the expression for the uplink received signal vector in (8.10) to account for the channel estimation error. Note that the linear receiver matrix now utilizes the channel estimate $\hat{G}$ to diagonalize the channel. More specifically, we have that

$$\hat{r} = \hat{T}^H \left(\sqrt{p_u} \hat{G} + \sqrt{p_u} E + n \right), \tag{8.30}$$

which can be decomposed further to get the received signal at the kth user,

$$\hat{r}_k = \underbrace{\sqrt{p_u}\hat{t}_k^H\hat{g}_k x_k}_{\text{desired signal}} + \underbrace{\sqrt{p_u}\sum_{j=1,k\neq j}^{K}\hat{t}_k^H\hat{g}_j x_j}_{\text{interuser interference}} + \underbrace{\sqrt{p_u}\sum_{j=1}^{K}\hat{t}_k^H\hat{e}_j x_j}_{\text{estimation error}} + \underbrace{\hat{t}_k^H n}_{\text{effective noise}}, \tag{8.31}$$

where $\hat{t}_k$ and $\hat{e}_k$ are the kth columns of $\hat{T}$ and E, respectively. The achievable spectral efficiency of this MIMO system can be found in [11, Eq. (37)]. By elaborating on the massive MIMO regime when $M \to \infty$, it was found in [11] that the power scaling should not be as aggressive as in the case of perfect CSI. In fact, the power scaling can be at most $1/\sqrt{M}$, or the achievable spectral efficiency will converge to zero. An intuitive explanation is that as p_u decreases and, hence, p_p decreases too, the quality of the channel estimate deteriorates, which leads to a "squaring effect" on the sum spectral efficiency. This phenomenon was originally observed in the seminal paper [17]. Mathematically speaking, we have that the total spectral efficiency of all three reception schemes for $p_u = E_u/\sqrt{M}$ converges to

$$\bar{R}^u \to \frac{\tau_c - L}{\tau_c}\sum_{k=1}^{K}\log_2\left(1 + LE_u^2\beta_k^2\right), \quad M \to \infty, \tag{8.32}$$

where the prelog factor $(\tau_c - L)/\tau_c$ accounts for the pilot overhead.

8.5.2   Downlink Transmission

In the downlink transmission, the precoding matrix $W = f(\hat{G})$ will be a function of the channel estimate $\hat{G}$ and can admit any of the types of linear precoding outlined in Section 8.4. By combining (8.4) and (8.22), we can express the noisy signal received by the kth user $(k = 1, \ldots, K)$ as

$$\tilde{y}_k = \sqrt{p_d\lambda}g_k^T w_k z_k + \sqrt{p_d\lambda}\sum_{j=1,k\neq j}^{K}g_k^T w_j z_j + \epsilon_k, \tag{8.33}$$

where w_k is the kth column of W. The channel model in (8.33) implies that the users do not have any sort of instantaneous CSI. To overcome this limitation, we can rewrite (8.33) in the following way:

$$\tilde{y}_k = \underbrace{\sqrt{p_d\lambda}\mathcal{E}\left\{g_k^T w_k\right\} z_k}_{\text{desired signal}} + \underbrace{\tilde{\epsilon}_k}_{\text{effective noise}}, \tag{8.34}$$

where the "effective noise" is equal to

$$\tilde{\epsilon}_k = \sqrt{p_d\lambda}\left(g_k^T w_k - \mathcal{E}\left\{g_k^T w_k\right\}\right) z_k + \sqrt{p_d\lambda}\sum_{j=1,k\neq j}^{K}g_k^T w_j z_j + \epsilon_k. \tag{8.35}$$

Now, the expectation in (8.34) is assumed to be known as it depends only on the channel statistics, which change over a long timescale, and not on the instantaneous channels. In the massive MIMO context, such an assumption is very realistic owing to the channel-hardening effect, which renders $\mathbf{g}_k^{\mathrm{T}} w_k$ nearly deterministic. Moreover, the desired signal and effective noise are uncorrelated. Using this observation, combined with the technique developed in [17] (and used in the massive MIMO context in [20]), which states that the worst-case uncorrelated additive noise is independent Gaussian noise of the same variance, we can get the following achievable spectral efficiency of the kth user:

$$\tilde{R}_k = \log_2 \left(\frac{p_{\mathrm{d}}\lambda \left| \mathcal{E}\{\mathbf{g}_k^{\mathrm{T}} w_k\} \right|^2}{p_{\mathrm{d}}\lambda \, \mathrm{var}\left(\mathbf{g}_k^{\mathrm{T}} w_k\right) + p_{\mathrm{d}}\lambda \sum_{j=1,k\neq j}^{K} \mathcal{E}\{|\mathbf{g}_k^{\mathrm{T}} w_j|^2\} + 1} \right). \qquad (8.36)$$

8.6   Applications of Massive MIMO Technology

We will now describe two practical applications of massive MIMO topologies that boost the performance of full-duplex relaying [22] and joint wireless information and energy transfer (JWIET) [23] systems. These disruptive wireless technologies have been intensively researched over the past decade, since they can deliver substantial gains in data rates.

8.6.1   Full-Duplex Relaying with Massive Antenna Arrays

We consider the system model in Figure 8.2, where K communication pairs (S_k, D_k), $k = 1, \ldots, K$, share the same time–frequency resource and a common relay station, R. The kth source, S_k, communicates with the kth destination, D_k, via the relay station, which operates in a full-duplex mode. All source and destination nodes are equipped with a single antenna, while the relay station is equipped with N_{rx} receive antennas and N_{tx} transmit antennas.

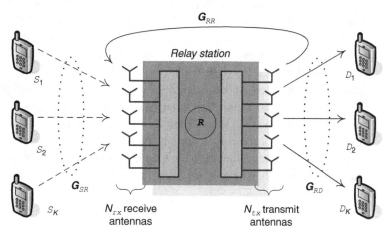

Figure 8.2 Multipair full-duplex relaying system with massive antenna arrays at the BS. Adapted from [22].

At time instant i, all K sources S_k, $k = 1, \ldots, K$, simultaneously transmit their signals $\sqrt{p_S}x_k[i]$ to the relay station, while the relay station broadcasts $\sqrt{p_R}s[i] \in \mathbb{C}^{N_{tx} \times 1}$ to all K destinations. Since the relay station receives and transmits at the same frequency, the received signal at the relay station is interfered with by its own transmitted signal, $s[i]$. This is called *loop interference*. We define $x[i] \triangleq [x_1[i], x_2[i], \ldots, x_K[i]]^\mathsf{T}$. The received signals at the relay station and the K destinations are respectively given by

$$y_R[i] = \sqrt{p_S}G_{SR}x[i] + \sqrt{p_R}G_{RR}s[i] + n_R[i], \qquad (8.37)$$

$$y_D[i] = \sqrt{p_R}G_{RD}^T s[i] + n_D[i], \qquad (8.38)$$

where $G_{SR} \in \mathbb{C}^{N_{rx} \times K}$ and $G_{RD}^T \in \mathbb{C}^{K \times N_{tx}}$ are the channel matrices from the K sources to the relay station's receive antenna array and from the relay station's transmit antenna array to the K destinations, respectively. The channel matrices account for both small-scale fading and large-scale fading. More precisely, G_{SR} and G_{RD} can be expressed as $G_{SR} = H_{SR}D_{SR}^{1/2}$ and $G_{RD} = H_{RD}D_{RD}^{1/2}$, where the small-scale fading matrices H_{SR} and H_{RD} have i.i.d. distributed $\mathcal{CN}(0,1)$ elements, while D_{SR} and D_{RD} are the large-scale fading diagonal matrices, whose kth diagonal elements are denoted by $\beta_{SR,k}$ and $\beta_{RD,k}$, respectively. Also, in (8.37), $G_{RR} \in \mathbb{C}^{N_{rx} \times N_{tx}}$ is the channel matrix between the relay's transmit and receive arrays, which represents the loop interference. We model the loop interference channel via the Rayleigh fading distribution, under the assumption that any line-of-sight component is efficiently reduced by antenna isolation and the major effect comes from scattering. Therefore, the elements of G_{RR} can be modeled as i.i.d. $\mathcal{CN}(0, \sigma_{LI}^2)$ random variables, where σ_{LI}^2 can be understood as the level of loop interference, which depends on the distance between the transmit and receive antenna arrays and/or the capability of the hardware loop interference cancellation technique.

In our analysis in [22], we considered MMSE channel estimation at the relay with $2K$ symbols along with two types of linear processing at the relay, namely ZF/ZF and MRT/MRC. The main idea in [22] was to investigate the potential of using massive antenna arrays at the relay station to eliminate the effects of loop interference. The loop interference can be canceled out by projecting it onto its orthogonal complement. However, this orthogonal projection may harm the desired signal. Yet, when N_{rx} is large, the subspace spanned by the loop interference is nearly orthogonal to the desired signal's subspace and, hence, the orthogonal projection scheme will perform very well. The next question is, "How to project the loop interference component?" It is interesting to observe that, when N_{rx} grows large, the channel vectors of the desired signal and the loop interference become nearly orthogonal. Therefore, a ZF or MRC receiver can act as an orthogonal projection of the loop interference. As a result, the loop interference can be reduced significantly by using a large N_{rx} together with a ZF or MRC receiver.

Lemma 8.2 *Assume that the number of source–destination pairs, K, is fixed. For any finite N_{tx} or for any N_{rx}, such that N_{rx}/N_{tx} is fixed, as $N_{rx} \to \infty$, the received signal* *at the relay station for decoding the signal transmitted from* S_k *is given by*

$$r_k[i] \overset{\text{a.s.}}{\to} \sqrt{p_S} x_k[i] \ \text{for ZF,} \tag{8.39}$$

$$\frac{r_k[i]}{N_{rx}\sigma^2_{SR,k}} \overset{\text{a.s.}}{\to} \sqrt{p_S} x_k[i] \ \text{for MRC, MRT,} \tag{8.40}$$

where $\sigma^2_{SR,k}$ *is the variance of the kth column of the channel estimate* $\hat{G}_{SR}$.

The result in the above lemma indicates that when N_{rx} grows to infinity, the loop interference can be canceled out. Furthermore, the interpair interference and noise effects also disappear. The received signal at the relay station after using a ZF or MRC receiver includes only the desired signal and, hence, the capacity of the communication link $S_k \to R$ grows without bound. As a result, the system performance is limited only by the performance of the communication link $R \to D_k$, which does not depend on the loop interference. In Figure 8.3, we compare the performance of half-duplex and full-duplex relaying as a function of the number of antennas at the relay station. The simulation parameters resembled those of [22, p. 1732], namely $K = 10$, $p_R = p_S = p_P = 10$ dB (where p_P is the pilot power), the duration of the training interval was $2K$, and the loop interference level was $\sigma^2_{LI} = 10$ dB. The large-scale fading coefficients were taken as $\beta_{SR,k} = \beta_{RD,k} = 1, \forall k = 1, \ldots, K$.

Recall that half-duplex relaying requires two orthogonal time slots for two transmissions: from the sources to the relay station and from the relay station to the destinations. The half-duplex mode does not induce loop interference, at the cost of imposing a prelog factor $(1/2)$ on the spectral efficiency. The results show that by using large antenna arrays at the relay station, the effect of loop interference can be effectively suppressed and the performance of full-duplex relaying is better than that of half-duplex relaying in the vast majority of cases. More specifically, we can see that, for MRC/MRT, the full-duplex mode is always better than the half-duplex mode, while for ZF, depending on the large-scale fading, full-duplex can be better than half-duplex relaying or vice versa. Another important observation is that in both cases of linear processing, the achievable spectral efficiency scales logarithmically with the number of receive (or transmit) antennas. This is consistent with the classical results on massive MIMO [5, 11, 18].

8.6.2   Joint Wireless Information Transfer and Energy Transfer for Distributed Massive MIMO

In this section, we first describe the general architecture of the distributed massive MIMO system employed for JWIET, which is illustrated on the left-hand side of Figure 8.4. Multiple antennas of a BS, called remote antenna units (RAUs), are distributed over the cell. All RAUs within a cell are connected to a baseband processing unit (BPU) using high-quality bidirectional wired (e.g., radio over fiber) or wireless (e.g., microwave repeater) links. Usually, the RAUs are functionally simple and represent the transceivers of the BSs, whereas the centralized BPU has all the baseband

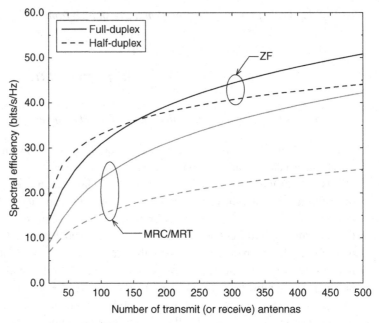

Figure 8.3 Sum spectral efficiency versus the number of transmit (or receive) antennas for half-duplex and full-duplex relaying. Adapted from [22].

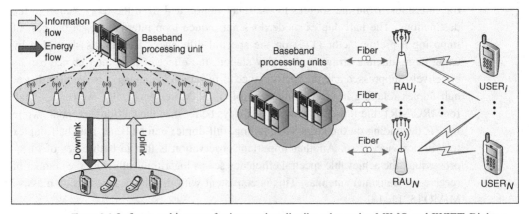

Figure 8.4 Left: an architecture for integrating distributed massive MIMO and JWIET. Right: a multiuser system model of downlink wireless energy transfer for distributed massive MIMO.

processing capability of a typical BS. The UEs can access energy and/or information services upon request.

We then describe a practical application of JWIET which uses a single-cell system model for a multiuser downlink wireless energy transfer (WET) scenario, as shown on the right-hand side of Figure 8.4. It is assumed that the distributed massive MIMO system has $M > 1$ RAUs and $K > 1$ UTs, which are all equipped with a single antenna. We denote the channel between the mth RAU and the kth UT by $g_{m,k} = h_{m,k}\sqrt{\beta_{m,k}}$. For the kth user, the channel vector is given by $\mathbf{h}_k = \left[h_{1,k},\ldots,h_{M,k}\right]^{\mathrm{T}}$, indicating the small-scale fading, with each entry assumed to have zero mean and unity variance, and $\beta_{m,k}$ accounts for the large-scale fading and can be expressed as

$$\beta_{m,k} = \frac{1}{d_{m,k}^v}, \tag{8.41}$$

where $d_{m,k}^v$ is the distance between the mth RAU and the kth UT, and v is the path loss exponent. The transmitted signal is given by $\mathbf{x} = \sum_{k=1}^{K} \mathbf{w}_k s_k$, where the transmitted data symbol s_k for UT_k is an i.i.d. Gaussian random variable with zero mean and unit variance, i.e., $s_k \sim \mathcal{N}(0,1), \forall k$. Since electromagnetic energy is propagated isotropically if the transmit antenna is isotropic, the UT harvests only a portion of the transmitted energy without specific control, resulting in a low WET efficiency. Distributed energy beamforming enables a cluster of distributed energy sources to cooperatively emulate an antenna array by transmitting RF energy simultaneously in the same direction to an intended energy harvester for improved diversity gain. Thus, we define $\mathbf{w}_k$ here as an $M \times 1$ transmit beamforming vector for UT_k. Moreover, we define $R_{\mathbf{w}} = \sum_{k=1}^{K} \mathbf{w}_k \mathbf{w}_k^{\mathrm{H}}$, such that the per-RAU power constraint corresponds to $[R_{\mathbf{w}}]_{m,m} = Q_m \leq p_{\max}$, $m = 1,\ldots,M$. Hence, the received signal at UT_k can be expressed as

$$y_k = \mathbf{g}_k^{\mathrm{H}} \sum_{j=1}^{K} \mathbf{w}_j s_j + n_k, \tag{8.42}$$

where n_k is the AWGN term, with power σ_k^2. As a result, the achievable spectral efficiency of the kth UT is given by

$$R = \log_2 \left(1 + \frac{\left| \mathbf{g}_k^{\mathrm{H}} \mathbf{w}_k \right|^2}{\sum_{j \neq k} \left| \mathbf{g}_k^{\mathrm{H}} \mathbf{w}_j \right|^2 + \sigma_k^2} \right). \tag{8.43}$$

However, for WET, the interuser interference can be taken as a kind of energy harvesting source, leading to a totally different metric compared with wireless information transfer (WIT). Thus, the harvested power at UT_k is given as

$$Q_k = \xi \left(\sum_{j=1}^{K} \left| \mathbf{g}_k^{\mathrm{H}} \mathbf{w}_j \right|^2 + \sigma_k^2 \right), \tag{8.44}$$

where $\xi \in (0,1]$ stands for the energy conversion efficiency. We compare the performance of four different scenario designs in Figure 8.5:

- *CAS-MFBF.* Traditional centralized antenna systems (CAS) with matched-filter beamforming (MFBF).
- *DAS-MFBF.* Distributed antenna systems (DAS) with MFBF.
- *NRS-3-MFBF.* The cost of the RF chains, which include power amplifiers, microprocessors, and analog-to-digital converters (ADCs), associated with the RAUs might make distributed massive MIMO an expensive technology. Moreover, a large backhaul infrastructure between the cells is required to collect the large amount of channel information, which represents another fundamental limitation of distributed

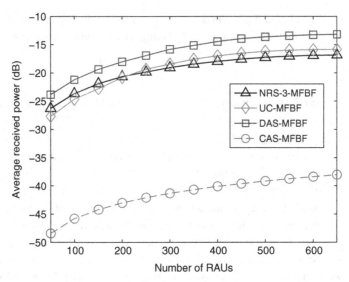

Figure 8.5 WET performance versus the number of RAUs for four different scenario designs.

massive MIMO. RAU selection is an effective approach to solving these problems
while preserving the diversity and multiplexing gains. NRS-3-MFBF is a scheme
which selects the nearest three RAUs that have MFBF.
- *UC-MFBF.* There may be advantages if some of the transmission points collaboratively serve a single UT, while others collaborate to serve different UTs. This
is termed a user-centric (UC) architecture. UC-MFBF denotes selecting an area of
RAUs (which support the best energy services for users) that have MFBF.

The RAUs were uniformly distributed in the cell and their number varied from 50
to 600, serving users placed in a series of circles of radius 15 m, 40 m, 65 m, and
90 m: each circle contained three users at equal distances, so that there was a total of
12 users. The carrier frequency was 5 GHz, and the noise power was -70 dBm. We
assumed a Rayleigh fading model for the propagation paths, such that $h_{m,k} \sim \mathcal{CN}(0,1)$,
and set the large-scale fading to be $\beta_{m,k} = 10^{-3}d_{m,k}^{-3}$, where the path loss exponent was
3. For fair comparison, the total number of RAUs, total transmit power, and coverage
area were assumed to be the same in all cases. As can be seen from the figure, a large
gain in the average received power could be achieved by deploying DASs rather than
CASs, which means that the transmit power efficiency of DASs is larger than that of
CASs as we expected. Although NRS-3 is not the optimal scheme, it greatly reduces
the hardware costs. The UC approach performs poorly for small numbers of RAUs and
improves when the number of RAUs becomes large; this is due to the fact that the RAUs
which can be selected by the users are few or even nonexistent at the very start, and start
increasing with the total number of RAUs. Overall, the energy harvested in the DASs is
higher than in the CASs as we expected, demonstrating that distributed massive MIMO
is more suitable for WET than is CAS.

8.7   Open Future Research Directions

The area of massive MIMO has been intensively explored over the past five years, ever since the pioneering work of Marzetta was published [5]. However, there is a plethora of research challenges that require further investigation in the coming years before massive MIMO is incorporated into practical wireless systems. In this section, we will set forth the most formidable challenges and review the first related efforts in the literature.

1. *Massive MIMO with frequency division duplex (FDD) operation.* The development of massive MIMO has been intrinsically based on TDD operation in order to make use of the uplink–downlink reciprocity. This makes the amount of pilot overhead proportional to the number of UTs and not to the number of BS antennas. Yet, there are practical system imperfections that limit the performance gains of TDD massive MIMO systems, such as calibration errors in the uplink/downlink RF chains. Moreover, FDD dominates current wireless cellular systems. Unfortunately, in FDD massive MIMO systems, the overhead of downlink training with orthogonal pilots and also the number of feedback bits increase linearly with the number of BS antennas. This renders the pilot overhead of FDD systems prohibitive for practical massive MIMO operation. Consequently, there have been some tentative steps toward minimizing the pilot overhead of FDD massive MIMO by exploiting the sparse structure of the propagation channels [24]–[26]. In essence, the key idea is to partition users with similar spatio-temporal characteristics (similar covariance matrices) and split the beamforming into a pre-beamforming and a multiuser MIMO precoder. After the pre-beamforming stage, the training dimensions are significantly reduced. In a recent study [26], a complete transmission scheme suitable for massive FDD-based topologies with only statistical CSI at the transmitter was proposed.

2. *Antenna design for massive MIMO systems.* Massive MIMO induces new challenges in the design and fabrication of antenna arrays, since a large number of antennas needs to be packed within a finite volume. Such topologies will suffer from increased electromagnetic mutual coupling and spatial correlation. To this end, a preliminary study [27] has demonstrated that for a fixed total physical space for the massive MIMO array, the assumption of favorable propagation is violated and the channel vectors do not become orthogonal. As a consequence, and contrary to ongoing studies, interuser interference does not vanish in the massive MIMO regime, thereby creating a saturation effect in the achievable rate with MRT precoding. The design of two- and three-dimensional massive antenna arrays, which also make use of propagation in the elevation domain, is a viable candidate for overcoming such limitations but this requires intensive research activity and real-time channel measurements [28].

3. *Massive MIMO at millimeter wave (mmWave) frequencies.* The technological combination of massive MIMO and mmWave systems has given birth to mmWave massive MIMO [3], which offers a broad spectrum of advantages: (a) delivery of multiplexing and array gains due to multiple transmit and receive antennas, (b) very high data rates due to the vast bandwidth that is available at millimeter frequencies,

and (c) reduced interference due to narrow beamforming. Yet, the practical design of mmWave massive MIMO faces many fundamental challenges. To begin with, with the current 4G system architectures, future massive MIMO BSs will deploy hundreds of antenna units, each connected to a dedicated RF chain. Typically, a transmit RF chain consists of a digital-to-analog converter, a modulator, an upconverter, a frequency synthesizer, mixers, and a power amplifier, with similar circuitry on the receive side. Such topologies are likely to scale badly into the mmWave regime, thereby creating new challenges with respect to energy consumption, electronic-circuit area, and signal-processing algorithms, and requiring much closer collaboration between communications and microwave engineers. Furthermore, the advances in mmWave chip fabrication have significantly reduced the cost of the electronics but the prohibitive power consumption of mixed-signal hardware is still the biggest bottleneck; for instance, today's commercial high-speed (e.g., more than 1 GSample/s), high-precision (e.g., 8–12 bit) ADCs are costly and power-hungry components, with a power consumption of few watts per unit [29].

4. *Hybrid processing in massive MIMO.* One solution to keeping the energy consumption, implementation cost, and complexity of massive MIMO at affordable levels is to perform analog beamforming, operating in the RF domain, which requires a much smaller number of baseband ADCs, instead of power-hungry digital signal processing. Unfortunately, complete analog beamforming still has some critical shortcomings, since it lacks flexibility and adaptability, sacrifices the overall system performance, and entails reliability issues in hardware design, especially for mmWave signals. A viable alternative is to perform a portion of the processing in the baseband (digital) and the remaining portion in the RF band (analog) with a reduced number of RF chains (compared with conventional MIMO systems). Such topologies can (a) offer a new hardware dimension for finding the optimal balance between analog and digital processing; (b) reduce the complexity and power consumption of the signal-processing algorithms for precoding/decoding, since the dimensions of the matrices will be smaller; and (c) reduce the complexity of the circuitry and the implementation cost as the number of RF chains will be much less than the number of antennas. For all these reasons, the area of hybrid processing for massive MIMO has received increasing interest over the past years from academia and industry (see [30–32] and references therein, among others).

8.8   Conclusion

This chapter has provided a general description of distributed massive MIMO, which is a very promising candidate for 5G wireless networks. The extremely high number of antennas offers an equivalently high number of degrees of freedom, which can be used to focus energy into desired directions and improve the end-to-end energy efficiency. Our analysis covered the uplink (and downlink) of single-cell massive MIMO and showed that the use of very simple linear receivers (or precoders) achieves very satisfactory performance. This is due to the channel-hardening effect, which effectively makes the channel vectors from (or to) different UTs nearly orthogonal as the number of BS antennas grows large. The acquisition of CSI in TDD massive MIMO was also presented in detail, followed by two practical applications of massive MIMO, namely full-duplex relaying with massive antenna arrays and joint wireless information and energy transfer. In both scenarios, we observed that the large number of antennas offers an unprecedented boost in spectral efficiency and remarkable power savings. Finally, we described some directions for future research in the context of massive MIMO that require close collaboration between academia and industry in the foreseeable future.

References

[1] Cisco, "Cisco visual networking index: Global mobile data traffic forecast update, 2014–2019," White paper.

[2] J. G. Andrews, S. Buzzi, W. Choi, S. V. Hanly, A. Lozano, A. C. K. Soong, and J. C. Zhang "What will 5G be?" *IEEE J. Sel. Areas Commun.*, vol. 32, no. 6, pp. 1065–1082, Jun. 2014.

[3] A. L. Swindlehurst, E. Ayanoglu, P. Heydari, and F. Capolino, "Millimeter-wave massive MIMO: The next wireless revolution?" *IEEE Commun. Mag.*, vol. 52, no. 9, pp. 56–62, Sep. 2014.

[4] M. Matthaiou, G. K. Karagiannidis, E. G. Larsson, T. L. Marzetta, and R. Schober, "Guest editorial: Large-scale multiple antenna wireless systems," *IEEE J. Sel. Areas Commun.*, vol. 31, no. 2, pp. 113–116, Feb. 2013.

[5] T. L. Marzetta, "Noncooperative cellular wireless with unlimited numbers of base station antennas," *IEEE Trans. Wireless Commun.*, vol. 9, no. 11, pp. 3590–3600, Nov. 2010.

[6] F. Rusek, D. Persson, B. K. Lau, E. G. Larsson, T. L. Marzetta, O. Edfors, and F. Tufvesson, "Scaling up MIMO: Opportunities and challenges with very large arrays," *IEEE Signal Process. Mag.*, vol. 30, pp. 40–60, Jan. 2013.

[7] E. G. Larsson, O. Edfors, F. Tufvesson, and T. L. Marzetta, "Massive MIMO for next generation wireless systems," *IEEE Commun. Mag.*, vol. 52, pp. 186–195, Feb. 2014.

[8] M. Matthaiou, C. Zhong, M. R. McKay, and T. Ratnarajah, "Sum rate analysis of ZF receivers in distributed MIMO systems," *IEEE J. Sel. Areas Commun.*, vol. 31, no. 2, pp. 180–191, Feb. 2013.

[9] H. Q. Ngo, E. G. Larsson, and T. L. Marzetta, "Aspects of favorable propagation in massive MIMO," in *Proc. of European Signal Processing Conf. (EUSIPCO)*, Sep. 2014.

[10] T. L. Marzetta and B. M. Hochwald, "Capacity of a mobile multiple-antenna communication link in Rayleigh flat fading," *IEEE Trans. Inf. Theory*, vol. 45, no. 1, pp. 139–157, Jan. 1999.

[11] H. Q. Ngo, E. G. Larsson, and T. L. Marzetta, "Energy and spectral efficiency of very large multiuser MIMO systems," *IEEE Trans. Commun.*, vol. 61, no. 4, pp. 1436–1449, Apr. 2013.

[12] P. A. Dighe, R. K. Mallik, and S. S. Jamuar, "Analysis of transmit–receive diversity in Rayleigh fading," *IEEE Trans. Commun.*, vol. 51, no. 4, pp. 694–703, Apr. 2003.

[13] H. Q. Ngo, M. Matthaiou, and E. G. Larsson, "Massive MIMO with optimal power and training duration allocation," *IEEE Wireless Commun. Lett.*, vol. 3, no. 6, pp. 606–608, Dec. 2014.

[14] M. Matthaiou, C. Zhong, and T. Ratnarajah, "Novel generic bounds on the sum rate of MIMO ZF receivers," *IEEE Trans. Signal Process.*, vol. 59, no. 9, pp. 4341–4353, Sep. 2011.

[15] M. R. McKay, I. B. Collings, and A. M. Tulino, "Achievable sum rate of MIMO MMSE receivers: A general analytic framework," *IEEE Trans. Inf. Theory*, vol. 56, no. 1, pp 396–410, Jan. 2010.

[16] T. Guess and M. K. Varanasi, "An information-theoretic framework for deriving canonical decision-feedback receivers in Gaussian channels," *IEEE Trans. Inf. Theory*, vol. 51, no. 1, pp. 173–187, Jan. 2005.

[17] B. Hassibi and B. M. Hochwald, "How much training is needed in multiple-antenna wireless links?" *IEEE Trans. Inf. Theory*, vol. 49, no. 4, pp. 951–963, Apr. 2003.

[18] J. Hoydis, S. ten Brink, and M. Debbah, "Comparison of linear precoding schemes for downlink massive MIMO," in *Proc. of IEEE International Conf. on Communications (ICC)*, Jun. 2012.

[19] H. Q. Ngo, "Massive MIMO: Fundamentals and system designs," Ph.D. dissertation, Department of Electronic Engineering, Linköping University, Linköping, Sweden, 2015.

[20] J. Jose, A. Ashikhmin, T. L. Marzetta, and S. Vishwanath, "Pilot contamination and precoding in multi-cell TDD systems," *IEEE Trans. Wireless Commun.*, vol. 10, no. 8, pp. 2640–2651, Aug. 2011.

[21] H. Q. Ngo and E. G. Larsson, "Blind estimation of effective downlink channel gains in massive MIMO," in *Proc. of IEEE International Conf. on Acoustics, Speech, and Signal Processing (ICASSP)*, Apr. 2015.

[22] H. Q. Ngo, H. A. Suraweera, M. Matthaiou, and E. G. Larsson, "Multipair full-duplex relaying with massive arrays and linear processing," *IEEE J. Sel. Areas Commun.*, vol. 32, no. 9, pp. 1721–1737, Oct. 2014.

[23] F. Yuan, S. Jin, Y. Huang, K.-K. Wong, Q. T. Zhang, and H. Zhu, "Joint wireless information and energy transfer in massive distributed antenna systems," *IEEE Commun. Mag.*, vol. 53, no. 6, pp. 109–116, Jun. 2015.

[24] J. Choi, D. J. Love, and P. Bidigare, "Downlink training techniques for FDD massive MIMO systems: Open-loop and closed-loop training with memory," *IEEE J. Sel. Top. Signal Process.*, vol. 8, no. 5, pp. 802–814, Mar. 2014.

[25] J. Nam, A. Adhikary, J.-Y. Ahn, and G. Caire, "Joint spatial division and multiplexing: Opportunistic beamforming, user grouping and simplified downlink scheduling," *IEEE J. Sel. Top. Signal Process.*, vol. 8, no. 5, pp. 876–890, Mar. 2014.

[26] C. Sun, X. Gao, S. Jin, M. Matthaiou, Z. Ding, and C. Xiao, "Beam division multiple access transmission for massive MIMO communications," *IEEE Trans. Commun.*, vol. 63, no. 6, pp. 2170–2184, Jun. 2015.

[27] C. Masouros and M. Matthaiou, "Space-constrained massive MIMO: Hitting the wall of favorable propagation," *IEEE Commun. Lett.*, vol. 19, no. 5, pp. 771–774, May. 2015.

[28] X. Gao, O. Edfors, F. Rusek, and F. Tufvesson, "Massive MIMO performance evaluation based on measured propagation data," *IEEE Trans. Wireless Commun.*, vol. 14, no. 7, pp. 3899–3911, Jul. 2015.

[29] C. H. Doan, S. Emami, D. A. Sobel, A. M. Niknejad, and R. W. Brodersen, "Design considerations for 60 GHz CMOS radios," *IEEE Commun. Mag.*, vol. 42, no. 12, pp. 132–140, Dec. 2004.

[30] S. Hur, T. Kim, D. J. Love, J. V. Krogmeier, T. A. Thomas, and A. Ghosh, "Millimeter wave beamforming for wireless backhaul and access in small cell networks," *IEEE Trans. Commun.*, vol. 61, no. 10, pp. 4391–4403, Oct. 2013.

[31] O. E. Ayach, S. Rajagopal, S. Abu-Surra, Z. Pi, and R. W. Heath, Jr., "Spatially sparse precoding in millimeter wave MIMO systems," *IEEE Trans. Wireless Commun.*, vol. 13, no. 3, pp. 1499–1513, Mar. 2014.

[32] T. E. Bogale and L. B. Le, "Beamforming for multiuser massive MIMO systems: Digital versus hybrid analog–digital," in *Proc. of IEEE Global Communications Conf. (GLOBECOM)*, Dec. 2014.

9 Full-Duplex Protocol Design for 5G Networks

Taneli Riihonen and Risto Wichman

9.1 Introduction

This chapter concerns the usage of *in-band full-duplex* transceivers in the emerging fifth generation (5G) wireless systems. The specific focus of the treatment of the topic is on the design and analysis of wideband physical layer and link layer data transmission protocols for heterogeneous networks, i.e., how to take advantage of the capability for simultaneous transmission and reception on the same frequency band provided by recent developments in self-interference cancellation.

First of all, we modernize the basic definition of the term 'full-duplex' such that it is applicable to 5G systems and consistent with the research community's new conception of the terminology. The technical challenges of the full-duplex technology are the self-interference and co-channel interference that are the price of its main advantage: improved spectral efficiency by frequency reuse. Here, we characterize the fundamental rate–interference trade-off imposed by the choice between full- and half-duplex modes in a primitive three-terminal system.

Thereafter, our discussion proceeds to the design and analysis of full-duplex transmission protocols for 5G communication links. The benefits of spectrum reuse by employing the full-duplex mode in this case are reduced by any kind of asymmetry in the system. In particular, the requested input and output data rates to and from a transceiver are not necessarily equal, and there may be imbalances between corresponding channels or even within them in wideband transmission.

The remainder of this chapter is organized as follows. Section 9.2 serves as an introduction to the basics of in-band full-duplex operation, namely its definition, purpose, challenges, and advantages at large. Section 9.3 discusses the design of full-duplex transmission protocols for heterogeneous 5G networks, identifying the types of communication links for which the technology could be suitable. Section 9.4 analyzes the performance of full-duplex transmission over wideband fading channels in typical asymmetric communication scenarios. Finally, Section 9.5 concludes the discussion with some visions of prospective future research directions and the schedule for adopting full-duplex technology for commercial use.

9.2   Basics of Full-Duplex Systems

9.2.1   In-Band Full-Duplex Operation Mode

Over the last few years, the research community working on state-of-the-art wireless concepts has been realigning its conception of some very basic terminology related to bidirectional communication and the operation modes of transceivers. In particular, the notion of *duplex(ing)* has undergone a complete transformation. This terminology change stems from the need to resolve ambiguity problems that appear when using classic terms to discuss modern complex networks: When using the old terminology, one needs always to specify explicitly, for instance, which Open Systems Interconnection (OSI) layer an operation mode corresponds to, and whether the two communication directions may use different transmission media. In fact, even the Radio Communication sector of the International Telecommunication Union (ITU-R) characterizes some of the related key terms as "deprecated" in the integrated database of ITU terms and definitions.

The core of the new emerging terminology for duplexing is the following adaptation of the ITU's original classic definition (cf. Figure 9.1).

Definition 9.1 *Full-duplex (FD) designates or pertains to a mode of operation by which information is transmitted to and from a transceiver* in two directions simultaneously on the same physical channel. [1]*

Conversely, in the half-duplex (HD) mode, information is transmitted to and from a transceiver in two directions as well, but not simultaneously on the same physical channel. To avoid confusion, one needs to recognize that conventional time division duplexing (TDD) and, especially, frequency division duplexing (FDD) are both actually regarded as HD modes in this new terminology. During the transition period, some authors have thus included the adjunct "in-band" before the term "full-duplex" in order to emphasize that they are considering the "new" full-duplex mode in contrast to, for example, FDD which could be classified as an FD mode according to the old terminology. We will consistently use the new terminology in the remainder of this chapter, and hence it is unnecessary to repetitiously emphasize the physical layer in-band perspective.

The new duplex terminology is unambiguous and is suitable for discussing modern wireless concepts. Firstly, the physical layer perspective establishes an inherent link to spectral efficiency by closing a "loophole" in the old definition: true FD operation as per the new definition compels one to measure the genuine aggregate data rate of a communication system per total bandwidth of allocated frequencies. For example, FDD links that use separate frequency bands, or even totally different transmission schemes, for bidirectional communication are subject to spectral efficiency loss owing to the fact that information flows in one direction only over each allocated band. Secondly, the focus of the terminology is now on the operation mode of any transceiver

---

* The definition in [1] actually uses the word "relay" instead of the word "transceiver" owing to the scope of the document, whereas the current definition is valid for all kinds of applications.

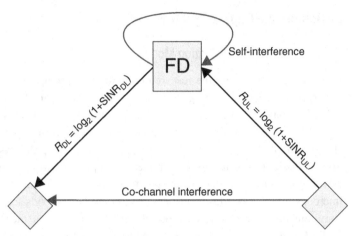

Figure 9.1 According to the emerging modern conception, the terms "FD" and "HD" specify the operation mode of a radio transceiver rather than that of a bidirectional link between two terminals. In particular, an FD transceiver transmits downlink (DL) signals and receives uplink (UL) signals simultaneously on the same channel, which causes interference, whereas a corresponding HD transceiver separates its DL and UL signals in time and/or in frequency to avoid interference.

instead of on bidirectional communication between exactly two points, and for this reason the definition is applicable to modern communication systems that are large, heterogeneous networks by their nature. For example, the old duplex terminology becomes cumbersome to use even in the case of the simple three-terminal network shown in Figure 9.1 because, strictly speaking, each individual point-to-point link in this network operates in a simplex, i.e., unidirectional, mode (as opposed to a duplex, i.e., bidirectional, mode).

9.2.2   Self-Interference and Co-channel Interference

We can summarize the discussion in the above section by remarking that, in plain words, a full-duplex transceiver is an advanced radio device that is (made) capable of simultaneous transmission and reception (STAR) on the same center frequency, or band. This progressive and valuable STAR capability does not come for free by any means, but at the cost of inevitable interference, though, as shown in Figure 9.1. Namely, transceivers operating in networks that take advantage of the FD mode need to implement special transmitter and receiver processing for interference mitigation and be robust to interference or its residue by design.

Firstly, STAR operation inevitably causes a feedback loop where signals transmitted by an FD radio leak back to its own receiver front end (see Figure 9.1); this phenomenon is referred to as self-interference (SI). In principle, cancellation is a straightforward task: the FD transceiver should know what it is transmitting and the loopback channel can be estimated, so that the SI component could be mathematically just subtracted away from the received sum signal. However, in progressing from this initial conception toward a more practical direction, researchers in the field soon noticed that the cancellation task is not so easy in reality but actually is the main technical problem to be solved first. In particular, the transceiver electronics are imperfect, such that a terminal always transmits a distorted version of what it ideally intended to transmit, and, at the receiver, the sum signal is subject to nonlinear distortion before digital processing can take place. Since the research on SI cancellation has now reached a mature state, this chapter focuses on generic scenarios where FD transceivers operate under residual SI without assuming any specific implementation. For more details, see, for example, [2, 7].

Secondly, FD operation creates new types of co-channel interference [8] at the network level because more transmitters are simultaneously active than in conventional HD systems. For example, in Figure 9.1, the uplink transmission from one terminal interferes with the downlink transmission to the other terminal. It may turn out that a significant problem in 5G systems will be powerful "cross-slot" interference between base stations in different cells, which is avoided with FDD (and TDD as well, if the downlink and uplink slots are synchronized throughout the network). Except for their existence, these phenomena are not physically different from other co-channel interference issues in conventional wireless systems, though. Moreover, if an FD transceiver operates as a relay that forwards information between terminals, the co-channel crosstalk may actually become useful.

9.2.3   Full-Duplex Transceivers in Communication Links

In essence, the full-duplex operation mode is no more than one spectrum-reuse concept among many others: it allows a system to schedule two information flows simultaneously on the same time–frequency channel resource at the cost of extra interference. Let us consider next the simple system of Figure 9.1 with a generic wireless transceiver that is capable of both FD and HD operation, to highlight the potential of the concept in comparison with conventional TDD and FDD. In particular, the transceiver device at hand is transmitting data to some receiver device and receiving data from some transmitter device; these communication directions can be referred to as the DL and UL, respectively, of the system from the transceiver's perspective.

Let SNR_{DL} and SNR_{UL} denote the signal-to-noise ratios (SNRs) achieved in each communication direction in the HD operation mode, where the corresponding data signals are separated in time or in frequency. In the FD operation mode, both the DL and the UL of the system are subject to (residual) self-interference and/or co-channel interference because the data signals overlap in time and in frequency. The signal-to-interference-plus-noise ratios (SINRs) achieved in each direction in the FD mode are denoted by $SINR_{DL}$ and $SINR_{UL}$. It is reasonable to presume that $SINR_{DL} <$ SNR_{DL} and $SINR_{UL} < SNR_{UL}$, or even that $SINR_{DL} \ll SNR_{DL}$ and $SINR_{UL} \ll SNR_{UL}$, since the signal overlap is mostly harmful and cannot be eliminated perfectly in practice.

For the purpose of illustration, let us then idealize the system and translate the S(I)NRs to corresponding transmission rates (bit/s/Hz) achieved by the links using the

Shannon capacity formula as follows:

$$R_{DL} = \begin{cases} \log_2(1 + SNR_{DL}) & \text{in the HD mode,} \\ \log_2(1 + SINR_{DL}) & \text{in the FD mode,} \end{cases} \tag{9.1a}$$

$$R_{UL} = \begin{cases} \log_2(1 + SNR_{UL}) & \text{in the HD mode,} \\ \log_2(1 + SINR_{UL}) & \text{in the FD mode.} \end{cases} \tag{9.1b}$$

Consequently, the spectral efficiency (bit/s/Hz) achieved by the system, i.e., the sum of the DL and UL rates, can be expressed as follows for each operation mode:

$$R_{HD} = \delta_{DL} \log_2(1 + SNR_{DL}) + \delta_{UL} \log_2(1 + SNR_{UL}), \tag{9.2a}$$

$$R_{FD} = \log_2(1 + SINR_{DL}) + \log_2(1 + SINR_{UL}), \tag{9.2b}$$

where δ_{DL} and δ_{UL} denote the relative amounts of time–frequency channel resources allocated to each communication direction when the HD mode is used, such that $0 \leq \delta_{DL} \leq 1$, $0 \leq \delta_{UL} \leq 1$, and $0 \leq \delta_{DL} + \delta_{UL} \leq 1$; the FD mode is free from such resource division, owing to spectrum reuse.

Remark 9.1. The ultimate potential of FD operation can be highlighted by considering the ideal symmetric case where $SNR = SNR_{DL} = SNR_{UL}$ and $SINR = SINR_{DL} = SINR_{UL}$, as well as $\delta_{DL} = \delta_{UL} = \frac{1}{2}$. In particular, we see that the spectral efficiency gain of the FD mode over the HD mode converges as follows in the extreme cases:

$$\frac{R_{FD}}{R_{HD}} = \frac{\log_2(1 + SINR)}{\frac{1}{2}\log_2(1 + SNR)} \rightarrow \begin{cases} 2 & \text{when } SINR \rightarrow SNR, \\ 0 & \text{when } SINR \rightarrow 0. \end{cases} \tag{9.3}$$

In plain words, the FD mode could even double the spectral efficiency of the system in comparison with conventional TDD or FDD operation if the residual SI could be suppressed down to a negligible level, whereas it is totally useless if the interference situation is prohibitive.

Although the literature bangs the drum for the FD concept by advertising its impressive 100% gain in system capacity or, equivalently, 50% savings in allocated spectrum resources, practical scenarios are likely to be situated somewhere between the extremes, and asymmetry in the system changes the comparison significantly.

Continuing with the symmetric case, the spectral efficiency ratio in (9.3) exhibits a fundamental rate–interference trade-off imposed by the choice between the FD and HD operation modes: should the system choose to eliminate the interference and achieve higher SNR or to avoid the rate loss due to the prelog factor of a half but achieve lower SINR owing to the interference? The two extreme cases in (9.3) obviously favor different modes and the ratio is continuously nonincreasing in terms of the amount of interference, so that there must be a crossover point where the modes are equal in terms of spectral efficiency.

Condition 9.2 *The FD mode outperforms the HD mode when*

$$\frac{R_{FD}}{R_{HD}} > 1 \Leftrightarrow SINR > \sqrt{SNR + 1} - 1 \approx \sqrt{SNR}, \tag{9.4}$$

for which the inequality is replaced by an equality in the exact crossover case.

In plain words, the HD mode needs to achieve an SNR that is *squared on a linear scale*, i.e., doubled on a decibel scale, compared with the SINR of the FD mode in order to compensate for the spectrum loss due to reserving separate channels for DL and UL communication. This aspect offers rather a large amount of headroom for interference in the FD mode because the effect of the prelog factor of a half on the HD rates is substantial. For example, if $\text{SNR} = 10^2 = 100 = 20$ dB, FD operation may degrade the signal quality down to $\text{SINR} \approx 10 = 10$ dB (by a factor of ten) before it offers worse spectral efficiency than HD operation.

The SINR is related to the SNR and the signal-to-interference ratio (SIR) as follows:

$$\text{SINR} = \frac{1}{1/\text{SIR} + 1/\text{SNR}} \Leftrightarrow \text{SIR} = \frac{1}{1/\text{SINR} - 1/\text{SNR}}, \tag{9.5}$$

from which the interference-to-noise ratio[†] can be expressed as

$$\frac{\text{SNR}}{\text{SIR}} = \frac{\text{SNR}}{\text{SINR}} - 1. \tag{9.6}$$

Therefore, the condition (9.4) can be equivalently restated as

$$\frac{\text{SNR}}{\text{SIR}} < \frac{\text{SNR}}{\sqrt{\text{SNR} + 1} - 1} - 1 \approx \sqrt{\text{SNR}} - 1, \tag{9.7}$$

and the fact that the right-hand side of this inequality is always greater than one (because SNR must obviously be positive) proves the following condition.

Condition 9.3 *The FD mode outperforms the HD mode irrespective of the SNR or SINR level if the residual interference power can be suppressed below the noise power, i.e., SNR < SIR. Otherwise, the SNR specifies an upper limit as per (9.7) for the interference level at which the FD mode is superior.*

Hence, suppressing the residual SI and co-channel interference levels below the receiver noise floor represents the ultimate target for the design of FD transceivers and protocols, although this is very challenging to achieve in general.

The plain sum rate in symmetric scenarios does not tell the full story, though, because the channels and rates can be asymmetric between the DL and UL directions. As an extreme example, FD operation is obviously pointless if information needs to or can be transmitted in the simplex mode only. Consequently, the comparison between the FD and HD modes should be performed instead in terms of achievable-rate regions or the sum rate subject to some constant ratio between the DL and UL rates. The concept of a flexible HD mode has been developed for this purpose, allowing the system to optimize the rates by tuning δ_{DL} and δ_{UL}.

When either communication direction is a bottleneck owing to asymmetry, it is optimal to use a hybrid FD/HD mode that allocates channel resources for both HD

[†] Neither of the corresponding abbreviations INR ("interference-to-noise ratio") and NIR ("noise-to-interference ratio") have been commonly used in the literature, so we refrain from using these abbreviations as well.

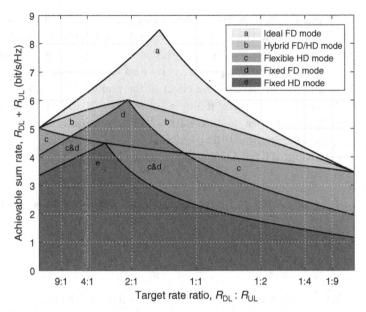

Figure 9.2 The maximum achievable sum rates in terms of the DL-to-UL rate ratio constraint when $\text{SNR}_{\text{DL}} = \text{SIR}_{\text{DL}} = 15$ dB, $\text{SNR}_{\text{UL}} = 10$ dB, $\text{SIR}_{\text{UL}} = 6$ dB, and the baseline reference case of fixed HD operation uses $\delta_{\text{DL}} = 1 - \delta_{\text{UL}} = 2/3$.

and FD operation simultaneously so that the spectral efficiency becomes

$$R = \delta_{\text{DL}} \log_2(1 + \text{SNR}_{\text{DL}}) + \delta_{\text{FD}} \log_2(1 + \text{SINR}_{\text{DL}}) + \delta_{\text{UL}} \log_2(1 + \text{SNR}_{\text{UL}}), \quad (9.8)$$

where δ_{FD} denotes the relative amount of time–frequency channel resources allocated for bidirectional FD communication, such that $0 \leq \delta_{\text{FD}} \leq 1$ and $0 \leq \delta_{\text{DL}} + \delta_{\text{FD}} + \delta_{\text{UL}} \leq 1$ in addition to $0 \leq \delta_{\text{DL}} \leq 1$ and $0 \leq \delta_{\text{UL}} \leq 1$ as above. In fact, one can easily show that always either $\delta_{\text{DL}} = 0$ or $\delta_{\text{UL}} = 0$ when the spectral efficiency is maximized, which means that it is beneficial to complement FD operation with additional simplex transmission in order to facilitate a higher data rate in the bottleneck direction.

Figure 9.2 illustrates the achievable sum rates subject to the ratio between the required DL and UL rates. Compared with an ideal FD system without any interference, not only the residual SI but also asymmetry in the communication scenario significantly reduces the benefits of the FD mode. For example, when $R_{\text{DL}}:R_{\text{UL}} = 9:1$, FD operation can yield only a 45% gain, which is quite far from the hyped 100% gain, and almost half of it is already achieved with flexible TDD. The channels are also unbalanced in this example, namely, the UL direction suffers from strong SI, which partially compensates for a similar rate imbalance.

9.2.4   Other Applications of Full-Duplex Transceivers

Above and in the remainder of this chapter, we assume that the FD mode is utilized solely for communication purposes, such that the information flowing in and out of the FD transceivers is actual data to be received from and transmitted to other transceivers.

However, it should be noted that the general STAR capability provided by the FD concept can be very useful beyond the pure communication context as well.

In some prospective applications of FD radio transceivers, one or other of the two directions of information flow is not utilized to transfer data, but instead is used to receive or transmit some other type of information or signal. In particular, cognitive radios could benefit greatly from the capability of receiving while transmitting [9], i.e., sensing and simultaneously using the spectrum whenever it is vacant. Conventional HD secondary users need to take regular sensing breaks from transmission. Thus, they lose communication opportunities whenever the channel still remains available, and the collision probability is increased because the primary user may become active during the secondary user's transmission, which cannot be detected until a sensing break takes place.

As a converse case to the above, physical layer security concepts can benefit greatly from the STAR capability as well, namely the possibility of transmitting while receiving. For example, a terminal may transmit jamming signals to hinder eavesdropping at the same time as it is receiving from a legitimate transmitter some information to be kept secure. Related conventional HD schemes require that the original transmitter itself sends the jamming signals (usually additive noise that is orthogonal to the data signals), which reduces the transmission capacity of the actual payload, or another third-party cooperative node must perform the jamming operation while avoiding disturbing the legitimate receiver.

Finally, it is also worth mentioning that FD radios are already used in applications where neither direction of information flow is utilized for actual communication data. Namely, continuous-wave radar systems are based on STAR operation, such that they transmit radio energy continuously while receiving reflections from the objects to be detected [3]. FD radars are somewhat different from communication terminals, though, in the sense that the signal looping back to the receiver from the transmitter is pure self-interference in communication but it is actually the signal of interest for detection and ranging in radar systems.

9.3   Design of Full-Duplex Protocols

9.3.1   Challenges and Opportunities in Full-Duplex Operation

The industry is anticipated to offer a thousand times higher throughput per total coverage area in 5G systems than in current 4G systems [10–12]. This network-level requirement is directly proportional to the spectral-efficiency targets for the physical and link layers of the systems, but FD technology may only double the spectral efficiency at best, and the gain is much smaller in practice, as discussed above. Yet, we could think that "great, we are almost halfway through," [4] and the potential gain from FD technology is actually still substantial compared with new advanced modulation and coding schemes. Thus, the remaining 500-fold throughput increase needs to come from the allocation of new frequencies in the physical layer [13, 14] and from network densification [15].

All networks that exploit FD technology in at least one transceiver can be divided into subparts, each of which can be classified as a conventional HD network or one of three basic FD link setups. The system in Figure 9.1 can be interpreted to represent any of them. With two FD transceivers, point-to-point communication can take place between them simultaneously in both directions; this scenario could be relevant to the wireless backhaul connections in heterogeneous 5G networks. With one FD transceiver, simultaneous transmission and reception can take place so that one HD terminal is receiving while another HD terminal is transmitting; this scenario could be relevant to a form of 5G network where a base station serves two FD mobile users or an access point relays signals between a base station and users in a two-hop manner.

The main challenge for the efficiency of FD operation is the fact that the simultaneous transmission and reception need to use the band for the same amount of time to make the most of the concept. If one direction needs more time–frequency channel resources, then it is better to switch to hybrid FD/HD operation because the double spectral efficiency can be achieved only in the ideal case in symmetric systems. The asymmetry is caused by the system's traffic pattern, i.e., the requested rates in the two simultaneous directions, and the channel quality, i.e., the achieved rates in the two simultaneous directions. In particular, the DL traffic from base stations to users is much more abundant than the UL traffic.

There are also some other emerging technologies which are somewhat competing with the FD technology by offering similar benefits. Firstly, flexible (or dynamic) TDD deals with the problem of unbalanced channels and rates in bidirectional communication by allowing control over the relative transmission rates in each direction such that the bottleneck link can be boosted. Flexible TDD also shares some implementation problems with the FD mode, such as those related to co-channel interference. Furthermore, the purpose of the FD technology overlaps somewhat with that of the emerging millimeter wave transmission concepts when it comes to wireless backhaul connections between macrobase stations and small-cell access points.

9.3.2   Full-Duplex Communication Scenarios in 5G Networks

The main conceivable scenarios in which FD technology may be applied in emerging 5G networks are illustrated in Figure 9.3. The systems considered comprise macrobase stations (BSs), small-cell access points (APs), and mobile users (MUs), which are situated in a heterogeneous multicell network. In principle, the three basic link setups discussed above can also be identified in these systems, but there are now multiple different variations because the terminals may be of various types.

Firstly, the application scenarios (a)–(c) correspond to point-to-point communication between two FD transceivers, namely (a) between an MU and a BS or an AP, (b) between a BS and an AP, and (c) between two MUs within a cell. The scenarios (a), (c), and (e) have two variations since the communication can take place within a macrocell (indicated by a subscript 1) or a small cell (subscript 2). The scenario (b) is characteristic

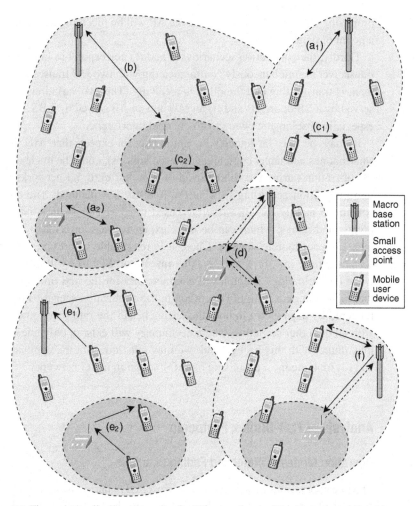

Figure 9.3 The main application scenarios for FD operation in 5G cellular networks: (a) point-to-point communication between an MU and a BS or an AP; (b) point-to-point backhaul communication between a BS and an AP; (c) point-to-point communication between two MUs within a cell; (d) two-hop DL or UL relaying through an AP, i.e., self-backhauling; (e) two MUs (one in DL and the other in UL) served by a BS or an AP; (f) an MU and an AP (one in DL and the other in UL) served by a BS.

of the prospective FD 5G systems that are envisioned to employ wireless backhaul connections to serve small-cell APs. We may assume that the bidirectional channels are approximately reciprocal in these scenarios.

Secondly, the application scenario (d) represents two-hop decode-and-forward relaying [16] through an AP between a BS and an MU, which has recently been referred to as "self-backhauling" instead, although the concept is still exactly the same. The communication can take place in either the DL or the UL direction. We have omitted cooperative communication scenarios where an MU acts as a relay for another MU, although these are conceivable as well, at least in theory. The scenario (d) looks very promising for FD operation because the AP will be receiving and transmitting the same amount of data by design.

Thirdly, the application scenarios (e) and (f) correspond to link setups where an FD transceiver is simultaneously communicating with two HD transceivers such that one of them is transmitting and the other is receiving. The two variations cover (e) two MUs served by a BS or an AP, and (f) an MU and an AP served by a BS. These scenarios are especially problematic owing to co-channel interference.

It may be too far-fetched or optimistic to expect that MU devices, such as smartphones and tablet computers (or even laptops), could be made capable of efficient FD operation during the short period of time left until 5G networks will be launched into production according to what the industry has planned. The current technology is suitable mainly for infrastructure-based transceivers, such as base stations, access points, and relays, which can be more expensive and have much bigger form factors than compact mobile devices. Thus, it is reasonable to presume that the application scenarios (a_1), (a_2), (c_1), and (c_2) that involve an FD user will not be in the front line when the technology is employed commercially for the first time. Likewise, achieving the maximum gain from FD operation in the application scenarios (e_1), (e_2), and (f) is hindered by asymmetry in both the channels and the rate requirements in comparison with all the other scenarios, where asymmetry will exist in only one of these aspects at maximum. With this background, we may conclude that the application scenarios (b) and (d) look the most promising ones for adoption in 5G networks.

9.4   Analysis of Full-Duplex Protocols

9.4.1   Operation Modes in Wideband Fading Channels

The rate expressions in Section 9.2.3 correspond to narrowband frequency-flat channels. In principle, this analysis can be translated to the case of wideband fading channels by assuming that the rate expressions in (9.1), (9.2), and (9.8) represent individual subchannels in orthogonal frequency division multiplexing (OFDM) transmission. However, there are certain fundamental restrictions: it is not possible to choose the duplex mode separately for each subchannel or to control δ_{DL}, δ_{FD}, and δ_{UL} in (9.2) and (9.8) on a per-subchannel basis.

In the following, the random subchannels are modeled by Rayleigh fading such that the probability density function (PDF) of the SNR is given by

$$f_{SNR}(x) = \frac{1}{\mathcal{E}\{SNR\}} \exp\left(-\frac{x}{\mathcal{E}\{SNR\}}\right), \tag{9.9}$$

where $\mathcal{E}\{SNR\}$ denotes the average of the random variable SNR. For the (residual) self-interference and co-channel interference channels, the interference-to-noise ratio, SNR/SIR, is modeled for simplicity as a constant, which means that the receiver noise floor is increased by this constant factor.

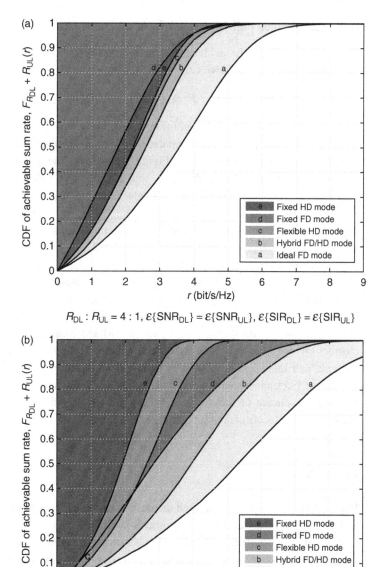

Figure 9.4 Comparison of operation modes in terms of sum rates achieved in transmission over wideband fading channels: (a) asymmetric DL-to-UL rate ratio but symmetric channels; (b) symmetric rate ratio but asymmetric channels.

9.4.2 Full-Duplex Versus Half-Duplex in Wideband Transmission

Let us now analyze the achievable transmission rates in terms of their cumulative distribution functions (CDFs), namely

$$F_{R_{DL}+R_{UL}}(r) = P(R_{DL} + R_{UL} \le r \,|\, R_{DL}:R_{UL} \text{ constant}), \qquad (9.10)$$

where $P(\Phi_1 \,|\, \Phi_2)$ denotes the conditional probability of the event Φ_1 given the event Φ_2. The numerical CDFs were computed in a Monte Carlo manner.

The numerical results shown in Figure 9.4 assume the following system setups: $\mathcal{E}\{\text{SNR}_{UL}\} = 10$ dB in both subfigures, and $\mathcal{E}\{\text{SNR}_{DL}\} = \mathcal{E}\{\text{SNR}_{UL}\}$ and $\mathcal{E}\{\text{SNR}_{DL}\} = 15$ dB in Figures 9.4(a) and (b), respectively; $\text{SNR}_{UL}/\text{SIR}_{UL} = 4$ dB in both subfigures, and $\text{SNR}_{DL}/\text{SIR}_{DL} = \text{SNR}_{UL}/\text{SIR}_{UL}$ and $\text{SNR}_{DL}/\text{SIR}_{DL} = 0$ dB in Figures 9.4(a) and (b), respectively; and the downlink-to-uplink rate ratios were chosen as $R_{DL}:R_{UL} = 4:1$ and $R_{DL}:R_{UL} = 1:1$ in Figures 9.4(a) and (b), respectively. In plain words, Figure 9.4(a) corresponds to a scenario where the transmission rates are unequal in the two communication directions while the corresponding average channel gains are equal, and vice versa in Figure 9.4(b). However, there is still an imbalance within each wideband channel due to fading in both setups.

In comparison with the case of static narrowband channels, fading in a wideband channel changes the trade-off between FD and HD operation quite drastically. For example in Figure 9.4(a), fixed HD operation outperforms fixed FD operation. The benefits of FD transmission over a wideband fading channel are limited mainly by the fact that the choice of the duplex mode is highly suboptimal for many subchannels because the protocols cannot be optimized individually for them.

9.5 Conclusion

In this chapter, we have considered the design and analysis of wideband physical layer and link layer data transmission protocols for heterogeneous networks where some transceivers are capable of wireless FD operation, i.e., simultaneous transmission and reception at the same center frequency. Let us conclude the treatment of the subject by envisioning the future research directions and discussing the prospects for FD technology in commercial 5G systems.

9.5.1 Prospective Scientific Research Directions

The scientific research on self-interference cancellation techniques for FD transceivers has now reached quite a mature state. In particular, academia has generated extensive theoretical understanding of the phenomena that limit cancellation performance and cause residual self-interference, and has developed advanced techniques to counteract these phenomena. In addition, there are already many convincing (in the authors' opinion) research prototypes that demonstrate the viability of FD operation at large. Designing and implementing FD radio devices in isolation from their surrounding networks can be regarded from now on as a task of product development for the industry rather than any major scientific problem. Consequently, it seems that the major open research directions in the subject are related to how to take advantage most efficiently of the progressive capabilities of FD devices when they are deployed as an integral part of full-scale heterogeneous networks in different roles.

9.5.2   Full-Duplex in Commercial 5G Networks

As of today, it seems that FD technology will not, at least not yet, be adopted in the first phase of global commercial 5G standards. This is because the concept has not so far been discussed in any noteworthy meeting contribution document ("TDoc") from the leading companies in RAN WG1 (the radio access network layer 1 working group) of the 3rd Generation Partnership Project (3GPP) since the group started the actual 5G standardization process at its meeting in Busan, Korea, in April 2016. RAN WG1 is responsible for the specification of the physical layer of the radio interface for the emerging terrestrial radio access network and covers both FDD and TDD operation modes. Thus, the FD mode would have already been a serious subject in the working group among candidate waveforms, multiple access techniques, symbol numerology, frame structures, channel coding, modulation, etc. if it were to be introduced in 3GPP Release 15, which is expected to be the first 5G standard and is expected to be completed quite soon, in the latter half of 2018.

On the other hand, we cannot rule out the possibility that FD technology will be incorporated into some proprietary or regional 5G wireless systems. Especially, some markets in Asia, for example China, have a history of deploying TDD systems where the transition to FD operation via flexible TDD concepts seems to be easier to approve than when it is suggested that both the downlink and the uplink bands in FDD systems should be opened up to bidirectional transmission. Thus, it is not any coincidence that there has also been significant research interest in developing FD technology in these specific nations. It is not unlikely that the significant amount of time and money invested in scientific research will eventually materialize as compelling technical solutions which regional industries will not mind adopting in some form or another in their 5G systems. Furthermore, 3GPP Release 15 is actually just the first phase of the 5G standard, which will evolve over the next decade or so with future 3GPP releases. The global industry aims to launch commercial 5G networks around the year 2020, so that there is still enough time pushing FD technology through 3GPP standardization, and more than a slim chance that this will happen.

Finally, one should bear in mind that transparent FD wireless operation has already been exploited in all generations of cellular mobile communications and will be available for 5G systems as well. In particular, devices such as on-channel repeaters, i.e., gap fillers, can be deployed more or less transparently when it comes to the communication standards and transmission protocols of the actual network. Their spread to wide commercial use in 5G systems would result mainly in network design problems and spectrum regulation questions rather than be complicated because of heavy requirements for standardization.

A relevant question, then, is whether standardization of FD technology is actually needed in 5G at all, or whether it is merely an implementation issue. Downlink traffic is expected to continue dominating uplink traffic, so most of the gains of FD technology are squeezed from relaying applications, where the traffic is inherently symmetric irrespective of uplink/downlink ratio requirements. As already mentioned, the deployment of relays working as on-channel repeaters is a network-planning issue, requiring no standardization activities. At the other extreme, an out-of-band FD relay specified in Long Term Evolution (LTE)-Advanced creates its own cell and generates its own control plane signals. A decode-and-forward relay that demodulates, decodes, and then encodes and modulates the signal lies between these two extremes. Although relays have not been widely deployed in LTE networks thus far, there is no reason to believe that the relay option will be left out of the new radio specifications. Wireless relays require mainly the specification of resource allocation, as well as control and data structures for backhaul and access links. For optimal performance, the allocation between the links should be dynamic and fully flexible. When these structures are specified for a conventional TDD relay, the resource allocation is rather unconstrained by the standard and is driven by a vendor-specific scheduler such that the relay may operate in the FD mode as well. FD relays do change the statistics of cross-link and cross-cell interference when compared with HD TDD relays, though, but when flexible TDD is allowed, mechanisms to deal with this interference already exist in the network. Thus, the adoption of FD technology in 5G systems may indeed be an implementation issue, and this may speed up its deployment.

References

[1] T. Riihonen, "Design and analysis of duplexing modes and forwarding protocols for OFDM(A) relay links," D.Sc. thesis, Aalto University School of Electrical Engineering, Helsinki, Finland, Jul. 2014.

[2] S. Hong, J. Brand, J. I. Choi, M. Jain, J. Mehlman, S. Katti, and P. Levis, "Applications of self-interference cancellation in 5G and beyond," *IEEE Commun. Mag.*, vol. 52, no. 2, pp. 114–121, Feb. 2014.

[3] A. Sabharwal, P. Schniter, D. Guo, D. W. Bliss, S. Rangarajan, and R. Wichman, "In-band full-duplex wireless: Challenges and opportunities," *IEEE J. Sel. Areas Commun.*, vol. 32, no. 9, pp. 1637–1652, Sep. 2014.

[4] M. Heino, D. Korpi, T. Huusari, E. Antonio-Rodríguez, S. Venkatasubramanian, T. Riihonen, L. Anttila, C. Icheln, K. Haneda, R. Wichman, and M. Valkama, "Recent advances in antenna design and interference cancellation algorithms for in-band full duplex relays," *IEEE Commun. Mag.*, vol. 53, no. 5, pp. 91–101, May 2015.

[5] Z. Zhang, X. Chai, K. Long, A. V. Vasilakos, and L. Hanzo, "Full duplex techniques for 5G networks: Self-interference cancellation, protocol design, and relay selection," *IEEE Commun. Mag.*, vol. 53, no. 5, pp. 128–137, May 2015.

[6] T. Riihonen and R. Wichman, "Full-duplex in wireless communications," in *Wiley Encyclopedia of Electrical and Electronics Engineering*, J. G. Webster, ed., Wiley, pp. 1–24, Aug. 2016.

[7] E. Antonio-Rodríguez, K. Haneda, D. Korpi, T. Riihonen, and M. Valkama, "Design and implementation of full-duplex transceivers," in *Signal Processing for 5G: Algorithms and Implementations*, F.-L. Luo and C. J. Zhang, eds., Wiley/IEEE Press, ch. 17, pp. 402–428, 2016.

[8] R. Pitaval, O. Tirkkonen, R. Wichman, K. Pajukoski, E. Lähetkangas, and E. Tiirola, "Full-duplex self-backhauling for small-cell 5G networks," *IEEE Commun. Mag.*, vol. 22, no. 5, pp. 83–89, Oct. 2015.

[9] Y. Liao, L. Song, Z. Han, and Y. Li, "Full duplex cognitive radio: A new design paradigm for enhancing spectrum usage," *IEEE Commun. Mag.*, vol. 53, no. 5, pp. 138–145, May 2015.

[10] Q. C. Li, H. Niu, A. Papathanassiou, and G. Wu, "5G network capacity: Key elements and technologies," *IEEE Veh. Technol. Mag.*, vol. 9, no. 1, pp. 71–78, Mar. 2014.

[11] S. Chen and J. Zhao, "The requirements, challenges, and technologies for 5G of terrestrial mobile telecommunication," *IEEE Commun. Mag.*, vol. 52, no. 5, pp. 36–43, May 2014.

[12] E. Dahlman, G. Mildh, S. Parkvall, J. Peisa, J. Sachs, Y. Selén, and J. Sköld, "5G wireless access: Requirements and realization," *IEEE Commun. Mag.*, vol. 52, no. 12, pp. 42–47, Dec. 2014.

[13] C. Dehos, J. González, A. De Domenico, D. Kténas, and L. Dussopt, "Millimeter-wave access and backhauling: The solution to the exponential data traffic increase in 5G mobile communications systems?" *IEEE Commun. Mag.*, vol. 52, no. 9, pp. 88–95, Sep. 2014.

[14] R. Taori and A. Sridharan, "Point-to-multipoint in-band mmWave backhaul for 5G networks," *IEEE Commun. Mag.*, vol. 53, no. 1, pp. 195–201, Jan. 2015.

[15] N. Bhushan, J. Li, D. Malladi, R. Gilmore, D. Brenner, A. Damnjanovic, T. Sukhavasi, C. Patel, and S. Geirhofer, "Network densification: The dominant theme for wireless evolution into 5G," *IEEE Commun. Mag.*, vol. 52, no. 2, pp. 82–89, Feb. 2014.

[16] T. Riihonen and X. Wang, "Relaying in full-duplex radio communication systems," in *Advanced Relay Technologies in Next Generation Wireless Communications*, I. Krikidis and G. Zheng, eds., Institution of Engineering and Technology, pp. 129–173, 2016.

10 Millimeter Wave Communications for 5G Networks

Jiho Song, Miguel R. Castellanos, and David J. Love

10.1 Motivations and Opportunities

Wireless broadband systems, such as cellular and Wi-Fi systems, are now a ubiquitous presence in our lives around the world. Today, these systems operate using carrier frequencies below 6 GHz. The fifth generation (5G) systems must be designed to meet future demands for wireless data, which are projected to continue growing exponentially with time. This has motivated researchers to look at the underutilized bands available at higher frequencies [1–5]. Current research has shown that millimeter wave systems could provide the capacity enhancements needed for serving future wireless data [6–11].

Bands at millimeter wave frequencies, which can roughly be defined as bands with carrier frequencies between 30 and 100 GHz, have received little commercial attention in the past. To meet the demands for new spectrum, the Federal Communications Commission (FCC) has explored the deployment of new services at millimeter wave frequencies [12]. The potential for commercial success of millimeter wave is mainly due to the potential for very large bandwidths. For example, the 60 GHz band, which has received much discussion for potential unlicensed wireless broadband access, has a bandwidth of 7 GHz.

Despite this promise, there are many challenges to be overcome. Communication at millimeter wave frequencies must occur under relatively demanding propagation conditions. High-frequency systems suffer from increased path loss, additional channel losses, and more costly radio frequency (RF) technology [13–17]. Compared with today's systems, the path loss alone can be tens of dB more at millimeter wave frequencies. Luckily, advances in hardware and communication theory are making widespread use of millimeter communication more likely. Device technology has continued to evolve, with advances in the fabrication of complementary metal–oxide–semiconductor (CMOS) compatible millimeter wave radios becoming more and more cost-effective [6, 18]. There have been accompanying advances in RF technology that may make the generation, reception, and analog processing of millimeter wave signals less challenging [8, 9, 19].

Communication researchers have continued to look at multiple-antenna technology, which can be readily employed at millimeter wave. Multiple-antenna radio links in the millimeter wave frequencies are inherently directional, which could enable denser wireless broadband deployments, generating less interference. Research has focused on millimeter wave systems using a multitude of antennas in a small form factor utilizing uniform linear arrays (ULAs) and uniform planar arrays (UPAs). These large-array communication systems can be combined with advances in signal processing under sparse channel conditions [20–24].

10.2 Millimeter Wave Radio Propagation

To understand the signal-processing and communication techniques used at millimeter wave frequencies, it is important to discuss the challenges of high-frequency propagation. We shall compare the characteristics of millimeter wave channels with those of lower-frequency channels. We shall also discuss the millimeter wave channel model used as a baseline model for the development of advanced signal-processing techniques.

10.2.1 Radio Attenuation

Systems operating in the millimeter wave bands suffer from path loss, which results in severe link quality degradation and can hinder reliable communication. Signal attenuation can be understood by taking a more detailed look at some of the major causes.

We first consider the attenuation due to atmospheric gases. Transmission losses are incurred when gaseous atmospheric particles absorb energy from millimeter waves. Interactions between millimeter waves and most gases do not cause significant attenuation; however, water vapor and oxygen molecules are well known to absorb significant amounts of energy from higher-frequency radio waves [26]. Peak attenuation occurs when the transmission frequency coincides with the resonant frequencies of the gas molecules: 24 GHz for water vapor and 60 GHz for oxygen in the sub-100 GHz range. Atmospheric attenuation can cause severe losses when millimeter waves propagate through a medium with a large water vapor concentration, such as settings with heavy fog, smoke, or high humidity.

In the same way that attenuation occurs due to water vapor absorption, millimeter wave radio links are seriously affected by rain. To study the impact of rainfall on millimeter wave links, the measured loss due to rainfall is summarized in Table 10.1 [25]. The loss can be estimated based on the attenuation model

$$(L_{\mathrm{r}})_{\mathrm{dB}} = kR^{\alpha}, \tag{10.1}$$

where R is the rainfall rate (mm/hour). The two parameters in (10.1) k and α are summarized in the ITU recommendation [26].

The predicted rain attenuation in scenarios with different of rainfall rates, i.e., 22, 63, 120, and 163 mm/hour, is depicted in Figure 10.1(a). For example, when it rains at a rate of 22 mm/hour, a millimeter wave transmission at 90 GHz will be diminished by approximately 10.72 dB per kilometer, while a microwave transmission in the 3 GHz

Table 10.1 Attenuation by rainfall in 120 GHz band [25].

Rainfall rate (mm/hour)	Attenuation (dB/km)
22	10.9
63	23.2
120	36.9
165	40.5

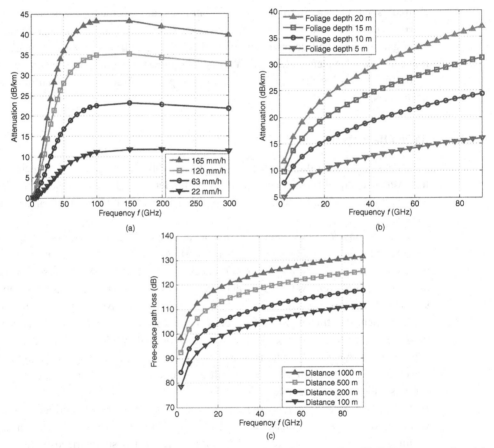

Figure 10.1 Millimeter wave propagation losses. (a) Rain attenuation; (b) Foliage attenuation; (c) Free-space path loss.

band will be attenuated by only around 0.0053 dB per kilometer. In strong rainfall of 120 mm/hour, radio links at 90 GHz frequency can be seriously degraded owing to an attenuation of approximately 34.41 dB per kilometer. Lower-frequency links are not as susceptible to variations in rainfall rates (e.g., note the low attenuation of a transmission at 3 GHz). Millimeter wave signals can be scattered by rain because the diameters of raindrops, distributed between approximately 0.5 and 5 mm, are similar to the wavelength of the radio waves [8, 27]. To guard against significant outages during strong rainfall, systems can leverage heterogenous structures with multiple frequency bands of operation.

Attenuation due to foliage is also more of a concern at millimeter wave. In the spectrum between 0.2 and 95 GHz, the channel loss (dB/km) can be estimated based on empirical studies [27], for example

$$(L_f)_{dB} = 2 \times 10^{-0.1} f^{0.3} D^{0.6}, \tag{10.2}$$

where f is the center frequency (GHz) and D is the depth of foliage (m). The predicted attenuation is depicted in Figure 10.1(b). Note that the foliage loss increases substantially as the depth of foliage increases. As expected from Figure 10.1(b), foliage blockage can impede millimeter wave radio links severely.

10.2.2   Free-Space Path Loss

Assuming that the propagation distance is much longer than the radio wavelength, the received power at each antenna can be defined for free-space propagation using the Friis transmission equation as

$$P_r = \frac{P_t G_t A_r L}{4\pi d^2} = P_t G_t G_r L L_F, \tag{10.3}$$

where d is the distance between the transmitter and receiver. In the Friis equation, P_t is the transmit signal power, P_r is the received signal power, G_t is the transmit antenna gain, G_r is the receive antenna gain, and $L \doteq L_a L_r L_f L_m$ is the additional path loss, which is defined based on the atmospheric loss L_a, rainfall loss L_r, foliage loss L_f, and miscellaneous loss L_m (e.g., cable loss). In addition, the effective aperture of a single-antenna element is

$$A_r = \frac{G_r}{4\pi} \left(\frac{c}{f} \right)^2,$$

and the free-space path loss is defined as

$$L_F = \left(\frac{c}{4\pi d f} \right)^2, \tag{10.4}$$

where c is the speed of light.

An antenna with a larger effective aperture captures more energy from incident radio waves. According to the Friis equation, the free-space path loss increases with the square of the center frequency because the path loss is proportional to the size of the antenna aperture. For a fixed additional path loss L, the path loss of 90 GHz millimeter wave signals is 29.5 dB higher than that of 3 GHz microwave signals, computed as

$$20 \log_{10} \left(\frac{90 \times 10^9}{3 \times 10^9} \right).$$

In millimeter wave systems, the small wavelengths allow large-sized arrays to offer directional gains comparable to their large single-antenna counterparts. An example of

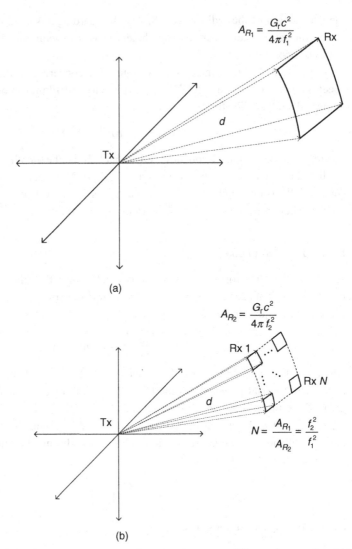

$$A_{R_1} = \frac{G_r c^2}{4\pi f_1^2}$$

(a)

$$A_{R_2} = \frac{G_r c^2}{4\pi f_2^2}$$

$$N = \frac{A_{R_1}}{A_{R_2}} = \frac{f_2^2}{f_1^2}$$

(b)

Figure 10.2 Effective antenna apertures for operating at two different frequencies. (a) Effective antenna aperture of lower-frequency system; (b) effective antenna aperture of higher-frequency system.

a millimeter wave array and a single antenna of the same physical size are shown in Figure 10.2. For example, a transceiver operating at 90 GHz can place a factor of

$$N \doteq \frac{G_r c^2 / (4\pi \times 3 \times 10^9)}{G_r c^2 / (4\pi \times 90 \times 10^9)} = 30^2$$

more antennas in the same form factor than a 3 GHz transceiver in order to provide the same gain in ideal conditions [1, 8, 9]. Although millimeter wave systems can capture more energy from incident waves by employing large antenna arrays, joint processing using techniques such as beamforming and combining must be utilized in order to reach the performance levels of microwave systems. Beamforming and combining will be discussed in detail in later sections.

10.2.3   Severe Shadowing

Electromagnetic waves can diffract when they encounter obstacles that are comparable in size to their wavelength. Radio signals in current cellular systems experience shadowing when line-of-sight (LOS) paths are obstructed by diffracting objects. Diffraction can pose significant problems during millimeter wave transmissions since smaller objects can cause shadowing. The larger propagation distances of millimeter waves diffracting around objects also result in path losses larger than those seen in microwave systems.

The penetration loss of many materials increases gradually with carrier frequency. However, the penetration loss of some materials, for example brick, tinted glass, and the human body, increases considerably as the frequency increases. Compared with microwave signals, millimeter wave signals do not penetrate these materials well. For example, the penetration losses of tinted glass and brick in the 28 GHz frequency band have been measured as 40.1 and 28.3 dB, respectively [28]. Also, experiments have shown that the human body results in an 18−36 dB penetration loss [29]. This means that humans could be a major cause of blockages in millimeter wave links.

The shorter wavelength and higher penetration loss give rise to serious shadowing effects. As well, millimeter wave signals may arrive at a receiver from a number of discrete directions owing to the diffraction-limited characteristics of millimeter wave channels [19]. The intermittent connectivity due to blockages can significantly degrade the quality of service. Supporting seamless radio connections in the presence of moving objects, for example humans, is a serious challenge in developing reliable millimeter wave communications.

10.2.4   Millimeter Wave Channel Model

As discussed earlier in the chapter, radio signals at millimeter wave frequencies are severely attenuated so that only a small number of dominant paths may exist. This causes many lower-frequency channel models, which assume a multitude of signal paths, to be rendered unusable at higher frequencies.

The small number of dominant paths can usually be modeled using a ray-like propagation assumption [16]. Assuming a ULA structure in a system with M_t and M_r transmit and receive antennas, respectively, a millimeter wave channel **H** can be modeled as [14, 30]

$$\mathbf{H} = \sqrt{\frac{KM_rM_t}{1+K}}\alpha_0\mathbf{d}_{M_r}(\psi_{r0}^v)\mathbf{d}_{M_t}^H(\psi_{t0}^v) + \sqrt{\frac{M_rM_t}{L_P(1+K)}}\sum_{\ell=1}^{L_P}\alpha_\ell\mathbf{d}_{M_r}(\psi_{r\ell}^v)\mathbf{d}_{M_t}^H(\psi_{t\ell}^v), \quad (10.5)$$

where K is the Ricean K-factor and L_P is the number of non-line-of-sight (NLOS) paths. Each path is defined by a complex channel gain $\alpha_\ell \sim \mathcal{CN}(0,1)$ and an array response vector

$$\mathbf{d}_{M_b}(\psi) = \frac{1}{\sqrt{M_b}}\left[1, e^{j\psi} \cdots, e^{j(M_b-1)\psi}\right] \in \mathbb{C}^{M_b}.$$

For a UPA structure, each ray-like beam can be defined by

$$\alpha_\ell \sqrt{M_r M_t} \mathbf{d}_{M_r}(\psi_{r\ell}^h, \psi_{r\ell}^v) \mathbf{d}_{M_t}^H(\psi_{t\ell}^h, \psi_{t\ell}^v),$$

using multivariate (azimuth and elevation) array response vectors defined by the Kronecker product of the array response vectors in the horizontal and vertical domains,

$$\mathbf{d}_{M_b}(\psi^h, \psi^v) = \mathbf{d}_{M_b^h}(\psi^h) \otimes \mathbf{d}_{M_b^v}(\psi^v) \in \mathbb{C}^{M_b},$$

with $\psi_{r\ell}^h = (2\pi d_h/\lambda)\sin\theta_{r\ell}^h\cos\theta_{r\ell}^v$ in the horizontal domain and $\psi_{r\ell}^v = (2\pi d_v/\lambda)\sin\theta_{r\ell}^v$ in the vertical domain [31]. Here, $M_r = M_r^h M_r^v$ is the number of receive antennas and $M_t = M_t^h M_t^v$ is the number of transmit antennas. $\theta_{r\ell}^a$ is the angle of arrival (AoA) and $\theta_{t\ell}^a$ is the angle of departure (AoD) of the ℓth ray-like beam in the domain $a \in \{h, v\}$.

The blockage-limited nature of millimeter wave channels gives rise to a large Ricean K-factor [16]. In many cases, this means that millimeter wave channels can be approximated by the most dominant path, for example an LOS path. Thus, using a rank-one channel matrix is often a decent approximation.

A virtual channel model can be used to map spatial dimensions to orthogonal beams at the transmitter and receiver as in [32, 33]. This model can be a useful representation when millimeter wave channels are dominated by a single LOS path. The beamspace channel model can be expressed as [34]

$$\mathbf{H} = \sum_{i=1}^{M_r}\sum_{j=1}^{M_t} h_v(i,j)\mathbf{d}_{M_r}(\psi_{r,i})\mathbf{d}_{M_t}^H(\psi_{t,j}) \tag{10.6}$$

$$= \mathbf{D}_{M_r}\mathbf{H}_v\mathbf{D}_{M_t}^H,$$

where the unitary matrices $\mathbf{D}_{M_b}$ are equal to $1/\sqrt{M_b}[\mathbf{d}_{M_b}(\psi_1),\ldots,\mathbf{d}_{M_b}(\psi_M)]^T$ with the array response vectors $\{\mathbf{d}_{M_b}(\psi_i)\}$ defined as above, and the entry $h_v(i,j)$ of the matrix $\mathbf{H}_v$ is the virtual complex channel gain corresponding to the fixed virtual AoA and AoD $\psi_{r,i}$ and $\psi_{j,t}$, respectively. The virtual angles $\{\psi_{r,i}\}$ and $\{\psi_{j,t}\}$ are chosen such that the matrices $\mathbf{D}_{M_r}$ and $\mathbf{D}_{M_t}$ are of full rank, and a uniform spacing between virtual angles is common since this results in the matrices becoming discrete Fourier transform (DFT) matrices. This linear representation of the millimeter wave channel is practical owing to the inherent low-dimensional structure of LOS links.

10.2.5   Link Budget Analysis

To maintain good wireless links for reliable communications and develop insights into network design parameters, for example cell size and transmit power, we conducted a simple link budget analysis based on [36].

We considered two channel scenarios, i.e., a rainfall channel condition with a rate of 22 mm/hour and a 5 m foliage blockage channel condition. The additional losses due

Table 10.2 System parameters for link budget analysis.

Parameter	Value			
Maximum transmit power P_T	30 dBm			
Thermal noise N_t	−74 dBm			
Link margin L_M	10 dB			
Miscellaneous loss L_m	8 dB			
Required received power P_R	$N_t + L_M + L_m$ dBm			
Transmit antenna gain G_T	28 dBi			
Receive antenna gain G_R	15 dBi			
Beamforming gain G_{BF}	$10 \log_{10} M$ dB			
Free-space path loss L_f	$20 \log_{10} (c/4\pi d f)$ dB			
Center frequency f (GHz)	28	60	72	94
Atmospheric loss L_a (dB/km) [26]	0.06	18	0.3	0.2
Rain loss L_r (dB/km) [35]	3.46	8.61	9.69	10.87
Foliage loss L_f (dB/km) [27]	11.34	14.25	15.05	16.30

to rainfall and foliage blockage were estimated based on (10.1) and (10.2), respectively. The required received signal power P_R was set to −56 dBm, which is $L_M + L_m$, i.e., 18 dB, higher than the power of thermal noise. A 2 dB cable loss at both ends and a 4 dB system implementation loss were considered in order to define the miscellaneous loss L_m. The system parameters for the link budget analysis are summarized in Table 10.2. The maximum transmit power, and transmit and receive antenna gains were taken from [3].

In Figure 10.3, the link budget results for the rain and foliage scenarios are plotted by solid lines and dotted lines, respectively. It is clear that the cell radius decreases as the center frequency increases. One exception is that the recommended cell size of a 60 GHz millimeter wave system is smaller than that of other systems because millimeter wave signals in the 57−64 GHz oxygen absorption band suffer most severely from atmospheric loss.

10.3   Beamforming Architectures

It is expected that the beamforming gain will determine the level of link reliability available in millimeter wave networks. Advanced signal-processing techniques need to be developed with the purpose of facilitating highly directional communications. The following beamforming architectures offer design criteria for practical millimeter wave systems with large beamforming gains.

At millimeter wave, the analog or digital processing in the array must be explicitly modeled. Most multiple-input multiple-output (MIMO) work has focused on sub-6 GHz systems employing only a small number of antennas. Most of this work implicitly assumes that one RF chain and one analog-to-digital/digital-to-analog converter (ADC/DAC) are available for each antenna. Owing to cost and power concerns, millimeter wave systems have instead focused on alternative architectures. Analog

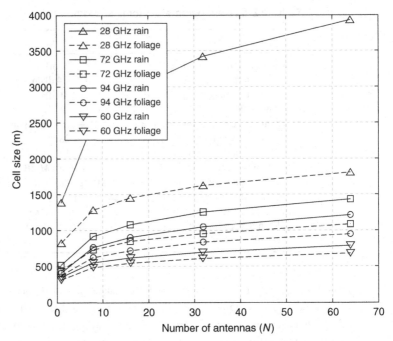

Figure 10.3 Recommended cell size based on the link budget analysis.

and hybrid beamforming techniques have been extensively studied as a solution for millimeter wave implementations with only moderate hardware complexity. In an analog-based beamforming architecture, beamforming/combining is accomplished by using a single RF chain coupled to a set of RF phase shifters [37–42]. A hybrid beamforming architecture, which uses a mixture of analog and digital beamforming solutions, generalizes analog beamforming to allow more than one RF chain, combined with sets of RF phase shifters [20, 21, 42–45]. We overview several beamforming solutions that have moderate hardware complexity.

10.3.1   Analog Beamforming Solutions

10.3.1.1   Phase-Shifter-Based Analog Beamforming Architecture

We first consider analog beamforming, which is implemented by using a single RF chain coupled to a set of phase shifters. Figure 10.4(a) shows an example of this architecture. Beamforming is implemented in the analog domain and usually modeled as a per-antenna phase shift. In this model, the antennas cannot generate different levels of output gain.

In a block fading channel model, consider the input–output expression

$$y[v] = \sqrt{\rho}\mathbf{z}^{H}[v]\mathbf{H}[v]\mathbf{f}[v]s[v] + n[v], \tag{10.7}$$

where $\mathbf{z}[v] \in \mathbb{C}^{M_r}$ is the receive combiner, ρ is the signal-to-noise ratio (SNR), $\mathbf{H}[v] \in \mathbb{C}^{M_r \times M_t}$ is the block fading channel matrix, $\mathbf{f}[v] \in \mathbb{C}^{M_t}$ is the transmit beamformer,

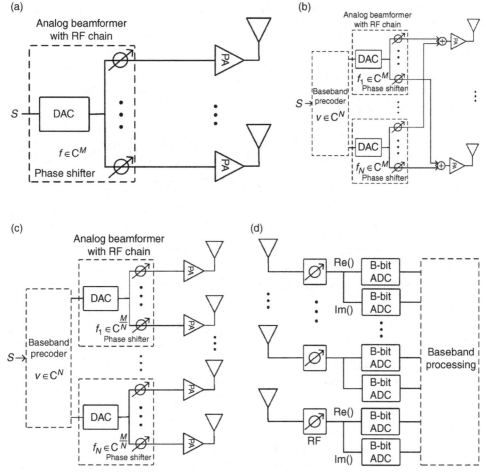

Figure 10.4 Beamforming and combining architectures for millimeter wave systems. (a) Phase-shifter-based analog beamforming architecture; (b) Hybrid beamforming structure 1; (c) Hybrid beamforming structure 2; (d) Receiver with low-resolution ADCs.

$s[v]$ is the transmitted signal, subject to $E[|s[v]|^2] \leq 1$, and $n[v]$ is the normalized effective noise after combining, modeled as additive white Gaussian noise (AWGN) with a per-sample distribution $\mathcal{CN}(0,1)$. For simplicity, the $[v]$ denoting each frequency subcarrier is dropped in the following sections.

In the beamforming/combining literature, the standard convention is to impose the restriction $\|\mathbf{z}\|_2 = \|\mathbf{f}\|_2 = 1$. Assuming that both beamforming and combining are realized using RF phase shifters, the transmit beamformer $\mathbf{f} \in \mathcal{B}_{M_t}$ and the receive combiner $\mathbf{z} \in \mathcal{B}_{M_r}$ are restricted to lie in an equal-gain set of unit vectors denoted by

$$\mathcal{B}_M = \left\{ \mathbf{a} \in \mathbb{C}^M : \left(\mathbf{a}\mathbf{a}^H\right)_{m,m} = \frac{1}{M}, \ m \in \{1, \cdots, M\} \right\}. \tag{10.8}$$

Before discussing practical analog beamforming techniques, we review equal-gain beamforming and combining under perfect CSI conditions. For a given input–output expression for the system as expressed in (10.7), the optimization problem is defined as

$$(\hat{\mathbf{z}}, \hat{\mathbf{f}}) = \underset{(\mathbf{z}, \mathbf{f}) \in \mathcal{B}_{M_r} \times \mathcal{B}_{M_t}}{\arg\max} |\mathbf{z}^H \mathbf{H} \mathbf{f}|^2, \tag{10.9}$$

where $\mathbf{z}$ is the equal-gain combining (EGC) vector and $\mathbf{f}$ is the equal-gain transmission (EGT) vector. To obtain more insight into the optimization problem, we first compute the upper bound of the maximizer

$$|\mathbf{z}^H \mathbf{H} \mathbf{f}|^2 \le \frac{1}{M_r} \left(\sum_{m=1}^{M_r} |(\mathbf{H}\mathbf{f})_m| \right)^2 = \frac{1}{M_r} \|\mathbf{H}\mathbf{f}\|_1^2, \tag{10.10}$$

where the equality holds when the optimal combiner is

$$\hat{\mathbf{z}} = \frac{1}{\sqrt{M_r}} e^{j(\varphi + \mathrm{phase}(\mathbf{H}\hat{\mathbf{f}}))} \tag{10.11}$$

for an arbitrary phase $\varphi \in [0, 2\pi)$ [46]. Here, $e^{j\theta} \doteq [e^{j\theta_1}, \cdots, e^{j\theta_M}]^T$ with $\theta \doteq [\theta_1, \cdots, \theta_M]^T$, and $\|\mathbf{a}\|_1$ is the 1-norm of the vector $\mathbf{a}$.

Without loss of generality, we assume $\varphi = 0$. Then, we can compute the optimal beamformer $\hat{\mathbf{f}} = (1/\sqrt{M_t}) e^{j\theta}$, where

$$\boldsymbol{\theta} = \underset{\vartheta \in [0, 2\pi)^{M_t}}{\arg\max} \|\mathbf{H} e^{j\vartheta}\|_1^2. \tag{10.12}$$

By substituting the optimal combiner in (10.11) and beamformer in (10.12) into the maximizer, the optimization problem can be rewritten as

$$(\hat{\mathbf{z}}, \hat{\mathbf{f}}) = \left(\frac{1}{\sqrt{M_r}} e^{j\mathrm{phase}(\mathbf{H} e^{j\theta})}, \frac{1}{\sqrt{M_t}} e^{j\theta} \right),$$

$$\boldsymbol{\theta} = \underset{\vartheta \in [0, 2\pi)^{M_t}}{\arg\max} \|\mathbf{H} e^{j\vartheta}\|_1^2. \tag{10.13}$$

In the special case where the channel has a dominant LOS path, the channel can be approximated as the rank-one matrix $\mathbf{H} = \mathbf{g}_{rx} \mathbf{g}_{tx}^H$. Then, the optimization problem can be divided into two optimization problems seeking AoA and AoD independently,

$$\hat{\mathbf{z}} = \frac{1}{\sqrt{M_r}} e^{j\theta}, \tag{10.14}$$

$$\boldsymbol{\theta} = \underset{\vartheta \in [0, 2\pi)^{M_r}}{\arg\max} |\mathbf{g}_{rx}^H e^{j\vartheta}|^2,$$

and

$$\hat{\mathbf{f}} = \frac{1}{\sqrt{M_t}} e^{j\phi}, \tag{10.15}$$

$$\boldsymbol{\phi} = \underset{\varphi \in [0, 2\pi)^{M_r}}{\arg\max} |\mathbf{g}_{tx}^H e^{j\varphi}|^2.$$

However, many NLOS millimeter wave channels cannot be approximated by a single dominant path, and therefore the optimization problem in (10.13) cannot be solved in a closed form.

The implementation of the analog beamformer may also add additional constraints to the optimization problem in (10.13). For example, some practical phase shifters can only generate phases that lie in a discrete set, for example

$$\mathcal{W}_{2^B} = \left\{0, \frac{2\pi}{2^B}, \cdots, \frac{2\pi(2^B-1)}{2^B}\right\}. \tag{10.16}$$

Lastly, it is not feasible to assume that the transmitter has perfect CSI. For these reasons, practical beamformers/combiners for beam alignment need to be computed using a brute force search and codebook-based subspace sampling approaches. The practical solutions will be discussed in Section 10.4.1.

10.3.1.2 Lens-based Analog Beamforming Architecture

We also consider lens-based analog beamforming in a continuous aperture phased (CAP) MIMO architecture as in [22, 32, 33]. A cost-effective and less complex alternative to analog beamforming using RF phase shifters is to use an architecture based on a discrete lens array (DLA). The DLA acts as a convex lens that maps transmitted beamspace signals to analog signals. A DLA can also be implemented at the receiver to map analog signals back to beamspace signals for processing. We view the transmitter and receiver as critically sampled n-dimensional ULAs and can thus express the LOS link as a MIMO system

$$\mathbf{y} = \mathbf{H}\mathbf{x} + \mathbf{n}, \tag{10.17}$$

where $\mathbf{x}$ is the transmitted M_t-dimensional aperture domain signal, $\mathbf{r}$ is the received M_r-dimensional aperture domain signal, $\mathbf{H}$ is the $M_r \times M_t$ aperture domain channel matrix, and $\mathbf{n}$ is an M_r-dimensional AWGN vector. We can also represent the link in the beam space domain as

$$\mathbf{y}_v = \mathbf{H}_v \mathbf{x}_v + \mathbf{n}_v = \mathbf{D}_{M_r}^H \mathbf{H} \mathbf{D}_{M_t} \mathbf{x}_v + \mathbf{n}_v, \tag{10.18}$$

where $\mathbf{H}_v$ is modeled as in (10.6) and $\mathbf{r}_v$, $\mathbf{x}_v$, and $\mathbf{n}_v$ are the beamspace representations of their aperture domain counterparts. In this case, the matrices $\mathbf{D}_{M_r}$ and $\mathbf{D}_{M_t}$ are the receive combiner and transmit beamformer matrices, respectively.

The rank of $\mathbf{H}_v$ is related to the number of orthogonal beams p_{max} coupled from the transmitter to the receiver, which generally satisfies $p_{max} \ll M_t, M_r$. From the low-dimensional structure of the virtual channel, we have that there is only a relatively small $p_{max} \times p_{max}$ submatrix $\tilde{\mathbf{H}}_v$ of nonzero elements. This implies that the beamforming and combiner matrices approximate the singular vectors of the millimeter wave channel and that the optimal DLA beamformer (or combiner) $\mathbf{F}_{DLA}$ ($\mathbf{F}_{DLA}^H$) should approximate a DFT (or IDFT) matrix. The DLA matrix can modeled as $\mathbf{F}_{DLA} = \mathbf{P}\mathbf{F}_{FA}$, where $\mathbf{P}$ represents the aperture phase profile and $\mathbf{F}_{FA}$ represents the paths from the focal points of the array to the DLA aperture elements. The goal of lens-based beamforming is to design the DLA using $\mathbf{P}$ and $\mathbf{F}_{FA}$ to best approximate a DFT matrix, thus achieving communication of p_{max} independent data streams. An optimal DLA is able to exploit the low number of spatial communication modes in LOS millimeter wave links with a significantly lower complexity than that of traditional MIMO architectures.

10.3.2   Hybrid Beamforming Solutions

Analog beamforming solutions are cost-effective and simple to implement, but they are inherently suboptimal owing to architecture limitations and single-stream transmission. Hybrid beamforming solutions have been proposed in order to achieve larger beamforming gains than are possible with analog solutions while reducing the complexity of digital precoding and combining [20, 21, 45]. Two hybrid precoding architectures are depicted in Figure 10.4(b) and (c). We consider a millimeter wave hybrid precoding system transmitting M_s data streams using a symbol vector $\mathbf{x} \in \mathbb{C}^{M_s}$. The transmitter and receiver are assumed to have M_t^{RF} and M_r^{RF} RF chains, respectively, where $M_s \leq M_t^{RF} \leq M_t$ and $M_s \leq M_r^{RF} \leq M_r$. Precoding and combining are divided between the analog and digital domains and the received signal can be expressed as

$$\mathbf{y} = \mathbf{Z}_{BB}^H \mathbf{Z}_{RF}^H \mathbf{H} \mathbf{F}_{RF} \mathbf{F}_{BB} \mathbf{x} + \mathbf{n}, \tag{10.19}$$

where $\mathbf{F}_{BB}$ and $\mathbf{F}_{RF}$ are the $M_t^{RF} \times M_s$ digital and $M_t \times M_t^{RF}$ analog precoding matrices, and $\mathbf{Z}_{BB}$ and $\mathbf{Z}_{RF}$ are the $M_r^{RF} \times M_s$ digital and $M_r \times M_r^{RF}$ analog combining matrices. The AWGN vector $\mathbf{n}$ has elements distributed as $\mathcal{CN}(0,1)$ after normalization, and we assume $E[\|\mathbf{x}\|_2^2] = \rho/M_s$. The analog matrices are implemented using RF phase shifters, and thus we require that their columns lie in the set $\mathcal{B}_M$. The digital matrices are not subject to any structural constraints but are assumed to be normalized such that $\|\mathbf{F}_{RF}\mathbf{F}_{BB}\|_F^2 = M_s$ and $\|\mathbf{Z}_{RF}\mathbf{Z}_{BB}\|_F^2 = M_s$. The digital precoding and combining matrices give the system more degrees of freedom for processing the transmitted and received signals.

Joint optimization of the transmitter and receiver can be significantly difficult, and thus we consider finding the optimal precoder in a decoupled setting assuming perfect CSI. The optimization problem is then defined as

$$(\mathbf{F}_{RF}, \mathbf{F}_{BB}) = \underset{\mathbf{F}_{RF}, \mathbf{F}_{BB}}{\arg\max} \|\mathbf{F}_{opt} - \mathbf{F}_{RF}\mathbf{F}_{BB}\|_F \tag{10.20}$$

subject to

$$\mathbf{F}_{RF}^{(i)} \in \mathcal{B}_M$$

$$\|\mathbf{F}_{RF}\mathbf{F}_{BB}\|_F^2 = M_s,$$

where $\mathbf{F}_{opt}$ is the unconstrained optimal unitary precoder. This is equivalent to finding a baseband precoder that corresponds to the projection of the optimal precoder onto the subspace spanned by the RF precoder. The complex nature of the equal-gain subset makes the optimization problem in (10.20) mathematically intractable. However, near-optimal solutions can be obtained via orthogonal matching pursuit by restricting the columns of the analog precoder to be in the transmitter array manifold.

Although the performance of hybrid precoding is limited by the number of RF chains so that it is suboptimal compared with digital precoding, results show that the sparsity of the millimeter wave channel bridges the gap between the two architectures. While hybrid precoding provides larger beamforming gains than analog beamforming and is able to support multistream transmission, the freedom provided by the digital precoder also allows designs that reduce interference in multiuser systems. In [43], hybrid precoder optimization was considered in systems with only partial CSI at the transmitter. Training of hybrid precoder designs for millimeter wave channel estimation was studied in [44].

10.3.3   Low-Resolution Receiver Architecture

One of the main allures of millimeter wave communication is the availability of large-bandwidth communication, allowing millimeter wave systems to support multigigabit rates. However, a wide bandwidth imposes constraints on sampling. Using a large number of high-rate ADCs generating many bits per sample would be expensive and consume too much power. Generally, the power consumption of ADCs increases with the resolution [47]. For example, it is reported that a 14-bit ADC operating at a rate of 2.5 GS/s consumes almost 2.39 W [48], making it unfit for mobile devices.

MIMO system architectures with low-resolution ADCs have been studied in recent years [49–52]. Unfortunately, receiver design with low-resolution ADCs is much more complicated in millimeter wave systems. The large number of antennas in millimeter wave systems has generated research on new beamforming architectures [53–55]. In the low-resolution receiver architecture, baseband beamforming/combining techniques are utilized by using one RF phase shifter per antenna. Sampling is conducted by using a pair of low-resolution ADCs (a single ADC is assigned to sampling each of the in-phase and quadrature demodulated signal components) connected to each of the antenna elements. The sampled signals are then reprocessed, for example by combining or beamforming, at baseband. An overview of a low-resolution receiver architecture is depicted in Figure 10.4(d).

A low-precision, for example $B = 1-3$ bit, ADC allows millimeter wave systems to conduct combining with substantially reduced power. For example, it has been noted that a 1-bit ADC operating at a rate of 240 GS/s consumes only 10 mW [56]. However, low-precision ADCs could be a bottleneck in millimeter wave systems and hinder any capacity enhancement. For this reason, physical layer techniques must be reevaluated to consider the additional hardware constraints. Beamforming techniques, for example analog and hybrid beamforming solutions, in millimeter wave systems need to be redesigned to lessen the system throughput degradation due to the low-precision ADCs. In addition, channel-training techniques need to be revisited by considering the constraint due to the low-precision ADCs and the sparse characteristics of millimeter wave links [53, 54]. The reevaluation of physical layer technologies and the analysis of the capacity of these links are promising research topics.

10.4   Channel Acquisition Techniques

Maximizing the performance of MIMO communication requires accurate CSI at the transmitter. This transmit-side CSI is obtained either through reciprocity or by feedback techniques. In today's fourth generation (4G) systems, feedback is the dominant method for obtaining CSI at the transmitter. This is enabled by having the transmitter send a pilot or reference signal from each transmit antenna [57–59]. The receiver then uses its knowledge of the reference signal to estimate the channel, quantize the channel, and convey it back to the transmitter through a small number of control information bits [60–62].

Having CSI at the transmitter is also necessary for millimeter wave systems to compute the relevant transmit beamformers and receive combiners [63, 64]. Furthermore, some MIMO techniques in multiuser settings, for example multibeam transmission for multiuser MIMO (MU-MIMO) systems, will require highly accurate CSI. Inaccurate CSI results in beam misalignment. In a single-user link, beam misalignment causes a performance loss due to an SNR drop or other interstream interference. In a multiuser system, the misaligned beams causes interference between users and lower the signal-to-interference-plus-noise-ratio (SINR).

The channel acquisition problem is much more complicated in millimeter wave systems. As mentioned earlier, beamforming will likely have to be facilitated by at least some amount of analog processing [8, 9, 20, 21, 37–40, 44]. The receiver will likely not be able to sample the received signal of each antenna element directly. Instead, CSI will have to be estimated from a postprocessed received signal that has been subjected to some level of combining or to passage through a low-rate/low-resolution ADC.

The other complication is training overhead. A large number of transmit antennas could necessitate a large training overhead to enable accurate, explicit channel estimation. Using more time for channel training can increase the quality of the estimated CSI. However, it is critical in most systems to limit the training overhead to improve overall user data throughput. Furthermore, this training overhead limitation is combined with the higher expected path loss and the small effective aperture of each antenna element to make channel estimation an important research topic in the millimeter wave field.

10.4.1   Subspace Sampling for Beam Alignment

Linear channel estimation methods will necessitate sending a large number of channel training signals [65–67]. The large size of training sequences imposes a heavy burden on the downlink and, potentially, the uplink control channels.

The channel estimation problem in millimeter wave systems can be replaced by one of beam alignment. This means that we exploit the subtle points that (i) the receiver needs only to acquire the effective channel $\mathbf{z}^H \mathbf{H} \mathbf{f}$ for coherent demodulation, not the entire matrix $\mathbf{H}$, and (ii) the transmitter needs only to acquire the beamformer $\mathbf{f}$. In a beam alignment approach to channel estimation, subspace sampling is conducted by using a finite number of training vectors to choose the proper combiner and beamformer pair. The transmitter must send pilot sequences on different beams, for example scan different AoDs, while the receiver switches between different receive beams, for example it scans different AoAs.

10.4.1.1   Nonadaptive Subspace Sampling

The most simplistic method of subspace sampling is implemented by having the transmitter send beamformed training vectors and the receiver process an effective receive signal, obtained by performing combining across the receive array. This approach effectively sounds the channel with one beamformer–combiner pair at a time. For a point-to-point transmission scenario, one possible approach is to choose the final beamformer–combiner pair as

$$(\hat{\mathbf{z}}, \hat{\mathbf{f}}) = \arg\max_{(\mathbf{z}, \mathbf{f}) \in \mathcal{C}_r \times \mathcal{C}_t} \left| \sqrt{\rho} \mathbf{z}^H \mathbf{H} \mathbf{f} + n \right|^2, \tag{10.21}$$

where $\mathcal{C}_r$ and $\mathcal{C}_t$ are predefined sets used for the subspace sampling. The millimeter wave channel subspace can be scrutinized carefully because all pairs of beamformers and combiners are fully intertwined.

10.4.1.2   Adaptive Subspace Sampling with Hierarchical Subcodebooks

The total channel-sounding time required for joint subspace sampling is $Q = \text{card}(\mathcal{C}_r) \cdot \text{card}(\mathcal{C}_t)$. If we scan the AoD and AoA with codebooks having $\text{card}(\mathcal{C}_r) = M_r$ and $\text{card}(\mathcal{C}_t) = M_t$ codewords, the channel-training time increases with the product $M_t \cdot M_r$. Millimeter wave systems require a large number of antennas to facilitate directional transmission, so that the joint subspace-sampling approach entails a heavy burden on the downlink, for example a training overhead, and on the uplink control channels, for example feedback links. For example, if $M_t = 128$ and $M_r = 64$, the required sounding time will be $Q = 8192$, which is not feasible.

Most channel acquisition techniques allow the transmitter to utilize the observed training data to update channel-training packets adaptively. Adaptive subspace-sampling approaches have been proposed to reduce the higher expected beam search complexity and the total training time [37, 38]. We now overview one ping-pong adaptive sampling algorithm designed for millimeter wave MIMO systems [30].

Assuming R rounds of ping-pong sounding, both the transmitter and the receiver conduct subspace sampling $Q/2R$ times in each round. The abilities to properly align the transmit beam and receiver beam are clearly closely linked. For this reason, the AoA at the receiver and the AoD at the transmitter are estimated sequentially back and forth. In the adaptive subspace-sampling algorithm, high-resolution codebooks are used in the last few sampling rounds in order to refine the selected beamformers and combiners by using narrow beams with a higher beamforming gain. The ping-pong sounding algorithm is summarized in Algorithm 10.1, where $\mathcal{C}_{r,k}$ and $\mathcal{C}_{t,k}$ are the multiresolution codebooks designed for the kth round of sampling. Each codebook has $Q/2R$ beamformers.

10.4.1.3   TDD- and FDD-Based Beam Alignment

To facilitate the adaptive subspace sampling, the CSI (each selected beamformer $\mathbf{f}_k$ in Algorithm 10.1) needs to be supplied to the transmitter. In 5G networks, techniques that supply CSI to the transmitter must rely upon frequency division duplexing (FDD) or time division duplexing (TDD).

Algorithm 10.1 Ping-pong adaptive sampling.

Initialization

1: Define an initial omnidirectional combiner $\mathbf{z}_0 \in \mathcal{B}_{M_r}$

Iterative update

2: **for** $1 \leq k \leq R$

3:   Update a beamformer $\mathbf{f}_k = \hat{\mathbf{f}}$, where
  $\hat{\mathbf{f}} = \arg\max_{\mathbf{f} \in \mathcal{C}_{t,k}} \left| \sqrt{\rho} \mathbf{z}_{k-1}^H \mathbf{H} \mathbf{f} + n \right|^2$

4:   Update a combiner $\mathbf{z}_k = \hat{\mathbf{z}}$, where
  $\hat{\mathbf{z}} = \arg\max_{\mathbf{z} \in \mathcal{C}_{r,k}} \left| \sqrt{\rho} \mathbf{z}^H \mathbf{H} \mathbf{f}_k + n \right|^2$

5: **end for**

Final update

6: Final beamformer $\mathbf{f}_R$

7: Final combiner $\mathbf{z}_R$

In feedback-assisted FDD systems, the transmitter sends pilot beams $\mathbf{f} \in \mathcal{C}_{t,k}$ to scan different AoDs and the receiver selects the beamformer $\mathbf{f}_k$. In each round, the index of $\mathbf{f}_k$ is then communicated through a feedback link incorporating a feedback overhead of $\log_2 |\mathcal{C}_{t,k}|$ bits. The total feedback overhead of $\sum_{k=1}^{R} \log_2 |\mathcal{C}_{t,k}|$ bits must be assigned in a feedback-assisted beam alignment architecture.

TDD systems can leverage channel reciprocity to allow the transmitter to access the updated beamformers $\mathbf{f}_k$. In the kth round, the receiver sends the pilot beam, i.e., the combiner $\mathbf{z}_{k-1}$ that is selected in the $(k-1)$th round, via a reverse link and the transmitter then switches between different beamformers to scan different AoDs. In a TDD-based beam alignment architecture, the transmitter can be informed of $\mathbf{f}_k$ without the assistance of feedback links. However, time synchronization and calibration issues between the forward and reverse links need to be resolved to facilitate TDD-based approaches.

10.4.1.4   Beam Alignment Performance

The practical deployment of millimeter wave systems requires analyses of the beam alignment performance in order to determine the time to be allocated to channel training. In realistic millimeter wave channel environments, such as street geometry scenarios [14], millimeter wave channels with a LOS path can be approximated by using a single ray-like beam, owing to the high Ricean K-factors [16, 37]. In codebook-based beamforming strategies, the transmit beamformer is chosen from a predefined codebook. Beam misalignment occurs when the optimally aligned beamformers found under noise-free and noisy conditions differ.

Under the conditions described and assuming the ideal, but unachievable, beam pattern described in [68], the probability of beam misalignment found in [40] can be used to show that the beam alignment performance increases as the codebook size increases. As the number of antennas increases, millimeter wave systems need to utilize more time for channel training because ray-like beams become narrower. Beam pattern design for subspace sampling needs to be extensively studied to ensure reliable beam alignment with the limited total channel-training time.

10.4.1.5 Beam Pattern Design

Codebooks need to be well designed because the beamformers in the codebook are used for both channel sounding and transmit beamforming. To sample the channel subspace accurately, beamformers have to be able to generate a beam pattern with high beamforming gain. Also, the sidelobe beam pattern that appears in undesired sections needs to be fully suppressed. Furthermore, different levels of codebooks need to be considered to exploit hierarchical beam alignment approaches.

We now overview a few codebooks that were designed for both analog and hybrid beamforming strategies. In [37, 38], multiresolution codebooks were developed for an analog beamforming architecture. In that work, the beamformers in the codebooks were designed to maximize the minimum beamforming gain in their own quantized sector. To achieve a favorable beam pattern, each beamformer was computed by combining a few subbeams based on a subarray method and beam-spoiling techniques. The normalized beamforming gain of the beamformer $\mathbf{f} \in \mathbb{C}^{M_t}$,

$$G_{BF} = \left| \mathbf{d}_{M_t}^H \left(\frac{2\pi d_h}{\lambda} \sin\theta \right) \mathbf{f} \right|^2, \tag{10.22}$$

is plotted in Figure 10.5(a), over a range of beam directions $\theta \in [-\pi/2, \pi/2]$ for $d_h = \lambda/2$.

Codebook designs for hybrid beamforming architectures have also been developed. In [44], a multiresolution codebook was designed by considering a hybrid beamforming architecture based on the orthogonal matching pursuit (OMP) algorithm. Furthermore, in [68], OMP-based hybrid codebooks were improved by minimizing the mean square error between each beamformer and the corresponding ideal beam pattern. The normalized beamforming gain of the hybrid beamformers presented in [68] is depicted in Figure 10.5(b). Note that the codebooks mentioned so far were designed by considering ULA structures. Designing codebooks for diverse antenna structures and channel conditions is still a topic of research.

10.4.2 Compressed Channel Estimation Techniques

Present-day wireless systems are able to reap the benefits of rich multipath channels by effectively utilizing the increased number of degrees of freedom (DoF) to obtain higher rates and/or a lower outage probability. Owing to the LOS properties of millimeter wave propagation, the channels encountered in millimeter wave systems often exhibit sparse multipath efforts. Because of this, traditional channel estimation methods are not as effective and different techniques must be used in order to achieve a target reconstruction error.

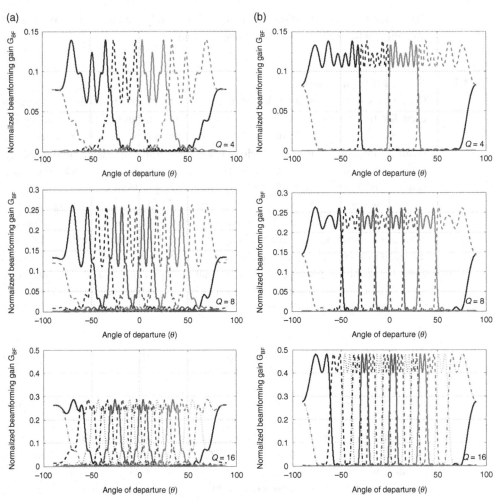

Figure 10.5 Beam patterns of multiresolution codebooks ($M_t = 32$, $Q = 4, 8, 16$). (a) Analog beamforming codebook with single RF chain [38]; (b) hybrid beamforming codebook with four RF chains [40].

Estimation of sparse channels using compressed sensing techniques has been a topic of interest in recent years [69–74]. Consider the linear training model

$$y[k] = \mathbf{H}x[k] + \mathbf{n}[k], \quad k = 0, \dots, N_{tr} - 1, \tag{10.23}$$

where $\mathbf{x}[k] \in \mathbb{C}^{M_t}$ is the kth column of the training sequence, $\mathbf{y}[k] \in \mathbb{C}^{M_t}$ is the kth column of the received training sequence, and $\{\mathbf{n}[k]\} \in \mathbb{C}^{M_r}$ is a sequence of independent and identically distributed (i.i.d.) AWGN vectors. The main premise of compressed channel sensing (CCS) is that if the parameter matrix is sparse, then $\mathbf{H}$ can be reliably estimated from observations and measurements in the case where $N_t \ll M_t$. The conditions for reliable estimation depend on the sparsity of the parameter matrix as well as on the structure of the measurement matrix. For a narrowband MIMO system, we can describe the channel $\mathbf{H}$ using a beamspace channel representation as in (10.6). The methods for reconstruction include the use of mixed-norm optimization, efficient greedy algorithms, and fast iterative thresholding methods [73]. We first premultiply the received training sequences $\mathbf{y}_n$ by $\mathbf{D}_{M_r}^{\mathrm{H}}$ and form the matrix $\mathbf{Y} \in \mathbb{C}^{N_{\mathrm{tr}} \times M_r}$ by row-wise stacking of the resulting vectors. We can then form the standard matrix training model

$$\mathbf{Y} = \mathbf{X}\mathbf{H}_{\mathrm{v}} + \mathbf{N}, \tag{10.24}$$

where $\mathbf{X} = [\mathbf{x}_0 \cdots \mathbf{x}_{N_{\mathrm{tr}}-1}]^{\mathrm{T}} \mathbf{D}_{M_t}^{\mathrm{H}}$, and $\mathbf{N}$ comprises i.i.d. entries distributed as $\mathcal{CN}(0,1)$.

Given this model, most traditional least-squares channel estimation approaches would require N_{tr} to be on the order of M_t. The advantage of CCS over other channel estimation methods is in the number of training dimensions needed for success. CCS requires the number of training vectors to scale as a function of the number of nonzero entries in the matrix. Since the channels are assumed to be sparse, CCS performs better than least-squares algorithms for a relatively small number of training dimensions.

10.5   Deployment Challenges and Applications

10.5.1   EM Exposure at Millimeter Wave Frequencies

As with other frequency bands, the transmission and reception of millimeter waves exposes users to electromagnetic radiation. There is an ongoing debate over the effects of long-term exposure, but the principal short-term effect of radiation at common wireless frequencies is superficial tissue heating [75]. In the millimeter wave bands, the three major dosimetric quantities are incident power density, specific absorption rate (SAR), and steady-state and/or transient temperature. Although all wireless devices must comply with the regulations of the FCC, most research in communications ignores RF exposure constraints. In [76], a SAR constraint was developed and incorporated into the system design to jointly optimize far-field system performance and near-field SAR performance.

Consideration of RF exposure limitations becomes more critical when one is transmitting in the millimeter wave bands. SAR increases rapidly as a function of frequency owing to a concurrent decrease in penetration depth and increase in the power-coupling to the skin. These factors result in SAR measurements that are significantly higher than those for lower-frequency bands with identical incident power densities [77]. In addition, millimeter wave systems are often implemented using phased arrays owing to the need for a large directional gain to combat the high path loss. The use of multiple antennas needed for satisfactory performance in these systems thus increases the amount of electromagnetic radiation absorbed by users. EM exposure needs to be considered in system design to ensure user safety and to satisfy FCC regulations, but it is currently unclear how to model millimeter wave EM exposure and how to incorporate exposure models into millimeter wave system design. Thorough surveys of research on the interactions between millimeter waves and humans have been presented in [78, 79].

10.5.2   Heterogeneous and Small-Cell Networks

The success of future wireless networks is dependent on the development of techniques and technologies with increased spectrum efficiency and on the design of innovative wireless networks. A solution to maximizing the efficiency of spectrum usage in a given area is to increase frequency reuse. Geographical spectrum efficiency can be achieved through the implantation of small-cell (picocell and femtocell) networks, bringing users closer to higher-rate backhaul links.

Deploying a large number of small-cell networks in a small geographic area promises to provide multigigabit rates [80]. Among various physical layer technologies, millimeter wave is a prime candidate for facilitating small-cell networks with cells of radius 200 m or less [6, 11].

Heterogeneous networks are also expected to be widely used in 5G networks. In heterogeneous networks, varying carrier frequencies and access technologies can be deployed together and they can all be jointly managed to control intercell and interuser interference, which severely degrades network quality. In addition, it is likely that mobile users will be simultaneously connected on both macrocell networks operating in sub-6 GHz bands and small-cell networks, for example below 200 m radius, in the millimeter wave band. The advanced signal-processing techniques needed for dual-connectivity and multiuser systems necessitate reliable, flexible, and scalable backhaul networks in order to be successful.

The highly sophisticated network architecture significantly increases the backhaul complexity. One possible solution is to connect these heterogeneous networks by using a fiber-based wired backhaul. However, wired backhaul, while effective, will be expensive and difficult to install for dense cell networks. A more cost-effective solution is wireless backhaul, which is more flexible and easier to install. However, the saturation of the sub-3 GHz spectrum renders in-band backhaul impractical. Another option is to provide backhaul through millimeter wave links, thus avoiding traditional wireless bands. In addition, the large amount of available spectrum in the millimeter band gives users access to high-data-rate access points [37].

However, millimeter wave backhaul deployment gives rise to challenges such as beam alignment with limited resources, and decreased link performance caused by environmental factors. The path loss of outdoor millimeter wave systems is considerably higher than that of in-band wireless systems, necessitating the use of large arrays to provide large beamforming gains. Beam alignment becomes difficult owing to the narrow width of the beam and constraints such as limited channel knowledge and small alignment times. A MIMO system could potentially have antennas numbering in the hundreds or thousands in order to combat the severe path loss. These systems could be mounted on traffic control structures such as lampposts, road signs, and stop lights, providing users in urban deployments with both backhaul and broadband access. These structures are subject to movement from wind and cars, among other factors, disturbing the propagation geometry of the arrays. Perturbations to these systems can cause undesirable outages if beam alignment is not performed frequently. These challenges are discussed further in [81].

Acknowledgments

Miguel R. Castellanos and David J. Love would like to thank the National Science Foundation for support from grant CCF-1403458. David J. Love would also like to thank the National Science Foundation for support from grant CCF-1141868.

References

[1] W. Roh, "Performances and feasibility of mmwave beamforming prototype for 5G cellular communications," in *Proc. of IEEE International Conf. on Communications (ICC)*, Jun. 2013.

[2] A. Ghosh, "Can mmWave wireless technology meet the future capacity crunch," in *Proc. of IEEE International Conf. on Communications (ICC)*, Jun. 2013.

[3] F. Khan and J. Pi, "Millimeter-wave mobile broadband: Unleashing 3–300 GHz spectrum," in *Proc. of IEEE Wireless Communications and Networking Conf. (WCNC)*, Mar. 2011.

[4] Qualcomm, "Leading the world to 5G," Feb. 2015. Available at www.qualcomm. com/media/documents/files/qualcomm-5g-vision-presentation.pdf.

[5] Ericsson, "Microwave towards 2020," Sep. 2015. Available at www.ericsson.com/res/docs/2015/microwave-2020-report.pdf.

[6] T. S. Rappaport, S. Sun, R. Mayzus, H. Zhao, Y. Azar, K. Wang, G. N. Wong, J. K. Schulz, M. Samimi, and F. Gutierrez, "Millimeter wave mobile communications for 5G cellular: It will work!," *IEEE Access*, vol. 1, pp. 335–349, May 2013.

[7] A. Ghosh, T. A. Thomas, M. C. Cudak, R. Ratasuk, P. Moorut, F. W. Vook, T. S. Rappaport, G. R. MacCartney, S. Sun, and S. Nie, "Millimeter-wave enhanced local area systems: A high-data-rate approach for future wireless networks," *IEEE J. Sel. Areas Commun.*, vol. 32, no. 6, pp. 1152–1163, Jun. 2014.

[8] Z. Pi and F. Khan, "An introduction to millimeter-wave mobile broadband systems," *IEEE Commun. Mag.*, vol. 49, no. 6, pp. 101–107, Jun. 2011.

[9] W. Roh, J.-Y. Seol, J. Park, B. Lee, J. Lee, Y. Kim, J. Cho, K. Cheun, and F. Aryanfar, "Millimeter-wave beamforming as an enabling technology for 5G cellular communications: Theoretical feasibility and prototype results," *IEEE Commun. Mag.*, vol. 52, no. 2, pp. 106–113, Feb. 2014.

[10] R. W. Heath, N. Gonzalez-Prelcic, S. Rangan, W. Roh, and A. Sayeed, "An overview of signal processing techniques for millimeter wave MIMO systems," *IEEE J. Sel. Top. Signal Process.*, vol. 10, no. 3, pp. 436–453, Apr. 2016.

[11] Z. Pi, J. Choi, and R. W. Heath, "Millimeter-wave Gbps broadband evolution towards 5G: Fixed access and backhaul," *IEEE Commun. Mag.*, vol. 54, no. 4, pp. 138–144, Apr. 2016.

[12] FCC, "Allocations and service rules for the 71–76 GHz, 81–86 GHz, and 92–95 GHz bands," Memorandum opinion and order, Mar. 2005.

[13] E. Ben-Dor, T. S. Rappaport, Y. Qiao, and S. J. Lauffenburger, "Millimeter-wave 60 GHz outdoor and vehicle AoA propagation measurements using a broadband channel sounder," in *Proc. of IEEE Global Communications Conf. (GLOBECOM)*, Dec. 2011.

[14] H. Zhang, S. Venkateswaran, and U. Madhow, "Channel modeling and MIMO capacity for outdoor millimeter wave links," in *Proc. of IEEE Wireless Communications and Networking Conf. (WCNC)*, Apr. 2010.

[15] E. Torkildson, H. Zhang, and U. Madhow, "Channel modeling for millimeter wave MIMO," in *Proc. of Information Theory and Applications Workshop (ITA)*, Jan. 2010.

[16] Z. Muhi-Eldeen, L. Ivrissimtzis, and M. Al-Nuaimi, "Modelling and measurements of millimeter wavelength propagation in urban environments," *IET Microw. Antennas Propag.*, vol. 4, no. 9, pp. 1300–1309, Sep. 2010.

[17] T. S. Rappaport, F. Gutierrez, E. Ben-Dor, J. N. Murdock, Y. Qiao, and J. I. Tamir, "Broadband millimeter-wave propagation measurements and models using adaptive-beam antennas for outdoor urban cellular communications," *IEEE Trans. Antennas Propag.*, vol. 61, no. 4, pp. 1850–1859, Apr. 2013.

[18] F. Gutierrez, S. Agarwal, K. Parrish, and T. S. Rappaport, "On-chip integrated antenna structures in CMOS for 60 GHz WPAN systems," *IEEE J. Sel. Areas Commun.*, vol. 27, no. 8, pp. 1367–1378, Oct. 2009.

[19] S. Rangan, T. S. Rappaport, and E. Erkip, "Millimeter-wave cellular wireless networks: Potentials and challenges," *Proc. IEEE*, vol. 102, no. 3, pp. 366–385, Mar. 2014.

[20] O. E. Ayach, R. W. Heath, S. Abu-Surra, S. Rajagopal, and Z. Pi, "Low complexity precoding for large millimeter wave MIMO systems," in *Proc. of IEEE International Conf. on Communications (ICC)*, Jun. 2012.

[21] O. E. Ayach, S. Rajagopal, S. Abu-Surra, Z. Pi, and R. W. Heath, "Spatially sparse precoding in millimeter wave MIMO systems," *IEEE Trans. Wireless Commun.*, vol. 13, no. 3, pp. 1499–1513, Mar. 2014.

[22] J. Brady, N. Behdad, and A. M. Sayeed, "Beamspace MIMO for millimeter-wave communications: System architecture, modeling, analysis, and measurements," *IEEE Trans. Antennas Propag.*, vol. 61, no. 7, pp. 3814–3827, Jul. 2013.

[23] C.-S. Choi, M. Elkhouly, E. Grass, and C. Scheytt, "60-GHz adaptive beamforming receiver arrays for interference mitigation," in *Proc. of IEEE Personal Indoor and Mobile Radio Communications (PIMRC)*, Sep. 2010.

[24] D. Ramasamy, S. Venkateswaran, and U. Madhow, "Compressive adaptation of large steerable arrays," in *Proc. of Information Theory and Applications Workshop (ITA)*, Feb. 2012.

[25] A. Hirata, T. Kosugi, H. Takahashi, R. Yamaguchi, F. Nakajima, T. Furuta, H. Ito, H. Sugahara, Y. Sato, and T. Nagatsuma, "120-GHz-band millimeter-wave photonic wireless link for 10-Gb/s data transmission," *IEEE Trans. Microw. Theory Tech.*, vol. 54, no. 5, pp. 1937–1944, May 2006.

[26] ITU-R, "Attenuation by atmospheric gases," Recommendation ITU-R P.676-10, 2013.

[27] FCC, "Millimeter wave propagation: Spectrum management implications," Bulletin 70, 2007.

[28] H. Zhao, R. Mayzus, S. Sun, M. Samimi, J. K. Schulz, Y. Azar, K. Wang, G. N. Wong, F. Gutierrez, and T. S. Rappaport, "28 GHz millimeter wave cellular communication measurements for reflection and penetration loss in and around buildings in New York City," in *Proc. of IEEE International Conf. on Communications (ICC)*, Jun. 2013.

[29] S. Collonge, G. Zaharia, and G. E. Zein, "Influence of the human activity on wide-band characteristics of the 60 GHz indoor radio channel," *IEEE Trans. Wireless Commun.*, vol. 3, no. 6, pp. 2396–2406, Oct. 2004.

[30] S. Hur, T. Kim, D. J. Love, J. V. Krogmeier, T. A. Thomas, and A. Ghosh, "Millimeter wave beamforming for wireless backhaul and access in small cell networks," *IEEE Trans. Commun.*, vol. 61, no. 10, pp. 4391–4403, Oct. 2013.

[31] R. C. Hansen, *Phased Array Antennas*, 2nd edn, Wiley-Interscience, 2009.

[32] A. M. Sayeed and N. Behdad, "Continuous aperture phased MIMO: Basic theory and applications," in *Proc. of Allerton Conf. on Communication, Control, and Computing*, Oct. 2010.

[33] A. M. Sayeed and N. Behdad, "Continuous aperture phased MIMO: A new architecture for optimum line-of-sight links," in *Proc. of IEEE International Symposium on Antennas and Propagation (ISAP)*, Jul. 2011.

[34] A. M. Sayeed, "Deconstructing multiantenna fading channels," *IEEE Trans. Signal Process.*, vol. 50, no. 10, pp. 2563–2579, Oct. 2002.

[35] ITU-R, "Specific attenuation model for rain for use in prediction methods," Recommendation ITU-R P. 838-3, 2005.

[36] J. G. Proakis, *Digital Communications*, 4th edn, McGraw-Hill, 2001.

[37] S. Hur, T. Kim, D. J. Love, J. V. Krogmeier, T. A. Thomas, and A. Ghosh, "Multilevel millimeter wave beamforming for wireless backhaul," in *Proc. of IEEE Global Communications Conf. Workshops (GLOBECOM)*, Dec. 2011.

[38] S. Hur, T. Kim, D. J. Love, J. V. Krogmeier, T. A. Thomas, and A. Ghosh, "Millimeter wave beamforming for wireless backhaul and access in small cell networks," *IEEE Trans. Commun.*, vol. 61, no. 10, pp. 4391–4403, Oct. 2013.

[39] J. Song, S. G. Larew, D. J. Love, T. A. Thomas, and A. Ghosh, "Millimeter wave beam-alignment for dual-polarized outdoor MIMO systems," in *Proc. of IEEE Global Communications Conf. Workshops (GLOBECOM)*, Dec. 2013.

[40] J. Song, J. Choi, S. G. Larew, D. J. Love, T. A. Thomas, and A. Ghosh, "Adaptive millimeter wave beam alignment for dual-polarized MIMO systems," *IEEE Trans. Wireless Commun.*, vol. 14, no. 11, pp. 6283–6296, Nov. 2015.

[41] IEEE, "PHY/MAC complete proposal specification (TGad D0.1)," IEEE 802.11-10/0433r2 Std., 2012.

[42] S. Sun, T. S. Rappaport, R. W. Heath, A. Nix, and S. Rangan, "MIMO for millimeter-wave wireless communications: Beamforming, spatial multiplexing, or both?" *IEEE Commun. Mag.*, vol. 52, no. 12, pp. 110–121, Dec. 2014.

[43] A. Alkhateeb, O. E. Ayach, G. Leus, and R. W. Heath, "Hybrid precoding for millimeter wave cellular systems with partial channel knowledge," in *Proc. of Information Theory and Applications Workshop (ITA)*, Feb. 2013.

[44] A. Alkhateeb, O. E. Ayach, G. Leus, and R. W. Heath, "Channel estimation and hybrid precoding for millimeter wave cellular systems," *IEEE J. Sel. Top. Signal Process.*, vol. 8, no. 5, pp. 831–846, Oct. 2014.

[45] A. Alkhateeb, G. Leus, and R. W. Heath, "Limited feedback hybrid precoding for multi-user millimeter wave systems," *IEEE Trans. Wireless Commun.*, vol. 14, no. 11, pp. 6481–6494, Nov. 2015.

[46] D. J. Love, R. W. Heath, and T. Strohmer, "Grassmannian beamforming for multiple-input multiple-output wireless systems," *IEEE Trans. Inf. Theory*, vol. 49, no. 10, pp. 2735–2747, Oct. 2003.

[47] R. H. Walden, "Analog-to-digital converter survey and analysis," *IEEE J. Sel. Areas Commun.*, vol. 17, no. 4, pp. 539–550, Apr. 2009.

[48] B. Murmann, "ADC performance survey," in *Proc. of ISSCC and VLSI Symposium*, vol. 2013, 1997.

[49] O. Dabeer, J. Singh, and U. Madhow, "On the limits of communication performance with one-bit analog-to-digital conversion," in *Proc. of IEEE International Workshop on Signal Processing Advances in Wireless Communications (SPAWC)*, Dec. 2006.

[50] M. T. Ivrlac and J. A. Nossek., "On MIMO channel estimation with single-bit signal-quantization," in *International ITG Workshop on Smart Antennas (WSA)*, Feb. 2007.

[51] A. Mezghani and J. A. Nossek, "On ultra-wideband MIMO systems with 1-bit quantized outputs: Performance analysis and input optimization," in *Proc. of IEEE International Symposium on Information Theory (ISIT)*, Jun. 2007.

[52] A. Mezghani, F. Antreich, and J. A. Nossek, "Multiple parameter estimation with quantized channel output," in *Proc. of International ITG Workshop on Smart Antennas (WSA)*, Feb. 2010.

[53] O. E. Ayach, S. Rajagopal, S. Abu-Surra, Z. Pi, and R. W. Heath, "Analog-to-digital converter survey and analysis," *IEEE Trans. Wireless Commun.*, vol. 13, no. 3, pp. 1499–1513, Jan. 2014.

[54] J. Mo and R. W. Heath, "High SNR capacity of millimeter wave MIMO systems with one-bit quantization," in *Proc. of Information Theory and Applications Workshop (ITA)*, Feb. 2014.

[55] D. Palguna, D. J. Love, T. Thomas, and A. Ghosh, "Millimeter wave receiver design using parallel delta sigma ADCs and low precision quantization," in *Proc. of IEEE Global Communications Conf. Workshops (GLOBECOM)*, Dec. 2015.

[56] G. P. Fettweis, "HetNet wireless fronthaul: The challenge missed," in *Proc. of Information Theory and Applications Workshop (ITA)*, Feb. 2014.

[57] H. Q. Ngo and E. G. Larsson, "EVD-based channel estimation in multicell multiuser MIMO systems with very large antenna arrays," in *Proc. of IEEE International Conf. on Acoustics, Speech and Signal Processing (ICASSP)*, Mar. 2012.

[58] H. Yin, D. Gesbert, M. Filippou, and Y. Liu, "A coordinated approach to channel estimation in large-scale multiple-antenna systems," *IEEE J. Sel. Areas Commun.*, vol. 31, no. 2, pp. 264–273, Feb. 2013.

[59] A. Soysal and S. Ulukus, "Joint channel estimation and resource allocation for MIMO systems – part II: Multi-user and numerical analysis," *IEEE Trans. Wireless Commun.*, vol. 9, no. 2, pp. 632–640, Feb. 2010.

[60] C.-B. Chae, D. Mazzarese, N. Jindal, and R. W. Heath, "Coordinated beamforming with limited feedback in the MIMO broadcast channel," *IEEE J. Sel. Areas Commun.*, vol. 26, no. 8, pp. 1505–1515, Oct. 2008.

[61] R. Bhagavatula and R. W. Heath, "Adaptive limited feedback for sum-rate maximizing beamforming in cooperative multicell systems," *IEEE Trans. Signal Process.*, vol. 59, no. 2, pp. 800–811, Feb. 2011.

[62] P. Kerret and D. Gesbert, "CSI feedback allocation in multicell MIMO channels," in *Proc. of IEEE International Conf. on Communications (ICC)*, Jun. 2012.

[63] D. Gesbert, S. Hanly, H. Huang, S. S. Shitz, O. Simeone, and W. Yu, "Multi-cell MIMO cooperative networks: A new look at interference," *IEEE J. Sel. Areas Commun.*, vol. 28, no. 9, pp. 1380–1408, Dec. 2010.

[64] J. Zhang, R. Chen, J. G. Andrews, A. Ghosh, and R. W. Heath, "Networked MIMO with clustered linear precoding," *IEEE Trans. Wireless Commun.*, vol. 8, no. 4, pp. 1910–1921, Apr. 2009.

[65] B. Hassibi and B. M. Hochwald, "How much training is needed in multiple-antenna wireless links?" *IEEE Trans. Inf. Theory*, vol. 49, no. 4, pp. 951–963, Apr. 2003.

[66] W. Santipach and M. L. Honig, "Asymptotic performance of MIMO wireless channels with limited feedback," in *Proc. of IEEE Military Communcations Conf. (MILCOM)*, Oct. 2003.

[67] W. Santipach and M. L. Honig, "Optimization of training and feedback overhead for beamforming over block fading channels," *IEEE Trans. Inf. Theory*, vol. 56, no. 12, pp. 6103–6115, Dec. 2010.

[68] J. Song, J. Choi, and D. J. Love, "Codebook design for hybrid beamforming in millimeter wave systems," in *Proc. of IEEE International Conf. on Communcations (ICC)*, Jun. 2015.

[69] J. L. Paredes, G. R. Arce, and Z. Wang, "Ultra-wideband compressed sensing: Channel estimation," *IEEE J. Sel. Top. Signal Process.*, vol. 1, no. 3, pp. 383–395, Oct. 2007.

[70] G. Taübock and F. Hlawatsch, "A compressive sensing technique for OFDM channel estimation in mobile environments: Exploiting channel sparsity for reducing pilots," in *Proc. of IEEE International Conf. on Acoustics, Speech and Signal Processing (ICASSP)*, Mar. 2008.

[71] G. Taübock, F. Hlawatsch, D. Eiwen, and H. Rauhut, "Compressive estimation of doubly selective channels in multicarrier systems: Leakage effects and sparsity-enhancing processing," *IEEE J. Sel. Top. Signal Process.*, vol. 4, no. 2, pp. 255–271, Apr. 2010.

[72] C. R. Berger, S. Zhou, J. C. Preisig, and P. Willett, "Sparse channel estimation for multicarrier underwater acoustic communication: From subspace methods to compressed sensing," *IEEE Trans. Signal Process.*, vol. 58, no. 3, pp. 1708–1721, Mar. 2010.

[73] W. U. Bajwa, J. Haupt, A. M. Sayeed, and R. Nowak, "Compressed channel sensing: A new approach to estimating sparse multipath channels," *Proc. IEEE*, vol. 98, no. 6, pp. 1058–1076, Jun. 2010.

[74] C. R. Berger, Z. Wang, J. Huang, and S. Zhou, "Application of compressive sensing to sparse channel estimation," *IEEE Commun. Mag.*, vol. 48, no. 11, pp. 164–174, Nov. 2010.

[75] World Health Organization, "Electromagnetic fields and public health: Mobile phones," Fact Sheet 193, Oct. 2014. Available at www.who.int/mediacentre/factsheets/fs193/en/.

[76] B. Hochwald, D. J. Love, S. Yan, P. Fay, and J.-M. Jin, "Incorporating specific absorption rate constraints into wireless signal design," *IEEE Commun. Mag.*, vol. 52, no. 9, pp. 126–133, Sep. 2014.

[77] O. P. Gandhi and A. Riazi, "Absorption of millimeter waves by human beings and its biological implications," *IEEE Trans. Microw. Theory Technol.*, vol. 34, no. 2, pp. 228–235, Feb. 1986.

[78] M. Zhadobov, N. Chabat, R. Sauleau, C. L. Quement, and Y. L. Drean, "Millimeter-wave interactions with the human body: State of knowledge and recent advances," *Int. J. Microw. Wireless Technol.*, vol. 3, pp. 237–247, Mar. 2011.

[79] T. Wu, T. S. Rappaport, and C. M. Collins, "Safe for generations to come: Considerations of safety for millimeter waves in wireless communications," *IEEE Microw. Mag.*, vol. 16, no. 2, pp. 65–84, Mar. 2015.

[80] A. Ghosh, T. A. Thomas, M. C. Cudak, R. Ratasuk, P. Moorut, F. W. Vook, T. S. Rappaport, G. R. MacCartney, S. Sun, and S. Nie, "Millimeter wave enhanced local area systems: A high data rate approach for future wireless networks," *IEEE J. Sel. Areas Commun.*, vol. 32, no. 6, pp. 1152–1163, Jun. 2014.

[81] Y. Niu, Y. Li, D. Jin, L. Su, and A. V. Vasilakos, "A survey of millimeter wave communications for 5G: Opportunities and challenges," *Wireless Netw.*, vol. 21, no. 8, pp. 2657–2676, Sep. 2014.

11 Interference Mitigation Techniques for Wireless Networks

Koralia N. Pappi and George K. Karagiannidis

11.1 Introduction

Mobile broadband communications based on fourth generation (4G) Long Term Evolution (LTE) services are currently being deployed worldwide and are increasingly expanding across global markets, providing a user experience that was previously possible only through wired connections. The evolving fifth generation (5G) of wireless communication systems is expected to cope with a thousand-fold increase in total mobile broadband data and a hundred-fold increase in connected devices. It will also be required to overcome various challenges affecting current cellular networks, and provide higher data rates, improved end-to-end performance and coverage, low latency, and low energy consumption at low cost per transmission [1]. These challenges are expected to be addressed in 5G wireless networks by adopting a multi-tier heterogeneous architecture, since the capacity of the current macrocellular network cannot be increased infinitely. Future mobile networks will comprise macrocells and small cells, relays, and device-to-device (D2D) links, while they will be accessed by a large number of smart and heterogeneous devices.

This chapter discusses the management of interference in wireless networks of the 5G era. In the following, the goals of 5G networks and the interference management challenges they pose are first presented. Furthermore, the characteristics of future mobile networks which offer new tools for combating interference are also summarized. In this framework, coordinated multipoint techniques for both the uplink and the downlink are revisited, with special focus on the improvement of the cell-edge user experience. Finally, two indicative techniques for interference management are discussed, namely interference alignment and the compute-and-forward protocol.

11.2 The Interference Management Challenge in the 5G Vision

11.2.1 The 5G Primary Goals and Their Impact on Interference

The advance from 4G to 5G networks is based on some primary visions and goals, which are summarized in the following [2]:

- *High data rates and low latency.* 5G networks are envisioned to enable data rates of 300 Mb/s and 60 Mb/s in the downlink and uplink, respectively, and end-to-end latency between 2 and 5 milliseconds.
- *Machine-type communications (MTC).* The Internet of Things (IoT) is expected to introduce a large number of self-organized MTC devices, such as home appliances, and surveillance and sensor devices.
- *Millimeter wave communications.* Owing to the scarcity of spectral resources and the increasing demand for better service and more connected devices, the millimeter wave frequency band is a potential candidate for utilization in 5G networks, since it allows transmission over a wide bandwidth, with channels exceeding 20 MHz.
- *Multiple radio access technologies (RATs).* 5G is expected to enhance the existing RATs, with the aim of improving the system performance.
- *Prioritized spectrum access.* Users will be prioritized into high- and low-priority groups, according to both the network tiers they are connected to and their different requirements in terms of quality of service (QoS). Cognitive radio technology, which enables the simultaneous operation of high- and low-priority user equipment, is a promising option in this regard.
- *Network-assisted D2D communications.* In 5G systems, other network nodes, apart from the macrocell base station (BS), will be able to perform control signaling, thereby controlling D2D links, when the user equipment (UE) does not have a good-quality connection to the macrocell BS.
- *Energy harvesting for energy-efficient communications.* One of the main goals of future networks will be to prolong the battery lifetime of wireless devices, which can be improved by harvesting energy from various energy sources, including ambient radio signals. Hence, simultaneous wireless information and power transfer (SWIPT) is a candidate technique for adoption in 5G systems.

From the visions above, some of the main characteristics of 5G cellular networks can be inferred. These networks will be multitier, comprising different types of cells; they will serve a huge number of devices with high QoS demands, making more efficient spectrum utilization necessary; different radio access technologies will coexist in the same network and also in the same UE; the users will be prioritized; and the processing on the network side will be decentralized.

Multiple challenges regarding interference management arise from the requirements of 5G wireless networks, and thus radio resource and interference management is a key research challenge for moving toward 5G. Interference restricts the reusability of the spectral resources in space, time, frequency slots, and codes, thus limiting the overall spectral efficiency. Interference is stronger for cell-edge users, since the spectral resources dedicated to a cell-edge user might be reused by a neighboring cell.

The multitier network architecture also introduces interference between overlapping macrocells and small cells, in addition to neighboring cells. Moreover, since 5G envisions an extensive deployment of small cells, reusability of resources becomes more challenging. Interference between macrocells and small cells depends on the handoff boundary between cells. For example, when the boundary is based on the received signal power at the UE, a UE device close to a small cell may be served by the macrocell. In this case, this may cause interference in the uplink of the small cell, while a high macrocell transmission power shrinks the coverage of the small cell.

In order to avoid uplink interference and small-cell underutilization, the technique of cell range expansion (CRE) can be used [3]. In CRE, the handoff boundaries are shifted in favor of small cells, so that most UEs are served by small cells, effectively expanding their service area. However, in this case, although uplink interference is mitigated, CRE creates downlink interference for users which are associated with a small cell, but actually receive a stronger signal from a macrocell. It is thus evident that intercell interference is a major challenge for multitier wireless networks.

Another key goal of 5G networks is extended D2D communication, which can potentially improve user experience by reducing latency and power consumption. It is also expected to improve further the overall spectral efficiency of the network, since it will enable dense spectrum reuse. However, this comes at the cost of irregular interference, due to the ad hoc nature of D2D deployment. When multiple D2D pairs coexist in space, the transmitter of one pair might be closer to the receiver of another pair than to its intended destination, but still transmit to the intended destination, causing strong interference to the second receiver. This phenomenon is called restricted association [3]. Furthermore, interference is also caused not only between multiple D2D pairs but also between D2D pairs and cellular communications.

Interference management is also a major challenge for the employment of prioritized spectrum access. Users with a high priority must operate with adequate QoS, while the operation of low-priority users over the same spectrum must be completely transparent to the high-priority users, i.e., without causing severe interference. This is usually addressed through distributed power control algorithms with different objectives and interference constraints for each group of users.

From all the above, we conclude that the interference management challenges in 5G networks arise from the heterogeneity of the network, the dense deployment of devices, the diverse interference levels in different tiers due to the varying transmit powers of different BSs, and the priorities for accessing different spectrum resources.

11.2.2   Enabling Technologies for Improving Network Efficiency and Mitigating Interference

The architecture of 5G heterogeneous networks offers the opportunity of utilizing new technologies which enable the improvement of network efficiency and the mitigation of interference. One main characteristic is the connection of macrocells and small cells via ideal or nonideal backhaul links, which allows coordination inside the network for seamless handoff and interference mitigation.

One of these new enabling technologies is the cloud radio access network (C-RAN) [4]. The C-RAN provides low-latency high-rate ("ideal") backhaul links, and thus the baseband signals of multiple cells can be received and processed at a centralized server platform. Thus, this creates a centralized BS with distributed antennas, which may support the processing of multiple RAN protocols and varying traffic loads, since the pool of resources can be dynamically allocated. The C-RAN architecture saves the operational cost, since it enables the centralized processing of signals from multiple distributed antennas, simplifying the implementation of interference mitigation techniques, which require coordination and centralized baseband processing. C-RAN may also enable joint resource allocation across multiple RATs, increasing the capacity of 5G networks. C-RAN technology is implemented mainly through optical-fiber backhaul links.

When ideal backhaul links are not available, an anchor–booster architecture may be employed [5]. In this scheme, the macrocell acts as an anchor BS, controlling and regulating the mobility of the users. The small cell acts as a booster BS, where its main responsibility is to offload data traffic. This architecture is also very attractive for easier integration of different RATs, since the booster cells may employ different access schemes.

Millimeter wave (mmWave) technology and massive multiple-input multiple-output (MIMO) transceivers enable the utilization of large chunks of bandwidth, enabling high throughput, but also a large number of small antennas in handheld devices [3]. Although mmWave technology presents many challenges, including large path loss, signal absorption by obstacles, and low efficiency of amplifiers in this frequency range, its combination with large antenna arrays enables the implementation of interference-aware beamforming. Highly directional beams will improve the link budget, while at the same time providing spatial orthogonalization, enabling dense cell deployment with low intercell interference.

Another key characteristic of 5G networks is backhaul densification [3], where small cells are expected to be deployed in various indoor and also outdoor locations. Some BSs might be in positions where wired backhaul links are not possible, and wireless backhaul is necessary, with mmWave technology being a viable solution for line-of-sight (LoS) aggregation links (i.e., toward a node with fiber backhaul). The dense deployment of small cells enables the implementation of distributed/multiuser MIMO techniques, the improvement of the quality of wireless links, and an increase in the channel coherence time, which reduces the need for pilot overhead and feedback.

In modern cellular networks, optimizing the cell association and power control (CAPC) [2], i.e., the assignment of a UE to a BS and the selection of the transmit power both for the uplink and for the downlink, is vital to the system performance in various respects, such as interference mitigation, throughput maximization, and reduction in power consumption. However, based on the above enabling technologies, 5G wireless networks will allow partial or full cooperation between transmitters for the downlink, and between receivers for the uplink. A single UE will be associated with multiple BSs, enabling the utilization of interference coordination techniques, which promise a linear scaling of capacity with the number of network nodes.

11.3   Improving the Cell-Edge User Experience: Coordinated Multipoint

Coordinated multipoint (CoMP) [6], or cooperative MIMO, is a concept which is intended to improve the data rate for cell-edge users and the spectral efficiency for LTE-Advanced (LTE-A) and beyond. With CoMP, interference can be exploited or mitigated by cooperation between different sectors of the same cell or between different cells, either macrocells or small cells. In this section, some basic CoMP schemes for the uplink and the downlink will be described, while the technical challenges regarding backhaul traffic, synchronization, feedback, etc. will be identified and discussed.

11.3.1   Deployment Scenarios and Network Architecture

Future networks will be characterized mainly by heterogeneity and the coexistence of different types of cells within the network. Furthermore, the concept of C-RAN provides a concise description of the main parts of such a network [7], which are depicted in Figure 11.1:

- *The macrocell BS*, which is also referred to as an eNodeB. Communication between different eNodeBs is usually performed utilizing the X2 interface [7]. The X2

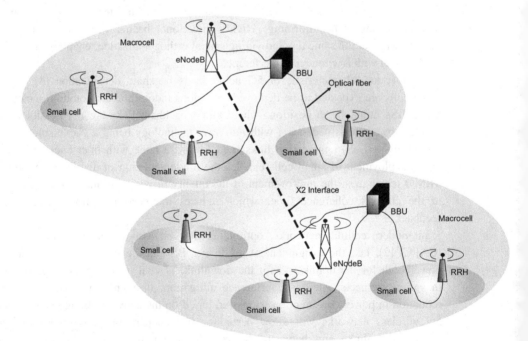

Figure 11.1 Illustration of the heterogeneous network architecture.

interface is a protocol stack used by LTE, including various functions such as end-user packet tunneling between eNodeBs, packet loss management, exchange of overload and traffic load information between eNodeBs, handover of a UE between eNodeBs, and error handling.

- *Multiple small-cell BSs*, which may belong to a C-RAN distributed antenna system, and are called *remote radio heads* (RRHs). These units are mainly responsible for the radio frequency (RF) operations (e.g., carrier frequency conversion, filtering, and power amplification).
- *The baseband unit* of each distributed antenna system. In these systems, the BS functionalities are split between the BBU and the RRHs, with the BBU performing the scheduling and the baseband processing. The BBU is usually placed in a special room, while the RRH is collocated with the antenna. The BBU and multiple RRHs are connected via optical fiber, which allows baseband signal transmission with high capacity and low latency.

Taking the above characteristics into account, four different CoMP deployment scenarios can be considered for homogeneous and heterogeneous networks, including only macrocells, or both macrocells and small cells, respectively, as follows [8]:

1. In the first scenario, CoMP is implemented within a macrocell, which is attractive mainly for homogeneous networks. The macrocells are divided into sectors, which are controlled by the same macro-BS. In this case, no backhaul connection is needed for the communication between the macro-BS antennas of multiple sectors.
2. In the second scenario, CoMP is performed by coordinating multiple macrocells or sectors that belong to different macrocells. Again, this scenario targets mainly homogeneous networks. The coordination between macro-BSs requires communication through the backhaul, which is performed mainly through the X2 interface. The performance of this scenario, depending on the network topology and on the number of BSs that are involved, is affected by the latency of communication between the eNodeBs.
3. In the third scenario, a macrocell and multiple RRHs, each one forming its own small cell with its own physical cell identity, are coordinated. The macrocell BS is usually connected to the RRHs via high-rate optical fiber.
4. The last scenario is similar to the third one; however, the coverage area of the RRHs has the same physical cell identity as the macrocell. Thus, the macro-BS together with the RRHs acts as a distributed antenna system, while there is no need for mobility support and handover procedures among the areas covered by the connected RRHs.

In current networks which employ MIMO transmission in each cell, intercell interference coordination (ICIC) is employed, which provides coordination over an X2 interface for message exchange between BSs [9]. The role of ICIC is to communicate information between cells about current and prospective interference between neighboring macro-BSs. However, the backhaul technology which implements the X2 interface imposes limitations in terms of latency, which cannot be guaranteed to be low. Thus, ICIC is not intended for real-time dynamic coordination between cells, but for semistatic coordination. CoMP envisions dynamic coordination, but also requires very low latency, and possibly high data rates over the backhaul links.

On the other hand, heterogeneous networks are characterized by intense intercell interference, especially between macrocells and small cells [9]. This is caused mainly by the overlapping areas of these cells, but also by the different power classes of their antennas. Thus, in the case of a heterogeneous network, a more dynamic interference mitigation strategy is needed, making CoMP a prominent technique for achieving this goal.

In conventional networks, a set of collocated antennas corresponds to a single cell, despite the application of sectorization [7]. A UE terminal is assigned to a single cell, based on received signal strength. However, in the fourth deployment scenario, a cell may employ multiple antennas which are not collocated. Thus, in CoMP, the term *transmission point* (TP) is used, in order to refer to collocated antennas. Using this term, multiple TPs may coexist within one cell. Furthermore, multiple TPs may correspond to multiple sectors when sectorization is applied.

11.3.2   CoMP Techniques for the Uplink

Uplink coordination is more straightforward than downlink coordination, in the sense that the UE terminals do not need any modification for uplink CoMP to be implemented. Thus, BS cooperation for the uplink may be easier to implement than downlink CoMP. Joint reception by BSs requires a high backhaul capacity and low latency, for which the X2 interference may be a bottleneck. The following cases of uplink CoMP can be discerned [10]:

- *Interference-aware detection.* This technique can be achieved without cooperation between BSs. The receiving node estimates the interference links, not just the desired signal links, and decides accordingly which receive filter should be applied, performing interference rejection combining.
- *Joint scheduling across multiple cells.* This technique requires information exchange between BSs regarding the channel state and scheduling decisions, which is performed over the X2 interface for the communication between different macro-BSs. Furthermore, it is based on interference prediction, so that the scheduling of users is done with the aim of mitigation of the interference between cells.
- *Joint signal processing across multiple cells.* This technique provides the greatest advantage in terms of performance, and can be applied in various forms, with different CoMP gains and different requirements for backhaul capacity, depending on the implementation. For example, techniques which are mostly based on distributed interference subtraction at different TPs require the exchange of decoded signals. On the other hand, techniques which form a distributed antenna system require centralized decoding of messages, and thus quantized received signals need to be exchanged, increasing the requirements for backhaul capacity, but also the gain from

CoMP. In general, there is a trade-off between the required backhaul capacity and the gains that a joint processing technique can provide.

The above uplink CoMP techniques are sorted by increasing backhaul requirements for high data rates and low latency. In the case of coordination between macrocell BSs, the X2 interface becomes the bottleneck. However, when the coordination takes place between different sectors of one cell, or between distributed RRHs which are connected via optical fiber to a BBU, the information exchange does not involve the X2 interface; the TPs share quantized baseband samples of the received signal, channel state information (CSI), and resource allocation decisions. Owing to the challenge of intercell cooperation through the X2 interface, combining joint detection for intracell CoMP and interference prediction for intercell CoMP is considered a viable scheme which balances backhaul communication requirements and CoMP gains.

Uplink CoMP presents more challenges in its implementation, with clustering, synchronization, and channel estimation being the most important ones, apart from backhaul requirements. Clustering refers to the grouping of the TPs that will participate in the uplink CoMP scheme, which may be a static or dynamic process, based on the interference conditions of the cell-edge users. Synchronization is vital, so that intersymbol and intercarrier interference are avoided, while propagation delays for different terminals must be acceptable.

11.3.3   CoMP Techniques for the Downlink

According to the amount of information transferred between nodes, downlink CoMP techniques range from simple interference avoidance schemes to complex joint transmission techniques from multiple TPs. The downlink CoMP techniques can be divided into three categories, depending on the availability of the transmitted message at different TPs and the information exchange between TPs [7, 8, 10]:

- *Coordinated scheduling/coordinated beamforming (CS/CB).* In this technique, the message intended for a UE is available at and transmitted from only one TP. However, multiple TPs exchange information regarding the CSI of multiple UE terminals. This is done in order to select appropriate beamforming weights for interfering TPs and to reduce the interference directed toward these terminals by neighboring cells. To this end, the interference is steered toward the null space of the interfered UE. Coordinated scheduling also refers to the division of the network into clusters, in order to determine which TPs in a cluster are allowed to transmit in a specific resource slot and toward a specific UE. An example of CS/CB is the case of two UE terminals and two TPs which are co-scheduled with specific precoders in order to form nulls toward the UE terminal which receives its message from the opposite TP. CS/CB can be performed in a semistatic manner, thus making it suitable for application in cases with a nonperfect backhaul link.
- *Dynamic point selection (DPS).* DPS takes advantage of a fast backhaul link in order to coordinate scheduling in real time, for realistic traffic conditions. More specifically, the message intended for a UE is available at multiple TPs at any time,

but it is transmitted in each subframe only by one TP. The serving TP of a UE can be dynamically switched, depending on the availability of wireless resources and the channel state information. Although the information which is exchanged between TPs does not contain UE messages, it requires low-latency backhaul for fast decisions.

- *Joint transmission (JT)*. In JT, the message intended for a specific UE is transmitted by multiple TPs to coherently or noncoherently increase the signal strength received at the UE. In coherent JT, the signals transmitted by multiple TPs are jointly precoded so that they can be coherently combined through the wireless channel at the receiver end. In noncoherent JT, the UE receives multiple transmissions from various TPs with individual precoding, without prior consideration of coherent combining. JT utilizes time and frequency resources at multiple TPs which are intended for a single UE, and thus JT will provide more gain in lightly loaded networks. In more heavily loaded networks, multiuser JT must be used, which allows transmission to multiple UEs using the same resources and the same set of TPs. In this case, a combination of JT-CoMP and multiuser MIMO (MU-MIMO) is necessary, but this approach is also more sensitive to channel estimation errors. Finally, JT can also be combined with DPS, where the set of transmitting TPs is dynamically selected.

The information exchange and scheduling varies according to the downlink CoMP scheme selected. For example, one method belonging to the CS/CB category is precoding matrix indicator (PMI) coordination, which is aimed at inter-TP interference mitigation [7]. More specifically, a fixed codebook of quantized precoders used by the TPs is assumed to be known at the UEs. Thus, a cell-edge UE calculates and reports to the serving TP its recommended or restricted PMIs, that is, the precoding matrices which create the least or the most interference at the cell-edge UE due to the interfering TPs. These preferred or restricted PMIs are then reported by the serving TP to the interfering TPs. In this way, the interfering TPs are restricted to using only a subspace of the codebook of precoding matrices, in order to avoid interference at the receiver of the cell-edge UE, achieving coordinated scheduling.

Another coordinated beamforming technique for interference suppression uses a coordinated selection of transmit precoders at each TP, which is based not on PMIs but on channel state information [7]. This technique relies on zero-forcing beamforming (ZFBF), i.e., a transmit filter requiring CSI feedback which forces the interference caused by a TP to a UE served by another TP to zero. This technique is more flexible than PMI coordination, but it presents increased requirements for CSI accuracy, feedback, and backhaul overhead.

Downlink JT CoMP is mostly designed for use with a set of TPs which are directly connected to a BBU via low-latency, high-capacity links. Coherent schemes are especially dependent on spatial CSI feedback, since they employ MIMO techniques for the transmission. Theoretical schemes with ideal CSI feedback set the information-theoretical bounds for JT CoMP, predicting high gains with multi-TP coordination. However, in practical scenarios, perfect CSI at the transmitter is not available with the currently supported feedback schemes through the uplink. More specifically, the current schemes are focused mainly on linear precoding techniques, where the UE reports a PMI and a corresponding channel quality indicator (CQI). Furthermore, there is a trade-off between the flexibility of using a larger number of cooperating TPs and the overhead needed for it. Channel estimation becomes a challenge when multiple TPs transmit to a single UE. The number of TPs involved defines the number of orthogonal pilot sequences that are necessary for channel estimation. A large number of cooperating TPs increases the CoMP gain, but also increases the overhead required for channel estimation and feedback, since the UE will have to transmit PMIs/CQIs for multiple TPs [7]. Finally, multiuser JT schemes can also be used where multi-TP ZFBF is performed, which is calculated by a centralized scheduler that obtains the PMI/CQI from the UEs regarding all TPs in a cluster.

An ideal backhaul connection between multiple TPs and ideal CSI feedback are assumed for various schemes, both for the uplink and for the downlink, which provide some fundamental bounds on the performance of coordinated transmission and reception in heterogeneous networks. In the following sections, two representative techniques are briefly presented, namely interference alignment and the compute-and-forward protocol, while the challenges and requirements of their practical implementation are discussed.

11.4   Interference Alignment: Exploiting Signal Space Dimensions

In this section, we present the fundamental concepts of interference alignment (IA), a technique which, although it requires many idealized assumptions about the network, provides theoretical insight into the limits of the interference channel. A representative setup in which IA can be applied is a wireless interference channel with K transmitter–receiver pairs, where each transmitter simultaneously transmits its message toward the corresponding receiver. With the application of IA, a data rate equal to half of the interference-free channel capacity can be achieved, irrespective of the number of user pairs K.

The implementation of IA requires many idealized assumptions [11]: global channel knowledge, bandwidth expansion, unlimited resolution, and high signal strengths compared to the additive noise. However, the IA technique gives surprising and fundamental insights into the number of accessible signaling dimensions in wireless interference networks. The IA technique has attracted the attention of the research community, leading to the proposal of a variety of IA schemes [11], such as spatial alignment, lattice alignment, asymptotic alignment, asymmetric complex signal alignment, opportunistic alignment, ergodic alignment, aligned interference neutralization, blind alignment, and retrospective alignment.

In this section, the concept of IA will be described, including some basic setups as examples, while the scheme's possible impact on cellular networks will also be discussed.

11.4.1   The Concept of Linear Interference Alignment

The concept behind all IA schemes, even if they are sophisticated in their implementation, is the access of a receiver to multiple linear equations for the transmitted messages, through the signaling dimensions that are provided for the communication. More specifically, let us consider a system of linear equations

$$y_1 = h_{11}x_1 + h_{12}x_2 + \ldots + h_{1K}x_K,$$

$$y_2 = h_{21}x_1 + h_{22}x_2 + \ldots + h_{2K}x_K,$$

$$\vdots$$

$$y_B = h_{B1}x_1 + h_{B2}x_2 + \ldots + h_{BK}x_K.$$

(11.1)

The above equations represent B observations, each in the form of a linear combination of K information symbols $x_1, x_2, \ldots, x_K$, with coefficients h_{ij}. If these equations model an interference network, then K is the number of transmitters, each trying to send one information symbol, h_{ij} are the channel coefficients, and B may be the bandwidth, the number of antennas, or, in general, the number of signaling dimensions which are accessible at a receiver through a linear channel.

If the channel coefficients are independently drawn from a continuous distribution, then all the symbols can be recovered, given that the number of observations is at least equal to the number of unknown information symbols. However, in an interference network with transmitter–receiver pairs, at each receiver only one information symbol is desired. Thus, the receiver needs to be able to extract this specific information symbol from the $K - 1$ interfering symbols.

The simplest solution to this problem is that, in general, each receiver needs K signaling dimensions in order to extract the desired message. This is known as *cake-cutting bandwidth allocation*. However, IA schemes are based on the observation that it is possible to recover the desired information symbol even when the number of the available linear equations is much smaller.

In order to demonstrate this, let us consider the following observed equations [11]:

$$y_1 = 3x_1 + 2x_2 + 3x_3 + x_4 + 5x_5,$$

$$y_2 = 2x_1 + 4x_2 + x_3 - 3x_4 + 5x_5,$$

(11.2)

$$y_3 = 4x_1 + 3x_2 + 5x_3 + 2x_4 + 8x_5.$$

In the case where the receiver desires all transmitted messages, then the above equations are not enough; five independent equations are needed. However, the receiver requires the decoding of only one message, for example x_1. In this case, we notice that the vector $\mathbf{U} = [17, -1, -10]^{\mathrm{T}}$ is orthogonal to all the interference vectors, while the projection $\mathbf{U}^{\mathrm{T}}\mathbf{Y}$ eliminates all interference when $\mathbf{Y} = [y_1, y_2, y_3]^{\mathrm{T}}$. Thus,

$$\mathbf{U}^{\mathrm{T}}\mathbf{Y} = 17y_1 - y_2 - 10y_3 = 9x_1,$$

(11.3)

and therefore message x_1 can be extracted.

Based on the above observations, the IA scheme consolidates the space spanned by the interference at each receiver into a small number of dimensions, while the desired signals are projected into the null space of the interference and thereby recovered free from interference. This requires appropriate beamforming/precoding at the sources, of course. One key principle for achieving this goal is relativity of alignment, i.e., the alignment of the signals is different at each receiver, since the observed equations are different, but also the desired signals are different.

11.4.2 The Example of the X-Channel

One of the first examples of IA was the X-Channel [12], where spatial beamforming is performed. A MIMO X-channel is considered here, where all nodes are equipped with three antennas each, and the sources transmit one message for each receiver, that is, four independent messages in total. The channel between transmitter t and receiver r is assumed to be a generic 3×3 linear transformation, denoted by $\mathbf{H}^{[rt]}$, with $r,t \in \{1,2\}$. Accordingly, the transmitted messages are denoted by x_{rt}. Furthermore, two generic 3×1 vectors, $\mathbf{S}^{[i]}$, denote the dimensions in which each receiver anticipates the interfering signals, that is $\mathbf{S}^{[1]}$ are the dimensions to which the interference signals which are not intended for receiver 1 are to be steered.

The channel matrices are considered to be known at the transmitters, and thus they are utilized for selecting the corresponding beamforming vectors. The transmitted vectors and their alignment at the receiver end are illustrated in Figure 11.2. The vectors transmitted by the first source are denoted by circles, while the vectors transmitted by the second source are denoted by squares. The vectors denoted by white symbols are intended for the first receiver, while the vectors denoted by black symbols are intended for the second receiver. It is evident from Figure 11.2 that the undesired signals at the first receiver are aligned along the direction of the vector $\mathbf{S}^{[1]}$, while those at the second receiver are aligned along the direction of vector $\mathbf{S}^{[2]}$. Thus, the receivers can eliminate the signals along those directions, effectively canceling the interference.

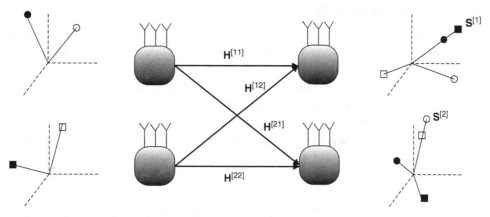

Figure 11.2 Interference alignment in the X-channel.

It is evident from the above that, even though there are a total of four messages and only three signaling dimensions, IA enables one symbol per message to be delivered free from interference. As a result, the desired symbols occupy linearly independent signal dimensions and can be isolated for a total of four degrees of freedom (DoFs). Note that the corresponding two user interference channel setting allows only three DoFs. This can be achieved with the following beamforming scheme.

The first source multiplies its transmitted messages by beamforming vectors depending on the channel matrices and the vectors $\mathbf{S}^{[i]}$, and sends the following sum:

$$\left(\mathbf{H}^{[21]}\right)^{-1}\mathbf{S}^{[2]}x_{11} + \left(\mathbf{H}^{[11]}\right)^{-1}\mathbf{S}^{[1]}x_{21}. \tag{11.4}$$

Similarly, the second source transmits the following sum:

$$\left(\mathbf{H}^{[22]}\right)^{-1}\mathbf{S}^{[2]}x_{12} + \left(\mathbf{H}^{[12]}\right)^{-1}\mathbf{S}^{[1]}x_{22}. \tag{11.5}$$

At the side of the first receiver, after the signals transmitted by the two sources have been propagated through the wireless medium and multiplied by the corresponding channel matrices, the resulting received signal is as follows:

$$\mathbf{H}^{[12]}\left(\mathbf{H}^{[22]}\right)^{-1}\mathbf{S}^{[2]}x_{12} + \mathbf{H}^{[11]}\left(\mathbf{H}^{[21]}\right)^{-1}\mathbf{S}^{[2]}x_{11} + \mathbf{S}^{[1]}\left(x_{22}+x_{21}\right). \tag{11.6}$$

It is evident from the equation above that the undesired signals x_{21} and x_{22} are aligned along the direction of $\mathbf{S}^{[1]}$, and thus they can be removed. Similarly, the received signal at the second receiver is

$$\mathbf{H}^{[21]}\left(\mathbf{H}^{[11]}\right)^{-1}\mathbf{S}^{[1]}x_{21} + \mathbf{H}^{[22]}\left(\mathbf{H}^{[12]}\right)^{-1}\mathbf{S}^{[1]}x_{22} + \mathbf{S}^{[2]}\left(x_{11}+x_{12}\right). \tag{11.7}$$

It can be observed that, in the above equations for the received signals, the additive noise has been neglected, since the IA scheme targets the high signal-to-noise ratio (SNR) regime. However, in practical systems, additive noise will lead to imperfect alignment of the interference, increasing the error rate.

11.4.3   The K-User Interference Channel and Cellular Networks: Asymptotic Interference Alignment

The IA scheme was generalized to $K > 2$ user pairs in [13]. In a K-user interference channel, $K/2$ DoFs can be achieved, that is, *half the cake* is available for each user at every time instant, but only when the available bandwidth is made very large. This phenomenon is called *bandwidth scaling*. The IA scheme in [13] is asymptotic, which limits its practical use. However, it allows the satisfaction of arbitrary large numbers of alignment constraints, without needing to sacrifice a fraction of the available signal space in order to achieve IA. Furthermore, it is applicable also in nonlinear forms, and it can also be applied to a variety of setups apart from the K-user channel, including cellular networks.

The DoFs of cellular networks were characterized in [14], where K cells were considered, each containing a BS and M users that were served by that BS, while all the nodes were equipped with a single antenna. For the uplink, the DoF result was that each user achieved a fraction $1/(M + 1)$ and each cell was able to access a fraction $M/(M + 1)$ of the cake. Thus, interestingly, for a large number of users, the number of DoFs per cell approaches the interference-free setting.

The downlink was discussed in [15], where the BSs were equipped with M antennas and transmitted to N single-antenna users in each cell. It was shown that a total of $MN/(M + N)$ DoFs was achievable per cell. Thus, here also the number of DoFs per cell approaches the interference-free setting, i.e., each cell achieves close to M DoFs when the number of users per cell becomes large.

11.4.4   Cooperative Interference Networks

Cooperation among nodes can alleviate interference. Regarding IA, if the channels are of comparable strength, then there is no DoF benefit from cooperation [11]. However, when heterogeneous links exist, such as backhaul links that do not compete for wireless spectrum (e.g., optical fibers), these ideal links may enable cognitive settings, i.e., where some messages can be shared between nodes. The presence of cognitive and cooperative transmitters or receivers leads to new aspects of interference alignment.

An application of IA which takes advantage of cooperation between receiving nodes was introduced in [16], namely interference alignment and cancellation (IAC). In a setup similar to that in Figure 11.2, with two transmitters and two receivers equipped with three antennas, the transmitters desire to transmit five messages, namely a, b, c, d, e. These messages are desired by both receivers. At the receiver side, these signals are coupled over the three available signaling dimensions. For example, at the first receiver, a arrives in the first dimension, b and c are aligned in the second, and d and e are aligned in the third. Accordingly, at the second receiver, a and b arrive aligned in the first dimension, c and d in the second, and e in the third. Now suppose that there is a mechanism that allows the receivers to share their decoded messages. Then, the first receiver decodes a from the observation of the first dimension and sends it over to the second receiver; the second receiver cancels a from the observation of its first dimension and then extracts b, which it sends over to the first receiver. In the same manner, all the messages are extracted one by one, by the cancellation of already extracted messages from the rest of the available signaling dimensions.

In such schemes, the number of achievable DoFs is $(2M - 1)/M$. Note that this number of DoFs approaches the value 2, which is the maximum number possible with full cooperation, that is, joint signal processing at the two receivers.

11.4.5   Insight from IA into the Capacity Limits of Wireless Networks

Interference alignment is an information-theoretic scheme which has enabled significant progress in the understanding of the capacity limits of wireless communication networks. It has shown that the theoretical limits can be orders of magnitude higher than previously thought possible.

IA targets asymptotically high SNRs, refers mainly to the DoF characterizations of various setups, and may seem of limited practical relevance. However, although the

DoF characterization cannot always provide a capacity-achieving scheme, it identifies new ways to exploit symbol extensions, channel variations, structured codes, etc.

Many challenges arise in the practical application of IA, with the most important ones being the overhead for acquiring enough channel knowledge, the penalty of residual channel uncertainty at the transmitters, the impact of channel correlations, and the solutions for time-varying channels.

11.5   Compute-and-Forward Protocol: Cooperation at the Receiver Side for the Uplink

In this section, we describe the compute-and-forward (CoF) protocol, first proposed in [17], which exploits interference to achieve significantly higher rates between users in a network. The key idea is that relays should decode linear functions of transmitted messages according to their observed channel coefficients. After decoding these linear equations, the relays simply send them toward the destinations, which, given enough equations, can recover their desired messages. This strategy relies on codes with a linear structure, specifically, nested lattice codes [18]. The linearity of the codebook ensures that integer combinations of codewords are themselves codewords. A relay is free to determine which linear equation to recover, but those closer to the channel's fading coefficients are available at higher rates. Encoders map messages from a finite field to a lattice, and decoders recover equations of lattice points which are then mapped back to equations over the finite field. This scheme is applicable even if the transmitters lack CSI.

More specifically, the CoF protocol finds application in a system with M distributed receiving antennas and L transmitting sources, where $M \geq L$. The distributed antennas are connected to a centralized decoder (CD). If we compare this system to the 5G network described in previous sections, the antennas can be mapped to the RRHs, while the CD can be located at the BBU. The decoded codeword is communicated to the CD, through ideal backhaul links. The CD, having enough linearly independent integer equations, decodes all messages transmitted by the UE terminals. The selection of integer equations at the antennas is performed through physical layer network coding (PNC) [19].

The CoF protocol requires decentralized decoding at each antenna, transmission of decoded equations toward the CD, and local CSI at each antenna regarding only the links between all UE terminals and that specific antenna. However, it may also require coordination between all antennas through the CD regarding the selection of the integer equations which are to be decoded by each antenna, so that they are linearly independent. On the other hand, in the original form of the protocol as presented in [17], no CSI at the transmitters is needed.

11.5.1   Encoding and Decoding of the CoF Protocol

A general CoF network comprises L sources, M relays, and one destination, namely the CD. Each source needs to transmit a message vector whose elements are drawn from a finite field of size p, denoted by $\mathbb{F}_p$, where p is a prime number. More specifically, the field $\mathbb{F}_p$ will be considered from now on as the subset of integers $\{0, 1, \ldots, p-1\}$, while addition over the field is denoted by $\oplus$ and is the modulo-p addition.

Each transmitter, indexed by $l = 1, 2, \ldots, L$, has a length-k_l message vector, denoted by $\mathbf{w}_l \in \mathbb{F}_p^{k_l}$. The CoF protocol allows nonsymmetric transmission rates of the sources, and thus the message vectors of all sources are zero-padded to the maximum vector length among all sources, which is $k = \max_l k_l$. These messages are encoded and mapped to length-n real-valued codewords $\mathbf{x}_l \in \mathbb{R}^n$, where the power constraint is

$$||\mathbf{x}_l||^2 \leq nP. \tag{11.8}$$

Here, we focus on the case of real-valued signaling at the sources. The CoF relaying protocol can be described for complex signaling as well. The codewords $\mathbf{x}_l$ belong to nested lattice codes, which are constructed based on Construction A in [20], and subtractive dithering is also performed, that is,

$$\mathbf{x}_l = [\mathbf{t}_l - \mathbf{d}_l] \mod \Lambda, \tag{11.9}$$

where $\mathbf{d}_l$ is a subtractive dither, $\mathbf{t}_l$ is the lattice point, and Λ is the coarse lattice of the code.

The transmission rates of each user, measured in bits per channel use, are given by

$$\mathcal{R}_1^s = \frac{k_l}{n} \log_2 p, \tag{11.10}$$

where the $(\cdot)^s$ denotes a quantity measured at the source.

The signal received at a relay, indexed by $m = 1, 2, \ldots, M$, is given by

$$\mathbf{y}_m = \sum_{l=1}^{L} h_{ml} \mathbf{x}_l + \mathbf{z}_m, \tag{11.11}$$

where $h_{ml} \in \mathbb{R}$ are the channel coefficients and $\mathbf{z}$ is independent and identically distributed (i.i.d.) additive white Gaussian noise (AWGN), i.e., $\mathbf{z} \sim \mathcal{N}(\mathbf{0}, \mathbf{I}_{n \times n})$, where $\mathbf{I}_{n \times n}$ is the identity matrix of dimension n. The channel coefficients are considered fixed over the duration of one codeword. Each relay tries to reliably recover a linear combination of the messages, given by

$$\mathbf{u}_m = \bigoplus_{l=1}^{L} q_{ml} \mathbf{w}_l, \tag{11.12}$$

where the $q_{ml} \in \mathbb{F}_p$ are coefficients which are chosen by the relay based on the maximization of the achievable rates, as it will be described below. The decoder at each relay maps the received signal to an estimated codeword $\hat{\mathbf{u}}_m$.

Although the desired equations are codewords, i.e., they belong to the finite field $\mathbb{F}_p$, the channel operates over the field of real numbers, and thus the relay maps the received signal to an integer equation for codewords. That is, the relay tries to decode an integer equation for the codewords with equation coefficients $\mathbf{a}_m = [a_{m1}, a_{m2}, \ldots, a_{mL}]^\mathrm{T}$, $\mathbf{a}_m \in \mathbb{Z}^L$. This integer equation of codewords corresponds to the equation of the messages

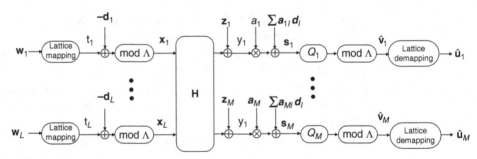

Figure 11.3 System diagram of compute-and-forward protocol.

$\mathbf{u}_m$, for which

$$q_{ml} = [a_{ml}] \quad \mod p. \tag{11.13}$$

Corresponding to the message equation in (11.12), the lattice equation which is desired by a relay is given by

$$\mathbf{v}_m = \left[\sum_{l=1}^{L} a_{ml} \mathbf{t}_l \right] \quad \mod \Lambda. \tag{11.14}$$

In order to decode such an equation, the relay amplifies the received signal $\mathbf{y}_m$ by α_m and computes the following:

$$\mathbf{s}_m = \alpha_m \mathbf{y}_m + \sum_{l=1}^{L} a_{ml} \mathbf{d}_l. \tag{11.15}$$

In the equation above, the signal $\mathbf{y}_m$ is shifted toward an integer combination of lattice points by the use of the amplification factor α_m, and the subtractive dithers are re-added to the received signal.

The estimate of the desired lattice equation is given by

$$\hat{\mathbf{v}}_m = [Q_m(\mathbf{s}_m)] \quad \mod \Lambda, \tag{11.16}$$

where $Q_m(\cdot)$ is the lattice quantizer of the finest lattice among the sources with index l, for which $a_{ml} \neq 0$.

The system diagram of the CoF protocol is depicted in Figure 11.3.

11.5.2   Achievable-Rate Region and Integer Equation Selection

In [17], it was proven that the equivalent channel induced by the modulo-Λ transformation can be represented by an AWGN channel with additive noise $\mathbf{z}_{\text{eq},m}$, whose variance is

$$N_{\text{eq},m} = \alpha_m^2 + P \, ||\alpha_m \mathbf{h}_m - \mathbf{a}_m||^2, \tag{11.17}$$

as $n \to \infty$. Here, $\mathbf{h}_m = [h_{m1}, \ldots, h_{mL}]^{\mathrm{T}}$.

From the equivalent noise above, a very interesting conclusion can be drawn: the CoF system is equivalent to an AWGN channel, where the noise comprises two sources, the actual thermal noise, amplified by the coefficient α_m, and the mismatch between the linear combination that the wireless channel provides and the integer equation that the relay chooses to decode. Furthermore, it is evident that the minimization of the above noise depends on the selection of α_m and the coding vector $\mathbf{a}_m$. Their optimization depends on the power constraint at the sources and the observed channel vector.

Thus, the *achievable computation rate* at the relay, i.e., the rate at which a relay can decode a desired equation of messages, is given by

$$\mathcal{R}_m^r\left(\mathbf{h}_m, \mathbf{a}_m, \alpha_m\right) = \frac{1}{2}\log_2^+\left(\frac{P}{\alpha_m^2 + P\|\alpha_m\mathbf{h}_m - \mathbf{a}_m\|^2}\right) \qquad (11.18)$$

for any choice of α_m and $\mathbf{a}_m$, where $\log_2^+(\cdot) = \max\left(\log_2(\cdot), 0\right)$ and $(\cdot)^r$ denotes a quantity measured at the relay. The optimal choice for the amplification factor α_m, in terms of maximizing the achievable computation rate, is the minimum mean squared error (MMSE) coefficient, denoted by β_m, given by

$$\beta_m = \frac{P\mathbf{h}_m^T\mathbf{a}_m}{1 + P\|\mathbf{h}_m\|^2}, \qquad (11.19)$$

which results in a computation rate region of

$$\mathcal{R}_m^r\left(\mathbf{h}_m, \mathbf{a}_m\right) = \frac{1}{2}\log_2^+\left(\left(\|\mathbf{a}_m\|^2 - \frac{P\left(\mathbf{h}_m^T\mathbf{a}_m\right)^2}{1 + P\|\mathbf{h}_m\|^2}\right)^{-1}\right). \qquad (11.20)$$

The best choice of the vector of coefficients of the equation, in terms of maximizing the achievable computation rate at the relay with index m, is given by [21, 22]

$$\mathbf{a}_m = \arg\min_{\mathbf{a}\in\mathbb{Z}^L, \mathbf{a}\neq\mathbf{0}} \mathbf{a}^T\mathbf{G}\mathbf{a}, \qquad (11.21)$$

where

$$\mathbf{G} = \mathbf{I}_{L\times L} - \frac{P\mathbf{h}_m\mathbf{h}_m^T}{1 + P\|\mathbf{h}_m\|^2}. \qquad (11.22)$$

The transmission rate of each source cannot exceed the computation rate of the relays which receive its signal, that is,

$$\mathcal{R}_l^s \leq \min_{m, a_{ml}\neq 0}\left(\mathcal{R}_m^r\right). \qquad (11.23)$$

The CD requires L equations for the received messages, which are decoded and forwarded to the CD by the relays. However, when these selected L relays choose their equation coefficient vectors independently, the coding matrix in the field $\mathbb{F}_p$,

$$\mathbf{Q} = [\mathbf{q}_1, \mathbf{q}_2, \ldots, \mathbf{q}_L], \qquad (11.24)$$

where $\mathbf{q}_m = [q_{m1}, q_{m2}, \ldots, q_{mL}]$, must be a *full-rank matrix*, so that the equations for the messages are independent. When p is assumed to grow with n, and $n \to \infty$, the same property holds for the coding matrix

$$\mathbf{A} = [\mathbf{a}_1, \mathbf{a}_2, \ldots, \mathbf{a}_L]. \qquad (11.25)$$

Thus, apart from maximizing the achievable computation rates, the equation coefficient vectors which are chosen by the relays must satisfy the need for independent equations at the CD, so that the original messages can be decoded.

11.5.3   Advantages and Challenges of the CoF Protocol

The CoF relaying protocol is an attractive choice for cloud networks where one aims to reduce the complexity of the backbone by decentralized processing at the relaying nodes:

- *Low complexity at the receiving antennas.* Each node performs lattice decoding, a rather simple decoding technique, which does not require a complex node architecture (e.g., a quantizer or a maximum-likelihood decoder). Thus, minimal signal processing without noise amplification is performed at the relaying nodes, while digital information is forwarded to the backbone, usually through a fiber optical network.
- *No CSI at the transmitters is necessary.* Another important advantage of the CoF protocol is the fact that it does not require CSI at the sources, but only at the relays, for the optimization of the choice of the MMSE coefficient β_m and the equation coefficient vector $\mathbf{a}_m$. This is in contrast to other techniques which have been proposed for multiuser multiple access networks and have been the focus of recent research, such as the interference alignment scheme, described in Section 11.4, where CSI is required at the sources.
- *The CoF protocol can be combined with other techniques.* For example, the IA scheme is known to offer $K/2$ DoFs, which means that each of the K users uses half of the resources at each instant. When CSI is available at the sources, the CoF scheme offers K degrees of freedom [23], combining the benefits of structured coding and signal alignment [24, 25].

Apart from the simple and efficient decoding and exploitation of the interference that CoF offers, however, several practical implementation issues of the CoF protocol still remain unsolved, some of which are described below:

- *Imperfect CSI.* The implementation of the CoF protocol requires CSI only at the relays. In practical communication systems which employ channel estimation, the acquired CSI is usually imperfect, since it is acquired through pilot symbols which are perturbed by AWGN or is acquired with the use of a finite impulse response filter, where the estimate may be outdated or may resemble the actual channel according to a specific correlation factor. CSI affects the computation of the MMSE coefficient and the search for the best equation coefficient vector, and thus imperfect CSI estimation has a direct impact on the performance of the protocol.
- *Physical layer network coding challenges.* The choice of equation coefficient vectors by the relays is done through the use of PNC, which exploits the linear combination of messages received through the wireless medium. Most techniques seek to maximize the achievable computation rate at a specific relay. However, this does not simply

translate to achieving the maximum throughput in a CoF network with multiple users. Several challenges arise, the most important of which are the probability of rank failure of the coding matrix, the computational complexity of the calculation of coding vectors, and the complex optimization of the sum-throughput in the case of nonsymmetric rates. More specifically, the protocol requires the cooperation of relays in order for the coding matrix to be of full rank, so that the number of independent equations at the CD is sufficiently large for calculating the original messages. The selection of the coding vector at each relay is a computationally demanding problem, and the case of nonsymmetric rates does not have a closed-form solution for the selection of the PNC vectors when the maximization of the sum-throughput of the network is desired.

Despite the challenges in practical implementation that the CoF protocol presents, it is an example of uplink cooperative reception which proves that interference can be harnessed, offering the possibility of multiple cell-edge UE terminals using the same frequency and time resources and transmitting their messages toward multiple receiving antennas, using simple code structures and allowing distributed processing of the received signal at the RRHs.

11.6   Conclusion

In this chapter, we have outlined the challenges and opportunities for interference management in 5G multitier networks, with special focus on the key enabling technologies for advanced interference mitigation techniques. We described the concept of coordinated multipoint, both for the uplink and for the downlink, its variations, and the corresponding requirements. We then proceeded to a brief presentation and discussion of two information-theory-oriented techniques, namely the interference alignment scheme and the compute-and-forward protocol, which provide interesting solutions for both the uplink and the downlink, accommodating multiple users over the same time and frequency resources. The challenges of the practical implementation of such techniques were discussed in relation to the anticipated network architecture and key features of 5G networks.

References

[1] Ericsson,"5G radio access, research and vision," White paper, Jun. 2013.
[2] E. Hossain, M. Rasti, H. Tabassum, and A. Abdelnasser, "Evolution toward 5G multi-tier cellular wireless networks: An interference management perspective," *IEEE Wireless Commun.*, vol. 21, no. 3, pp. 118–127, Jun. 2014.
[3] N. Bhushan, J. Li, D. Malladi, R. Gilmore, D. Brenner, A. Damnjanovic, R. T. Sukhavasi, C. Patel, and S. Geirhofer, "Network densification: The dominant theme for wireless evolution into 5G," *IEEE Commun. Mag.*, vol. 52, no. 2, pp. 82–89, Feb. 2014.

[4] China Mobile Research Institute, "C-RAN: The road towards green RAN," White paper, ver. 2.5, Oct. 2011.

[5] B. Bangerter, S. Talwar, R. Arefi, and K. Stewart, "Networks and devices for the 5G era," *IEEE Commun. Mag.*, vol. 52, no. 2, pp. 90–96, Feb. 2014.

[6] M. Sawahashi, Y. Kishiyama, A. Morimoto, D. Nishikawa, and M. Tanno, "Coordinated multipoint transmission/reception techniques for LTE-Advanced [Coordinated and distributed MIMO]," *IEEE Wireless Commun.*, vol. 17, no. 3, pp. 26–34, Jun. 2010.

[7] D. Lee, H. Seo, B. Clerckx, E. Hardouin, D. Mazzarese, S. Nagata, and K. Sayana, "Coordinated multipoint transmission and reception in LTE-Advanced: Deployment scenarios and operational challenges," *IEEE Commun. Mag.*, vol. 50, no. 2, pp. 148–155, Feb. 2012.

[8] J. Lee, Y. Kim, H. Lee, B. L. Ng, D. Mazzarese, J. Liu, W. Xiao, and Y. Zhou, "Coordinated multipoint transmission and reception in LTE-Advanced systems," *IEEE Commun. Mag.*, vol. 50, no. 11, pp. 44–50, Nov. 2012.

[9] V. Jungnickel, K. Manolakis, W. Zirwas, B. Panzner, V. Braun, M. Lossow, M. Sternad, R. Apelfrojd, and T. Svensson, "The role of small cells, coordinated multipoint, and massive MIMO in 5G," *IEEE Commun. Mag.*, vol. 52, no. 5, pp. 44–51, May 2014.

[10] R. Irmer, H. Droste, P. Marsch, M. Grieger, G. Fettweis, S. Brueck, H.-P. Mayer, L. Thiele, and V. Jungnickel, "Coordinated multipoint: Concepts, performance, and field trial results," *IEEE Commun. Mag.*, vol. 49, no. 2, pp. 102–111, Feb. 2011.

[11] S. A. Jafar, *Interference Alignment: A New Look at Signal Dimensions in a Communication Network*, NOW Publishers, 2011.

[12] S. A. Jafar and S. Shamai, "Degrees of freedom region for the MIMO X channel," *IEEE Trans. Inf. Theory*, vol. 54, no. 1, pp. 151–170, Jan. 2008.

[13] V. Cadambe and S. Jafar, "Interference alignment and degrees of freedom of the K-user interference channel," *IEEE Trans. Inf. Theory*, vol. 54, no. 8, pp. 3425–3441, Aug. 2008.

[14] C. Suh and D. Tse, "Interference alignment for cellular networks," in *Proc. of Allerton Conf. on Communications, Control and Computing,* Sep. 2008.

[15] A. Ghasemi, A. Motahari, and A. Khandani, "Interference alignment for the K user MIMO interference channel," in *Proc. of IEEE International Symposium on Information Theory (ISIT)*, Jun. 2010.

[16] S. Gollakota, S. Perli, and D. Katabi, "Interference alignment and cancellation," in *Proc. of ACM Special Interest Group on Data Communication (SIGCOMM)*, Aug. 2009.

[17] B. Nazer and M. Gastpar, "Compute-and-forward: Harnessing interference through structured codes," *IEEE Trans. Inf. Theory*, vol. 57, no. 10, pp. 6463–6486, Oct. 2011.

[18] U. Erez and R. Zamir, "Achieving $\frac{1}{2}\log(1+\text{SNR})$ on the AWGN channel with lattice encoding and decoding," *IEEE Trans. Inf. Theory*, vol. 50, no. 10, pp. 2293–2314, Oct. 2004.

[19] S. Zhang, S. Liew, and P. P. Lam, "Hot topic: Physical-layer network coding," in *Proc. of ACM International Conf. on Mobile Computing and Networking (MobiCom)*, Sep. 2006.

[20] J. H. Conway and N. J. A. Sloane, *Sphere Packings, Lattices and Groups*, 3rd edn, Springer, 1998.

[21] C. Feng, D. Silva, and F. Kschischang, "Design criteria for lattice network coding," in *Proc. of 45th Annual Conf. on Information Sciences and Systems (CISS)*, Mar. 2011.

[22] C. Feng, D. Silva, and F. Kschischang, "An algebraic approach to physical-layer network coding," *IEEE Trans. Inf. Theory*, vol. 59, no. 11, pp. 7576–7596, Nov. 2013.

[23] U. Niesen and P. Whiting, "The degrees of freedom of compute-and-forward," *IEEE Trans. Inf. Theory*, vol. 58, no. 8, pp. 5214–5232, Aug. 2012.

[24] U. Niesen, B. Nazer, and P. Whiting, "Computation alignment: Capacity approximation without noise accumulation," *IEEE Trans. Inf. Theory*, vol. 59, no. 6, pp. 3811–3832, Jun. 2013.

[25] A. S. Motahari, S. O. Gharan, M.-A. Maddah-Ali, and A. K. Khandani, "Real interference alignment: Exploiting the potential of single antenna systems," *IEEE Trans. Inf. Theory*, vol. 60, no. 8, pp. 4799–4810, Aug. 2014.

12 Physical Layer Caching with Limited Backhaul in 5G Systems

Vincent Lau, An Liu, and Wei Han

12.1 Introduction

It is envisioned that the capacity demand in fifth generation (5G) wireless networks will increase by 1000 times before the year 2020, and this huge demand will be fueled by high-definition video and content streaming applications, which not only consume immense wireless bandwidth but also have stringent real-time quality of service (QoS) requirements. Small-cell dense wireless networks are regarded as a key candidate 5G technology to meet such aggressive demand. By dense deployment of small-cell base stations (BSs), the network becomes closer to mobile users, and thus the spectral efficiency per unit area can be significantly improved. Moreover, relay stations (RSs) without wired backhaul can also be deployed to enhance the coverage area and provide signal-to-noise ratio (SNR) gain.

While there are many potential opportunities associated with dense wireless networks, the potential spectrum efficiency gain they can provide is significantly limited by increasingly severe interference due to the densification of BSs. There are several common schemes to mitigate interference. In the strong-interference case, channel orthogonalization, such as frequency division multiple access (FDMA) or time division multiple access (TDMA), is used to avoid interference [1], and advanced schemes, such as interference alignment (IA) [2, 3] and interference coordination [4, 5], have been proposed to increase the spectral efficiency of interference channels by jointly mitigating the interference using shared channel state information (CSI). However, these approaches may lead to inefficient use of channel resources because the transmitters do not share payload information and therefore are not fully cooperative. To improve the spectrum efficiency further, coordinated multipoint (CoMP) transmission has been proposed as one of the most important core technologies for Long Term Evolution-Advanced (LTE-A) and future 5G wireless networks [6]. By sharing both CSI and payload data among the BSs concerned, CoMP transmission can transform the wireless network from an unfavorable interference topology to a favorable broadcast topology, where the interference can be mitigated much more efficiently. However, the conventional CoMP scheme is quite costly because it requires a high-capacity backhaul for payload exchange between the BSs, and this poses a huge challenge for practical applications of CoMP, especially for small-cell dense wireless networks.

In this chapter, we show that it is possible to achieve the CoMP gain without a high cost of the backhaul between the BSs. We introduce a new wireless caching technique called *physical layer* (PHY) *caching* [7–10], which can exploit BS-level caching to mitigate interference and improve the degrees of freedom (DoF) of wireless networks. In PHY caching, if the content accessed by several users exists simultaneously at nearby BSs/RSs, it induces various types of *side information* at the BSs. As a result, the nearby BSs/RSs can exploit this side information from the PHY cache and engage in CoMP to serve these users, achieving a CoMP (spatial multiplexing) gain without consuming backhaul bandwidth. In this way, we can opportunistically transform the *interference/relay topology* into a more favorable *multiple-input multiple-output* (MIMO) *broadcast channel topology*, as illustrated in Figure 12.2(b) later in this chapter. This is referred to as *cache-induced opportunistic CoMP*.

This proposed PHY caching is fundamentally different from traditional applications of caches in fixed-line networks [11] or peer-to-peer (P2P) content distribution networks [12]. For example, in [13], the primary role of the cache is to bring the content closer to the content consumer, and the role of the cache in [14] is for load balancing. In addition to caching in the core network, there have also been some recent studies advocating caching at the BS. In [15], femto-caching was proposed for increasing the spectral efficiency of a wireless distributed caching network by using small BSs, called helpers, with high storage capacity to cache popular video files. The authors of [16, 17] studied the benefits of caching at the BSs as well as caching at the devices for device-to-device (D2D) communications. However, in all of these studies, the cache state (hit or miss) does not affect the underlying physical layer topology of the wireless network, and hence the use of caching does not contribute to resolving the aforementioned interference issues.

As discussed above, the interplay between the PHY topology and the PHY cache at the BS provides unique benefits in terms of interference mitigation and DoF gain. However, it also brings several new technical challenges:

- *Increasing the CoMP opportunity with limited PHY cache size.* One important benefit of PHY caching is captured by the cache-induced CoMP gain, which depends heavily on the CoMP opportunity (or probability). However, if naive random caching is used, the CoMP probability is exponentially small with respect to the number of users involved in the CoMP transmission. Hence, the conventional cache design for fixed-line networks cannot be used.

- *Mathematical model of PHY caching.* To understand the fundamental limits of PHY caching, we need to construct a mathematical model that embraces all possible caching strategies and all PHY transmission strategies with different possible types of side information induced by the dynamic cache state. The traditional cache model for fixed-line content delivery networks (CDNs) cannot be used because it does not exploit the side information induced by the dynamic cache state to change the underlying PHY topology.

- *DoF of cached wireless networks.* In addition to understand the fundamental limits of PHY caching, we also need to have a low-complexity and implementation-friendly

PHY caching scheme and understand the optimality of such a scheme. For analytical tractability, we shall consider DoF as the performance metric. However, we need to extend the definition of DoF for traditional uncached wireless networks to embrace the content-centric scenarios of PHY caching.

- *Cache content placement algorithm.* One important component of the PHY caching design is how to initialize the cache content at each BS. Given that the cache storage capacity is not usually large enough to store the entire content library, it is important to have an algorithm to determine the priority of caching one packet over others. This is referred to as the *cache content placement algorithm.* In many cases, the BSs and content gateway do not have knowledge of the popularity of the content files. How to design an online cache content placement algorithm to optimize the DoF of the cached wireless network without explicit knowledge of the content popularity is a challenge.

The rest of this chapter is organized as follows. In Section 12.2, we elaborate on the basic concepts associated with PHY caching. In particular, in Section 12.2.4, we discuss the key design challenges of PHY caching and outline the potential solutions. In Section 12.3, we introduce a generic cache model and study the DoF upper bound. In Section 12.4, we propose a maximum-distance separable (MDS) coded PHY caching scheme and study the associated achievable DoF. A cache content placement algorithm is proposed in Section 12.5 to maximize the DoF of the MDS-coded PHY caching scheme. Finally, the system performance of PHY caching is analyzed in Section 12.6, and our conclusion is presented in Section 12.7, together with some interesting future research directions for PHY caching.

12.2   What Is PHY Caching?

In this section, we first introduce some typical PHY topologies in radio access networks (RANs). Then we elaborate on what PHY caching is and how it induces favorable PHY topology changes.

12.2.1   Typical Physical Layer Topologies

In wireless access networks, there are various *canonical topologies* depending on the amount of payload information available at the transmitters. When the transmitters have sufficient payload information, they can utilize this side information to enhance the spectral efficiency, and the resulting topologies are called *favorable PHY topologies.* On the other hand, when side information is not available, the canonical topologies may have spectral efficiency overheads (which may occur in either the same or different spectral resources used for payload data transmission between the transmitters and receivers) for either cooperating or mitigating interference; these topologies are called *unfavorable topologies.* Below, we elaborate on various canonical topologies with different amounts of payload side information available to the transmitters.

12.2.1.1 Cooperative Relay Topology

One common canonical PHY topology in cellular networks is the *cooperative relay topology*. Figure 12.1(a) illustrates a simple three-node relay channel where there is a BS, an RS, and two users. The RS helps the BS to serve the users without prior knowledge of the information transmitted by the BS. There are various relay schemes to utilize the RS nodes to enhance the signal-to-interference-plus-noise ratio (SINR) of the users. Among them, the amplify-and-forward (AF) relay scheme, which applies a linear transformation to the received signal, is attractive in practical systems owing to its simplicity. However, noise will also be amplified at each hop in AF. As a result, the performance of AF degrades as the number of hops increases. This has motivated the development of decode-and-forward (DF), which decodes the received signal for relay transmission so that the transceiver is free from the superposition of relay noise [18]. Another efficient relay strategy is compress-and-forward (CF) [18]. Owing to the absence of side information at the RS, there is only a marginal spectral efficiency gain as a result of the cut-set bound [19], and hence the cooperative relay topology is regarded as an unfavorable topology. For example, in Figure 12.1(a), the total number of independent data streams that can be supported is 2, which is the same with or without the RS node.

12.2.1.2 Interference Topology

Another common PHY topology is the *interference channel*, where there are multiple BS–user pairs and the transmission of each pair interferes with the transmissions of the other pairs, as illustrated in Figure 12.1(b). Owing to the absence of knowledge about the data symbols transmitted by the other BSs, the capacity is fundamentally limited by the interference between multiple transmitter–receiver pairs. There are several common schemes to mitigate interference. Traditionally, in the weak-interference case, we may treat interference as noise and focus on detecting the desired signals. In the strong-interference case, channel orthogonalization, such as FDMA or TDMA, is used to avoid interference. Advanced schemes such as IA [2] and interference coordination [4] have been proposed to increase the spectral efficiency of interference channels by jointly mitigating the interference using shared CSI. However, these approaches may lead to inefficient use of channel resources because the transmitters do not share payload information and therefore are not fully cooperative. Hence, the interference topology is also regarded as an unfavorable topology.

12.2.1.3 Cooperative MIMO Broadcast Topology

In the interference channel example, if the transmitters are allowed to share the payload data of all the receivers (e.g., via a backhaul network between the transmitters), the PHY topology between these transmitters and users can be transformed to a cooperative MIMO broadcast channel, where all transmitters form a virtual transmitter with more antennas, as illustrated in Figure 12.1(c). In contrast to relay and interference channels, the cooperative MIMO broadcast topology can deliver a maximum of four data streams owing to the payload sharing among the transmitters. The traditional approach to sharing payload is via a backhaul between the transmitters. However, this approach

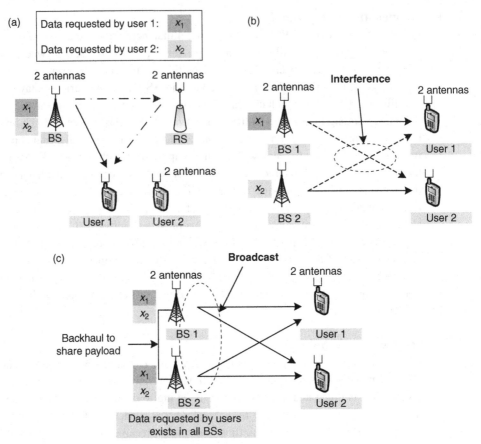

Figure 12.1 Illustration of some typical physical layer topologies. (a) Relay channel; (b) interference channel; (c) cooperative MIMO broadcast channel.

requires a high-capacity backhaul for payload sharing, which is expensive, especially for small-cell networks.

12.2.2   Basic Components of PHY Caching

PHY caching is a caching strategy at the BS to promote MIMO cooperation between peer transmitters in wireless networks [7–10]. As a result, the transmitters can occasionally engage in MIMO cooperation if the requested packets exist in the BS caches simultaneously. There are two major components in the PHY caching scheme: *PHY cache initialization and update* and *cache-assisted PHY transmission* for content delivery, as illustrated in Figure 12.2(a). The PHY cache initialization determines *what content to cache* and *how to cache the content* at each of the caches of the transmitter. For practical reasons, the cache content initialization is adaptive to the content popularity distribution rather than to the instantaneous user requests. Once the cache content has been initialized, the same cache content is used to serve different online user requests until the popularity distributions of the content change. As a result, the PHY cache induces *dynamic side information* that is available to the transmitters in the wireless network. At each time slot, the side information may be random, depending on the instantaneous user requests as well as on how the cache content is stored (deterministic cache versus random cache) [20, 21].

The cache-assisted PHY transmission component exploits the aforementioned side information induced by the PHY cache at the transmitters. For example, as illustrated in Figure 12.2(b), part (ii), if the requested packets of user 1 and user 2 are in the caches of transmitters 1 and 2, then both transmitters have knowledge of x_1 and x_2, and hence they have full information and can engage in full CoMP transmission, achieving a significant gain in spectral efficiency. In this case, the PHY topology is transformed from an interference channel to a cooperative MIMO broadcast channel.

12.2.3 Benefits of PHY Caching

When PHY caching is used at the base stations and relay nodes, the physical layer topology of the radio interface changes dynamically between an unfavorable topology (when there is no PHY cache hit) and a favorable topology (when there is a PHY cache hit). As a result, the benefits of the PHY caching depend on how the cache-assisted PHY transmission can exploit the side information and translate it into performance benefits in the radio interface. We summarize below the associated benefits depending on the availability of channel state information at the transmitter (CSIT).

12.2.3.1 PHY with Global CSIT: Opportunistic Cooperative Spatial Multiplexing Induced by PHY Caching

When there is global CSIT shared between the transmitters, cooperative spatial multiplexing between the users concerned becomes possible. Specifically, when there is a PHY cache hit, the content requested by the users exists simultaneously in the cache of the transmitters. In this case, the PHY topology in the current time slot becomes a cooperative MIMO broadcast topology with CSIT. As a result, the transmitters can cooperatively transmit multiple data streams to multiple users in the network simultaneously. For example, consider the cached interference channel in Figure 12.2(b), part (ii). When the packets requested by user 1 and user 2 exist simultaneously in the PHY cache at the BSs, they can be served cooperatively, delivering four streams of data to the users. On the other hand, when there is no PHY cache hit, the system can only deliver two streams of data to user 1 and user 2 because of the interference.

12.2.3.2 PHY without CSIT: Opportunistic Cooperative Spatial Diversity Induced by PHY Caching

When global CSIT is not available, it is more difficult to mitigate the interference between multiple users to achieve spatial multiplexing gain. In this case, we can

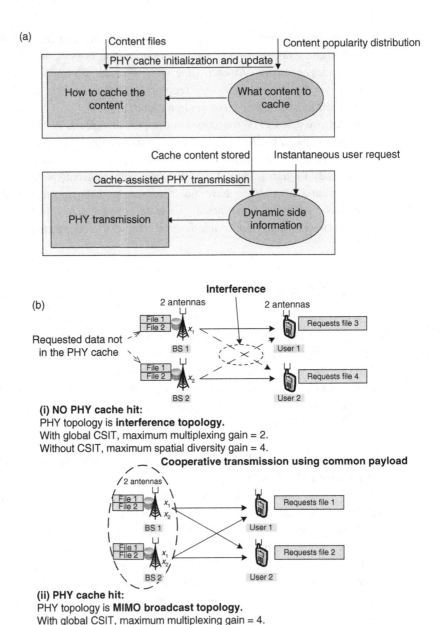

Figure 12.2 (a) Summary and interrelationship of the components of PHY caching. (b) Benefits of PHY caching.

focus on improving the reliability of the wireless communication measured by the *spatial diversity* gain. The spatial diversity is used to characterize how the probability of a transmission error declines with increasing SNR. Using PHY caching, it is possible to significantly improve the spatial diversity of the relay/interference topology via opportunistic cooperative spatial diversity. For example, consider the cached interference channel in Figure 12.2(b), part (ii). When the packets requested by the user in the current time slot exist simultaneously in the PHY cache at the BSs, the PHY topology becomes a cooperative broadcast channel without CSIT. In this case, the maximum spatial diversity per user is given by $(2+2) \times 2 = 8$. On the other hand, when there is no PHY cache hit, the maximum spatial diversity gain is limited by the number of BS antennas times the number of antennas at a user [22], given by $2 \times 2 = 4$.

12.2.4   Design Challenges and Solutions in PHY Caching

The benefits of PHY caching depend heavily on the probability of inducing various types of side information at the transmitters of the wireless network. Since the role of the PHY cache is to *promote* MIMO cooperation opportunities among peer transmitters, the design of PHY caching is quite different from that of traditional caching in fixed-line networks or CDNs. We shall elaborate on these design challenges below.

12.2.4.1   Exponentially Small MIMO Cooperation Probability with Asynchronous Access

We first use an example to show that, with a naive random cache scheme, the MIMO cooperation probability can be exponentially small, even if a significant portion of the content files are stored at the BS cache.

Example 12.1 (Naive random caching with asynchronous access) Consider a two-BS–user-pair interference channel with $L = 2$ content files (where each has size F), as illustrated in Figure 12.3. Each file consists of 12 packets, and suppose every node stores the same half of each file (packets 1–3 and 7–9) without source coding. User 1 requests file 1 at time slot t_1 and user 2 requests file 2 at time slot t_2. Without coding, the packets have to be sent sequentially to the users, from the first packet to the last packet, as illustrated in Figure 12.3. Note that the starting times of the transmission of file 1 and file 2 are random (since user requests are usually asynchronous with each other). As a result, the MIMO cooperation probability (the probability that the requested packets of both users exist in the cache of the two BSs) is given by 0.5^2 (averaged over all random request timings t_1 and t_2).* This will be the case no matter how we select the packets to be cached at the BSs. Hence, the MIMO cooperation probability decreases exponentially with respect to the number of users. This problem may be solved if we add a unique sequence number (ID) to each information packet and schedule the

---

* With random request timings t_1 and t_2, for a given time slot t, the probability that a requested packet of file 1 is in the cache is 0.5, and, similarly, the probability that a requested packet of file 2 is in the cache is 0.5. Since the request timings t_1 and t_2 are independent, the MIMO cooperation probability is 0.5^2.

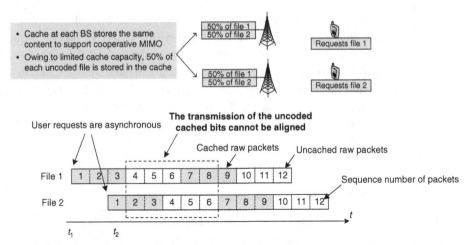

Figure 12.3 Illustration of the exponentially small MIMO cooperation probability under naive random caching. Note that file accesses are completely asynchronous between users, and hence t_1 and t_2 are random.

transmission order to align the transmission of the cached packets. However, this will impose a large segmentation and reassembly overhead and delay on the user.

Note that even if the user requests happen to be synchronized, the cached bits still cannot always be aligned, owing to the different instantaneous transmission rates caused by the fading wireless channels, and hence the transmission time of each packet between users will be different. As a result, the problem cannot be resolved merely by synchronizing the user requests. Such an exponentially small probability of MIMO cooperation (PHY cache hits) will seriously jeopardize the overall spectral-efficiency gain of the PHY cache in a wireless network. Later, in Section 12.4, we will propose an MDS-coded PHY caching scheme to address this issue.

12.2.4.2   PHY-Cache-Induced Partial MIMO Cooperation

The cache state in the PHY cache induces some side information in the BSs which the BSs can exploit to engage in partial cooperation [23], resulting in enhanced DoF in the radio interface. However, there is an exponentially large number of cache states and associated partial cooperation patterns (transmission modes), and thus a complicated MIMO precoding and transmission design is required to mitigate interference under conditions of partial side information. For example, consider an interference network with only two BS–user pairs. The cache state can be represented by a 4×1 binary vector, with each element indicating whether the content requested by user k is in the PHY cache of BS n, for $k, n = 1, 2$. As a result, there are $2^4 = 16$ possible cache states and associated partial cooperation patterns. For a general cached wireless network with K users and N BSs, there are about 2^{KN} possible cache states and associated partial cooperation patterns. In Section 12.4, we will propose a simple cache-assisted MIMO cooperation scheme with only two transmission modes to capture the first-order gain of PHY caching. There are already some studies of partial cooperation in cached wireless networks. For example, in [24], a cache-aided transmission scheme for a three-user interference channel was proposed, and it was shown that a DoF gain can be achieved by caching different parts of each content file at the transmitters. However, this study only provided an achievable scheme for a three-BS–user-pair single-antenna interference channel. The caching and transmission scheme and the upper bound for general cached MIMO wireless networks are still unknown. Joint data assignment and partial cooperative beamforming for backhaul-limited caching networks was also studied in [20]. However, to the best of our knowledge, the joint optimization of cache placement and cache-induced partial cooperative beamforming has not been considered before.

12.2.4.3   Cache Content Placement Algorithm Design

The cache content placement algorithm in Figure 12.2(a) determines the dynamic priority of caching of one file over the others, and this is a critical component that affects the spectral-efficiency gain of the PHY caching scheme. The design of the cache content placement algorithm is a complex large-scale *stochastic optimization* problem, with intrinsic coupling between the dynamic physical layer topology and the cache placement vector [7, 21]. Furthermore, the solution needs to be adaptive to the content access popularity as well as to the channel statistics. However, in practice, such information is not easy to obtain and is also nonstationary over time. It is desirable but challenging to have a self-adaptive and self-learning algorithm without explicit knowledge of these statistics.

To address the cache content placement issue, we can first exploit the timescale separations of the optimization variables and stochastic decomposition techniques to decompose the two-timescale stochastic optimization problem into a short-term precoding problem and a long-term stochastic cache placement problem [7, 8, 10]. Then stochastic learning algorithms, such as stochastic subgradient or stochastic cutting-plane methods, can be used to solve the cache content placement problem without explicit knowledge of the popularity of the content files. An example of the design of a cache placement algorithm will be given in Section 12.5.

12.3   DoF Upper Bound for Cached Wireless Networks

In this section, we shall develop a theory of the upper bound on the DoF of general cached wireless networks. We would like to investigate the maximum benefit we can expect when we introduce caches into wireless networks. One challenge we need to overcome is to embrace all possible caching strategies in developing the upper-bound results. We start with a generic cached wireless network model, followed by a generic cache model and a generic cache-assisted PHY model.

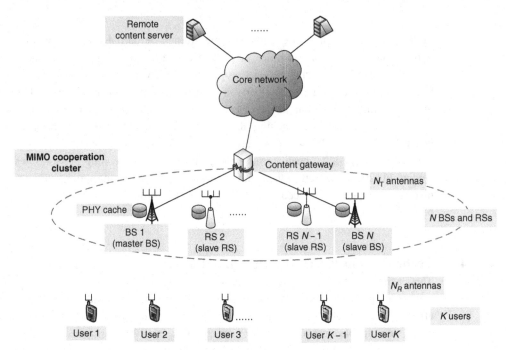

Figure 12.4 Architecture of a cached wireless network.

12.3.1   Architecture of Cached Wireless Networks

Consider a general cached wireless network, as illustrated in Figure 12.4. There are a total of N BSs and RSs, and K users. The RSs can be viewed as BSs without wired backhaul and are also treated as BSs when there is no ambiguity. Each BS is equipped with N_T transmitter antennas, and each user is equipped with N_R receiver antennas. Each BS/RS is equipped with a cache with a capacity of B_C bits. Let $\mathcal{U}_n$ denote the set of users associated with BS n. The data requested by the users in $\mathcal{U}_n$ can always be reached at their serving BS n through backhaul. For RS n, we have $\mathcal{U}_n = \emptyset$. Furthermore, there is a *master BS* responsible for beamforming and resource control of all the *slave BSs* and *slave RSs*, as illustrated in Figure 12.4. Such a cluster architecture appears in practical Long Term Evolution (LTE) networks. The time is partitioned into time slots indexed by t, with duration τ. There are L content files on the content server, indexed by $l \in \{1, \ldots, L\}$. The size of the lth content file is F_l bits. We assume the cache capacity is not large enough to store the entire content library, i.e., $B_C < \sum_{l=1}^{L} F_l$. The index of the content file requested by the kth user is denoted by π_k. We define $\pi = [\pi_1, \pi_2, \ldots, \pi_K]$ as the user request vector (URV), which is an ergodic random process and changes on a timescale much slower than the time slots.

12.3.2   Generic Cache Model

In a cached wireless network, each content file is first encoded at the content server, and then each BS stores part of the encoded content packets from the server. Collectively, the caches of all the BSs can be modeled by a *generic cache model* which consists of

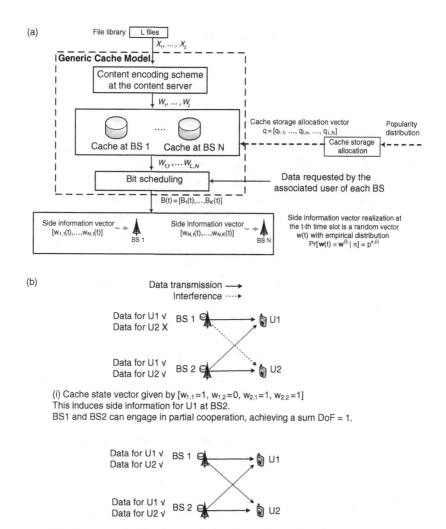

(a)

Figure 12.5 Illustration of (a) generic cache model, and (b) the relationship between the side information vector and PHY topology.

a content encoding scheme, a cache content placement scheme and a bit scheduling scheme, as illustrated in Figure 12.5(a).

Content Encoding Scheme: The content encoding scheme at the content server is defined as a collection of L mappings $\theta_l : \{0,1\}^{F_l} \rightarrow \{0,1\}^{F_l'}$ from content file X_l to a set of coded information bits $W_l \triangleq \theta_l(X_l)$, where $F_l' \in \mathbb{N}$ with $F_l' \geq F_l$ is the length of the lth coded content file.

Cache Content Placement Scheme: The cache content placement scheme at the BSs is defined as a collection of NL mappings $\phi_{l,n} : \{0,1\}^{F_l'} \rightarrow \{0,1\}^{q_{n,l}F_l'}, \forall n \in \{1,...,N\}, \forall l \in \{1,...,L\}$ from the coded content file W_l to its subset $W_{l,n} \triangleq \phi_{l,n}(W_l) \subseteq W_l$ cached at BS

n, where $q_{l,n}$ is the cache storage allocation factor of the lth content file at BS n. Denote $\mathbf{q} = [q_{1,1}, \ldots, q_{l,n}, \ldots q_{L,N}]$ as the *cache storage allocation vector*. It is clear that $\mathbf{q}$ must satisfy the cache storage capacity constraint:

$$\mathbf{q} \in \Lambda = \left\{ [q_{1,1}, \ldots, q_{l,n}, \ldots, q_{L,N}] \in \mathbb{R}_+^{LN} \,\middle|\, \sum_{l=1}^{L} q_{l,n} F_l \leq B_C, \forall n \in \{1, \ldots, N\} \right\}. \quad (12.1)$$

Bit Scheduling Scheme: Each user k requests files one after another. Let T_i^k denote the time slot when $\pi_k(t)$ changes for the ith time. In other words, user k starts to request file $\pi_k(T_i^k)$ at time slot T_i^k and the delivery of file $\pi_k(T_i^k)$ to user k is finished at time slot $T_{i+1}^k - 1$. In the following, we will focus on the delivery of file $\pi_k(T_i^k)$ and use the simplified notation π_k to denote $\pi_k(T_i^k)$. Let $B_k(t) \subseteq W_{\pi_k}$ denote the set of bits scheduled for delivery to user k in the tth time slot. After obtaining all the scheduled bits $\bigcup_{t \in [T_i^k, T_{i+1}^k - 1]} B_k(t)$ for file π_k at time $T_{i+1}^k - 1$, user k will apply a decoding function $\hat{X}_{\pi_k} = \psi_i^k \left(\bigcup_{t \in [T_i^k, T_{i+1}^k - 1]} B_k(t) \right)$ to obtain the estimated file $\hat{X}_{\pi_k}$. A bit scheduling scheme is *feasible* if

$$\lim_{F_{\pi_k} \to \infty} \Pr \left[\psi_i^k \left(\bigcup_{t \in [T_i^k, T_{i+1}^k - 1]} B_k(t) \right) \neq X_{\pi_k} \right] = 0, \forall k, i.$$

For convenience, let $B(t) = \{B_1(t), \ldots, B_K(t)\}$ denote the collective sets of bits scheduled for delivery to all K users in tth time slot.

Define $w_{n,k}(t) \in \{0, 1\}$ as the side information state of BS n in the tth time slot conditional on URV π, given by

$$w_{n,k}(t) = 1, \forall k \in \mathcal{U}_n,$$

$$w_{n,k}(t) = \mathbf{1}\left[B_k(t) \subseteq W_{\pi_k,n} \right], \forall k \notin \mathcal{U}_n,$$

where $w_{n,k}(t) = 1$ indicates the set of bits scheduled for delivery to user k can be obtained from cache or backhaul at BS n, and $w_{n,k}(t) = 0$ indicates $B_k(t)$ cannot be obtained from cache or backhaul at BS n. The terms $w_{n,k}(t) = 1, \forall k \in \mathcal{U}_n$ since the data requested by user k can always be reached at its serving BS n through backhaul. Collectively, the overall side information state of the network is denoted by the *side information vector* $\mathbf{w}(t) = [w_{1,1}(t), \ldots, w_{n,k}(t), \ldots, w_{N,K}(t)]$. Note that there are a total of $N_c \triangleq 2^{NK - \sum_n |\mathcal{U}_n|}$ possible side information patterns. We tabulated the N_c side information vectors by a codebook $\mathcal{W}$, and let $\mathbf{w}^{(i)} = [w_{1,1}^{(i)}, \ldots, w_{n,k}^{(i)}, \ldots, w_{N,K}^{(i)}]$ denote the ith side information vector in $\mathcal{W}$, where $i \in \mathcal{I} = \{1, \ldots, N_c\}$. Conditioned on URV π, the empirical probability that the ith side information pattern $\mathbf{w}^{(i)}$ occurs is given by

$$p^{\pi,(i)} = \lim_{T \to \infty} \frac{\sum_{t \in [1,T]} \mathbf{1}\left(\mathbf{w}(t) = \mathbf{w}^{(i)}, \pi(t) = \pi\right)}{T}.$$

We define $\mathbf{p}^\pi = [p^{\pi,(1)}, \ldots, p^{\pi,(N_c)}]$ as the cache state probability vector under URV π. The cache state probability vector $\mathbf{p}^\pi$ depends on the parameters $\tilde{B}_C$ and $\mathbf{q}$, and the feasible region of $\mathbf{p}^\pi$ will be characterized in Section 12.3.4. Note that the cache model does not specify the content encoding scheme $\{\theta_l\}$ and cache content placement scheme $\{\phi_{l,n}\}$ and therefore it embraces different possible caching schemes, including possibly random caching.

12.3.3   Cache-Assisted PHY Transmission Model

The availability of side information at the BSs is determined by the cache state vector $\mathbf{w}$, where $w_{n,k} = 1, k \notin \mathcal{U}_n$ means that the packets requested by user k are in the cache of BS n. As a result, the PHY transmission mode (or the PHY topology) depends on the dynamic cache state $\mathbf{w}$, as illustrated in Figure 12.5(b). The cache state vector $\mathbf{w}$ of the cache induces some side information in the BSs which the BS in an interference network can exploit to engage in partial cooperation [23], resulting in $k : (k,j) \in \mathcal{J}$ enhanced DoF for the radio interface.

Let $\mathbf{H}_{k,n}(t) \in \mathbb{C}^{N_R \times N_T}$ denote the channel matrix between BS n and user k. We assume that the master BS has knowledge of the global CSI $\mathbf{H}(t) = \{\mathbf{H}_{k,n}(t), \forall k,n\}$. For simplicity, we consider the case where no channel extension is allowed, since a large number of channel extensions is not possible in practical networks [3]. Note also that the global-CSI assumption is a common requirement for most interference mitigation schemes, such as joint beamforming, coordinated beamforming [4, 5], and cooperative MIMO [25, 26], with or without PHY caching. Furthermore, it is also implementable with a cluster architecture in LTE systems.[†] The goal of this chapter is to investigate how a PHY caching scheme can leverage the advanced PHY to achieve DoF gains with limited backhaul.

Given a certain side information vector $\mathbf{w}(t) \in \mathcal{W}$ at the tth time slot, the scheduled bits $B(t)$ are limited by the transmission data rate of the physical layer. Specifically,

$$|B_k(t)| = r_k(t)\,\tau,$$

where $r_k(t)$ is the received data rate of user k given $\mathbf{H}(t)$ and $\mathbf{w}(t)$. To achieve the data rate $r_k(t)$, a space-time block coding scheme $\beta_{n,k} : \{0,1\}^{r_k(t)\tau} \to \mathbb{C}^{N_T \times N^b}$ is applied at BS n for the content requested by user k, which maps the set of bits scheduled to transmit $B_k(t)$ to a space-time codeword of size $N_T \times N^b$

$$\mathbf{Z}_{n,k}(t) \triangleq \begin{cases} \beta_{n,k}\left(B_k^\pi(t)\right) & w_{n,k}(t) = 1 \\ \mathbf{0} & w_{n,k}(t) = 0 \end{cases}$$

$\forall k \in \{1, \ldots K\}$, where N^b is the number of symbol vectors in each time slot. Note that when $w_{n,k}(t) = 0$, BS n does not have the content requested by user k at the tth time slot, and thus we let $\mathbf{Z}_{n,k}(t) = \mathbf{0}$. The signal $\mathbf{Z}_n(t)$ transmitted at BS n in the tth time slot is a superposition of the codewords for all users: $\mathbf{Z}_n(t) = \sum_{k=1}^{K} \mathbf{Z}_{n,k}(t) \in \mathbb{C}^{N_T \times N^b}$, where each $N_T \times 1$ column vector in $\mathbf{Z}_n(t)$ is transmitted from the N_T antennas during a symbol period in time slot t. Moreover, $\mathbf{Z}_n(t)$ satisfies a power constraint $\mathrm{Tr}\left(\mathbf{Z}_n(t)\mathbf{Z}_n^H(t)\right)/N_b \leq P_\Sigma$. The received signal at user k in the tth time slot is denoted by $\mathbf{Y}_k(t) = \sum_{n=1}^{K} \mathbf{H}_{n,k}(t)\mathbf{Z}_n(t) + \mathbf{N}_k(t) \in \mathbb{C}^{N_R \times N^b}$, where $\mathbf{N}_k(t) \in \mathbb{C}^{N_R \times N^b}$ denotes the noise

---

[†] For example, Huawei has deployed a cloud-based LTE network in Japan and Chengdu with a CoMP cluster consisting of one master eNB and eight slave eNBs, supporting global CSI and CoMP.

matrix at user k in the tth time slot with i.i.d. entries of zero mean and unit variance. The rate $r_k(t)$ is achievable if the user k can decode the scheduled bits $B_k(t)$ from the received signal $\mathbf{Y}_k(t)$ successfully at the tth time slot. The set of all achievable rate vectors $\mathbf{r}(t) = [r_1(t),\ldots,r_K(t)]$ forms the capacity region $\mathcal{C}(\mathbf{H}(t),\mathbf{w}(t),P_\Sigma) \in \mathbb{R}^K$. For analytical tractability, we focus on the DoF instead of the absolute rate. Specifically, the corresponding DoF region of the PHY under given $\mathbf{H}$ and $\mathbf{w}^{(i)}$ is defined as follows.

Definition 12.1 (Instantaneous DoF region under given $\mathbf{H}$ and $\mathbf{w}^{(i)}$) *The instantaneous DoF region under the ith side information vector $\mathbf{w}^{(i)}$ and CSI $\mathbf{H}$ is defined as*

$$\mathcal{D}^{(i)}(\mathbf{H}) = \left\{ (d_1,\ldots,d_K) \in \mathbb{R}_+^K \middle| \exists (R_1(P_\Sigma),\ldots,R_K(P_\Sigma)) \in \mathcal{C}(\mathbf{H},\mathbf{w}^{(i)},P_\Sigma) \right.$$

$$\left. such\ that\ d_k = \lim_{\Delta\to\infty} \frac{R_k(P_\Sigma)}{\log_2 P_\Sigma}, \forall k \right\},$$

where $\Delta \to \infty$ means that $P_\Sigma \to \infty$, $B_C \to \infty$ and $F \to \infty$ such that $\frac{B_C}{\log_2(P_\Sigma)} = \tilde{B}_C$ and $\frac{F_l}{\log_2(P_\Sigma)} = \tilde{F}_l, \forall l$.

Note that DoF is an asymptotic metric for large SNR. To avoid trivial results, we need to scale the content file size F_l and the cache storage capacity B_C with the log SNR because, otherwise, the system will be able to deliver the entire content file in one time slot as SNR $\to \infty$. Hence, $\tilde{F}_l$ and $\tilde{B}_C$ are called the *file size scaling factor* and *cache storage capacity scaling factor*, respectively. For a given cache state (side information) $\mathbf{w}^{(i)}$, the corresponding DoF region $\mathcal{D}^{(i)}$ of the cached wireless network is in general unknown. However, there are outer bounds available for the DoF region $\mathcal{D}^{(i)}(\mathbf{H})$, as given in the following lemma.

Lemma 12.2 (DoF region for a given cache state vector $\mathbf{w}^{(i)}$) *An outer bound of the DoF region $\mathcal{D}^{(i)}$ for a given cache state (side information) $\mathbf{w}^{(i)}$ is given by*

$$\mathcal{D}^{(i)} \subseteq \hat{\mathcal{D}}^{(i)} = \left\{ [d_1,\ldots,d_K] \,\middle|\, M_k^{(i)} = N_T \sum_{n=1}^N w_{n,k}^{(i)}, \right.$$

$$\min\left\{ M_k^{(i)}, N_R \right\} \ge d_k, \forall k,$$

$$\max\left\{ M_k^{(i)}, N_R \right\} \ge d_k + d_j, \forall k,j, k \ne j,$$

$$\sum_{k:(k,j)\in\mathcal{J}} \left(M_k^{(i)} - d_k \right) d_k + \sum_{k:(k,j)\in\mathcal{J}} \left(N_R - d_j \right) d_j$$

$$\ge \sum_{k:(k,j)\in\mathcal{J}} d_k d_j, \forall \mathcal{J} \subseteq \{ (k,,j) \,|\, 1 \le k \ne j \le K \} \right\}. \tag{12.2}$$

Under side information induced by $\mathbf{w}^{(i)}$, the DoF region of the cached interference network is the same as that of a virtual interference channel with K transmitter–receiver pairs, where each receiver has N_R receive antennas and the kth transmitter has $N_T \sum_{n=1}^N w_{n,k}^{(i)}$ transmit antennas. As a result, the expression for $\hat{\mathcal{D}}^{(i)}$ can be obtained using the DoF bounds for general interference networks given in [3, Theorem 1]. Figure 12.6 illustrates the DoF regions of a two-user interference channel for different cache state vectors $\mathbf{w}^{(i)}$.

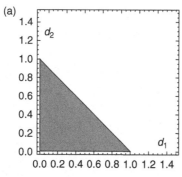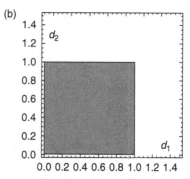

Figure 12.6 DoF regions of a two-user interference channel with $N = 2, K = 2, N_T = N_R = 1$ for different cache state vectors **w**. (a) Cache state vector given by $[w_{1,1} = 1, w_{1,2} = 0, w_{2,2} = 1]$; (b) cache state vector given by $[w_{1,1} = 1, w_{1,2} = 1, w_{2,1} = 1, w_{2,2} = 1]$.

12.3.4   Upper Bound of Sum DoF for Cached Wireless Networks

Since the capacity of the cache is not large enough to store the entire content library, for a given URV π, the cache state probabilities $\left[p^{\pi,(1)}, \ldots, p^{\pi,(N_c)}\right]$ are constrained by the number of information bits stored in the cache. Specifically, for a given URV π, the number of information bits transmitted from the nth BS cache to the kth user (which is $\sum_{i \in \mathcal{I}} w_{n,k}^{\pi,(i)} d_k^{\pi,(i)} p^{\pi,(i)} T^{\pi}$ in units of $\log_2(P_\Sigma)$ bits, where $T^{\pi} = T \Pr[\pi(t) = \pi]$ is the total transmission time under URV π) cannot exceed the total number of information bits obtained from the nth BS cache (which is $q_{\pi_k, n} \tilde{F}_{\pi_k} N^{\pi}$ in units of $\log_2(P_\Sigma)$ bits, where $N^{\pi} = \sum_{i \in \mathcal{I}} w_{n,k}^{\pi,(i)} d_k^{\pi,(i)} p^{\pi,(i)} T^{\pi} / \tilde{F}$ is the number of copies of content received at the kth user). This is illustrated in a toy example below.

Example 12.2 (Constraint on cache state probability) Consider a cached interference channel with one BS, one RS (i.e., $N = 2$) with $N_T = 1$ transmit antenna, $K = 1$ user with $N_R = 2$ receive antenna, and $L = 1$ content file with file size F. User 1 requests one copy of file 1, i.e., $\pi_1 = 1$ and $N^{\pi} = 1$. The cache placement is given by $q_{1,2} = 0.75$. There are two transmission modes, namely, relay mode $\left(w_{1,1} = 1, w_{2,1} = 0\right)$ and cooperation mode $\left(w_{1,1} = 1, w_{2,1} = 1\right)$. The transmission rate (in units of $\log_2(P_\Sigma)$ bits) in relay mode is given by $d_1^{\pi,(2)} = 1$ and in cooperation mode is given by $d_1^{\pi,(1)} = 2$, as illustrated in Figure 12.7. The total number of information bits obtained from the cache is $q_{1,2}F$. The total number of information bits transmitted from the cache of RS 2 is $p^{\pi,(1)} T^{\pi} d_1^{\pi,(1)}$, where T^{π} is the total transmit time, given by $T^{\pi} = F/(d_1^{\pi,(1)} p^{\pi,(1)} + d_1^{\pi,(2)} p^{\pi,(2)})$. Since the number of information bits transmitted from the cache of RS 2 cannot exceed the total number of information bits stored in the cache, any feasible cache state probability must satisfy $p^{\pi,(1)} T^{\pi} d_1^{\pi,(1)} \leq q_{1,2}F$, which indicates that the cache state probability must satisfy $p^{\pi,(1)} \leq 0.6$.

From Example 12.2, it can be seen that the feasible regions of the DoF and cache state probability are coupled together. For a given URV π, it is difficult to obtain an

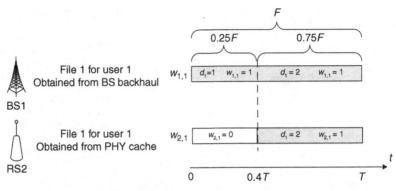

Figure 12.7 Illustration of the constraint on cache state probability.

exact expression for the feasible region Γ^π of the DoF and cache state probability tuple $(p^{\pi,(1)},\ldots,p^{\pi,(N_c)},d_1^{\pi,(1)},\ldots,d_K^{\pi,(N_c)})$, where $d_k^{\pi,(i)}$ is the DoF of user k under side information induced by $\mathbf{w}^{(i)}$ and URV π. Using Lemma 12.2, we obtain an outer bound of Γ^π, given in the following theorem.

Theorem 12.3 (Outer bound on the feasible region of the DoF and cache state probability tuple)   *Given a cache model with parameters $\tilde{B}_C$ and $\mathbf{q}$, and a PHY model with parameters N_R, N_T, N, and K, the feasible region for a given URV π is bounded by*

$$\Gamma^\pi \subseteq \hat{\Gamma}^\pi = \left\{ \left(p^{\pi,(1)},\ldots,p^{\pi,(N_c)},d_1^{\pi,(1)},\ldots,d_K^{\pi,(N_c)}\right) \mid \right.$$

$$0 \le p^{\pi,(i)} \le 1, \forall \in \mathcal{I}, \quad \sum_{i\in\mathcal{I}} p^{\pi,(i)} = 1, \tag{12.3}$$

$$\left[d_1^{\pi,(i)},\ldots,d_K^{\pi,(i)}\right] \in \hat{\mathcal{D}}^{(i)}, \forall i \in \mathcal{I}, \tag{12.4}$$

$$\left. \sum_{i\in\mathcal{I}} w_{n,j}^{\pi,(i)} d_j^{\pi,(i)} p^{\pi,(i)} \le q_{\pi_j,n} \sum_{i\in\mathcal{I}} d_j^{\pi,(i)} p^{\pi,(i)}, \forall n, \forall j \notin \mathcal{U}_n \right\}. \tag{12.5}$$

Proof Equation (12.3) is the conditions satisfied by any probability vector. Condition (12.4) means that a feasible DoF vector for a given cache state vector $\mathbf{w}^{(i)}$ must lie in the outer bound $\hat{\mathcal{D}}^{(i)}$ of the corresponding DoF region. Finally, the condition in (12.5) means that the cache state probabilities $\left[p^{\pi,(1)},\ldots,p^{\pi,(N_c)}\right]$ are constrained by the number of information bits stored in the cache. Specifically, for a given URV π, the number of information bits transmitted from the nth BS cache to the kth user, $\sum_{i\in\mathcal{I}} w_{n,k}^{\pi,(i)} d_k^{\pi,(i)} p^{\pi,(i)} T^\pi$, cannot exceed the total number of information bits obtained from the nth BS cache $q_{\pi_k,n} F_{\pi_k} N^\pi$, which implies that (12.5) must hold.

The average sum DoF over URV distribution is defined as

$$D = \mathbb{E}_\pi\left[D^\pi\right],$$

where $D^\pi = \sum_{k=1}^K \sum_{i\in\mathcal{I}} d_k^{\pi,(i)} p^{\pi,(i)}$ is the sum DoF conditional on URV π. The instantaneous DoF region $\mathcal{D}^{(i)}\left(\mathbf{H}(t)\right)$ captures the feasible DoF vector $\mathbf{d}^{\pi,(i)}$ for a given instantaneous side information vector $\mathbf{w}(t) = \mathbf{w}^{(i)}$ at time slot t. On the other hand, given a URV π, the instantaneous side information vector $\mathbf{w}(t)$ at the tth time slot is random, with empirical probability vector $\mathbf{p}^\pi$ constrained by region $\hat{\Gamma}^\pi$. Furthermore, for fairness consideration, we consider symmetric traffic model where all users have the same average DoF requirement (average over the random instantaneous side information vector $\mathbf{w}(t)$ under given URV π). As a result, with the outer bound on the instantaneous DoF in Lemma 12.2 and the outer bound on the side information probability vector in Theorem 12.3, the average sum DoF upper bound is given in the following theorem.

Theorem 12.4 (Sum DoF upper bound for cached wireless networks averaged over content popularity distribution) *Given a cache model with parameters $\tilde{B}_C$ and $\mathbf{q}$, and a PHY model with parameters N_R, N_T, N, and K, the sum DoF of a cached wireless network averaged over the distribution of URV π is upper bounded by*

$$D_{\max} \leq \max_{\mathbf{q}, \left[d_1^{\pi,(i)}, \ldots, d_K^{\pi,(i)}\right], \mathbf{p}^{\pi,(i)}} \mathbb{E}_\pi \left[\sum_{k=1}^{K} \sum_{i \in \mathcal{I}} d_k^{\pi,(i)} p^{\pi,(i)} \right] \tag{12.6}$$

such that. $\mathbf{q} \in \Lambda$, *(cache capacity constraint)*

$$\sum_{i \in \mathcal{I}} d_k^{\pi,(i)} p^{\pi,(i)} = \sum_{i \in \mathcal{I}} d_j^{\pi,(i)} p^{\pi,(i)}, \forall j \neq k,$$

(fairness constraint)

$$\left(p^{\pi,(1)}, \ldots, p^{\pi,(N_c)}, d_1^{\pi,(1)}, \ldots, d_K^{\pi,(N_c)} \right) \in \hat{\Gamma}^\pi, \forall \pi.$$

(Outer bound on the DoF and cache state probability region)

Note that the sum DoF upper bound increases with $\tilde{B}_C$, since the feasible regions Λ and $\hat{\Gamma}^\pi$ become larger for larger $\tilde{B}_C$. The DoF upper bound also increases with the number of antennas at each BS, N_T, and the number of antennas at each user, N_R, since the DoF region for each cache state becomes larger. In the following corollary, we give a closed-form upper bound for the special case of a cached single-input single-output (SISO) interference network under an independent reference model (IRM) popularity distribution.

Corollary 12.5 (Closed-form upper bound of sum DoF for cached SISO interference networks) *Consider a cached interference network with N BSs, K users, and L content files, where each BS has $N_T = 1$ antenna and a cache capacity $\tilde{B}_C$, each user has $N_R = 1$ antenna, and the size of each content file is $\tilde{F}$. Assume that the URV π follows the IRM, where consecutive user requests are mutually independent, and each user independently requests the lth content file with a probability ρ_l satisfying $\rho_l \in [0,1]$ and $\sum_{l=1}^{L} \rho_l = 1$. Without loss of generality, we sort the content file such that $\rho_l \geq \rho_{l+1}$ for $l = 1, \ldots, L-1$. Then the sum DoF averaged over the IRM popularity distribution is upper bounded by*

$$D_{\max} \leq \min \left[N \sum_{l=1}^{\tilde{B}_C/\tilde{F}} \rho_l + 1, K \right]. \tag{12.7}$$

Proof $T^\pi = T\Pr[\pi(t) = \pi]$ is the total transmission time under URV π. It can be shown that as the total transmission time $T \to \infty$, we have

$$T^\pi \geq \sum_{j=1}^{K}\sum_{i\in\mathcal{I}}\sum_{r=1}^{K}\frac{d_j^{\pi,(i)}p^{\pi,(i)}1\left(\sum_{n=1}^{N}w_{n,j}^{(i)}=r\right)T^\pi}{r}, \tag{12.8}$$

where $d_j^{\pi,(i)}p^{\pi,(i)}1\left(\sum_{n=1}^{N}w_{n,j}^{(i)}=r\right)T^\pi$ indicates the number of information bits where each of these bits exists simultaneously in r BSs, and these information bits can be transmitted with a maximum sum DoF of r. As a result, we have

$$1 \geq \sum_{j=1}^{K}\sum_{i\in\mathcal{I}}\sum_{r=1}^{K}\frac{d_j^{\pi,(i)}p^{\pi,(i)}1\left(\sum_{n=1}^{N}w_{n,j}^{(i)}=r\right)}{r} \tag{12.9}$$

$$\geq \frac{\left(\sum_{j=1}^{K}\sum_{i\in\mathcal{I}}d_j^{\pi,(i)}p^{\pi,(i)}\right)^2}{\sum_{j=1}^{K}\sum_{i\in\mathcal{I}}d_j^{\pi,(i)}p^{\pi,(i)}\sum_{n=1}^{N}w_{n,j}^{(i)}}, \tag{12.10}$$

where (12.10) is due to the Cauchy–Schwartz inequality. Hence, the average sum DoF is given by

$$\mathbb{E}_\pi\left[\sum_{j=1}^{K}\sum_{i\in\mathcal{I}}d_j^{\pi,(i)}p^{\pi,(i)}\right] \tag{12.11}$$

$$\leq \mathbb{E}_\pi\left[\frac{\sum_{j=1}^{K}\sum_{i\in\mathcal{I}}\sum_{n=1}^{N}d_j^{\pi,(i)}p^{\pi,(i)}w_{n,j}^{(i)}}{\sum_{j=1}^{K}\sum_{i\in\mathcal{I}}d_j^{\pi,(i)}p^{\pi,(i)}}\right] \tag{12.12}$$

$$\leq \mathbb{E}_\pi\left[\frac{\sum_{j=1}^{K}\sum_{i\in\mathcal{I}}d_j^{\pi,(i)}p^{\pi,(i)}\min\left(\sum_{n=1}^{N}q_{\pi j,n}+1,N\right)}{\sum_{j=1}^{K}\sum_{i\in\mathcal{I}}d_j^{\pi,(i)}p^{\pi,(i)}}\right] \tag{12.13}$$

$$= \mathbb{E}_\pi\left[\frac{\sum_{j=1}^{K}\min\left(\sum_{n=1}^{N}q_{\pi j,n}+1,N\right)}{K}\right] \tag{12.14}$$

$$\leq N\sum_{l=1}^{\tilde{B}_C/\tilde{F}}\rho_l+1, \tag{12.15}$$

where (12.13) indicates that the total number of bits used for transmission among N BSs, $\sum_{j=1}^{K}\sum_{i\in\mathcal{I}}\sum_{n=1}^{N}d_j^{\pi,(i)}p^{\pi,(i)}w_{n,j}^{(i)}T^\pi$ is no larger than the total number of information bits obtained from the caches at the BSs, $\sum_{n=1}^{N}q_{\pi k,n}\tilde{F}_{\pi k}N^\pi = \sum_{n=1}^{N}q_{\pi j,n}\sum_{i\in\mathcal{I}}d_j^{\pi,(i)}p^{\pi,(i)}T^\pi$, plus the number of information bits obtained from the backhaul, $\sum_{i\in\mathcal{I}}d_j^{\pi,(i)}p^{\pi,(i)}T^\pi$, and that the total number of bits used for transmission among N BSs cannot be larger than $N\sum_{i\in\mathcal{I}}d_j^{\pi,(i)}p^{\pi,(i)}T^\pi$. Together with the trivial bound $D_{\max} \leq K$, we conclude the proof.

In the following, we give one simple example to illustrate the sum DoF upper bound in Theorem 12.4.

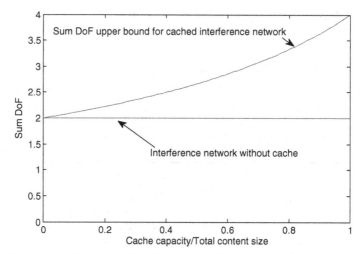

Figure 12.8 Illustration of DoF upper bound for a cached interference channel with $N = K = 2$ BS–user pairs and $N_T = N_R = 2$ antennas at each BS and user.

Example 12.3 (Upper bound of sum DoF for cached MIMO interference networks) Consider a two-user interference network with cache capacity $\tilde{B}_C$, $N = K = 2$ BS–user pairs, N_T antennas at each BS, and N_R antennas at each user satisfying $N_T \geq N_R$, and where the user association is given by $\mathcal{U}_1 = \{1\}$ and $\mathcal{U}_2 = \{2\}$. There are L content files with equal size $\tilde{F}$, and each file is requested independently with equal probability $1/L$. In this case, the average sum DoF upper bound in (12.6) can be simplified to

$$D_{\max} \leq D_0 + \frac{(D_1 - D_0)\,\tilde{B}_C/\tilde{F}L}{G\left(1 - \tilde{B}_C/\tilde{F}L\right) + \tilde{B}_C/\tilde{F}L}, \tag{12.16}$$

where $D_1 = 2N_R$, $D_0 = \min(2N_R, N_T)$, and $G = D_1/D_0$. In Figure 12.8, we illustrate the DoF upper bound for a cached interference network with $N = K = 2$ BS–user pairs and $N_T = N_R = 2$ antennas at each BS and user. Note that the sum DoF upper bound increases with the cache capacity, and this DoF upper bound embraces all possible caching strategies and cache-assisted PHY transmission schemes.

12.4   MDS-Coded PHY Caching and the Achievable DoF

From Figure 12.8, we observe that there are huge potential DoF gains in cached wireless networks. In this section, we outline a simple cache scheme, namely *MDS-coded PHY caching*, and discuss the achievable DoF. We shall illustrate that the scheme is DoF optimal in some special cases, and in general the gap between the DoF upper bound and achievable DoF is small for typical scenarios.

Table 12.1 Long-timescale cache initialization.

Step 1 (content file segmentation and MDS encoding). Each file is divided into segments of size L_S bits. Each segment is then encoded into $L_S + q_l L_S$ parity bits using an MDS rateless code, where the encoded parity bits of each segment are further divided into L_S parity bits stored in the content server and $q_l L_S$ parity bits cached in each BS cache, as shown in Figure 12.9.

Step 2 (offline cache initialization/updating). For $l = 1,..,L$, all the caches at the BSs are initialized with the $q_l L_S$ MDS-encoded parity bits for each segment of the lth file in an offline manner during off-peak hours. Note that the cached parity bits at each BS cache are identical to create cooperative MIMO opportunities.

12.4.1   MDS-Coded PHY Caching with Asynchronous Access

We propose an MDS-coded PHY caching scheme for cached wireless networks. The steps of the cache initialization scheme are summarized in Table 12.1. Specifically, BS n caches the same portion, of size $q_{l,n} F_l$ MDS-encoded bits, of the lth content file, where $q_{l,n} = q_l, \forall n$. Since all the BSs cache the same content, we use q_l to denote the portion of content l stored in each cache, for simplicity. An MDS rateless code generates an arbitrarily long sequence from an information sequence of L_S bits ($L_S \in \mathbb{Z}^+$) such that if the decoder receives any L_S parity bits, it can recover the original L_S information bits. The segment size L_S and the time slot duration τ must satisfy $L_S \gg \bar{r}_k \tau$, where $\bar{r}_k$ is the average rate of user k. Note that the segment size L_S is proportional to the decoding latency. Without segmentation, the decoding latency would become unacceptably large. In practice, the MDS rateless code can be implemented using Raptor codes [27] at the cost of a small redundancy overhead. The motivation for MDS encoding at the content server will be elaborated on in Section 12.4.3. The MDS-based PHY cache accepts a user request vector π at the beginning of a communication session,[‡] and the associated cache state at each time slot is given by the ith cache state $\mathbf{w}^{(i)}$ with probability $p^{\pi,(i)}$. Given $\tilde{B}_C$ and $\mathbf{q}$, the cache state probability vector $\mathbf{p}^\pi = \left[p^{\pi,(1)},...,p^{\pi,(N_c)} \right]$ can be determined from the given URV π, and this will be elaborated on in Section 12.4.3. Furthermore, as illustrated in Figure 12.9, there is an online algorithm for determining the *cache placement vector* $\mathbf{q}$ to maximize the DoF, and this will be elaborated on in Section 12.5.

12.4.2   Cache-Assisted MIMO Cooperation in the PHY

As described in Section 12.3.3, a given cache state vector $\mathbf{w}$ in the cache will induce corresponding side information at the BSs of a cached wireless network. In general, the PHY can exploit the side information induced by $\mathbf{w}$ to engage in partial cooperation to enhance the DoF. For practicality, we consider two transmission modes for online content delivery, namely, a *coordinated MIMO transmission mode* ($w_{n,k} = 0, \exists n, k$), and a *cooperative MIMO transmission (full cooperation) mode* ($w_{n,k} = 1, \forall n, k$), as illustrated in Figure 12.9. This is a special case of the cache-assisted PHY model in

---

[‡] Note that the user requests may arrive asynchronously at random times.

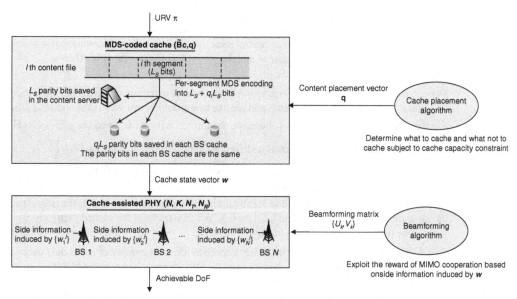

Figure 12.9 Components of the MDS-based PHY caching framework.

Section 12.3.3 in the sense that we have combined the $N_c - 1$ "partial cooperation" modes into a single coordinated MIMO mode, even though more complicated partial cooperation schemes are possible in these modes. For convenience, let $S(t) = 0$ denote the event $w_{n,k} = 0, \exists n, k$ at time t (i.e., the coordinated MIMO transmission mode), and let $S(t) = 1$ denote the event $w_{n,k} = 1, \forall n, k$ at time t (i.e., the cooperative MIMO transmission mode). When there is no ambiguity, we will drop the time index t and use S to denote the current transmission mode. The detailed steps of the cache-assisted PHY transmission scheme are summarized in Table 12.2, and we elaborate on them as follows:

- *Coordinated MIMO transmission mode (S = 0).* In this case, the BSs do not utilize any side information induced by the cache state vector **w** and the PHY operates in the coordinated MIMO transmission mode. We define $\mathbf{V}_k \in \mathbb{C}^{N_T \times d_k}$ and $\mathbf{U}_k \in \mathbb{C}^{N_R \times d_k}$ as the transmit and receive beamforming matrices for user k, where d_k is the number of data streams for user k. The beamforming matrices are chosen to satisfy the following interference-nulling condition:

$$\mathbf{U}_k^\dagger \mathbf{H}_{k,k'} \mathbf{V}_{k'} = \mathbf{0}, \quad \forall k' \neq k. \tag{12.17}$$

Many existing beamforming algorithms can be used to calculate beamforming matrices that satisfy the above conditions. For clarity, we shall use the simple block diagonalization (BD) beamforming algorithm [28, 29] as an example to calculate the beamforming matrices. In this case, the instantaneous sum DoF is given by $\sum_{k=1}^{K} d_k = \min(N_T, KN_R)$. Note, however, that the BD beamforming algorithm can be replaced with any other beamforming algorithm, such as the IA-based beamforming algorithm [3], without affecting the key results in this chapter.

Table 12.2 Cache-assisted PHY transmission.

Step 1 (transmission mode decision at central node). At the beginning of each time slot, the master BS first collects the *cache state* **w** and the global CSI **H** from the BSs. Then **w** determines the transmission mode S.

Step 2 (computation of beamformers at central node). After determining the transmission mode, the central node computes the transmit and receive beamforming matrices ($\{\mathbf{V}_k\}$ and $\{\mathbf{U}_k\}$ when $S = 0$ or $\{\tilde{\mathbf{V}}_k\}$ and $\{\tilde{\mathbf{U}}_k\}$ when $S = 1$, based on the transmission mode S and the global CSI **H**).

Step 3 (data transmission). The central node sends the beamforming matrices $\{\mathbf{V}_k\}/\{\tilde{\mathbf{V}}_k\}$, $\{\mathbf{U}_k\}/\{\tilde{\mathbf{U}}_k\}$ and the transmission mode S to the BSs and users. When $S = 0$, the BSs obtain some parity bits directly from the content server via the data backhaul, and transmit them to the associated users using the coordinated MIMO transmission mode. When $S = 1$, the BSs obtain the same cached parity bits from the local caches, and then transmit the cached parity bits to users using the cooperative MIMO transmission mode.

Step 4 (segment decoding at the users). User k receives a certain number of the cached parity bits of the currently requested file segment from the BSs using the cooperative MIMO transmission mode (if $S = 1$) or the uncached parity bits from the content server via BS k using the coordinated MIMO mode (if $S = 0$). These received parity bits are stored in a reassembling buffer until the total number of parity bits of the current segment is equal to L_S. The current segment is then decoded and the reassembling buffer is cleared for the next segment.

- *Cooperative MIMO transmission mode ($S = 1$).* In this case, the cache state $w_{n,k} = 1, \forall n, k$ induces full side information at each BS. Hence, the BSs can engage in full MIMO cooperation. We define $\tilde{\mathbf{H}}_k = [\mathbf{H}_{k,1}, \ldots, \mathbf{H}_{k,K}] \in \mathbb{C}^{N_R \times KN_T}$ as the composite channel matrix between the K users and K BSs. Denote $\tilde{\mathbf{V}}_k \in \mathbb{C}^{KN_T \times \tilde{d}_k}$ and $\tilde{\mathbf{U}}_k \in \mathbb{C}^{N_R \times \tilde{d}_k}$ as the transmit and receive beamforming matrices for user k, where $\tilde{d}_k$ is the number of data streams for user k. The beamforming matrices are chosen to satisfy the following interference-nulling condition:

$$\tilde{\mathbf{U}}_k^\dagger \tilde{\mathbf{H}}_{k,k'} \tilde{\mathbf{V}}_{k'} = \mathbf{0}, \quad \forall k' \neq k. \tag{12.18}$$

In this case, the instantaneous sum DoF is given by $\sum_{k=1}^K \tilde{d}_k = \min(NN_T, KN_R)$.

12.4.3   MIMO Cooperation Probability of MDS-Coded PHY Caching with Asynchronous Access

In this section, we shall first present the motivation for the MDS-based PHY caching design and derive the cache state probability $p^{\pi,(1)} = \Pr[S = 1|\pi]$ for the MDS-based PHY caching scheme.[§] The overall performance gain of the cache-assisted interference network depends heavily on the probability of an opportunity for the cooperative MIMO transmission mode, $p^{\pi,(1)}$. In Example 12.1 in Section 12.2.4.1, we showed that, with a naive random cache scheme, $p^{\pi,(1)}$ can be exponentially small for asynchronous access,

[§] Without loss of generality, we let the first cache state $\mathbf{w}^{(1)}$ represent the case where the BSs have sufficient side information for cooperative MIMO transmission (i.e., $w_{k,n}^{\pi,(1)} = 1, \forall k, n$).

even if a significant portion of the content files is stored in the BS cache. This is because, without coding, packets have to be sent sequentially to the users from the first packet to the last packet. As a result, we have no freedom to align the transmissions of cached packets of different users to improve the MIMO cooperation probability, as illustrated in Figure 12.10. In contrast, using the proposed MDS-based caching scheme, packets within a segment do not need to be delivered in sequence, and there is no need for packet reassembly at the user. As a result, the BSs are able to schedule and align the transmissions of the cached parity bits for different users so as to improve the probability of MIMO cooperation ($p^{\tau,(1)}$). This is illustrated in the following example.

Example 12.4 (Motivation for MDS-based PHY caching) We use the same setup as in Example 12.1. Each file is divided into two segments, and each segment consists of six packets. Suppose the BS cache stores the first three MDS-coded packets of each segment. Using the MDS-based PHY caching scheme with asynchronous access, the transmitters have the freedom to choose the transmission order of the parity bits in a segment to align the transmissions of the cached parity bits for different users, as illustrated in Figure 12.10. In this case, the probability $p^{\tau,(1)} = \Pr[S = 1|\pi] = 0.5$. Although the packets in a segment may not be delivered in sequence, the user is able to decode the segment as long as it receives enough parity bits, and there is no need for packet reassembly at the receiver.

The probability of cooperative MIMO transmission is the key performance indicator that captures the DoF and BS cache size trade-off for a cached wireless network. Compared with the naive random caching in Example 12.1, the probability of the cooperative MIMO transmission mode is greatly improved. The following theorem establishes the asymptotic probability of the cooperative MIMO transmission mode at high SNR.

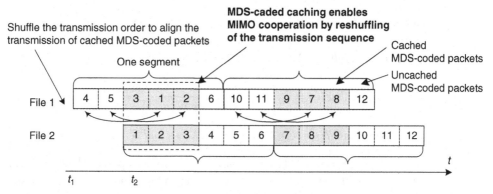

Proposed MDS based PHY caching with asynchronous access
The key to increasing the probability of cooperative MIMO transmission $p^{\tau,(1)}$ is to align the transmissions of **the cached MDS-coded packets** for different users as much as possible.

Figure 12.10 Illustration of MDS-based PHY caching for the system in Example 12.4.

Theorem 12.6 (Probability of cooperative MIMO transmission under MDS-coded PHY caching and asynchronous access)   *As the sum transmission power $P_\Sigma \to \infty$ and the normalized segment size $L_S / (\log(P_\Sigma)\tau) \to \infty$, we have*

$$p^{\pi,(1)} \to \frac{\min_k\{q_{\pi_k}\}}{G\left(1 - \min_k\{q_{\pi_k}\}\right) + \min_k\{q_{\pi_k}\}}, \tag{12.19}$$

where $G = \min(NN_T, KN_R) / \min(N_T, KN_R)$.

Proof Using the proposed MDS-coded caching scheme, the number of bits that can be transmitted in the cooperative MIMO mode is $\min_k\{q_{\pi_k}\}L_S$ in one segment. The transmission rate (in units of $\log_2(P_\Sigma)$ bits) is $D_0 = \min(N_T, KN_R)$ conditional on $S = 0$, and $D_1 = \min(NN_T, KN_R)$ conditional on $S = 1$. As a result, the average transmission time of coordinated MIMO transmission for a segment is given by $\bar{t}^B = \left(1 - \min_k\{q_{\pi_k}\}\right)L_S/D_0$, and the average transmission time of cooperative MIMO transmission for a segment is given by $\bar{t}^A = \min_k\{q_{\pi_k}\}L_S/D_1$. Hence, the probability of cooperative MIMO transmission is given by $p^{\pi,(1)} = \bar{t}^A / (\bar{t}^A + \bar{t}^B)$, which is then given by (12.19).

According to (12.19), the opportunity for cooperative MIMO is limited by $\min_k\{q_{\pi_k}\}$, which is the fraction of the least popular content requested by the users under the current URV realization. This is because cooperative MIMO transmission is only possible when the content requested by all users exists in the BS caches. As a result, a scheme that simply cached one hundred percent of the most popular content files would not always achieve the maximum DoF gain. Note that the cooperative-MIMO opportunity is not exactly equal to $\min_k\{q_{\pi_k}\}$, because the average transmission rates conditioned on the two transmission modes are different.

12.4.4   Achievable DoF for Cached Wireless Networks

The following theorem provides the achievable sum DoF for cached wireless networks.

Theorem 12.7 (Sum DoF achievable by MDS-coded PHY caching)   *For a given cache content placement vector $\mathbf{q}$, if $\tilde{F}_l \geq \min(NN_T, KN_R)\tau, \forall l$, and $\tilde{B}_C \geq \sum_{l=1}^L q_l\tilde{F}_l$, then the sum DoF achievable by MDS-coded PHY caching is given by*

$$\mathrm{DoF}(\mathbf{q}) = \mathbb{E}_\pi\left[D_0 + \frac{(D_1 - D_0)\min_k\{q_{\pi_k}\}}{G\left(1 - \min_k\{q_{\pi_k}\}\right) + \min_k\{q_{\pi_k}\}}\right], \tag{12.20}$$

where $D_1 = \min(NN_T, KN_R)$, $D_0 = \min(N_T, KN_R)$, and $G = D_1/D_0$.

Proof Under MDS-coded PHY caching, the network adopts coordinated transmission with a sum rate D_0 and probability $1 - p^{\pi,(1)}$, and adopts cooperative transmission with a sum rate D_1 and probability $p^{\pi,(1)}$. As a result, the DoF of the cached interference network is given by $\mathrm{DoF}(\mathbf{q}) = \mathbb{E}_\pi\left[D_0\left(1 - p^{\pi,(1)}\right) + D_1 p^{\pi,(1)}\right]$, which can then be expressed as (12.20) using the expression for the cooperative-transmission probability $p^{\pi,(1)}$ in (12.19).

Note that the sum DoF achievable by the proposed MDS-coded PHY caching scheme depends on the cache content placement vector $\mathbf{q}$. As a result, we have the freedom to choose $\mathbf{q}$ to maximize $\text{DoF}(\mathbf{q})$. Therefore, the achievable DoF after optimizing $\mathbf{q}$ is given by

$$\text{DoF}_{\text{MDS}} = \max_{\mathbf{q} \in \Lambda} \text{DoF}(\mathbf{q}), \tag{12.21}$$

where $\mathbf{q} \in \Lambda$ is the cache capacity constraint in (12.1).

Interestingly, the proposed MDS-coded PHY caching scheme is DoF-optimal for the two-user cached MIMO interference networks in Example 12.3. Specifically, if we set the cache placement vector to be $q_l = \tilde{B}_C / \tilde{F}L, \forall l$, we can immediately see that the proposed MDS-coded PHY caching scheme is DoF optimal in that example. Following similar analysis, it can be shown that the proposed MDS-coded PHY caching scheme is also DoF optimal for cached wireless networks with $N = K = 2$, $N_{\text{T}} \geq N_{\text{R}}$ under any general content popularity distribution.

Figure 12.11 illustrates the gap between the achievable DoF (after optimizing with respect to $\mathbf{q}$) and the DoF upper bound for a cached interference network with $K = N = 3$ BS–user pairs, $N_{\text{T}} = N_{\text{R}} = 2$ antennas at each BS and user, and $L = 50$ content files. The content file popularity follows a Zipf distribution with parameter γ [30, 31] (details of the Zipf distribution can be found in Section 12.6.1). As illustrated, the DoF gap is small for different cache capacities and content popularity distributions.

12.5   Cache Content Placement Algorithm for DoF Maximization

From (12.7), it can be observed that maximizing the DoF is equivalent to maximizing the probability of cooperative MIMO transmission. Note that q_l in the cache placement vector $\mathbf{q}$ determines the relative priority of caching of the lth content file in the BS caches during the offline cache initialization phase. In practice, we would prefer to find an optimizing cache placement vector $\mathbf{q}$ (and the associated cache content) based on the *content popularity statistics*, rather than on the instantaneous realization of π, to reduce the associated overhead. Hence, the optimal cache placement can be obtained from a DoF maximization problem formulated as

$$\mathcal{P}: \quad \min_{\mathbf{q} \in \Lambda} -\mathbb{E}_\pi \left[\frac{\min_k \{q_{\pi_k}\}}{G\left(1 - \min_k \{q_{\pi_k}\}\right) + \min_k \{q_{\pi_k}\}} \right], \tag{12.22}$$

where $\mathbf{q} \in \Lambda$ is the cache capacity constraint. The optimal solution of problem $\mathcal{P}$ for cache content placement is denoted by $\mathbf{q}^\star$.

Problem $\mathcal{P}$ is a stochastic optimization with a nonconvex and nonconcave objective function, and it is highly nontrivial to obtain the optimal solution. Using standard techniques [32], problem $\mathcal{P}$ can be transformed to an equivalent concave minimization

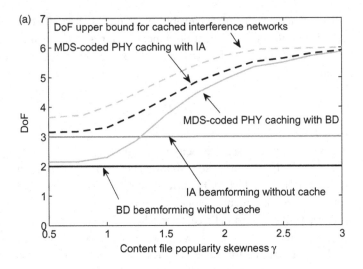

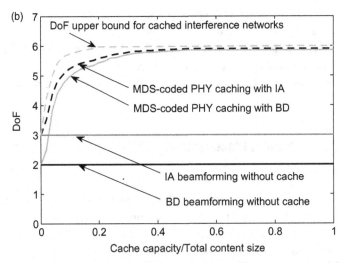

Figure 12.11 Illustration of achievable DoF (after optimizing with respect to **q**) and the DoF upper bound. (a) DoF versus content file popularity skewness γ. The cache capacity is equal to 10% of the total content size; (b) DoF versus cache capacity with $\gamma = 2$.

problem with linear constraints by introducing auxiliary variables.[¶] However, the number of constraints and auxiliary variables increases exponentially with the number of files, and hence the complexity of solving this concave minimization problem increases double-exponentially with the number of files. To address this problem, we propose an alternative transformation of $\mathcal{P}$. We define $x_l = q_l/(G - (G-1)q_l)$; then

---

[¶] By introducing auxiliary variables $\{u_1, u_2, \ldots, u_{L^K}\}$ satisfying $u_i \geq q_{\pi_k^i}$, $\forall k \in \{1, 2, \ldots, K\}$, and $\forall i \in \{1, 2, \ldots, L^K\}$, where $\{\pi^i\}$ is the set of possible realizations of π, the objective function can be written as $-\mathbb{E}_{\pi}[u_i/(G(1-u_i)+u_i)]$. The problem becomes a concave minimization problem with linear constraints.

Algorithm 12.1 Low-computational-complexity online cache content placement algorithm.

1. Start from the initial point $\mathbf{x}^0 = \mathbf{0}$ with iteration index $n = 0$.
2. Let $n+1 \rightarrow n$, calculate descent direction $\delta\mathbf{x}^{(n-1)}$ of $\psi^{(n)}$ at $\mathbf{x}^{(n-1)}$, and project $\delta\mathbf{x}^{(n-1)}$ to the tangent plane of Λ_x at $\mathbf{x}^{(n-1)}$.
3. Apply backtracking line search in the descent direction to find $\mathbf{x}^{(n)}$ satisfying $\psi^{(n)}\left(\mathbf{x}^{(n)}\right) \leq \psi^{(n)}\left(\mathbf{x}^{(n-1)}\right) + \sigma^{(n-1)}\delta\mathbf{x}^{(n-1)}$, where $\sigma^{(n-1)}$ is the step size.
4. Go to step 2 until convergence.

---

problem $\mathcal{P}$ can be rewritten as

$$\mathcal{P}' : \min_{\mathbf{x} \in \Lambda_x} \psi(\mathbf{x}) = -\mathbb{E}_\pi \left[\min_k \{x_{\pi_k}\} \right],$$

where $\mathbf{x} = \{x_l, l = 1, \ldots, L\}$ and $\Lambda_x = \left\{ x_l \in [0,1] \ \forall l, \sum_{l=1}^{L} x_l G\tilde{F}_l/((G-1)x_l + 1) \leq \tilde{B}_C \right\}$.

In the following, we propose a low-computational-complexity cache content placement algorithm based on a stochastic gradient method which converges to a stationary point of $\mathcal{P}'$ without knowing explicitly the popularity distribution of the content π. For notational convenience, let $\hat{\pi}^n$ denote the nth URV observed at the BS, and let $\psi^{(N)}(\mathbf{x}) = -(1/N)\sum_{n=1}^{N} \min_k \{x_{\hat{\pi}_k^n}\}$ denote the sample average of $\psi(\mathbf{x})$. The overall algorithm is summarized in Algorithm 12.1. In each iteration, the algorithm calculates an unbiased estimation of the descent direction[‖] at $\mathbf{x}^{(n-1)}$. If $\mathbf{x}^{(n-1)}$ is not on the boundary of the feasible region Λ_x, then backtracking line search is applied in the descent direction. Otherwise, if $\mathbf{x}^{(n-1)}$ is on the boundary of Λ_x, then the descent direction is projected to the tangent plane of Λ_x at $\mathbf{x}^{(n-1)}$ for a backtracking line search such that the neighborhood points of $\mathbf{x}^{(n-1)}$ along with the backtracking line search direction are all in the feasible region. The convergence of Algorithm 12.1 is established in the following theorem.

Theorem 12.8 (Convergence of Algorithm 12.1) *Algorithm 12.1 converges to a stationary point which satisfies the Karush–Kuhn–Tucker (KKT) conditions of problem $\mathcal{P}'$ with probability 1.*

Proof. According to the strong law of large numbers, the sample average of $\psi(\mathbf{x})$ satisfies

$$\lim_{n \to \infty} \psi^{(n)}\left(\mathbf{x}^{(n)}\right) - \psi^{(n)}\left(\mathbf{x}^{(n-1)}\right) = 0 \tag{12.23}$$

with probability 1. Define $\mathbf{x}^*$ as any accumulation point of $\mathbf{x}^{(n)}$, $n = 1, 2, \ldots$. Then there is a subsequence $\mathbf{x}^{(n_l)}$ such that $\lim_{l \to \infty} \mathbf{x}^{(n_l)} = \mathbf{x}^*$. Assume that $\mathbf{x}^*$ is not a stationary point of $\psi(\mathbf{x}^*)$. Since $\lim_{l \to \infty} \psi^{(n_l+1)}\left(\mathbf{x}^{(n_l)}\right) = \psi(\mathbf{x}^*)$, by Algorithm 12.1, there is a descent direction such that $\lim_{l \to \infty} \psi^{(n_l+1)}\left(\mathbf{x}^{(n_l+1)}\right) - \psi^{(n_l+1)}\left(\mathbf{x}^{(n_l)}\right) < 0$, which

---

[‖] The descent direction $\delta\mathbf{x}$ of f at $\mathbf{x}$ satisfies $f'(\mathbf{x}; \delta\mathbf{x}) = \arg\min_{\delta\mathbf{x}} \max_{\mathbf{g} \in \partial f(\mathbf{x})} \mathbf{g}^T \delta\mathbf{x} < 0$, where $\partial f(\mathbf{x})$ is the set of subgradients of f, and $f'(\mathbf{x}; \delta\mathbf{x})$ is the directional derivative of f at $\mathbf{x}$ in the direction $\delta\mathbf{x}$.

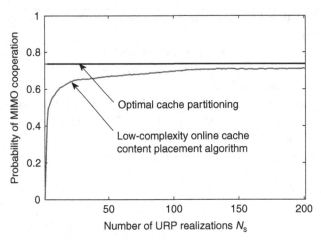

Figure 12.12 Probability of cooperative MIMO transmission $p^{\pi,(1)}$ versus number of URV samples N with $\gamma = 2$. The cache capacity was equal to 10% of the total content size.

contradicts (12.23). Hence, we can conclude that $\mathbf{x}^*$ is the stationary point of $\mathcal{P}'$ that satisfies the KKT conditions with probability 1 [33, 34].

In Figure 12.12, we plot the probability of cooperative MIMO transmission $p^{\pi,(1)}$ versus the number of URV samples N_s for a cached interference network with $K = N = 3$ BS–user pairs and $L = 50$ content files. The content file popularity followed a Zipf distribution with popularity skewness parameter $\gamma = 2$. The cache capacity was 10% of the total content size. The probability of cooperative MIMO transmission achieved by optimal cache content placement (obtained using exhaustive search) served as the upper bound on the performance. The simulation results indicate that Algorithm 12.1 converges quickly to a stationary-point solution, which is close to the optimal solution, as the number of URV realizations increases.

12.6   Closed-Form DoF Analysis and Discussion

In this section, we shall derive closed-form results for the achievable DoF of the MDS-coded PHY caching scheme. We shall show that a large DoF gain compared with a wireless network without a cache is possible, even if the number of files $L \to \infty$.

12.6.1   Content Popularity Model and Definition of DoF Gain

To study the scaling of the DoF, we make the following assumption about the popularity of the content files.

Assumption 12.9 (Content file popularity and size) *The URV follows the IRM, where consecutive user requests are assumed to be mutually independent, and each user independently requests the lth content file with probability ρ_l. The content popularity* *is modeled by the Zipf distribution, defined as follows:*

$$\rho_l = \frac{1}{Z_\gamma} l^{-\gamma}, \quad l = 1, 2, \dots, L, \tag{12.24}$$

where the parameter γ determines the rate of popularity decline as l increases, and $Z_\gamma = \sum_{l=1}^{L} l^{-\gamma}$ is a normalization factor. All files are of the same size, i.e., $\tilde{F}_l = \tilde{F}, \forall l$.

The IRM is a simple and widely studied model to analyze Internet traffic [35, 36]. The IRM assumption is known to be somewhat idealized when compared with real Internet traffic. However, it is mathematically tractable and is helpful for capturing first-order insight into the behavior of caching [37]. The Zipf distribution is a well-known law used to model Internet traffic [30, 31].

We define δDoF as the DoF gain achieved by PHY caching, given by

$$\delta\mathrm{DoF} = \mathbb{E}_\pi \left[\frac{(D_1 - D_0) \min_k \{q_{\pi_k}^\star\}}{(1 - \min_k \{q_{\pi_k}^\star\}) D_1 / D_0 + \min_k \{q_{\pi_k}^\star\}} \right], \tag{12.25}$$

where $\mathbf{q}^\star$ is the optimal cache content placement solution of problem (12.22). Note that in general cases, it is difficult to obtain a closed-form expression for the DoF gain δDoF since there is no closed-form solution for the optimal cache content placement $\mathbf{q}^\star$. In the following, we will consider several interesting asymptotic regimes in which it is possible to obtain an asymptotic closed-form expression for δDoF.

12.6.2   Asymptotic DoF Gain with Respect to the Number of Files

As the number of files L increases, one would expect the probability of a cache hit to go to zero, and hence the DoF gain compared with a wireless network without a cache to go to zero. However, we show below that this is not the case. In fact, a large DoF gain is also possible even when $L \to \infty$. To study the scaling of the DoF, we use Θ as a notation that describes the limiting behavior. Specifically, let f and g be real functions; then $f = \Theta(g)$ if there exist $k > 0$, $x_0 > 0$ such that for $x \geq x_0$, $|f(x)/g(x)| = k$.

Theorem 12.10 (Asymptotic DoF gain for large L)   *For a large number of content files L, the asymptotic scaling laws of the DoF are summarized below:*

• *Subcritical.*

  1. If $0 \leq \gamma < 1 - 1/K$, then $\delta\mathrm{DoF} = \Theta\left(L^{-1}\right)$.
  2. If $1 - 1/K \leq \gamma < 1$, then $\delta\mathrm{DoF} = \Theta\left(L^{-(1-\gamma)K}\right)$.

• *Critical. If $\gamma = 1$, then $\delta\mathrm{DoF} = \Theta\left(\ln^{-K}(L)\right)$.*
• *Supercritical. If $\gamma > 1$, then $\delta\mathrm{DoF} = \Theta(1)$.*

Proof It can be shown that for large L, when $\gamma > 1 - 1/K$, the content popularity is concentrated in the first few files, and the optimal cache placement is given by caching the L' most popular files. When $\gamma \leq 1 - 1/K$, the less popular files will be requested with higher probability, and the optimal cache content placement vector $\mathbf{q}$ becomes closer to uniform caching. As a result, the maximum probability of cooperative transmission and DoF gain are given by

- $0 \leq \gamma < 1 - 1/K$:

$$p^{\pi,(1)} = \frac{\tilde{B}_C/\tilde{F}L}{(D_1/D_0)\left(1 - \tilde{B}_C/\tilde{F}L\right) + \tilde{B}_C/\tilde{F}L}, \quad \text{DoF} = \Theta\left(L^{-1}\right).$$

- $1 - 1/K \leq \gamma < 1$:

$$p^{\pi,(1)} = \Theta\left(\frac{1}{\left((L+1)^{1-\gamma} - 1/(1-\gamma)\right)^K + 1}\right) = \Theta\left(L^{(\gamma-1)K}\right),$$

$$\delta\text{DoF} = \Theta\left(L^{(\gamma-1)K}\right).$$

- $\gamma = 1$:

$$p^{\pi,(1)} = \Theta\left(\frac{1}{\ln^K((L+1)/2)}\right) = \Theta\left(\ln^{-K}(L)\right), \quad \delta\text{DoF} = \Theta\left(\ln^{-K}(L)\right).$$

- $\gamma > 1$:

$$p^{\pi,(1)} = \left(\frac{\sum_{i=1}^{\lfloor \tilde{B}_C/\tilde{F}\rfloor} i^{-\gamma}}{\sum_{j=1}^{L} j^{-\gamma}}\right)^K = \Theta(1), \quad \delta\text{DoF} = \Theta(1).$$

This completes the proof.

For large L, the DoF gain depends heavily on the content popularity skewness, represented by the parameter γ. A larger γ means that the user requests are concentrated more on a few content files, and thus a larger probability of cooperative MIMO transmission and a large DoF gain can be achieved with a relatively small cache capacity $\tilde{B}_C$. Note that there is a *phase transition* behavior at the critical point $\gamma = 1$, as illustrated in Figure 12.13. In the supercritical case, where $\gamma > 1$, it is possible to achieve a nonzero

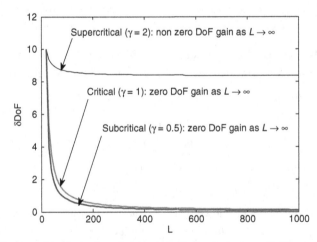

Figure 12.13 Phase transition behavior of asymptotic DoF gain for large L. The system parameters were configured as follows: $K = 6$, $N_T = 2$, $N_R = 2$, and $\tilde{B}_C/\tilde{F} = 20$.

DoF gain by caching the most popular files, even if the cache capacity $\tilde{B}_C$ is much smaller than the total content size $L\tilde{F}$. This result has an important impact on practical implementations of cached interference networks, where the popularity skewness γ can be large, especially for mobile applications [31]. On the other hand, in the subcritical case, where $\gamma < 1$, the popularity is more flat and the optimal cache content placement vector $\mathbf{q}$ becomes closer to uniform caching (i.e., $q_l = \tilde{B}_C/L\tilde{F}, \forall l$). In this case, if the cache capacity $\tilde{B}_C$ is not of the same order as the total content size $L\tilde{F}$, PHY caching does not provide an order-of-magnitude gain.

12.6.3 Asymptotic DoF Gain with Respect to the Number of Users

The scaling of the DoF gain for large K is summarized in the following theorem.

Theorem 12.11 (Asymptotic DoF gain for large K) *When $K \to \infty$, the DoF gain is given by*

$$\delta\mathrm{DoF} = \frac{(N-1)N_T\tilde{B}_C/\tilde{F}L}{\left(1-\tilde{B}_C/\tilde{F}L\right)N+\tilde{B}_C/\tilde{F}L}.$$

Proof When $K \to \infty$, it can be shown that the optimal solution is uniform caching (i.e., $q_l = \tilde{B}_C/L\tilde{F}, \forall l$). The cooperative MIMO transmission probability is given by

$$p^{\pi,(1)} = \frac{\tilde{B}_C/\tilde{F}L}{(D_1/D_0)\left(1-\tilde{B}_C/\tilde{F}L\right)+\tilde{B}_C/\tilde{F}L},$$

and thus the DoF gain $\delta\mathrm{DoF}$ is given by Theorem 12.11.

A larger K means a higher probability that the less popular files will be requested. As a result, the optimal cache content placement vector $\mathbf{q}$ becomes closer to uniform caching. When the cache capacity is less than the total content size, the DoF gain depends on the ratio of the cache capacity $\tilde{B}_C$ and the total content size $L\tilde{F}$ and does not scale with K.

12.7 Conclusion and Future Work

We have proposed PHY caching to improve spectral efficiency in wireless networks. After introducing the basic concepts of PHY caching, we discussed two fundamental roles/benefits of PHY caching, namely PHY-caching-induced opportunistic cooperative spatial multiplexing and spatial diversity. Then we pointed out the design challenges of PHY caching and outlined the potential solutions. After that, we provided a generic cache model and an upper bound on the DoF of cached wireless networks over all possible caching and transmission strategies. Then we proposed an MDS-coded caching scheme together with an online cache content placement algorithm to optimize the achievable DoF without explicit knowledge of the popularity of the content files. The proposed scheme is DoF optimal for some interesting special cases, and the gap between the upper bound on the DoF and the achievable DoF for general cases is small in some typical scenarios. Finally, we quantified the DoF gain of PHY caching with respect to some important system parameters, such as the cache size, number of users, and popularity of the content files. Both analysis and simulation show that the proposed scheme is able to achieve a large DoF gain, even when the cache capacity is much smaller than the total content size.

PHY caching has shown great potential to enhance spectral efficiency in future 5G wireless networks. In this chapter, we focused on a simple PHY caching scheme with either no cooperation or full cooperation. An interesting future study would be to consider the joint optimization of more flexible cache-induced partial cooperation and the associated cache placement to further enhance the performance of PHY caching. Other future work might be to extend PHY caching to D2D networks so that several users can exploit the cached content to simultaneously transmit multiple data streams to other users. In this case, PHY caching can also induce physical layer topology changes and provide a DoF gain in D2D networks. It is also important to study the impact of PHY caching on the capacity-scaling law of large-scale wireless networks, such as wireless ad hoc networks. There have been studies of the throughput-scaling laws of wireless ad hoc networks for a specific PHY caching scheme [38], as well as the fundamental capacity-scaling law of backhaul-limited dense wireless networks with PHY caching for some interesting special cases [39, 40]. However, the fundamental impact of PHY caching on the capacity-scaling law of wireless networks is still unknown for the general case.

References

[1] R. Ghaffar and R. Knopp, "Fractional frequency reuse and interference suppression for OFDMA networks," in *Proc. of International Symposium on Modeling and Optimization in Mobile, Ad Hoc and Wireless Networks (WiOpt)*, Jun. 2010.

[2] V. R. Cadambe and S. A. Jafar, "Interference alignment and degrees of freedom of the K-user interference channel," *IEEE Trans. Inf. Theory*, vol. 54, no. 8, pp. 3425–3441, Aug. 2008.

[3] M. Razaviyayn, G. Lyubeznik, and Z.-Q. Luo, "On the degrees of freedom achievable through interference alignment in a MIMO interference channel," *IEEE Trans. Signal Process.*, vol. 60, no. 2, pp. 812–821, Oct. 2011.

[4] L. Venturino, N. Prasad, and X. Wang, "Coordinated linear beamforming in downlink multi-cell wireless networks," *IEEE Trans. Wireless Commun.*, vol. 9, no. 4, pp. 1451–1461, Apr. 2010.

[5] H. Dahrouj and W. Yu, "Coordinated beamforming for the multicell multi-antenna wireless system," *IEEE Trans. Wireless Commun.*, vol. 9, no. 5, pp. 1748–1759, May 2010.

[6] R. Irmer, H. Droste, P. Marsch, M. Grieger, G. Fettweis, S. Brueck, H.-P. Mayer, L. Thiele, and V. Jungnickel, "Coordinated multipoint: Concepts, performance, and field trial results," *IEEE Commun. Mag.*, vol. 49, no. 2, pp. 102–111, Feb. 2011.

[7] A. Liu and V. K. N. Lau, "Mixed-timescale precoding and cache control in cached MIMO interference network," *IEEE Trans. Signal Process.*, vol. 61, no. 24, pp. 6320–6332, Dec. 2013.

[8] A. Liu and V. K. N. Lau, "Cache-enabled opportunistic cooperative MIMO for video streaming in wireless systems," *IEEE Trans. Signal Process.*, vol. 62, no. 2, pp. 390–402, Jan. 2014.

[9] A. Liu and V. K. N. Lau, "Exploiting base station caching in MIMO cellular networks: Opportunistic cooperation for video streaming," *IEEE Trans. Signal Process.*, vol. 63, no. 1, pp. 57–69, Jan. 2015.

[10] W. Han, A. Liu, and V. Lau, "Degrees of freedom in cached MIMO relay networks," *IEEE Trans. Signal Process.*, vol. 63, no. 15, pp. 3986–3997, Aug. 2015.

[11] U. C. Kozat, O. Harmanci, S. Kanumuri, M. U. Demircin, and M. R. Civanlar, "Peer assisted video streaming with supply-demand-based cache optimization," *IEEE Trans. Multimedia*, vol. 11, no. 3, pp. 494–508, Feb. 2009.

[12] B. Shen, S.-J. Lee, and S. Basu, "Caching strategies in transcoding-enabled proxy systems for streaming media distribution networks," *IEEE Trans. Multimedia*, vol. 6, no. 2, pp. 375–386, Apr. 2004.

[13] M. Hefeeda and O. Saleh, "Traffic modeling and proportional partial caching for peer-to-peer systems," *IEEE/ACM Trans. Netw.*, vol. 16, no. 6, pp. 1447–1460, Mar. 2008.

[14] G. Barish and K. Obraczke, "World wide web caching: Trends and techniques," *IEEE Commun. Mag.*, vol. 38, no. 5, pp. 178–184, May 2000.

[15] N. Golrezaei, K. Shanmugam, A. G. Dimakis, A. F. Molisch, and G. Caire, "Femtocaching: Wireless video content delivery through distributed caching helpers," in *Proc. of IEEE INFOCOM*, Mar. 2012.

[16] A. Altieri, P. Piantanida, L. Rey Vega, and C. Galarza, "On fundamental trade-offs of device-to-device communications in large wireless networks," *IEEE Trans. Wireless Commun.*, vol. 14, no. 9, pp. 4958–4971, Sep. 2015.

[17] M. Ji, G. Caire, and A. F. Molisch, "Fundamental limits of caching in wireless D2D networks," *IEEE Trans. Inf. Theory*, vol. 62, no. 2, pp. 849–869, Dec. 2015.

[18] G. Kramer, M. Gastpar, and P. Gupta, "Cooperative strategies and capacity theorems for relay networks," *IEEE Trans. Inf. Theory*, vol. 51, no. 9, pp. 3037–3063, Sep. 2005.

[19] T. M. Cover and J. A. Thomas, *Elements of Information Theory*, 2nd edn, Wiley, 2006.

[20] X. Peng, J.-C. Shen, J. Zhang, and K. B. Letaief, "Joint data assignment and beamforming for backhaul limited caching networks," in *Proc. of IEEE International Symposium on Personal, Indoor and Mobile Radio Communications (PIMRC)*, Sep. 2014.

[21] M. Tao, E. Chen, H. Zhou, and W. Yu, "Content-centric sparse multicast beamforming for cache-enabled cloud RAN," *IEEE Trans. Wireless Commun.*, vol. 15, no. 9, pp. 6118–6131, Sep. 2016.

[22] M. Yuksel and E. Erkip, "Multiple-antenna cooperative wireless systems: A diversity–multiplexing tradeoff perspective," *IEEE Trans. Inf. Theory*, vol. 53, no. 10, pp. 3371–3393, Oct. 2007.

[23] I. Marić, R. D. Yates, and G. Kramer, "Capacity of interference channels with partial transmitter cooperation," *IEEE Trans. Inf. Theory*, vol. 53, no. 10, pp. 3536–3548, Oct. 2007.

[24] M. A. Maddah-Ali and U. Niesen, "Cache-aided interference channels," in *Proc. of IEEE International Symposium on Information Theory (ISIT)*, Jun. 2015.

[25] D. Gesbert, S. Hanly, H. Huang, S. Shamai Shitz, O. Simeone, and W. Yu, "Multi-cell MIMO cooperative networks: A new look at interference," *IEEE J. Sel. Areas Commun.*, vol. 28, no. 9, pp. 1380–1408, Oct. 2010.

[26] O. Somekh, O. Simeone, Y. Bar-Ness, A. Haimovich, and S. Shamai, "Cooperative multicell zero-forcing beamforming in cellular downlink channels," *IEEE Trans. Inf. Theory*, vol. 55, no. 7, pp. 3206–3219, Jul. 2009.

[27] A. Shokrollahi, "Raptor codes," *IEEE Trans. Inf. Theory*, vol. 52, no. 6, pp. 2551–2567, Jun. 2006.

[28] Z. Shen, R. Chen, J. G. Andrews, R. W. Heath, and B. L. Evans, "Sum capacity of multiuser MIMO broadcast channels with block diagonalization," *IEEE Trans. Wireless Commun.*, vol. 6, no. 6, pp. 2040–2045, Jun. 2007.

[29] W. Li and M. Latva-aho, "An efficient channel block diagonalization method for generalized zero forcing assisted MIMO broadcasting systems," *IEEE Trans. Wireless Commun.*, vol. 10, no. 3, pp. 739–744, Dec. 2011.

[30] L. Breslau, P. Cao, L. Fan, G. Phillips, and S. Shenker, "Web caching and Zipf-like distributions: Evidence and implications," in *Proc. of IEEE INFOCOM*, Mar. 1999.

[31] T. Yamakami, "A Zipf-like distribution of popularity and hits in the mobile web pages with short life time," in *Proc. of International Conf. on Parallel and Distributed Computing, Applications and Technologies (PDCAT)*, Dec. 2006.

[32] A. Majthay and A. Whinston, "Quasi-concave minimization subject to linear constraints," *Discrete Math.*, vol. 9, no. 1, pp. 35–59, 1974.

[33] M. A. Abramson and C. Audet, "Convergence of mesh adaptive direct search to second-order stationary points," *SIAM J. Optim.*, vol. 17, no. 2, pp. 606–619, Aug. 2006.

[34] C. Audet and J. E. Dennis Jr., "Mesh adaptive direct search algorithms for constrained optimization," *SIAM J. Optim.*, vol. 17, no. 1, pp. 188–217, May 2006.

[35] S. Gitzenis, G. Paschos, and L. Tassiulas, "Asymptotic laws for joint content replication and delivery in wireless networks," *IEEE Trans. Inf. Theory*, vol. 59, no. 5, pp. 2760–2776, Dec. 2012.

[36] P. Olivier and A. Simonian, "Performance of a cache with random replacement and Zipf document popularity," in *Proc. of International Conf. on Performance Evaluation Methodologies and Tools (VALUETOOLS)*, Institute for Computer Sciences, Social-Informatics and Telecommunication Engineering (ICST), Jan. 2014.

[37] R. Fagin and T. G. Price, "Efficient calculation of expected miss ratios in the independent reference model," *SIAM J. Comput.*, vol. 7, no. 3, pp. 288–297, 1978.

[38] A. Liu and V. K. N. Lau, "Asymptotic scaling laws of wireless ad hoc network with physical layer caching," *IEEE Trans. Wireless Commun.*, vol. 15, no. 3, pp. 1657–1664, Mar. 2016.

[39] A. Liu and V. K. N. Lau, "Capacity gain of physical layer caching in backhaul-limited dense wireless networks," in *Proc. of IEEE INFOCOM*, Apr. 2015.

[40] A. Liu and V. K. N. Lau, "How much cache is needed to achieve linear capacity scaling in backhaul-limited dense wireless networks?" *IEEE/ACM Trans. Netw.*, in press.

13 Cost-Aware Cellular Networks Powered by Smart Grids and Energy Harvesting

Jie Xu, Lingjie Duan, and Rui Zhang

13.1 Introduction

To meet the dramatic growth in wireless data traffic driven by the popularity of new mobile devices and mobile applications, the fifth generation (5G) of cellular technology has recently attracted a lot of research interest from both academia and industry (see, e.g., [1]). As compared with its fourth generation (4G) counterpart, 5G is expected to achieve a roughly 1000 times data rate increase via dense base station (BS) deployments and advanced physical layer communication techniques [1]. However, the large number of BSs will lead to large energy consumption and high electricity bills for cellular operators, which amounts to a large portion of their operational expenditure. For example, China Mobile owned around 920 000 BSs in 2011 and the total energy cost per year was almost 3 billion US dollars, given that the annual cost for each BS is about 3000 US dollars [2]. Therefore, in the 5G era, it is becoming necessary for these cellular operators to reduce their energy costs by employing new cost-saving solutions in the design of cellular BSs, which are our main focus in this chapter.

In general, these cost-saving solutions can be categorized into two classes, which manage the energy supply and the communication demand of cellular BSs, respectively [2–5]. On the supply side, one commonly adopted solution is to use energy harvesting devices (e.g., solar panels and wind turbines) at cellular BSs, which can harvest cheap and clean renewable energy to reduce or even substitute for the energy purchased from the grid [5]. However, since renewable energy is often randomly distributed in both time and space and cellular BSs are very energy-hungry, it is very difficult to solely use different BSs' individually harvested energy to power their operation. As a result, the power grid is still needed to provide reliable energy to BSs. Besides serving as a reliable energy supply, the power grid also provides new opportunities for saving the BSs' costs with its ongoing paradigm shift from the traditional grid to the smart grid. Unlike the traditional grid, which uses a one-way energy flow to deliver power from central generators to electricity users, the smart grid deploys smart meters at end users to enable both two-way information and two-way energy flows between the grid and the end users [6, 7]. The two-way energy flow in the smart grid motivates a new idea of *energy cooperation* in cellular networks, as we will elaborate later in this chapter, which allows the BSs to trade and share their unevenly harvested renewable energy through the smart grid to support nonuniform wireless traffic in a cost-effective way.

On the demand side, various techniques have been proposed for cellular networks across different layers of communication protocols to reduce the energy consumption [2]. Among them, communication cooperation (e.g., traffic loading [8], spectrum sharing [9], and coordinated multipoint (CoMP) [10]) is particularly appealing; this allows the BSs to share wireless resources and shift traffic loads between them to achieve energy savings. However, the introduction of renewable energy at BSs imposes new challenges on the existing design of communication cooperation: the conventional energy-saving design may not be cost-effective any longer. This is due to the fact that renewable energy (though the supply is unreliable) is in general much cheaper than energy purchased from the grid and therefore BSs should use it maximally to save cost, whereas under the energy-saving design the harvested renewable energy at BSs may not be utilized efficiently when the BSs are serving time- and space-varying wireless traffic. To overcome this problem, it is desirable to design new *cost-aware communication cooperation* approaches, by taking into account the cost differences between renewable and conventional energy.

In this chapter, we first overview the recent advances in energy cooperation and cost-aware communication cooperation. Figure 13.1 illustrates the general concept of energy and communication cooperation for cellular networks in both the energy supply layer and the communication demand layer, respectively. Then we propose a new *joint energy and communication cooperation* technique to exploit both benefits. Specifically, we present the following three approaches [11]:

- *Approach I: Energy cooperation on the supply side.* Cellular systems or BSs use the two-way energy flow in a smart grid to trade or share renewable energy, taking the energy demands for communications as given.
- *Approach II: Communication cooperation on the demand side.* Cellular systems or BSs perform cost-aware communication cooperation to share wireless resources and reshape the wireless load over space and time, taking the energy supply (renewable and/or conventional) as given.
- *Approach III: Joint energy and communication cooperation on both sides.* Cellular systems or BSs jointly cooperate on both the supply and the demand sides to maximally reduce their total energy cost.

It is worth noting that there is another line of research, namely energy harvesting wireless communications, which has attracted a lot of attention recently. Its recent advances have been reviewed in, e.g., [12–14]. Specifically, the parallel studies [15, 16] first investigated the throughput maximization problem for point-to-point fading channels subject to energy harvesting constraints at the wireless transmitter, where the optimal offline transmission strategy was shown to be a "staircase" [15] (or so-called directional [16]) water-filling power allocation. The work in [15] was then extended to more practical cases with half-duplex energy storage constraints [17] and nonideal circuit power at the transmitter [18], and was also extended to other scenarios

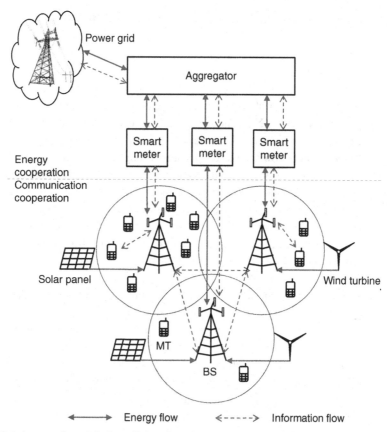

Figure 13.1 A general model of a cellular network with energy and communication cooperation among BSs [11]. Reproduced with permission from the IEEE. MT, mobile terminal.

such as relay channels [19]. Furthermore, [20] studied the average outage probability minimization problem for a point-to-point fading channel setup with energy harvesting constraints at the wireless transmitter, for which the optimal power profile was shown to be nondecreasing over time and have an interesting "save-then-transmit" structure. Nevertheless, the studies of conventional energy harvesting wireless communications assume that the transmitters are powered solely by renewable energy, while in this work we consider cellular networks where the BSs are powered by hybrid energy supplies with both smart grids and renewable energy, and thus we can explore the two-way energy flow in smart grids to enable energy cooperation among BSs for further cost saving.

In the rest of this chapter, we first introduce models of the energy supply and demand of cellular systems. Then we present the latest energy, communication, and joint cooperation approaches. Finally, we point out several future research directions and conclude the chapter.

13.2   Energy Supply and Demand of Cellular Systems

In this section, we introduce models of energy supply and demand for cellular systems. The models will be used to motivate the various cooperation schemes introduced later. For notational convenience, we focus on one particular time slot and thus skip the time indices of the variables (such as the energy demand and supply of BSs, and energy prices) used later. Note, however, that these variables can vary over time in practice. In addition, we normalize the length of each time slot to unity, and thus use the terms "energy" and "power" interchangeably throughout this chapter.

We consider a single cellular system with $N > 1$ BSs, in which each BS i is connected to a smart grid and also equipped with a renewable energy harvesting device that has a harvesting rate $E_i \geq 0, i = 1, \dots, N$. The value of E_i at a given time depends on the type of renewable energy source (e.g., solar or wind), the harvesting capacity of the device (e.g., the size of the solar panel), and the weather conditions at that location. As shown in the upper subfigure of Figure 13.2, the E_i's are generally different for BSs at different locations.

On the demand side, the power consumption of each cellular BS i, denoted by $Q_i \geq 0$, is composed of two main parts: a dynamic power consumption related to the transmission and reception of wireless signals for serving the mobile terminals (MTs), and a constant power consumption (e.g., at the circuits and air conditioners) for maintaining other operations. In reality, the value of Q_i varies according to the traffic load over the service coverage area of BS i. Owing to the mobility of MTs across cells and their time-varying service requests, the traffic loads (and thus the Q_i's) are different for different BSs and change over time, as shown in the middle subfigure of Figure 13.2.

By combining the supply and demand sides, we can denote the *net load* at BS i as $\delta_i = Q_i - E_i$, where $\delta_i > 0$ shows a deficit of renewable energy and $\delta_i < 0$ indicates an energy surplus. Since the Q_i's and E_i's are usually independent (see Figure 13.2), it is likely that some BSs have insufficient renewable energy to match the demand (i.e., $\delta_i > 0$), while the other BSs have adequate renewable energy (i.e., $\delta_i < 0$). Such a geographical diversity in net load requires some BSs to purchase energy from the grid (e.g., an amount δ_i of energy purchase for BS i with $\delta_i > 0$)* but the other BSs to waste their extra renewable energy (i.e., an amount $|\delta_j|$ of energy waste for BS j, with $\delta_j < 0$).[†] Overall, the total amount of energy purchased from the grid by all the N BSs is the total renewable energy deficit, denoted by $\Delta_+ := \sum_{i=1}^{N} [\delta_i]^+ \geq 0$, with $[x]^+ := \max(x, 0)$, while the total amount of renewable energy wasted by them is the total renewable energy surplus, given by $\Delta_- := -\sum_{j=1}^{N} [\delta_j]^- \geq 0$, with $[x]^- := \min(x, 0)$. If we denote the price at which BSs can purchase energy from the grid by $\pi > 0$, then the total energy cost of the cellular system is

$$C_1 = \pi \Delta_+, \qquad (13.1)$$

---

* By purchasing energy from the grid, all the BSs can maintain their routine operations all the time.
[†] One possible way to reduce such renewable energy waste is to install storage devices at BSs to store extra energy for future use. However, storage devices are currently expensive and capacity-limited, and commercial BSs are often not equipped with such devices for dynamic energy management.

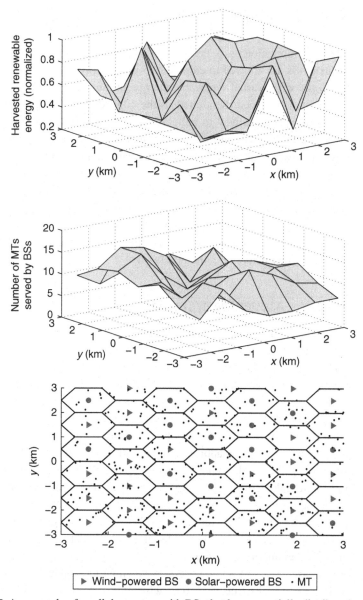

Figure 13.2 An example of a cellular system with BSs that have a spatially distributed traffic load and spatially distributed harvested energy at a given instance of time. It is assumed that the (normalized) energy harvesting capacity of all solar-powered BSs is 1, and that of all wind-powered BSs is 0.5 [11]. Reproduced with permission from the IEEE.

which is independent of Δ_-. This fact motivates us to use the wasted renewable energy surplus (Δ_-) to compensate the deficit (Δ_+) for the purpose of cost saving. To this end, we can implement energy and communication cooperation on the supply and demand sides, respectively, to reschedule and balance the E_i's and Q_i's.

13.3   Energy Cooperation

Energy cooperation is a cost-saving approach on the supply side, in which the cellular BSs are allowed to employ two-way energy trading or sharing to better utilize their otherwise wasted renewable energy surplus (Δ_-). Although the idea of energy cooperation has been mentioned in the context of smart grids for energy trading in microgrids [7], it is new to cellular networks. In particular, since it is too complex for a grid to directly control a large number of BSs, the energy trading and sharing in cellular networks needs to be enabled by using aggregators [21] (see, the upper energy cooperation layer in Figure 13.1). With aggregators, we can cluster BSs into a finite number of groups and an aggregator can serve as an intermediary party to control each group of BSs for the grid, thus helping realize a two-way energy flow between the grid and the BS groups. The implementation of energy cooperation is not difficult in a smart grid: it only requires a two-way energy flow and aggregators, and does not change the existing infrastructure of cellular networks.

13.3.1   Aggregator-Assisted Energy Trading

Aggregator-assisted energy trading is an energy cooperation scheme in which the aggregator performs two-way energy trading with the BSs by deciding buying and selling prices. In this scheme, the BSs adequate in renewable energy can sell their extra energy to the aggregator, from which sales revenue can be gained to offset the total energy cost; at the same time, the other BSs short of renewable energy can obtain cheap energy from the aggregator at a lower price than the regular price π for purchasing from the grid directly. As the coordinator in this trading market, the aggregator can also obtain some revenue by appropriately deciding the selling and buying prices of energy. Here, the selling and buying of energy at each BS is managed with the help of smart meters in real time, which can decide the amount of energy sold or purchased in any time slot a priori based on the energy harvesting rates, the power demand, and energy prices. It does not strictly require the BSs to deploy energy storage devices.

Let $\pi_{\text{buy}} > 0$ and $\pi_{\text{sell}} > 0$ denote the unit prices at which each BS can buy and sell energy from and to the aggregator, respectively.[‡] Here, $\pi_{\text{sell}} < \pi_{\text{buy}}$ holds to avoid the trivial case where a BS can benefit by reselling its bought energy from the aggregator; and $\pi_{\text{buy}} < \pi$ is also true, since otherwise all BSs short of energy can buy cheaper energy from the grid directly. With two-way energy trading, the BSs adequate in renewable energy will sell their total amount Δ_- of surplus energy to the aggregator at the price π_{sell}, and, accordingly, an energy quota Δ_- is set by the aggregator. The BSs short of renewable energy will first purchase an amount $\min(\Delta_+, \Delta_-)$ of cheap energy from the aggregator at the price π_{buy} (within the quota limitation of Δ_-) to use this resource maximally, and (if it is not enough) will buy an amount $\Delta_+ - \min(\Delta_+, \Delta_-)$ from the

---

[‡] The energy prices may vary over time based on the time-varying relationship between aggregate energy demand and supply at the aggregator. Depending on the information and energy exchange frequencies, the aggregator can decide such prices either a day ahead or in real time.

grid at the price π. Depending on the relationship between Δ_+ and Δ_-, the total cost for all N BSs is

$$C_2 = \begin{cases} \pi_{\text{buy}}\Delta_+ - \pi_{\text{sell}}\Delta_- & \text{if } \Delta_+ \leq \Delta_- \\ \pi_{\text{buy}}\Delta_- + \pi(\Delta_+ - \Delta_-) - \pi_{\text{sell}}\Delta_- & \text{if } \Delta_+ > \Delta_-. \end{cases} \quad (13.2)$$

Note that C_2 can even be negative, which is the case when Δ_- is sufficiently larger than Δ_+ such that $\pi_{\text{buy}}\Delta_+ < \pi_{\text{sell}}\Delta_-$. By comparing (13.1) and (13.2), it follows that $C_2 \leq C_1$.

13.3.2   Aggregator-Assisted Energy Sharing

Aggregator-assisted energy sharing is an energy cooperation scheme that allows BSs in a BS group to mutually negotiate and share renewable energy by simultaneously injecting and drawing energy into and from the aggregator, respectively. By matching the local renewable energy deficit (positive δ_i's) and surplus (negative δ_i's) between any two BSs, this scheme helps the group of BSs reduce their aggregate renewable energy deficit. The practical implementation of energy sharing requires this group of BSs to sign a *contract* with the aggregator at a contract fee that motivates the aggregator to support energy sharing. Note that the contract also requires the BSs to commit not to interfere with the overall operation of the aggregator, by equating their total injected energy (into the aggregator) to their total drawn energy (from the aggregator) at any given time instant.[§] Compared with the aggregator-assisted energy-trading scheme, the aggregator does not need to be actively involved in this energy-sharing scheme, and this ensures limited coordination complexity.

Specifically, suppose that BS i wants to transfer an amount $e_{ij} \geq 0$ of energy to BS j, $i \neq j$. This is accomplished at an appointed time by BS i injecting an amount e_{ij} of energy into the aggregator, and at the same time BS j drawing the same amount e_{ij} from the aggregator.[¶] Thanks to the mutual sharing of e_{ij}'s among the N BSs, the total energy deficit Δ_+ and surplus Δ_- can be effectively matched. When $\Delta_+ \leq \Delta_-$, the N BSs can maintain their operation without purchasing any energy from the grid; otherwise, a total amount $\Delta_+ - \Delta_-$ of energy must be purchased from the grid at the price π. If we denote the contract fee to the aggregator by $\bar{C}$, the total cost for all N BSs is given by

$$C_3 = \begin{cases} \bar{C} & \text{if } \Delta_+ \leq \Delta_- \\ \pi(\Delta_+ - \Delta_-) + \bar{C} & \text{if } \Delta_+ > \Delta_-. \end{cases} \quad (13.3)$$

By comparing (13.3) and (13.1), it generally follows that $C_3 \leq C_1$, i.e., the total energy cost is reduced, as long as $\bar{C}$ is sufficiently small.

[§] The aggregator can, alternatively, provide a long-term contract to the BSs, such that the BSs only need to ensure an energy-sharing balance over a longer time period (e.g., one day or even a month), thus offering more flexibility for the BSs' energy sharing. In this case, however, a higher contract fee may be required by the aggregator owing to the short-term perturbation to the aggregator.

[¶] The energy transfer between the two BSs through the aggregator may lead to a certain amount of energy loss [22]. Since this loss is very small (e.g., less than 5% of the transferred energy), its impact is neglected here, though our scheme also applies in that case.

The aggregator-assisted energy-trading and energy-sharing schemes may have different cost-saving performances depending on the buying and selling prices of energy in the former scheme, and on the contract fee in the latter, both of which incentivize the aggregator to help.

13.4   Communication Cooperation

Communication cooperation refers to a cost-saving approach on the demand side that exploits the broadcast nature of wireless channels and uses wireless resource sharing to reshape the BSs' wireless load and energy consumption. Differently from conventional communication cooperation (e.g., [8–10]), which aims to maximize data throughput or minimize energy consumption, the communication cooperation that we are interested in here seeks to minimize the total energy cost by optimally utilizing both cheap renewable energy and reliable on-grid energy. In so-called cost-aware communication cooperation, the rescheduling of the BSs' traffic load and energy consumption needs to follow their given renewable energy supply, such that the renewable energy can be maximally used to support the quality of service (QoS) requirements of the MTs, and the on-grid energy purchase is thus minimized. To implement this approach, BSs need to share with each other communication information (e.g., the channel state information and the QoS requirements of MTs) and energy information (e.g., the energy harvesting rates) through the backhaul links connecting them, as shown in Figure 13.1. This may require the cellular operator to install new infrastructure (e.g., high-capacity and low-latency backhaul links) and/or coordinate and standardize the communication protocols. It may involve more implementation complexity than the energy cooperation described in Section 13.3.

In this section, we discuss three different cost-aware communication cooperation schemes, namely, traffic offloading [8], spectrum sharing [9], and CoMP [10], which are implemented on different timescales. For the purpose of illustration, we consider a simple cellular system setup with two BSs as shown in Figure 13.3(a), in which BS 1 has sufficient harvested renewable energy but a light traffic load (serving two MTs), thus having a net load $\delta_1 < 0$, while BS 2 has insufficient renewable energy but a heavy traffic load (serving four MTs), leading to a net load $\delta_2 > 0$. The corresponding spectrum and power allocation and the user–BS association for the two BSs are shown in Figure 13.3(b)–(d) for several different cost-aware communication cooperation schemes.

13.4.1   Cost-Aware Traffic Offloading

Traffic offloading is traditionally designed to shift the traffic load (or served MTs) of heavily loaded BSs to lightly loaded ones for the purpose of avoiding traffic congestion and improving the QoS of the MTs. Differently, the cost-aware traffic offloading presented here focuses on the new issue of energy cost reduction, i.e., BSs short of renewable energy can offload their MTs to neighboring BSs with abundant renewable

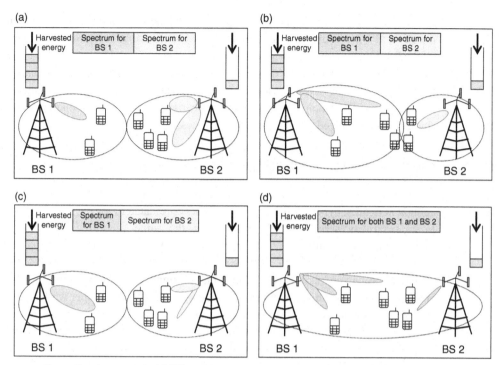

Figure 13.3 An example of different communication cooperation designs in a simple cellular network with two BSs [11]. (a) Conventional design; (b) cost-aware traffic offloading; (c) cost-aware spectrum sharing; (d) cost-aware CoMP. Reproduced with permission from the IEEE.

energy (even if they have greater or similar traffic loads), thus reducing the total amount of energy drawn from the gird to save costs. In the example in Figure 13.3(b), it is cost-effective for BS 2 to offload two MTs (at its cell edge) to BS 1, so that the renewable energy at BS 1 is better utilized. Traffic offloading is often employed on a timescale of several seconds.

13.4.2 Cost-Aware Spectrum Sharing

Besides energy, spectrum is another scarce resource in cellular networks, and spectrum sharing has been considered as a solution to improve the spectrum utilization efficiency [9]. Unlike conventional spectrum sharing, cost-aware spectrum sharing is based on the fact that energy and spectrum resources can partially substitute for each other to support wireless transmission, and sharing spectrum with a BS short of energy can help to reduce the energy cost of that BS.‖ In the example in Figure 13.3(c), BS 1

‖ Considering a point-to-point channel with additive white Gaussian noise (AWGN), the relationship between the transmit power $P \geq 0$ and the bandwidth $W \geq 0$ is given by $P = N_0 W \left(2^{r/W} - 1 \right)$, where r and N_0 denote the transmission rate and the noise power spectral density at the receiver, respectively. In this case, assuming fixed r, as the system bandwidth W increases, the transmit power P will decrease, and vice versa.

shares part of its available spectrum with BS 2. Under the same QoS requirements of the MTs, BS 2 can decrease the transmission power that it purchases from the grid, while BS 1 uses more renewable energy for transmission. Hence, the total cost is reduced. The implementation of spectrum sharing requires the BSs (and perhaps even the MTs) to have the capability to aggregate different frequency bands for transmission and reception, for example, by an advanced carrier aggregation technique. Spectrum sharing can be realized on a timescale of minutes.

13.4.3   Cost-Aware Coordinated Multipoint

Traditionally, CoMP is considered as a technique to improve spectral efficiency in cellular networks, by which BSs can implement coordinated baseband signal processing to cooperatively serve multiple MTs using the same time–frequency resources, transforming the harmful intercell interference (ICI) into useful information signals [10]. Differently, cost-aware CoMP is motivated by the following observation: since different BSs can cooperatively send information signals to the MTs (in the downlink), changes in their transmission power can be compensated by each other while satisfying the QoS requirements of the MTs. Therefore, by adaptively adjusting the BSs' transmit signals, cost-aware CoMP helps match the BSs' transmission power to their harvested renewable energy, thus minimizing the total energy drawn from the grid to save costs. For example, in Figure 13.3(d), BS 1, which has sufficient renewable energy, should use a high transmission power to provide strong wireless signals to the MTs, while BS 2, which is short of renewable energy, should transmit at a low power level in a CoMP transmission.

CoMP needs to be performed on a symbol or frame level on a timescale of microseconds or milliseconds, which is more complex than the aforementioned traffic offloading and spectrum sharing schemes, but it can achieve higher cost savings (as will be shown in the case study later). In practical cost-aware cellular networks, the three communication cooperation schemes can be employed depending on the trade-off between cost saving and implementation complexity.

13.5   Joint Energy and Communication Cooperation

Joint energy and communication cooperation can save cost maximally by applying both energy cooperation on the supply side and communication cooperation on the demand side. To achieve such joint cooperation, the BSs must share energy information by using the two-way information flow supported by the smart gird (through smart meters), and also exchange communication information through their backhaul connections (see Figure 13.1). Here, the exact required information sharing among BSs depends on the specific energy and communication cooperation schemes employed.

Joint energy and communication cooperation is more complex than energy or communication cooperation only, owing to the implementation complexity of solving the cost minimization problem by optimizing both the supply side (e.g., energy trading/sharing among BSs) and the demand side (e.g., spectrum and power allocations at BSs), as well as to the signaling overhead for sharing both the energy and the communication information among BSs. The complexity increases significantly as the network size or the number of BSs becomes larger. One potential solution to this problem is to dynamically group the huge number of BSs into clusters, where BSs within each cluster can implement joint cooperation in a centralized manner, and different clusters can perform limited coordination in a decentralized way. In this section, we focus our study on joint cooperation among a limited number of BSs in a single cluster.

As there are two energy cooperation schemes (aggregator-assisted energy trading and energy sharing) described in Section 13.3 and three communication cooperation schemes (traffic offloading, spectrum sharing, and CoMP) described in Section 13.4, there are in total six combinations that can be used in a joint cooperation design. In this section, we focus on three specific schemes: a joint energy and spectrum sharing design and two joint energy cooperation and CoMP designs. The ideas can be similarly extended to the other three combinations.

13.5.1   Joint Energy and Spectrum Sharing

Joint energy and spectrum sharing [23] is a scheme that allows neighboring BSs to share energy and spectrum with each other through the aggregator-assisted energy sharing described in Section 13.3.2 and the spectrum sharing described in Section 13.4.2, respectively. In this scheme, the BSs share their energy harvesting rates, energy prices, available bandwidth, and channel state information (e.g., channel gains), as well as the QoS requirements of the MTs, among each other. Accordingly, the BSs exchange energy and spectrum to take advantage of resource complementarity.

Building upon the spectrum sharing shown in Figure 13.3(c), the authors of [23] considered joint energy and spectrum sharing between two BSs to minimize their total energy cost, while ensuring the QoS requirements for all the MTs. It was shown that at optimality, it was possible that one BS having sufficient energy and spectrum would share these two resources with the other (in unidirectional cooperation), or one BS would exchange its energy for spectrum with the other (in bidirectional cooperation).

13.5.2   Joint Energy Cooperation and CoMP

In this scheme, different BSs implement the CoMP-based transmission/reception scheme presented in Section 13.4.3 to serve one or more MTs using the same time–frequency resources, and at the same time perform the aggregator-assisted energy trading described in Section 13.3.1 or the aggregator-assisted energy sharing described in Section 13.3.2. To implement this, the BSs need to share their energy harvesting rates, energy prices, and instantaneous channel state information (both channel gains and phases), as well as the QoS requirements of the MTs, among each other. For different types of energy cooperation schemes, [24] and [22] studied two different joint energy cooperation and CoMP designs for downlink transmission.

First, when the aggregator-assisted energy trading described in Section 13.3.1 is implemented among the BSs, the authors of [24] proposed to jointly optimize the BSs' cooperative transmit beamforming using CoMP-based communication and their two-way energy trading using the aggregator, so as to minimize the total energy cost. It was shown that by exploiting the nonuniform values of E_i for the harvested renewable energy over different BSs and the difference between the buying and selling prices π_{buy} and π_{sell} of energy, the joint energy trading and CoMP optimization achieved a significant cost reduction, as compared with a design that separately optimized the CoMP-based communication and the energy trading.

Next, the authors of [22] considered the use of the aggregator-assisted energy sharing described in Section 13.3.2 to enable a new, purely renewable-powered cellular system, in which the BSs do not purchase any energy from the grid but use harvested renewable energy together with energy sharing to maintain their operations. Taking into account the possible energy loss during the energy sharing, the weighted sum-rate for all served MTs in one particular CoMP cluster was maximized by jointly optimizing the cooperative BSs' zero-forcing beamforming design and the amounts of energy shared with each other.

13.5.3   A Case Study

We now present a case study to compare the energy cooperation presented in Section 13.3, the cost-aware spectrum sharing and CoMP presented in Section 13.4, and the three joint energy and communication cooperation schemes proposed in this section. Also, we consider a conventional design without energy or communication cooperation as a performance benchmark, where each BS first individually minimizes its energy consumption on the demand side while ensuring the QoS requirements at the MTs, and then (if the energy demand exceeds the renewable energy supply) purchases additional energy from the grid. For the purpose of illustration, as shown in Figure 13.4,

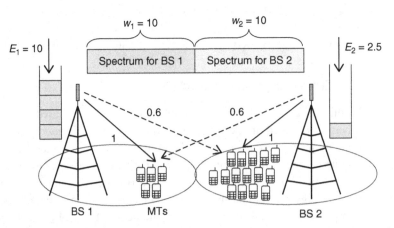

Figure 13.4 Case study model for comparing energy cooperation, communication cooperation, and joint cooperation [11]. Reproduced with permission from the IEEE.

we consider the downlink of a cellular system with two single-antenna BSs (i.e., BS 1 and BS 2), applying orthogonal frequency division multiple access (OFDMA) to serve $K_1 = 5$ and $K_2 = 15$ single-antenna MTs (denoted by the MT sets $\mathcal{K}_1$ and $\mathcal{K}_2$), respectively. Each BS uses an orthogonal frequency band with the same bandwidth ($W_1 = W_2 = 10$) initially. For simplicity, we generated the channels randomly based on independent and identically distributed (i.i.d.) Rayleigh fading with the average channel powers from each BS to its own associated MTs (i.e., from BS 1 to any MT in $\mathcal{K}_1$ and from BS 2 to any MT in $\mathcal{K}_2$) equal to 1, and the average channel powers from each BS to the other BS's associated MTs (i.e., from BS 1 to any MT in $\mathcal{K}_2$ and from BS 2 to any MT in $\mathcal{K}_1$) equal to 0.6. We set the noise power spectral density at each MT to be 1, and the QoS requirement of each MT to be a minimum data rate of 1. On the demand side, we set the power consumptions Q_1 and Q_2 of the two BSs to be their transmission power only; on the supply side, we considered their amounts of harvested renewable energy as constant over the time period of interest and set these to be $E_1 = 10$ and $E_2 = 2.5$, respectively. We set their buying price of energy from the grid to $\pi = 1$. Additionally, for aggregator-assisted energy trading, the BSs' buying and selling prices of energy from and to the aggregator were $\pi_{\text{buy}} = 0.5$ and $\pi_{\text{sell}} = 0.4$, respectively; and for aggregator-assisted energy sharing, the contract fee paid to the aggregator was $\bar{C} = 0.1$. In each scheme, the BSs employed an equal-bandwidth allocation among the MTs and only one MT was served in each subband. Note that all units are normalized for simplicity here.

We summarize the results obtained with these choices in Table 13.1, from which we have the following observations.

- For the conventional design, the two BSs' average energy demands for communications are computed to be $Q_1 = 4.14$ and $Q_2 = 18.28$. Their total energy cost is 15.78.
- For the two energy cooperation approaches, it is observed that the renewable energy supplies at BS 1 and BS 2 are changed to 4.14 and 8.36, respectively, by BS 1 trading or sharing its excess renewable energy with BS 2 through the aggregator. Since the new renewable energy supplies match better with the given energy consumptions of the two BSs, their total energy cost reduces to 10.51 and 10.03, respectively, where the different cost reductions are related to the different service fees charged by the aggregator.
- For communication cooperation, it is observed that the average energy consumptions of the two BSs are changed to $Q_1 = 10.00$ and $Q_2 = 14.04$ for the spectrum sharing scheme, and $Q_1 = 10.00$ and $Q_2 = 3.75$ for the CoMP scheme. Compared with the conventional design, communication cooperation increases the transmission power of BS 1 to partially substitute for that of BS 2 (together with certain wireless resource sharing) while satisfying the MTs' QoS requirements. The transmission power adaptation matches and better uses the given cheap renewable energy supplies at the two BSs. Consequently, the resulting total energy costs reduce to 11.54 and 1.25, respectively.

Table 13.1 Energy cost performance comparison.

	BS 1's renewable-energy supply	BS 2's renewable-energy supply	BS 1's average energy consumption	BS 2's average energy consumption	Total energy cost
Conventional design without energy or communication cooperation	10	2.5	4.14	18.28	15.78
Energy cooperation via aggregator-assisted energy trading	4.14	8.36	4.14	18.28	10.51
Energy cooperation via aggregator-assisted energy sharing	4.14	8.36	4.14	18.28	10.03
Communication cooperation via spectrum sharing	10	2.5	10.00	14.04	11.54
Communication cooperation via CoMP	10	2.5	10.00	3.75	1.25
Joint energy and spectrum sharing	5.00	7.50	5.00	15.00	7.60
Joint aggregator-assisted energy trading and CoMP	6.87	5.62	6.87	5.77	0.46
Joint aggregator-assisted energy sharing and CoMP	5.47	7.03	5.47	7.03	0.10

- For joint energy and communication cooperation, it is observed that by exploiting both supply and demand side management, each joint scheme outperforms the corresponding schemes with only energy or communication cooperation. For instance, the total energy cost with joint energy and spectrum sharing (7.6) is less than that with aggregator-assisted energy sharing only (10.03) and with spectrum sharing only (11.54). Furthermore, it is observed that the two joint energy cooperation and CoMP designs achieve the lowest total energy cost (0.46 and 0.10, respectively) and outperform all the other schemes. Therefore, joint energy cooperation and CoMP design is a promising way to achieve maximum cost saving.

13.6   Extensions and Future Directions

Despite the aforementioned studies on energy and communication cooperation, a lot of interesting topics remain unaddressed. We list several of them as follows for future study.

Practically, energy harvesting rates in general change slowly as compared with wireless channel and traffic load variations, and, as a consequence, the timescale of implementing energy cooperation is normally longer than that of communication cooperation [12]. However, the joint energy and communication cooperation described in this chapter requires energy cooperation to be realized on the same timescale as communication cooperation, thus needing more frequent decision making at the BSs and higher implementation complexity in smart meters. A promising way to overcome this issue is the multi-timescale implementation of joint energy and communication cooperation, for example by employing two-layer decision making with energy cooperation on a longer timescale and communications cooperation on a shorter timescale, so as to balance the trade-off between the cost-saving performance and the implementation complexity.

So far, we have focused on a single cellular system or multiple systems belonging to the same entity, aiming to minimize the total energy cost. In practice, however, multiple self-interested systems (owned by different operators) can coexist or be colocated and it is interesting to study their energy and/or communication cooperation. Unlike the energy trading considered in Section 13.3, on the energy supply side, more than one aggregator may be needed to facilitate trading across different BS groups. As for the mutual energy-sharing scheme, one selfish system may want to sell (or buy) renewable energy to (or from) another system at a high (or low) price. On the communication demand side, some systems (belonging to different operators, such as Verizon and T-Mobile) are sharing spectrum on a long-term basis. Yet enabling communication cooperation in the short term (as in Section 13.4) requires intersystem communication compatibility and more coordination. Moreover, to establish joint energy and communication cooperation, cellular systems may seek the advantages of resource complementarity. For example, in a preliminary study [23], it was shown that one system adequate in spectrum might be willing to cooperate with another that was adequate in energy, since both systems could efficiently reduce their individual costs by exchanging spectrum and energy with each other. Overall, the design of cooperation mechanisms is required to motivate or strengthen intersystem joint cooperation to realize a win–win situation for all systems involved.

Besides cellular networks, it is also appealing for heterogeneous communication networks (e.g., Wi-Fi, small cells, and even Long Term Evolution (LTE) in unlicensed spectrum (LTE-U)) to cooperate to reduce overall energy costs [25]. Offloading a mobile user from a macrocell to a small cell saves energy, and better utilizes the wireline backhaul resource to expand the limited wireless spectrum. But these networks are different in terms of service coverage, operating spectrum, and even energy harvesting availability (difficult indoors), and thus their joint energy and communication cooperation becomes more complicated than our design presented in Section 13.5. For example, scalability could be a problem, and one possible solution is to decompose the whole heterogeneous network into a number of micronetworks as in [6] with cooperation in each.

Up to now, we have focused on cases without the use of energy storage at BSs, owing to cost considerations. With the advancement of battery technologies, we envision that energy storage may be employed in future BSs and it is promising to study energy and communication cooperation jointly with the management of storage. In principle, storage devices handle the renewable energy fluctuations at the BSs to match the variations in energy demand over time [26], while the energy and communication cooperation approaches introduced in this chapter do that over space. Therefore, the two approaches may provide good complementarity. Nevertheless, such joint time and space domain optimization problems are very challenging to solve, since any decisions made by BSs in the present affect their storage status and traffic loads in the future. In an initial study, [27] has considered the joint energy cooperation and storage management problem to minimize the total energy cost in a simplified cellular system with given energy demands at the BSs.

Furthermore, it is worth noting that smart grids have recently also enabled time-varying energy prices to help stabilize energy generation and transmission. For example, a hybrid electricity market has been successfully implemented in the United States and Norway [28, 29], which combines a day-ahead energy market and a real-time energy market. In the day-ahead energy market, electricity consumers can make commitments about tomorrow's energy purchases at low prices, whereas in the real-time energy market, they are free from commitment and can flexibly buy energy at higher prices or sell back an excess energy commitment at prices lower than the day-ahead ones. In view of such a new hybrid electricity market, the energy purchased by cellular BSs in the day-ahead and real-time markets may lead to different unit costs at different times. This, together with the cost difference between renewable and on-grid energy, makes the cost-effective energy management of cellular networks more challenging. In this case, a cellular operator needs to jointly optimize many BSs' day-ahead and real-time energy cooperation (in terms of energy purchase) and their communication cooperation, based on the time-varying wireless traffic load and energy prices, for the purpose of minimizing energy costs [30].

13.7   Conclusion

This chapter has investigated novel energy and communication cooperation approaches for energy cost saving in cellular networks with BSs powered by renewable energy sources and smart grid. These approaches use both the two-way energy flow in the smart grid and communication cooperation in the cellular network to reshape the nonuniform energy supplies and energy demands of the cellular network to achieve cost savings. It is our hope that these new approaches can bring new insights into energy demand management in smart grids by considering the unique properties of the communication demands of cellular networks, and also into wireless resource allocation in cellular networks by taking into account the new characteristics of the emerging renewable and smart grid based energy supply.

References

[1] J. G. Andrews, S. Buzzi, W. Choi, S. V. Hanly, A. Lozano, A. C. K. Soong, and J. C. Zhang, "What will 5G be?" *IEEE J. Sel. Areas Commun.*, vol. 32, no. 6, pp. 1065–1082, Jun. 2014.

[2] Z. Hasan, H. Boostanimehr, and V. Bhargava, "Green cellular networks: A survey, some research issues and challenges," *IEEE Commun. Surv. Tutor.*, vol. 13, no. 4, pp. 524–540, Fourth Quarter 2011.

[3] J. Wu, S. Rangan, and H. Zhang, *Green Communications: Theoretical Fundamentals, Algorithms, and Applications*, CRC Press, 2012.

[4] S. Bu, F. R. Yu, Y. Cai, and X. P. Liu, "When the smart grid meets energy-efficient communications: Green wireless cellular networks powered by the smart grid," *IEEE Trans. Wireless Commun.*, vol. 11, no. 8, pp. 3014–3024, Aug. 2012.

[5] T. Han and N. Ansari, "Powering mobile networks with green energy," *IEEE Wireless Commun.*, vol. 21, no. 1, pp. 90–96, Feb. 2014.

[6] X. Fang, S. Misra, G. Xue, and D. Yang, "Smart grid – the new and improved power grid: A survey," *IEEE Commun. Surv. Tutor.*, vol. 14, no. 4, pp. 944–980, Fourth Quarter 2012.

[7] W. Saad, Z. Han, H. V. Poor, and T. Basar, "Game-theoretic methods for the smart grid: An overview of microgrid systems, demand-side management, and smart grid communications," *IEEE Signal Process. Mag.*, vol. 29, no. 5, pp. 86–105, Sep. 2012.

[8] Z. Niu, Y. Wu, J. Gong, and Z. Yang, "Cell zooming for cost-efficient green cellular networks," *IEEE Commun. Mag.*, vol. 48, no. 11, pp. 74–78, Nov. 2010.

[9] R. Zhang, Y. C. Liang, and S. Cui, "Dynamic resource allocation in cognitive radio networks," *IEEE Signal Process. Mag.*, vol. 27, no. 3, pp. 102–114, May 2010.

[10] D. Gesbert, S. Hanly, H. Huang, S. Shamai, O. Simeone, and W. Yu, "Multi-cell MIMO cooperative networks: A new look at interference," *IEEE J. Sel. Areas Commun.*, vol. 28, no. 9, pp. 1380–1408, Dec. 2010.

[11] J. Xu, L. Duan, and R. Zhang, "Cost-aware green cellular networks with energy and communication cooperation," *IEEE Commun. Mag.*, vol. 53, no. 5, pp. 257–263, May 2015.

[12] H. Li, J. Xu, R. Zhang, and S. Cui, "A general utility optimization framework for energy harvesting based wireless communications," *IEEE Commun. Mag.*, vol. 53, no. 4, pp. 79–85, Apr. 2015.

[13] D. Gunduz, K. Stamatiou, N. Michelusi, and M. Zorzi, "Designing intelligent energy harvesting communication systems," *IEEE Commun. Mag.*, vol. 52, no. 1, pp. 210–216, Jan. 2014.

[14] S. Ulukus, A. Yener, E. Erkip, O. Simeone, M. Zorzi, P. Grover, and K. Huang, "Energy harvesting wireless communications: A review of recent advances," *IEEE J. Sel. Areas Commun.*, vol. 33, no. 3, pp. 360–381, Mar. 2015.

[15] C. K. Ho and R. Zhang, "Optimal energy allocation for wireless communications with energy harvesting constraints," *IEEE Trans. Signal Process.*, vol. 60, no. 9, pp. 4808–4818, Sep. 2012.

[16] O. Ozel, K. Tutuncuoglu, J. Yang, S. Ulukus, and A. Yener, "Transmission with energy harvesting nodes in fading wireless channels: Optimal policies," *IEEE J. Sel. Areas Commun.*, vol. 29, no. 8, pp. 1732–1743, Sep. 2011.

[17] S. Luo, R. Zhang, and T. J. Lim, "Optimal save-then-transmit protocol for energy harvesting wireless transmitters," *IEEE Trans. Wireless Commun.*, vol. 12, no. 3, pp. 1196–1207, Mar. 2013.

[18] J. Xu and R. Zhang, "Throughput optimal policies for energy harvesting wireless transmitters with non-ideal circuit power," *IEEE J. Sel. Areas Commun.*, vol. 32, no. 2, pp. 322–332, Feb. 2014.

[19] C. Huang, R. Zhang, and S. Cui, "Throughput maximization for the Gaussian relay channel with energy harvesting constraints," *IEEE J. Sel. Areas Commun.*, vol. 31, no. 8, pp. 1469–1479, Aug. 2013.

[20] C. Huang, R. Zhang, and S. Cui, "Optimal power allocation for outage probability minimization in fading channels with energy harvesting constraints," *IEEE Trans. Wireless Commun.*, vol. 13, no. 2, pp. 1074–1087, Feb. 2014.

[21] L. Gkatzikis, I. Koutsopoulos, and T. Salonidis, "The role of aggregators in smart grid demand response markets," *IEEE J. Sel. Areas Commun.*, vol. 31, no. 7, pp. 1247–1257, Jul. 2013.

[22] J. Xu and R. Zhang, "CoMP meets smart grid: A new communication and energy cooperation paradigm," *IEEE Trans. Veh. Tech.*, vol. 64, no. 6, pp. 2476–2488, Jun. 2015.

[23] Y. Guo, J. Xu, L. Duan, and R. Zhang, "Joint energy and spectrum cooperation for cellular communication systems," *IEEE Trans. Commun.*, vol. 62, no. 10, pp. 3678–3691, Oct. 2014.

[24] J. Xu and R. Zhang, "Cooperative energy trading in CoMP systems powered by smart grids," *IEEE Trans. Veh. Tech.*, vol. 65, no. 4, pp. 2142–2153, Apr. 2016.

[25] S. Cai, Y. Che, L. Duan, J. Wang, S. Zhou, and R. Zhang, "Green 5G heterogeneous networks through dynamic small-cell operation," *IEEE J. Sel. Areas Commun.*, vol. 34, no. 5, pp. 1103–1115, May 2016.

[26] K. Rahbar, J. Xu, and R. Zhang, "Real-time energy storage management for renewable integration in microgrids: An off-line optimization approach," *IEEE Trans. Smart Grid*, vol. 6, no. 1, pp. 124–134, Jan. 2015.

[27] Y. K. Chia, S. Sun, and R. Zhang, "Energy cooperation in cellular networks with renewable powered base stations," *IEEE Trans. Wireless Commun.*, vol. 13, no. 12, pp. 6996–7010, Dec. 2014.

[28] A. L. Ott, "Experience with PJM market operation, system design, and implementation," *IEEE Trans. Power Syst.*, vol. 18, no. 2, pp. 528–534, May 2003.

[29] A. B. Philpott and E. Pettersen, "Optimizing demand-side bids in day-ahead electricity markets," *IEEE Trans. Power Syst.*, vol. 21, no. 2, pp. 488–498, May 2006.

[30] J. Xu, L. Duan, and R. Zhang, "Energy group buying with loading sharing for green cellular networks," *IEEE J. Sel. Areas Commun.*, vol. 34, no. 4, pp. 786–799, May 2016.

14 Visible Light Communication in 5G

Harald Haas and Cheng Chen

14.1 Introduction

Owing to the increasing demand for wireless data communication, the available radio spectrum below 10 GHz (centimeter wave communication) has become insufficient. The wireless communications industry has responded to this challenge by considering the radio spectrum above 10 GHz (millimeter-wave communication). However, the higher frequencies f mean that the path loss L increases according to the Friis free-space equation ($L \propto f^2$), i.e., moving from 3 to 30 GHz would add 20 dB signal attenuation or, equivalently, would require 100 times more power at the transmitter. In addition, blockages and shadowing in terrestrial communication are more difficult to overcome at higher frequencies. As a consequence, systems must be designed to enhance the probability of line-of-sight (LoS) communication, typically by using beamforming techniques and by using very small cells (about 50 m in radius). The requirement for smaller cells in cellular communication also benefits network capacity and data density. In fact, reducing cell size has without doubt been one of the major contributors to enhanced system performance in current cellular communications. The cell sizes in cellular communication have dramatically shrunk (35 km in the second generation (2G), 5 km in the third generation (3G), 500 m in the fourth generation (4G), and probably about 50 m in the fifth generation (5G) [1] and 5 m in the sixth generation (6G). This means that, contrary to general belief, using higher frequencies for terrestrial communication has become a practical option. However, there are some significant challenges associated with providing a supporting infrastructure for ever-smaller cells. One such challenge is the provision of a sophisticated backhaul infrastructure.

It is predicted that a capacity per unit area of 100 Mbps/m^2 will be required for future indoor spaces, primarily driven by high-definition video and billions of Internet of Things (IoT) devices. Achieving this with low energy consumption will be critical if the potential of "green" communication is to be realized. The goal of connectivity will require swathes of new spectrum, and energy harvesting will be needed to prevent exponentially increasing energy consumption for wireless communications. The available optical spectrum dwarfs that available in the radio frequency (RF) region, and can be accessed using low-cost optical components and simple (compared with radio frequency (RF) baseband processing. The introduction of Light Fidelity (LiFi) [2, 3] is, therefore, a natural consequence of the trend of moving to higher frequencies in the electromagnetic spectrum, and LiFi is part of the more general field of optical wireless communications (OWC) [4]. Specifically, it would be obvious to classify Light-Fidelity (LiFi) as nanometer wave communication. LiFi uses light-emitting diodes (LEDs) for high-speed communication, and speeds of over 3 Gbps from a single micro-LED [5] have been demonstrated using optimized direct current optical orthogonal frequency division multiplexing (DCO-OFDM) modulation [6]. Given that there is widespread deployment of LED lighting in homes, offices, and public places because of its energy-efficiency, there is an added benefit of LiFi cellular deployment in that it can build on existing lighting infrastructures. Moreover, the cell sizes can be reduced further beyond what is possible with millimeter wave communication, leading to the concept of LiFi attocells [7]. LiFi attocells constitute an additional wireless networking layer within existing heterogeneous wireless networks, and they have zero interference from, and add zero interference to, their RF counterparts such as femtocell networks. A LiFi attocell network uses the lighting system to provide fully networked (multiuser access and handover) wireless access. In 2030, the vision is that an LED infrastructure and distributed LED lights in appliances will form LiFi networks, and will provide illumination, secure communication, and energy to power network nodes via simple solar-cell energy harvesting [8]. A LiFi network will form a substantial part of the heterogeneous future networking landscape beyond 5G and 6G, and will carry a large share of wireless Internet connectivity. This chapter will demonstrate that visible light communications (VLC) has matured into a full wireless networking technology, moving from millimeter wave (mmWave) to nanometer wave communications for 5G and beyond. However, there are fundamental technical challenges to be overcome. Some of these challenges are addressed in this chapter.

14.2   Differences between Light-Fidelity and Visible Light Communication

VLC uses LEDs to transmit data wirelessly by using intensity modulation (IM). At the receiver, the signal is received by a photodiode (PD) and detected using the principle of direct detection (DD). visible light communication (VLC) was conceived as a point-to-point data communication technique – essentially as a cable replacement. This led to early VLC standardization activities as part of IEEE 802.15.7 [9]. This standard, however, has been revised to include LiFi. LiFi, in contrast, defines a mobile wireless system [10]. This includes bidirectional multiuser communication, i.e., point-to-multipoint and multipoint-to-point communication. By using LiFi, it is possible to create a network of multiple access points, each covering a very small LiFi attocell. Mobility is supported by seamless handover between LiFi-enabled luminaries. This means that LiFi enables full user mobility, and therefore establishes a new wireless networking layer within the existing RF heterogeneous wireless networks (Figure 14.1). Because LEDs are natural beamformers, excellent local containment of LiFi signals can be achieved, and because opaque walls block LiFi signals, co-channel

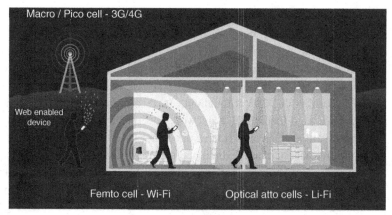

Figure 14.1 LiFi will create a step change improvement in the small-cell concept leading to area data densities of 1 Gbps/m^2. LiFi will create a new wireless networking layer, the optical attocell layer, which does does not cause interference to the RF layers, and is not susceptible to interference from RF layers. The optical attocell layer will support mobility and multiuser access. (Courtesy pureLiFi Ltd.)

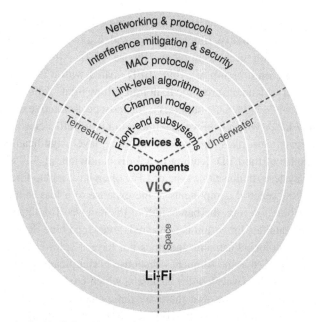

Figure 14.2 The principal building blocks of LiFi and its application areas.

interference (CCI) can be effectively managed. For the same reasons, physical layer security can be enhanced. Figure 14.2 illustrates the principal techniques that are needed to create optical attocell LiFi networks. At the core are novel devices such as gallium nitride (GaN) micro-LEDs and single-photon avalanche diodes (SPADs) [11]. These devices are embedded in optical front-ends and subsystems, which may include adaptive optics as well as analog circuitry to drive the LEDs and to filter the signals received by the optical detectors. In order to correctly model link margins, establish the coherence bandwidth of the channel, and correctly model CCI, precise channel models are required which take the spectral composition of the signal into account [12]. Link-level algorithms are required to optimally harness the signals to maximize the data throughput. In this context, because IM signals are strictly positive, new theoretical frameworks are needed to establish the channel capacity. The traditional Shannon framework is not strictly applicable [13]. In order to enable multiuser access, new medium access control (MAC) protocols are required that take the specific features of the LiFi physical layer into account. Similarly, interference mitigation techniques are needed to ensure fairness and high overall system throughput. Lastly, LiFi attocell networks need to be integrated into the emerging software-defined networks governed by separation of the control and data planes and by network virtualization [14].

14.3   LiFi LED Technologies

The achievable data rates in LiFi depend greatly on the LED technology used, and a number of 'hero' experiments have provided some rough practical maximum achievable data rates for each technology (Figure 14.3). Most commercial white LED lights are based on a high-power blue LED coated with a phosphor-based color-converting material which converts blue light into yellow light. Yellow and blue light combine and create white light. Currently this is the most cost-effective way to create high-brightness white LED, which achieve maximum luminous efficacies in the region of 160 lm/W.* However, the phosphor color-converting layer significantly reduces the frequency response of the LED, leading to a 3 dB bandwidth of 3–5 MHz. Therefore, VLC and LiFi systems that use this type of LED often have a blue filter at the receiver to remove the slow spectral components at the expense of a loss in signal-to-noise ratio (SNR). After blue filtering, the bandwidth of the received IM signal is about 20 MHz. A way to mitigate this limitation is to use red, green, and blue (RGB) LEDs. This type of LED enables color adjustment and does not have a color-converting phosphor layer. This means, first, that the frequency response is significantly higher and, second, that there are three color (wavelength) channels, which enable wavelength division multiplexing (WDM). This means that there are three channels for data transmission. It has been found that with red, green and blue (RGB) LEDs, data rates of about 5 Gbps can be achieved. It is important to note that if the 3 dB bandwidth of an LED is, say, 30 MHz, this does not mean that the channel bandwidth has to be limited to 30 MHz, since there are no channel regulations like those in the RF, where there exist well-defined spectrum masks for each channel. A 30 MHz LED may be modulated at up to 500 MHz using bit-loading and power-loading techniques. A third class of LEDs is GaN micro-LEDs

---

* As a rule of thumb, a 100 W incandescent light bulb produces about 1600 lm. This means that the energy savings are about 10 times with current commercial LEDs over incandescent lights.

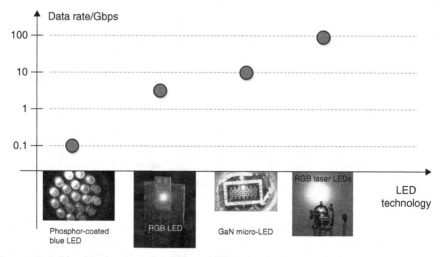

Figure 14.3 Achievable data rates for different LED technologies.

which operate at a very high current density and, owing to their small size, the 3 dB bandwidth is significantly improved and is in excess of 60 MHz. A WDM system with micro-LEDs can achieve 10 Gbps [15]. Lastly, white light can be generated by using multiple-wavelength laser LEDs at a wavelength spacing of about 10–15 nm. The bandwidth of a laser LED is about 1 GHz, and data rates of 100 Gbps are feasible [16]. Broad coverage is achieved by shining the white light beam onto a diffuser, and it is conceivable that future light bulbs will be made of lasers. The commercial introduction of a laser light bulb is primarily governed by commercial constraints, although there are also some technical issues that need to be overcome, such as the effect of speckle.

14.4 LiFi Attocell Networks

Typically, a cellular network is limited by the downlink capacity. We therefore focus our attention to the downlink of a LiFi attocell network [17]. A key metric for a wireless network is the achievable signal-to-interference-plus-noise ratio (SINR). Every LiFi access point needs a connection to the backbone which requires a backhaul connection. A unique advantage of LiFi is that it can piggyback on lighting infrastructures. For LiFi with lighting retrofit installations, the backhaul can be realized by using power line communication (PLC). For new installations, power over Ethernet (PoE) would be the preferred option to achieve gigabit connectivity. Alternatively, a high-throughput backhaul network can also be realized by optical fiber connections using, for example, passive optical networking or point-to-point wireless mmWave connections [18]. It is assumed that these backhaul connections can provide enough capacity and would not limit the performance of the LiFi attocell network. The uplink connection is typically achieved by using wavelength division duplex with wireless infrared or RF links [19].

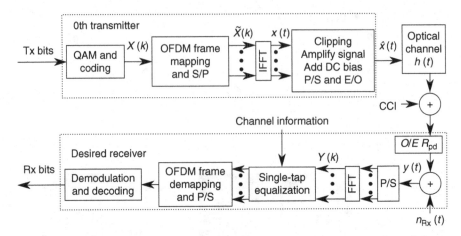

Figure 14.4 An LiFi attocell network DCO-OFDM downlink system. (S/P, serial to parallel; P/S, parallel to serial; E/O, electrical to optical; O/E, optical to electrical.)

It has been shown that LiFi can offload a large portion of data traffic from RF wireless networks [19]. Since there is sufficient spectral separation between the spectrum used for the uplink (infrared or RF) and the visible light spectrum, there is negligible interference between the uplink and the downlink.

14.4.1   Optical OFDM Transmission

High-data-rate transmission in IM/DD systems is achieved with optical orthogonal frequency division multiplexing (OFDM) and variants [2]. An in-depth treatment of OFDM for IM/DD and for general digital modulation techniques in VLC and LiFi systems is provided in [2]. Since LiFi offers the benefits of joint illumination and high-speed data communication, we assume spectrum-efficient DCO-OFDM in the sequel. However, the analysis can readily be extended to other types of optical OFDM systems. A block diagram of the downlink system is shown in Figure 14.4. For each OFDM frame, K quadrature amplitude modulation (QAM) symbols are fed into the modulator. Since the 0th and $K/2$th samples require no energy, this amount of energy is equally distributed among the remaining samples to ensure that the time domain signal is normalized. Therefore, the OFDM frame is normalized by a factor of $\xi = \sqrt{K/(K-2)}$. After an inverse fast Fourier transform (IFFT) operation and with a sufficiently large number of subcarriers, the real time domain signal $x(t)$ follows a Gaussian distribution with zero mean and unity variance. Next, a cyclic prefix with a length based on the length of the maximum delay of the channel is added to the frame. Furthermore, the time domain signal $x(t)$ is clipped, amplified by a factor of σ_x, and biased by a direct current (DC) component I_{DC} in order to modulate the intensity of the light with the signals. The optical-signal sample at time slot t can be written as

$$\hat{x}(t) = \eta_{led} \left(\sigma_x U(x(t)) + I_{DC} \right), \tag{14.1}$$

where η_{led} denotes the electrical-to-optical conversion coefficient. The output optical power is proportional to the input signal current, σ_x is equivalent to the standard deviation of the electrical signal, and the clipping function $U(v)$ is defined as

$$U(v) = \begin{cases} \lambda_t & : v > \lambda_t, \\ v & : \lambda_t \geq v \geq \lambda_b, \\ \lambda_b & : v < \lambda_b, \end{cases} \tag{14.2}$$

where λ_t and λ_b are the normalized top and bottom clipping levels, respectively [20]. According to the Bussgang theorem, the nonlinear clipped signal can be modeled as follows:

$$U(x(t)) = \rho x(t) + n_{\text{clip}}(t), \tag{14.3}$$

where ρ is an attenuation factor and $n_{\text{clip}}(t)$ is the time domain clipping noise.

To realize a LiFi attocell network, multiple optical base stations (BSs) have to be considered. The assignment of a user to a BS is based on the received signal strength. Note that because the detector size is much larger than the wavelength of the signal, VLC systems are not subject to fast fading effects. This circumstance will significantly aid in cell selection and resource allocation. However, as in RF networks, BSs that use the same transmission resources as neighboring BSs may cause CCI, which impacts SINR. To distinguish the signals from different BSs, a subscript $i = 0, 1, 2, \ldots$ is added. The case of $i = 0$ corresponds to the case of the desired BS, while $i \in \mathcal{I}$ corresponds to the case of the interfering BSs, in which $\mathcal{I}$ denotes the set of the BSs using the same transmission resources. Subsequently, the signals pass through free-space optical channels and are received by the receiver of the desired user. The received signal at time t can be written as follows:

$$y(t) = \eta_{\text{pd}} \left(\hat{x}_0(t) \otimes h_0(t) + \sum_{i \in \mathcal{I}} \hat{x}_i(t) \otimes h_i(t) \right) + n_{\text{Rx}}(t), \tag{14.4}$$

where η_{pd} is the photodiode (PD) responsivity, $h_i(t)$ denotes the channel impulse response between the ith BS and the desired user, $n_{\text{Rx}}(t)$ represents the noise at the receiver in the time domain, and $\otimes$ is the convolution operator. In conjunction with the clipping process (14.3), the frequency domain signal sample received on subcarrier k after the fast Fourier transform (FFT) operation can be written as follows:

$$Y(k) = \eta_{\text{pd}} \eta_{\text{led}} \sigma_x \left(\rho X_0(k) + N_{\text{clip},0}(k) \right) H_0(k) + N_{\text{Rx}}(k)$$
$$+ \eta_{\text{pd}} \eta_{\text{led}} \sigma_x \sum_{i \in \mathcal{I}} \left(\rho X_i(k) + N_{\text{clip},i}(k) \right) H_i(k), \tag{14.5}$$

where $H_i(k)$ is the channel transfer factor of the interference channel i for the kth subcarrier, and $N_{\text{Rx}}(k)$ corresponds to the frequency domain receiver noise, which follows a Gaussian distribution with zero mean and variance σ_{Rx}^2. Here, the power spectral density (PSD) of the noise is modeled by N_0. With a channel bandwidth of F_s, the receiver noise variance is $\sigma_{\text{Rx}}^2 = N_0 F_s / \xi^2$; and $N_{\text{clip},i}(k)$ represents the FFT of $n_{\text{clip},i}(t)$. According to the central limit theorem, $N_{\text{clip},i}(k)$ follows a Gaussian distribution with zero mean and a variance of σ_{clip}^2, assuming sufficiently many subcarriers. After the single-tap equalization, the desired signal can be recovered to the original QAM symbols $X_0(k)$.

The SINR is an important metric for evaluating the quality of the communication link and the system capacity. Based on (14.5), an expression for the SINR on subcarrier k can be written as

$$
\gamma(k) = \frac{\eta_{pd}^2 \eta_{led}^2 \sigma_x^2 \rho^2 \xi^2 |H_0(k)|^2}{\eta_{pd}^2 \eta_{led}^2 \sigma_x^2 \sigma_{clip}^2 |H_0(k)|^2 + \sum_{i \in \mathcal{I}} \eta_{pd}^2 \eta_{led}^2 \sigma_x^2 \left(\rho^2 + \sigma_{clip}^2\right) |H_i(k)|^2 + \sigma_{Rx}^2}
$$

$$
= \left(\left(\frac{\rho^2 \xi^2 |H_0(k)|^2}{\left(\rho^2 + \sigma_{clip}^2\right) \sum_{i \in \mathcal{I}} |H_i(k)|^2 + N_0 F_s / \xi^2 \eta_{pd}^2 \eta_{led}^2 \sigma_x^2}\right)^{-1} + \frac{\sigma_{clip}^2}{\rho^2 \xi^2}\right)^{-1}, \quad (14.6)
$$

where the channel gain $|H(k)|^2$, the electrical signal variance σ_x^2, the clipping-related parameters ρ, σ_{clip}^2, and the receiver noise PSD N_0 are analyzed in the following subsections.

14.4.2   Channel Model

As can be found by inspection of (14.6), the gain of the VLC channel $|H(k)|^2$ has a significant impact on the achievable SINR. The characteristics of $|H(k)|^2$ are influenced by the response of the front-end devices (LED, PD) and the propagation channel. Therefore, $|H(k)|^2$ can be modeled as

$$
|H(k)|^2 = |H_{fe}(k)|^2 |H_{fs}(k)|^2, \quad (14.7)
$$

where $H_{fe}(k)$ is the frequency response of the combined low-pass equivalent front-end devices, and $H_{fs}(k)$ represents the frequency response of the propagation channel.

Impact of Front-End Devices
The front-end devices exhibit a low-pass characteristic. The corresponding bandwidth is typically limited by the frequency responses of the LED and the PD. Four normalized transfer functions of real-world LED devices taken from [21–24] are plotted in Figure 14.5. The 3 dB bandwidth of these systems ranges from 10 to 60 MHz. Since the channel bandwidth can reach up to 1 GHz, this low-pass effect will significantly decrease the capacity. Therefore, the low-pass effect of the front-end devices is crucial to the performance of a LiFi attocell system. However, there have been remarkable developments in improving the bandwidth of LEDs during the past decade.

A first-order function (in the log domain) is adopted here to approximate the normalized transfer function of an LED device. This approximation has been shown to be accurate [21]:

$$
|H_{fe}(f)|^2 = \exp\left(-\frac{f}{F_{fe}}\right), \quad (14.8)
$$

where F_{fe} controls the frequency characteristics of the front-end device. The higher the value of F_{fe}, the larger the bandwidth. As shown in Figure 14.5, this approximation

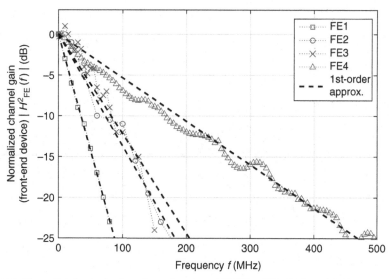

Figure 14.5 Normalized transfer functions of real-world LEDs: FE1, FE2, and FE3 correspond to devices taken from [21], [22], and [23], respectively. Commercially available white LEDs were used in these systems. The transfer function of FE4 is an experimental measurement of the system presented in [24], which used a 50 μm gallium nitride (GaN) micro-LED.

offers a good estimation of the low-pass characteristics. Converting (14.8) to the normalized transfer factor on subcarrier k yields

$$|H_{\text{fe}}(k)|^2 = \exp\left(-\frac{kF_{\text{s}}}{KF_{\text{fe}}}\right), \tag{14.9}$$

for $k = 1, 2, \ldots, K/2 - 1$. In the remainder of this chapter, systems with different front-end devices are considered. For convenience of description, the front-end device used in [21] is denoted as FE1, with a corresponding F_{fe} of 15.2 MHz. Similarly, the front-end devices used in [22–24] are denoted as FE2, FE3, and FE4, with corresponding F_{fe} values of 35.6 MHz, 31.7 MHz, and 81.5 MHz, respectively.

The Indoor Propagation Channel

In this study, a geometric ray-tracing method is used to analyze the channel characteristics. The wireless transmission geometry is given in Figure 14.6. The optical BS is installed on the ceiling of the room and oriented toward the floor. The receiver is at a given height from the floor, and the PD faces upward. This results in a vertical separation, h, between the terminal and the BS. Alignment of the PD detector can be achieved by using multiple PD detectors with different orientations at the terminal, and by applying adaptive selection and combining techniques [25, 26]. A possible receiver design is depicted in Figure 14.7. This is practically feasible because of the small wavelength, and consequently the small size of the PD components.

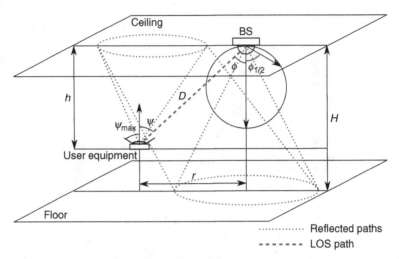

Figure 14.6 Transmission geometry in an LiFi attocell system.

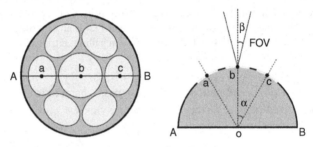

Figure 14.7 Receiver design with multiple PDs on a hemisphere.

The LoS path as shown in Figure 14.6 can be calculated using the DC channel gain H_{los} between the BS and the desired user [27]:

$$H_{\text{los}} = \frac{A_{\text{pd}}(m+1)}{2\pi D^2} \cos^m(\phi) \cos(\psi) \mathbf{1}_{\mathcal{D}_1}(\psi), \qquad (14.10)$$

$$H_{\text{los}}(r) = \frac{A_{\text{pd}}(m+1)h^{m+1}\mathbf{1}_{\mathcal{D}_1}(\psi)}{2\pi \left(r^2 + h^2\right)^{(m+3)/2}}, \qquad (14.11)$$

where m denotes the Lambertian emission order, which is given by $m = -1/\log_2(\cos(\phi_{1/2}))$, in which $\phi_{1/2}$ is the half-power semiangle of the LED. This quantity determines the beam width of the light source. A_{pd} is the physical area of the receiver PD, D is the Euclidean distance between the BS and the terminal, ϕ is the corresponding light radiance angle, ψ is the corresponding light incidence angle, and the function $\mathbf{1}_{\mathcal{D}}(x)$ is defined as

$$\mathbf{1}_{\mathcal{D}}(x) = \begin{cases} 1 & :x \in \mathcal{D}, \\ 0 & :x \notin \mathcal{D}. \end{cases} \qquad (14.12)$$

In (14.10), $\mathcal{D}_1 = [0, \psi_{max}]$, where ψ_{max} is the field of view (FoV) of the receiver. According to the geometry shown in Figure 14.6, H_{los} can be rewritten as a function of the horizontal offset r between the BS and the terminal, as suggested by (14.11).

As shown in Figure 14.6, there is also multipath propagation (MP), which results mainly from reflections from the internal surfaces of the room (walls, ceiling, and floor). Diffuse reflection can be assumed, as typically these internal surfaces are rough compared with the wavelength of light. A large office with side lengths of 5 m to 50 m was assumed for the multicell LiFi attocell network. Most of the users and BSs were away from the room edges. Consequently, in most cases (in the cells not near walls) the contribution to the signal from first-order reflections should be negligible. Therefore, the MP effect was caused mainly by second-order reflections bouncing off the floor and ceiling.

In order to evaluate the effect of the reflected signals on the channel, computer simulations based on [28] were carried out. To simplify the problem, a special (extreme) case was simulated with the floor and ceiling extending infinitely in all directions. A BS and user with a horizontal offset of r were considered. The transmission geometry is given in Figure 14.6. The simulation time bin width was 0.1 ns and the number of iterations was 5×10^5. The impulse responses and the frequency domain normalized channel gains of four example channels with $r = 0, 1, 2$, and 3 m are shown in Figure 14.8 . A dispersive source with a $\phi_{1/2}$ of 60° was simulated. The reflectivity of the ceiling was 0.7 and the reflectivity of the floor was 0.3. The height of the room was 3 m and the measurement plane was 0.75 m above the floor.

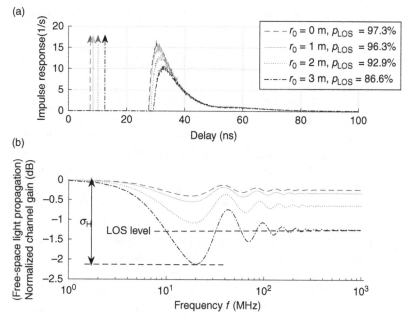

Figure 14.8 (a) Channel impulse response in time domain. (b) Normalized channel gain in frequency domain.

In Figure 14.8(a), the impulse responses show that there is a strong LoS component, followed by a gap period until the first reflected signal arrives. This is followed by a stream of closely spaced reflected signals. This is because the signal propagation delay of the LoS path is much shorter than the delay incurred by the reflected paths. To show the strength of the LoS component, a parameter p_{los} was defined; p_{los} represents the ratio of the received optical power of the LoS signal component to the total received optical power. The channels considered result in normalized channel gains $|H_{fs}(f)/H_{fs}(0)|^2$ as shown in Figure 14.8(b). The maximum channel gain always appears at DC. With increasing frequency, the channel gain decreases and reaches a minimum at about 20 MHz. With a further increase in the frequency, the channel gain oscillates around a constant level and the magnitude of this oscillation diminishes. The constant level corresponds to the channel gain considering the LoS component only. It can be observed that the maximum variance in the channel gain σ_H is less than 2.5 dB. This means that the variation of the channel gain due to the reflected signal is minor compared with the effect of the low-pass characteristic of the front-end devices.

It is known that the flatness of the channel gain is strongly related to the strength of the LoS component [29]. Therefore, the relationship between the maximum channel gain variance σ_H and the proportion of the LoS power p_{los} was evaluated for different transmitter half-power semiangles $\phi_{1/2}$. It was observed that σ_H was a decreasing function of p_{los}, as shown in Figure 14.9(a). As long as p_{los} is above 80%, σ_H can be kept below 4 dB. Instead of evaluating the channel gain directly, p_{los} can be evaluated to estimate the flatness of the channel gain. In addition to the case shown in Figure 14.8, the performance for all of the configurations of interest had to be investigated. Specifically, p_{los} was evaluated using different values of r and $\phi_{1/2}$. In conjunction with the analysis in Section 14.5, it was shown that in this study that p_{los} was always high in the region of interest, as shown in Figure 14.9(b). This is due to the longer travel distance ℓ in combination with the high electrical path loss in IM/DD systems, $L \propto (\ell)^4$, and the high absorption by the floor and ceiling.

The results shown in Figures 14.8 and 14.9 do not cover the performance of users near the edge of the room. Therefore, p_{los} was calculated considering reflections from all surfaces of the room, in order to validate the assumption that the first-order reflections are negligible in most cases. In this example, 23 hexagonal cells in a room of size of 26 m × 26 m × 3 m were considered. A half-power semiangle $\phi_{1/2}$ of 60° and a cell radius R of 3 m were used. The result is depicted in Figure 14.10(a). Users in the cells not close to the room edges have p_{los} above 85%. In addition, users in the center of the room edge cells also have high p_{los}. The remaining users in the cells close to the walls have a lower value of p_{los}, but generally above 50%. The results given in Figure 14.10(a) are also presented in the form of a cumulative density function (CDF), shown as the curve for setup 1 in Figure 14.10(b). Figure 14.10(b) shows that nearly 80% of the users experience p_{los} above 80%. Although the indoor signal propagation causes some considerable frequency selectivity in the channel for the remaining 20% of the low-p_{los} users, the effect of intersymbol interference (ISI) can easily be mitigated by the use of OFDM in conjunction with a well-designed cyclic prefix. In addition, the

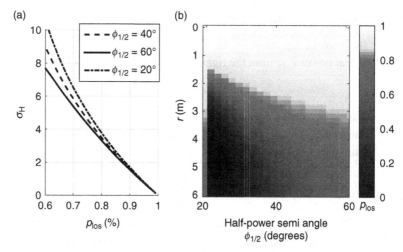

Figure 14.9 (a) Maximum channel gain variance σ_H against proportion of LoS power component p_{los}. (b) p_{los} against transmission horizontal offset r_0 and half-power semiangle $\phi_{1/2}$.

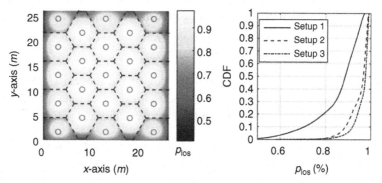

Figure 14.10 (a) Spatial distribution of p_{los} in setup 1: $\phi_{1/2} = 60°$, $R = 3$ m, $L_{room} = 26$ m. (b) p_{los} distribution in the form of a CDF. Setup 2: $\phi_{1/2} = 20°$, $R = 1$ m, $L_{room} = 8.5$ m. Setup 3: $\phi_{1/2} = 20°$, $R = 1$ m, $L_{room} = 14.5$ m.

p_{los} distribution for a smaller $\phi_{1/2}$ of 20° and $R = 1$ m is shown. When the number of cells was the same as in the $\phi_{1/2} = 60°$ case (setup 2), the room size was decreased to 8.5 m × 8.5 m × 3 m. The resulting p_{los} CDF shows that about 99% of the users have p_{los} above 80%. If the room size was increased to 14.5 m × 14.5 m × 3 m with 67 cells (setup 3), p_{los} was further improved. Therefore, it can be concluded that the reflected signal causes a negligible MP effect on the channel for the majority of users as long as the user experiences a dominant LoS signal component. Thus, $|H_{fs}(k)| \approx H_{los}$ for any k. Occasionally, the LoS path may be shadowed or completely blocked. In these cases, a user may need an alternative serving BS or a diffused link.

14.4.3   Light Source Output Power

The relationship between the electrical-signal standard deviation σ_x and the output optical power P_{opt} from the light source can be written as [20]

$$P_{\text{opt}} = \mathbb{E}\left[\hat{x}(t)\right] = \eta_{\text{led}}\left(\sigma_x \mathbb{E}\left[U(x(t))\right] + I_{\text{DC}}\right), \tag{14.13}$$

where $\mathbb{E}$ is the mathematical expectation operator. Generally, a fixed ratio of the DC bias level to the electrical-signal standard deviation is used, defined as $\zeta = I_{\text{DC}}/\sigma_x$. By combining (14.13) with ζ, it can be found that

$$\sigma_x^2 = \frac{P_{\text{opt}}^2}{\eta_{\text{led}}^2\left(\zeta + \mathbb{E}\left[U(x(t))\right]\right)^2}, \tag{14.14}$$

which represents the maximum possible σ_x^2 with a given P_{opt}. To get more electrical-signal power, more optical power is needed. However, the optical power is finite and is typically constrained by the illumination requirement. This requirement is specified by the European indoor lighting regulation [30], which requires an illuminance of 500 lux to be maintained in a typical indoor working environment for writing and reading purposes. To accommodate this requirement, the illuminance in the area below the luminaire (cell center) should be at least 500 lux. According to the analysis in Section 14.4.2, the illuminance at the cell center can be calculated as

$$E_v = \Phi_v \frac{H_{\text{los}}(0)}{A_{\text{pd}}} = \frac{(m+1)\Phi_v}{2\pi h^2}, \tag{14.15}$$

where Φ_v is the output luminous flux of the luminaire. This is the output power measured in photometry units which corresponds to the optical power in radiometry units [31]. The conversion between luminous flux and radiant optical power is given as follows:

$$\frac{\Phi_v}{P_{\text{opt}}} = K_{e/v} = \frac{683 \int V(\lambda)\Phi_e(\lambda)\,d\lambda}{\int \Phi_e(\lambda)\,d\lambda}, \tag{14.16}$$

where $K_{e/v}$ is referred to as the luminous efficacy, $V(\lambda)$ is the luminosity function with respect to the wavelength λ, and $\Phi_e(\lambda)$ is the spectral radiant power density function. The value of $K_{e/v}$ is determined by the characteristics of the specific LED used in the lighting system. Therefore, the value of P_{opt} required can be calculated as

$$P_{\text{opt}} = \frac{\Phi_v}{K_{e/v}} = \frac{2\pi E_v h^2}{(m+1)K_{e/v}}. \tag{14.17}$$

Considering a room height of 3 m and a $\phi_{1/2}$ of 20° to 45°, the required luminous flux for a minimum illuminance of 500 lux is in the range of 1300 to 5300 lumens. This amount of power agrees with the specifications of commercially available LED downlighters and LED panels for lighting in offices and public areas [32, 33]. Note that the output level of LED lamps for residential homes is typically lower than this level (< 1600 lumens). However, this does not necessarily affect the communication performance, as only a fraction of the optical output power, modeled by ζ, is used for the communication link. Moreover, by closer inspection of (14.3) and (14.14) it can be found that the performance also depends on the signal clipping, i.e., the linearity of the LED transfer characteristic.

14.4.4 Signal Clipping

The frameworks developed in [20] were used to determine the impact of signal clipping in [17]. Firstly, clipping affects the transfer relationship between the BS output optical power and electrical signal power, as shown in (14.14). The expectation of the clipped signal $\mathbb{E}[U(x(t))]$ in (14.14) can be calculated as follows [20]:

$$\mathbb{E}[U(x(t))] = (f_{\mathcal{N}}(\lambda_b) - f_{\mathcal{N}}(\lambda_t) + \lambda_t \mathcal{Q}(\lambda_t) + \lambda_b(1 - \mathcal{Q}(\lambda_b))), \qquad (14.18)$$

where $\mathcal{Q}(u) = (1/\sqrt{2\pi}) \int_u^\infty \exp(-v^2/2) \, dv$ is the Q-function and $f_{\mathcal{N}}(u) = (1/\sqrt{2\pi}) \exp(-u^2/2)$ is the probability density function (PDF) of the unit normal distribution. In addition, the transmitted signal is attenuated by a factor of ρ, which can be calculated as $\rho = \mathcal{Q}(\lambda_b) - \mathcal{Q}(\lambda_t)$. Finally, the clipping-noise variance σ_{clip}^2 yields [20]

$$\sigma_{\text{clip}}^2 = \mathcal{Q}(\lambda_b) - \mathcal{Q}(\lambda_t) + f_{\mathcal{N}}(\lambda_b)\lambda_b - f_{\mathcal{N}}(\lambda_t)\lambda_t + (1 - \mathcal{Q}(\lambda_b))\lambda_b^2 + \mathcal{Q}(\lambda_t)\lambda_t^2 - \rho^2$$
$$- (f_{\mathcal{N}}(\lambda_b) - f_{\mathcal{N}}(\lambda_t) + (1 - \mathcal{Q}(\lambda_b))\lambda_b + \mathcal{Q}(\lambda_t)\lambda_t)^2. \qquad (14.19)$$

14.4.5 Noise at Receiver

There are three main noise sources at the receiver, and this is reflected in the determination of the PSD, N_0, as follows:

$$N_0 = N_{0,s} + N_{0,ab} + N_{0,th}, \qquad (14.20)$$

where $N_{0,s}$ corresponds to the shot noise caused by the optical signal received from the BS, $N_{0,ab}$ corresponds to the shot noise caused by the received ambient light (mainly daylight), and $N_{0,th}$ corresponds to the thermal noise in the receiver circuit. The PSD of the shot noise caused by the signal can be calculated as follows [27]:

$$N_{0,s} = 2q P_{\text{opt,Rx}} \eta_{\text{pd}}, \qquad (14.21)$$

where q is the charge of an electron, 1.6×10^{-19} C, and $P_{\text{opt,Rx}}$ denotes the incident optical power on the PD detector in the receiver from the optical BS. Intuitively, the main contribution to the optical power is from the desired BS. To avoid unnecessary computational complexity, the incident optical power from the remaining BSs will be omitted. Thus, $P_{\text{opt,Rx}} = P_{\text{opt}} H_{\text{los}}(r_0)$, where r_0 denotes the horizontal offset between the desired BS and the user considered. The PSD of the shot noise caused by the ambient light can be computed as follows [27]:

$$N_{0,ab} = 2q E_{\text{r,ab}} A_{\text{pd}} \eta_{\text{pd}}, \qquad (14.22)$$

where $E_{\text{r,ab}}$ denotes the incident irradiance. Note that the actual effect of ambient light will be smaller, as only light in the relevant communication spectrum will cause distortions as long as appropriate optical filters are used. Finally, the PSD of thermal noise yields [34]

$$N_{0,\text{th}} = \frac{4\mathcal{K}_B T}{R_L}, \tag{14.23}$$

where $\mathcal{K}_B$ denotes Boltzmann's constant, which is 1.38×10^{-23} J/K; T denotes the absolute temperature, and R_L denotes the load resistance in the receiver circuit.

By inserting (14.7), (14.9), (14.11), (14.14), and (14.17) into the SINR expression (14.6), (14.6) can be modified to

$$\gamma(k) = \left(\left(\frac{\rho^2 \xi^2 \left(r_0^2 + h^2\right)^{-m-3} \mathbb{1}_{\mathcal{D}_2}(r_0)}{\left(\rho^2 + \sigma_{\text{clip}}^2\right) \sum_{i \in \mathcal{I}} \left(r_i^2 + h^2\right)^{-m-3} \mathbb{1}_{\mathcal{D}_2}(r_i) + \mathcal{Z}(k)} \right)^{-1} + \frac{\sigma_{\text{clip}}^2}{\rho^2 \xi^2} \right)^{-1}, \tag{14.24}$$

where $\mathcal{D}_2 = \left[0, h\sqrt{\sec^2(\psi_{\max}) - 1} \right]$ and

$$\mathcal{Z}(k) = \frac{K_{e/v}^2 N_0(r_0) F_s \exp\left(kF_s/KF_{\text{fe}}\right)\left(\zeta + \mathbb{E}\left[U(x(t))\right]\right)^2}{\left(\xi E_v A_{\text{pd}} \eta_{\text{pd}} h^{m+3}\right)^2}. \tag{14.25}$$

Note that N_0 is a function of r_0, as the shot noise varies with the user location owing to variation in the received signal strength.

14.4.6   Multiple Access and Spatial-Reuse Schemes

Since it is recognized that the magnitude response of the channel is affected mainly by the front-end devices, the magnitude response excluding the path loss changes little with the location of the user. In other words, there is little multiuser diversity, and therefore time division multiple access (TDMA) is used. In addition, in systems with a spatial reuse factor Δ larger than 1, the resources are also divided in the frequency domain.

14.5   Design of Key Parameters for LiFi Attocell Networks

The performance of a LiFi attocell network depends on many factors as implied by (14.24). Some of the parameters can be controlled by a predefined system configuration. In this section, two key parameters closely related to the network configuration are examined more closely. One of these parameters is the cell radius R or the BS density Λ of the network. This determines the number of users per cell and the number of cells in a room. The other parameter is the radiation pattern of the light source, which is controlled by the Lambertian emission order, m. This pattern determines the signal strength distribution within each cell and, also, importantly, the co-channel interference (CCI) experienced by the surrounding cells. Appropriate configuration of these two parameters offers a higher probability of achieving the desired system performance. Two configuration objectives are considered, namely the maximization of the desired signal strength and the minimization of CCI.

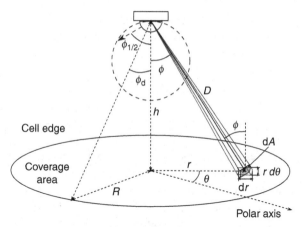

Figure 14.11 Geometry of radiation from optical base station.

14.5.1   Co-channel Interference Minimization

In this section, a mathematical analysis is used to determine the appropriate choices of R and m with the objective of minimizing CCI. If we consider an optical BS with an optical output of P_{opt} serving a cell underneath it, part of the radiated signal power will fall onto the desired coverage area, while the remaining part is incident on other cells, causing CCI. Figure 14.11 shows the geometry of this setup. The BS considered is h meters away from the cell center. In order to minimize CCI, it is preferred to allow more radiated signal power from the BS to stay within the coverage area of that BS, and to allow less signal power to leak into other cells. If the optical power reaching the desired coverage area is $P_{\text{opt,d}}$, the objective becomes maximizing $P_{\text{opt,d}}$. Firstly, $P_{\text{opt,d}}$ needs to be determined. If we assume the BS to be at the origin, the circular coverage area of the cell corresponds to a certain solid angle. Then, the desired signal power $P_{\text{opt,d}}$ for that solid angle can be computed as follows [35]:

$$P_{\text{opt,d}} = \int_{\text{cell}} P_{\text{opt}} \frac{m+1}{2\pi} \cos^m(\phi)\, d\Omega, \tag{14.26}$$

where Ω represents the solid angle of the radiation. The differential of this solid angle $d\Omega$ can be derived as follows, according to the geometry shown in Figure 14.11.

$$d\Omega = \frac{dA}{D^2} = \frac{r\, d\theta\, dr \cos(\phi)}{h^2 \sec^2(\phi)} = d\theta\, d\phi\, \sin(\phi). \tag{14.27}$$

By inserting (14.27) into (14.26), the two-dimensional integration can be decomposed into two one-dimensional integrations as

$$P_{\text{opt,d}} = P_{\text{opt}} \frac{m+1}{2\pi} \int_0^{2\pi} \int_0^{\phi_d} \cos^m(\phi) \sin(\phi)\, d\phi\, d\theta$$

$$= P_{\text{opt}} \left(1 - \left(\frac{h}{\sqrt{h^2 + R^2}} \right)^{m+1} \right), \tag{14.28}$$

where ϕ_d is defined as follows: $\phi_d = \arctan(R/h)$ and as shown in Figure 14.11. The partial derivatives of ϕ_d with respect to R and m are determined as follows:

$$\frac{\partial P_{\text{opt,d}}}{\partial R} = \frac{P_{\text{opt}}R(m+1)h^{m+1}}{\left(h^2+R^2\right)^{(m+3)/2}} > 0, \tag{14.29}$$

$$\frac{\partial P_{\text{opt,d}}}{\partial m} = P_{\text{opt}}\ln\left(\frac{\sqrt{h^2+R^2}}{h}\right)\left(\frac{h}{\sqrt{h^2+R^2}}\right)^{m+1} > 0, \tag{14.30}$$

which implies that $P_{\text{opt,d}}$ is a monotonically increasing function of R and m. This means that the CCI can be reduced by using a larger cell size. Naturally, a larger cell size will increase the distance between each neighboring interfering BS and the desired user. In addition, using a source with a narrower beam width will also decrease the level of CCI, as a smaller half-power semiangle leads to a more collimated beam, which ideally only illuminates the desired coverage area.

14.5.2 Maximization of Strength of Desired Signal

Since a Lambertian radiation pattern is used to model the light emission from the source, the further the user is away from the cell center, the weaker the received desired signal. Note that there are no fading effects as in RF communications. Consequently, a user at the cell edge receives the weakest signal from the BS. In other words, as long as the signal strength of a cell edge user is high enough, all of the users in the cell coverage area should have sufficient signal power. Therefore, the objective can be converted to maximizing the signal power received by a cell edge user that is R meters away from the cell center. According to the analysis in Section 14.4.2, $P_{\text{opt,e}}$ can be determined as follows:

$$P_{\text{opt,e}} = P_{\text{opt}}H_{\text{los}}(R) = \frac{P_{\text{opt}}A_{\text{pd}}(m+1)h^{m+1}}{2\pi\left(R^2+h^2\right)^{(m+3)/2}}. \tag{14.31}$$

Similarly, the partial derivatives of $P_{\text{opt,e}}$ with respect to R and m can be derived and yield

$$\frac{\partial P_{\text{opt,e}}}{\partial R} = -\frac{P_{\text{opt}}A_{\text{pd}}(m+1)(m+3)Rh^{m+1}}{2\pi\left(R^2+h^2\right)^{(m+5)/2}} < 0, \tag{14.32}$$

$$\frac{\partial P_{\text{opt,e}}}{\partial m} = \frac{P_{\text{opt}}A_{\text{pd}}h^{m+3}\left(1+\ln\left(h^{m+1}/\left(R^2+h^2\right)^{(m+1)/2}\right)\right)}{2\pi h^2\left(R^2+h^2\right)^{(m+3)/2}}, \tag{14.33}$$

which implies that $P_{\text{opt,e}}$ is a monotonically decreasing function of R. Therefore, for a source with a specified radiation pattern, a smaller cell offers higher received signal power for cell edge users. This is because a smaller cell size reduces the distance from the cell edge user to the cell center. On the other hand, $P_{\text{opt,e}}$ is a concave function of m, which means there is an optimal value for m that maximizes the cell edge user signal strength. By letting $\partial P_{\text{opt,e}}/\partial m = 0$, the optimal radiation pattern can be found as

$$\tilde{m} = 1 \Big/ \ln\left(\sqrt{R^2+h^2}/h\right) - 1. \tag{14.34}$$

In the case of a source with a narrower beam width ($m > \tilde{m}$), the beam is overconcentrated, which causes a significant signal strength variation between cell center users and cell edge users, and the signal strength for the cell edge users would eventually become too weak for reliable communication. In the case of a source with a wider beam width ($m < \tilde{m}$), the beam is overdiffused, which causes too much power leakage to other cells and the overall signal strength in the desired cell will become insufficient.

14.5.3   Parameter Configurations

From the analysis in Sections 14.5.1 and 14.5.2, it can be seen that there is a mutual dependence which needs to be considered (and exploited) when determining the parameters R and m. Apart from the requirement of reliable high-speed wireless communication, there are many possible other constraints which determine the cell size. For example, if the cell size is too large, the illumination performance may be compromised in an undesired fashion. An extremely small cell size leads to too many BSs (lights) being required in the room, which increases the installation cost and increases the handover overhead. It is therefore more convenient to fix the cell radius R and find the respective m to achieve the desired illumination *and* communication objectives.

According to (14.30) and (14.34), if m is smaller than $\tilde{m}$, both the CCI increases and the cell edge signal strength decreases. If m is equal to or greater than $\tilde{m}$, there is a trade-off between the two objectives. Therefore, $\tilde{m}$ can be considered as a lower bound for m. In a noise-limited system, a value of m closer to $\tilde{m}$ is preferred. In the case of a CCI-limited system, (14.30) shows that m should be maximized to minimize CCI. However, an upper bound must be set to allow the cell edge user signal strength to be high enough to achieve the minimum acceptable SNR. In order to find this upper bound, we define a simple metric: the ratio between the SNR of a cell center user ($r = 0$) and that of a cell edge user ($r = R$), denoted by σ_P. From the analysis in Section 14.4, it can be found that σ_P is proportional to the square of the ratio of the optical power received by the cell center user to that received by the cell edge user:

$$\sigma_P = \left(\frac{P_{opt} H_{los}(0)}{P_{opt} H_{los}(R)} \right)^2 = \frac{h^{-2(m+3)}}{(R^2 + h^2)^{-m-3}}. \tag{14.35}$$

For a fixed σ_P, the required lower bound on m can be found:

$$\hat{m} = \frac{\ln \sigma_P}{\ln(1 + R^2/h^2)} - 3. \tag{14.36}$$

According to [24, 23], the achievable cell center SNR is around 30 dB. For uncoded 4 QAM, the minimum required SNR is approximately 10 dB. Therefore, $\sigma_P = 20$ dB was chosen in the present study. The results for $\tilde{m}$, $\hat{m}$, and the corresponding $\phi_{1/2}$ against R based on (14.34) and (14.36) are plotted in Figure 14.12. It can be seen that the areas between the two curves define the appropriate configuration region, which includes the preferred choices of $\phi_{1/2}$. In the case where CCI is the main limiting factor, m can be

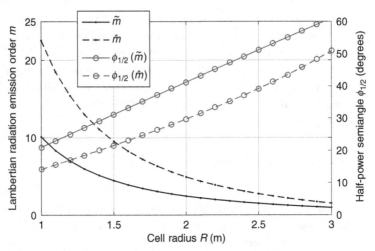

Figure 14.12 Half-power semiangle $\phi_{1/2}$ and the corresponding Lambertian emission order m as a function of the cell radius R.

set to a value that is close or equal to $\hat{m}$, which is determined using (14.36). In the case where receiver noise is the limiting factor, m can be configured to a value that is close or equal to $\tilde{m}$, which is determined using (14.34).

14.6   Signal-to-Interference-Plus-Noise Ratio in LiFi Attocell Networks

A user in a LiFi attocell network will experience variable levels of service quality based on its location. The SINR statistics are an important factor, as they directly determine the service quality of a LiFi attocell network, such as the achievable data rate and outage probability. The SINR varies with a large number of parameters, as discussed in Sections 14.4 and 14.5. In addition, it also depends on the cell deployment (i.e., lighting infrastructure). In order to provide a comprehensive characterization of the SINR statistics, two extreme cases are examined here. One case considers a network of hexagonal (HEX) cells. Some existing indoor lighting designs indeed follow HEX topologies. The second case considers a completely random placement of BSs. Specifically, the two-dimensional spatial distribution of BSs follows a homogeneous Possion point process (PPP). The irregular placement of luminaires is motivated mainly by the following considerations. Firstly, the placement of a luminaire may be limited by the wiring structure in the room. Secondly, in some cases, nonuniform illumination is required. Also, even for a uniform cell deployment, users may be absent from some cells. In that case, the downlink transmission can be switched off, which effectively results in a nonuniform cell deployment. Owing to these practical issues, the BS deployments corresponding to these two extreme cases (HEX and random Poisson point process (PPP)) are likely, to be very rare in practice. However, most practical scenarios will fall between these two extremes [36]. In different cell deployment scenarios, the shapes of the cells vary. In order to guarantee a fair comparison, the average cell size is scaled here to be the same as that of an equivalent circular cell with a radius of R.

14.6.1   System Model Assumptions

In order to simplify the analysis and to make the analysis tractable, the SINR derived in (14.24) has to be modified. Firstly, it is assumed that the nonlinear characteristic of the relationship between the input current and the output optical power is minimized by using predistortion techniques [37, 38]. Therefore, a linear dynamic range from 0 to $2I_{DC}$ is considered. This leads to a clipping level of $\lambda_t = -\lambda_b = \zeta$, and $\mathbb{E}[U(x(t))] = 0$. In addition, the function $\mathbf{1}_{\mathcal{D}_2}(x)$ makes (14.24) a piecewise function, which causes extra mathematical complexity in the analysis. Therefore, the worst case with a full FoV of $\psi_{\max} = 90°$ is assumed, thereby making $\mathbf{1}_{\mathcal{D}_2}(x)$ always equal to 1 in the region of interest. Then, the simplified SINR expression can be rewritten as follows:

$$\gamma(k) = \left(\frac{\left(\rho^2 + \sigma_{\text{clip}}^2 \right) \mathcal{I} + \mathcal{Z}(k)}{\rho^2 \mathcal{X}} + \frac{\sigma_{\text{clip}}^2}{\xi^2 \rho^2} \right)^{-1}, \tag{14.37}$$

where $\mathcal{X} = \xi^2 \left(r_0^2 + h^2 \right)^{-m-3}$ and $\mathcal{I} = \sum_{i \in \mathcal{I}} \left(r_i^2 + h^2 \right)^{-m-3}$.

14.6.2   Hexagonal Cell Deployment

Instead of considering a specified network in a room, an infinitely extended HEX network is considered in this analysis. There are two reasons for making this assumption. Firstly, the main concern in this study is the impact of CCI from neighboring BSs. The number of neighboring BSs causing CCI is maximized in an infinite network. Second, this removes the cell-boundary effect. However, it is unnecessary to consider neighboring BSs that are too far away from the cell of interest as they cause negligible CCI. Instead, a two-layer HEX cell deployment is considered in order to approximate the infinite network, as shown in Figure 14.13, and the user performance in the central cell is analyzed. In this investigation, all networks were assumed to be fully loaded. In addition, the cases with $\Delta = 1$ and $\Delta = 3$ are considered, since these cases are more likely to be used in practice. In our model, a polar coordinate system is used to represent the location of the users and BSs. Each two-dimensional location has a specified distance from the origin and a polar angle. A user at z is r_0 away from the origin and has a polar angle of θ, as shown in Figure 14.13. Similarly, the ith BS is located at (R_i, Θ_i). In order to ensure that the area of a cell in the HEX configuration is equal to that of the equivalent circular cell, the HEX cell radius satisfies $\tilde{R} \approx 1.1R$, as shown in Figure 14.13. Then the horizontal offset between the ith BS and the user at z is

$$r_i(z) = \sqrt{r_0^2 + R_i^2 - 2R_i r_0 \cos(\theta - \Theta_i)}. \tag{14.38}$$

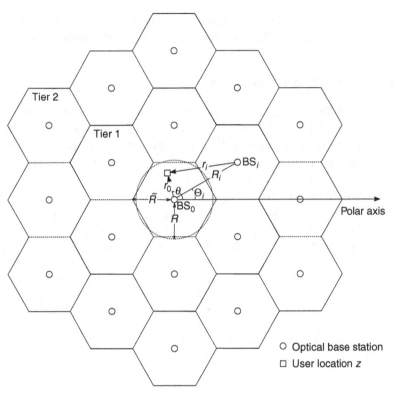

Figure 14.13 A two-layer HEX network model with polar coordinates.

The user at $z = (r_0, \theta)$ in the central cell is served by the 0th BS. The remaining BSs using the same transmission resource ($i \in \mathcal{I}$) cause CCI to the desired user at z. Since the coordinates of all BSs are known, by inserting (14.38) into (14.37), the SINR $\gamma(k)$ can be calculated as a function of the user location z. Since the users are assumed to be uniformly distributed in the cell, the PDF of r_0 and θ should follow $f_{r_0}(r_0) = 2r_0/R^2$ and $f_\theta(\theta) = 1/2\pi$. The objective is defined as $\mathbb{P}[\gamma(k) < T]$, which establishes the probability that the downlink SINR is less than a threshold T. In conjunction with (14.37) and letting the probability be conditioned on r_0, the following probability can be obtained:

$$\mathbb{P}[\gamma(k) < T|r_0] = \mathbb{P}\left[\mathcal{I} > \frac{\rho^2 \mathcal{X}\left(1/T - \sigma_{\text{clip}}^2/\rho^2\xi^2\right) - \mathcal{Z}(k)}{\rho^2 + \sigma_{\text{clip}}^2}\Bigg| r_0\right]. \tag{14.39}$$

Combining (14.38) and (14.37) results in $\mathcal{I}$ being an extremely complex function of θ that is unwieldy for carrying out a PDF transformation. Therefore, this relationship between $\mathcal{I}$ and θ need to be simplified in order to make the calculation tractable. Figure 14.14 shows the CCI term $\mathcal{I}$ for a HEX network against θ with a given r_0. It can

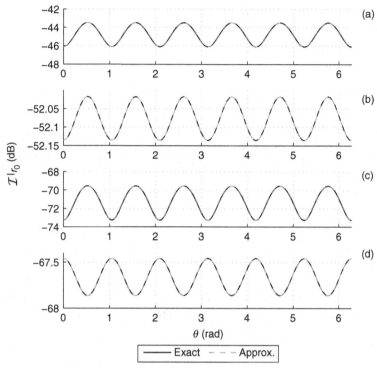

Figure 14.14 CCI approximation. In configuration (a), $R = 3$ m, $\Delta = 1$, and $r = R$ m; m is calculated using (14.36). Relative to configuration (a), configuration (b) changes r to $R/2$, configuration (c) changes R to 2 m, and configuration (d) changes Δ to 3.

be observed that with increasing θ, $\mathcal{I}(\theta|r_0)$ oscillates between two extreme values with a period of 60°. This is because of the centrally symmetric deployment of the interfering BSs. Therefore, an approach similar to the "flower" model introduced in [39] is used to simplify the relationship between $\mathcal{I}$ and θ. This approach uses a cosine function to approximate the oscillation of the function $\mathcal{I}(\theta|r_0)$. Firstly, $\mathcal{I}_{0°}(r_0)$ and $\mathcal{I}_{30°}(r_0)$ are calculated. These two values constitute the bounds of oscillation of the function $\mathcal{I}(\theta|r_0)$. Expressions for $\mathcal{I}_{0°}(r_0)$ and $\mathcal{I}_{30°}(r_0)$ can be derived in closed form, as shown in [17]. Then, the approximate CCI term can be found as follows:

$$\hat{\mathcal{I}} = \frac{\mathcal{I}_{30°}(r_0) + \mathcal{I}_{0°}(r_0)}{2} + \frac{|\mathcal{I}_{30°}(r_0) - \mathcal{I}_{0°}(r_0)|}{2}\cos(6\theta). \tag{14.40}$$

Figure 14.14 compares the exact conditional CCI term $\mathcal{I}(\theta|r_0)$ with the approximate term $\hat{\mathcal{I}}(\theta|r_0)$ for different system configurations. In all cases, the approximate model $\hat{\mathcal{I}}(\theta|r_0)$ matches well with the exact model $\mathcal{I}(\theta|r_0)$. The difference between the two curves is negligible, as shown in each plot in Figure 14.14. Thus, it is reasonable to replace $\mathcal{I}$ with $\hat{\mathcal{I}}$. By replacing $\mathcal{I}$ in (14.39) with (14.40), the conditional probability

$\mathbb{P}[\gamma(k) < T | r_0]$ can be written as follows:

$$\mathbb{P}[\gamma(k) < T | r_0]$$

$$= \mathbb{P}\left[\cos(6\theta) > \frac{2\rho^2 \mathcal{X}\left(1/T - \sigma_{\mathrm{clip}}^2 / \rho^2 \xi^2\right) - 2\mathcal{Z}(k)}{\left(\rho^2 + \sigma_{\mathrm{clip}}^2\right)|\mathcal{I}_{30^\circ} - \mathcal{I}_{0^\circ}|} - \frac{\mathcal{I}_{30^\circ} + \mathcal{I}_{0^\circ}}{|\mathcal{I}_{30^\circ} - \mathcal{I}_{0^\circ}|} \Big| r_0 \right]$$

$$= \frac{1}{2} - \frac{1}{\pi} \arcsin^\dagger \left(\frac{2\rho^2 \mathcal{X}\left(1/T - \sigma_{\mathrm{clip}}^2 / \rho^2 \xi^2\right) - 2\mathcal{Z}(k)}{\left(\rho^2 + \sigma_{\mathrm{clip}}^2\right)|\mathcal{I}_{30^\circ} - \mathcal{I}_{0^\circ}|} - \frac{\mathcal{I}_{30^\circ} + \mathcal{I}_{0^\circ}}{|\mathcal{I}_{30^\circ} - \mathcal{I}_{0^\circ}|} \right), \quad (14.41)$$

where

$$\arcsin^\dagger(x) = \begin{cases} 1 & : x > 1 \\ \arcsin(x) & : |x| \le 1 \\ -1 & : x < -1 \end{cases}. \quad (14.42)$$

The final CDF of the SINR can be determined by averaging (14.41) over r_0 as described in (14.43), which can be done efficiently by using numerical methods. In this integration, the range of r_0 is from 0 to R, which corresponds to an integration over the equivalent circular cell:

$$\mathbb{P}[\gamma(k) < T] = \int_0^R f_{r_0}(r_0) \mathbb{P}[\gamma(k) < T | r_0] dr_0$$

$$= \int_0^R \frac{r_0}{R^2} - \frac{2r_0}{\pi R^2} \arcsin^\dagger \left(\frac{2\rho^2 \mathcal{X}\left(1/T - \sigma_{\mathrm{clip}}^2 / \rho^2 \xi^2\right) - 2\mathcal{Z}(k)}{\left(\rho^2 + \sigma_{\mathrm{clip}}^2\right)|\mathcal{I}_{30^\circ} - \mathcal{I}_{0^\circ}|} - \frac{\mathcal{I}_{30^\circ} + \mathcal{I}_{0^\circ}}{|\mathcal{I}_{30^\circ} - \mathcal{I}_{0^\circ}|} \right) dr_0.$$

$$(14.43)$$

14.6.3   PPP Cell Deployment

Similar to the HEX network, an infinitely extended PPP network is considered, where the origin of the coordinates is placed at a random user [36]. The horizontal positioning of the nearby optical BSs follows a two-dimensional homogeneous PPP with a density of Λ, as shown in Figure 14.15. A method similar to that used before is used to retrieve the SINR statistics by calculating $\mathbb{P}[\gamma(k) < T | r_0]$ using (14.39). The exact distribution of $\mathcal{I}$ is difficult to obtain in closed form. However, the method presented in [40] can be used to determine the characteristic function (CF) of $\mathcal{I}$ conditioned on r_0. The details are as follows. Since there is no dependency between BSs, the only significant variable in this model is the Euclidean distance between a BS and the user D_i. According to the geometrical relationships shown in Figure 14.6, the CCI term $\mathcal{I}$ in (14.37) can be rewritten as $\mathcal{I} = \sum g(D_i)$, where the function $g(x) = x^{-2(m+3)}$. It is assumed that the furthest BS is a meters away from the user and the interfering BS is not closer than the desired BS, which is r_0 meters away from the user. As shown in Figure 14.15, r_i is within the range $[r_0, a]$. Since the interfering BSs are uniformly distributed, the PDF of

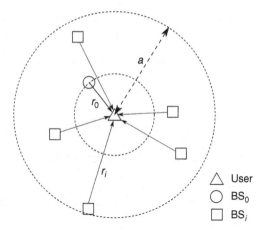

Figure 14.15 PPP network geometry.

r_i yields

$$f_{r_i}(r_i) = \frac{2r_i}{a^2 - r_0^2}, \quad r_0 \le r_i \le a. \tag{14.44}$$

Then the PDF of D_i can be calculated using standard PDF transformations from (14.44), resulting in

$$f_{D_i}(D_i) = \frac{2D_i}{a^2 - r_0^2}, \quad \sqrt{r_0^2 + h^2} \le D_i \le \sqrt{a^2 + h^2}. \tag{14.45}$$

The conditional CF of $\mathcal{I}$ is defined as $\varphi_{\mathcal{I}_a}(\omega|r_0) = \mathbb{E}\left[e^{j\omega \mathcal{I}_a}\right]$. Since the number of interfering BSs I is a nonnegative integer random variable, $\varphi_{\mathcal{I}_a}(\omega)$ can be extended as follows:

$$\varphi_{\mathcal{I}_a}(\omega|r_0) = \mathbb{E}_I\left[\mathbb{E}\left[e^{j\omega \mathcal{I}_a}|I = n\right]\right]. \tag{14.46}$$

Since all of the D_i in $\mathcal{I}$ are independent of each other, $\varphi_{\mathcal{I}_a}(\omega)$ conditioned on I can be factorized as follows:

$$\mathbb{E}\left[e^{j\omega \mathcal{I}_a}|I = n\right] = \prod_{i=1}^{n} \mathbb{E}\left[e^{j\omega g(D_i)}\right]$$

$$= \left(\int_{\sqrt{r_0^2 + h^2}}^{\sqrt{a^2 + h^2}} \frac{2De^{j\omega g(D)}}{a^2 - r_0^2} dD\right)^n. \tag{14.47}$$

Because I follows a Poisson distribution with a mean of Λ/Δ, the corresponding probability mass function of I can be written as follows:

$$\mathbb{P}[I = n] = \frac{e^{-(\Lambda\pi/\Delta)\left(a^2 - r_0^2\right)}\left((\Lambda\pi/\Delta)\left(a^2 - r_0^2\right)\right)^n}{n!}. \tag{14.48}$$

Next, (14.46) can be extended as follows:

$$\varphi_{\mathcal{I}_a}(\omega|r_0) = \sum_{n=0}^{\infty} \mathbb{P}[I=n]\,\mathbb{E}\left[e^{j\omega\mathcal{I}_a}\,|\,I=n\right]$$

$$= e^{-(\Lambda\pi/\Delta)\left(a^2-r_0^2\right)} \sum_{n=0}^{\infty} \frac{1}{n!}\left(\frac{\Lambda\pi}{\Delta}\int_{\sqrt{r_0^2+h^2}}^{\sqrt{a^2+h^2}} 2De^{j\omega g(D)}dD\right)^n$$

$$\stackrel{(a)}{=} e^{-(\Lambda\pi/\Delta)\left(a^2-r_0^2-\int_{\sqrt{r_0^2+h^2}}^{\sqrt{a^2+h^2}} 2De^{j\omega g(D)}dD\right)}, \tag{14.49}$$

where (a) uses the Taylor series for e^x. By taking the limit $a \to \infty$, the CF can be calculated as follows:

$$\varphi_{\mathcal{I}}(\omega|r_0) = \exp\left(\sum_{n=1}^{\infty} \frac{(j\omega)^n}{n!} \cdot \frac{\Lambda\pi\left(r_0^2+h^2\right)^{1-n(m+3)}}{\Delta(n(m+3)-1)}\right). \tag{14.50}$$

The proof of (14.50) can be found in [17]. Theoretically, (14.50) can be converted to the corresponding PDF. However, this operation is intractable. Therefore, an alternative approximation method is used to obtain the PDF of CCI. The cumulant generating function can be written as follows:

$$\ln\left(\varphi_{\mathcal{I}}(\omega)\right) = \sum_{n=1}^{\infty} \kappa_n(\mathcal{I})\frac{(j\omega)^n}{n!}. \tag{14.51}$$

By comparing (14.51) and (14.50), the nth cumulant of $\mathcal{I}$ conditioned on r_0 can be found:

$$\kappa_n^{\mathcal{I}} = \frac{\Lambda\pi(r_0^2+h^2)^{1-n(m+3)}}{\Delta(n(m+3)-1)}. \tag{14.52}$$

With all cumulants known, the corresponding n^{th} raw moment can be calculated recursively from the following set of equations:

$$\mu_n = \begin{cases} 1 & : n=0, \\ \kappa_1 & : n=1, \\ \kappa_n + \sum_{l=1}^{n-1}\binom{n-1}{l-1}\kappa_l\mu_{n-l} & : n \geq 2. \end{cases} \tag{14.53}$$

With all moments of the CCI distribution known, an expansion of the PDF as a sum of Gamma densities as proposed in [41] can be used. This expansion is based on the Gram–Charlier series and Laguerre polynomials. The Gamma density used in this expansion is $f_V(v) = v^{\alpha-1}e^{-v}/\Gamma(\alpha)$ for a random variable V. The expansion of the PDF is given as [41]

$$f_V(v) = \frac{v^{\alpha-1}e^{-v}}{\Gamma(\alpha)}\sum_{n=0}^{\infty} A_n\mathcal{L}_n^{\alpha}(v), \tag{14.54}$$

where the Laguerre polynomial $\mathcal{L}_n^\alpha(v)$ can be calculated as follows:

$$\mathcal{L}_n^\alpha(v) = (-1)^n v^{1-\alpha} e^v \frac{d^n}{dv^n}\left(v^{n+\alpha-1}e^{-v}\right)$$

$$= \sum_{l=0}^{n} \binom{n}{l}(-1)^{n-l}v^l S_l^n, \tag{14.55}$$

where l is a nonnegative integer and

$$S_l^n = \begin{cases} 1 & : l > n-1, \\ \prod_{\iota=l}^{n-1}(\alpha+\iota) & : l \le n-1. \end{cases} \tag{14.56}$$

The coefficients $\mathcal{A}_n$ in (14.54) can be computed using the following expression:

$$\mathcal{A}_n = \frac{\Gamma(\alpha)}{n!\Gamma(\alpha+n)}\int_0^\infty f_V(v)\mathcal{L}_n^\alpha(v)\,dv$$

$$= \frac{(-1)^n\Gamma(\alpha)}{n!\Gamma(\alpha+n)}\sum_{l=0}^{n}\binom{n}{l}(-1)^l S_l^n \int_0^\infty v^l f_V(v)\,dv$$

$$= \frac{(-1)^n\Gamma(\alpha)}{n!\Gamma(\alpha+n)}\sum_{l=0}^{n}\binom{n}{l}(-1)^l S_l^n \mu_l^V. \tag{14.57}$$

The expansion (14.54) requires the random variable V to have its mean and variance equal to α:

$$\mathbb{E}[V] = \sigma_V^2 = \alpha. \tag{14.58}$$

Therefore, the CCI random variable $\mathcal{I}$ has to be scaled to satisfy the condition in (14.58). So $V = \beta\mathcal{I}$ is defined, where β is the scaling factor. Then the cumulants and moments of V should follow

$$\kappa_n^V = \beta^n \kappa_n^\mathcal{I}, \tag{14.59}$$

$$\mu_n^V = \beta^n \mu_n^\mathcal{I}. \tag{14.60}$$

Note that $\kappa_1^\mathcal{I}$ and $\kappa_2^\mathcal{I}$ are equal to the mean and variance, respectively, of $\mathcal{I}$. Then the mean and variance of V should be $\beta\kappa_1^\mathcal{I}$ and $\beta^2\kappa_2^\mathcal{I}$, respectively. The values of α and β can be determined in conjunction with (14.52) and (14.58) as

$$\beta = \frac{(2m+5)\left(r_0^2+h^2\right)^{m+3}}{m+2}, \tag{14.61}$$

$$\alpha = \beta\kappa_1^\mathcal{I} = \beta^2\kappa_2^\mathcal{I} = \frac{A\pi(2m+5)\left(r_0^2+h^2\right)}{A(m+2)^2}. \tag{14.62}$$

By substituting $\beta\mathcal{I}$ for V in (14.54) and rearrangement, the conditional PDF $f_\mathcal{I}(\mathcal{I}|r_0)$ can be determined as follows:

$$f_\mathcal{I}(\mathcal{I}|r_0) = \sum_{n=0}^{\infty}\left(\sum_{l_1=0}^{n}\frac{\beta^\alpha C_{l_1}^n \mu_{l_1}^\mathcal{I}}{n!\Gamma(\alpha+n)}\right)\left(\sum_{l_2=0}^{n}\frac{C_{l_2}^n \mathcal{I}^{l_2+\alpha-1}}{e^{\beta\mathcal{I}}}\right), \tag{14.63}$$

where

$$C_l^n = \binom{n}{l} (-1)^{n-l} \beta^l S_l^n. \tag{14.64}$$

Next, the probability

$$\mathbb{P}\left[\mathcal{I} > \tilde{\mathcal{I}}|r_0\right] = \int_{\tilde{\mathcal{I}}}^{\infty} f_{\mathcal{I}}(\mathcal{I}|r_0) \, d\mathcal{I}$$

$$= \sum_{n=0}^{\infty} \left(\sum_{l_1=0}^{n} \frac{\beta^{\alpha} C_{l_1}^n \mu_{l_1}^{\mathcal{I}}}{n! \Gamma(\alpha+n)} \right) \left(\sum_{l_2=0}^{n} \frac{C_{l_2}^n}{\beta^{l_2+\alpha}} \Gamma\left(l_2+\alpha, \beta\tilde{\mathcal{I}}\right) \right), \tag{14.65}$$

where $\Gamma(\nu, \epsilon) = \int_{\epsilon}^{\infty} e^{-x} x^{\nu-1} \, dx$ is the upper incomplete Gamma function, and

$$\tilde{\mathcal{I}} = \frac{\rho^2 \mathcal{X}\left(1/T - \sigma_{\text{clip}}^2/\rho^2\xi^2\right) - \mathcal{Z}(k)}{\rho^2 + \sigma_{\text{clip}}^2}. \tag{14.66}$$

Because r_0 equals the distance between the user (the origin) and the serving BS (the closest node to the origin), the PDF of r_0 with a node density of Λ should be $f_{r_0}(r_0, \Lambda) = 2\pi \Lambda r_0 e^{-\Lambda \pi r_0^2}$ in a PPP [42]. The final CDF of the SINR can be calculated by combining (14.65) with (14.39) and averaging $\mathbb{P}[\gamma(k) < T|r_0]$ over r_0 as follows:

$$\mathbb{P}[\gamma(k) < T] = \int_0^{\infty} f_{r_0}(r_0, \Lambda) \mathbb{P}[\gamma(k) < T|r_0] \, dr_0$$

$$= \int_0^{\infty} \frac{2\pi \Lambda r_0}{e^{\pi \Lambda r_0^2}} \sum_{n=0}^{\infty} \left(\sum_{l_1=0}^{n} \frac{\beta^{\alpha} C_{l_1}^n \mu_{l_1}^{\mathcal{I}}}{n! \Gamma(\alpha+n)} \right) \left(\sum_{l_2=0}^{n} \frac{C_{l_2}^n \Gamma\left(l_2+\alpha, \beta\tilde{\mathcal{I}}\right)}{\beta^{l_2+\alpha}} \right) dr_0. \tag{14.67}$$

Note that there is a summation with an infinite upper bound in (14.67), which makes the calculation intractable. Therefore, the infinite upper bound of the summation is replaced by a finite integer number N. With increasing N, (14.67) quickly converges to the case of $N = \infty$. When calculating the results, $N = 10$ was found to be sufficient to provide accurate analytical results. With this approach, (14.67) can be solved using numerical methods.

14.6.4   SINR Statistics Results and Discussion

Figure 14.16 depicts the CDF of the achieved SINR at DC for different system setups with HEX and PPP cell deployments. The SINR at DC is used as an example here. The SINR at other frequencies decreases with increasing in frequency owing to the low-pass effect of the front-end devices. The values shown in Table 14.1 were used for system parameters not specified individually for each setup. The selection of F_s and F_{fe} was in accordance with the setup in [23]. The selection of σ_P is justified in Section 14.5.3. It can be seen that the numerical results obtained from (14.43) and (14.67) agree with the corresponding Monte Carlo simulations in the region of interest.

In setup 1, $R = 2.5$ m, $\phi_{1/2} = 40°$, and $\Lambda = 1$. The results for both the HEX and the PPP network are shown. It can be observed that with the same system configuration, a

Table 14.1 Simulation parameters.

Parameter	Symbol	Value
Vertical separation	h	2.25 m
Receiver field of view	$\psi_{\max}$	90°
Sampling frequency	F_s	360 MHz
Front-end device bandwidth factor	F_{fe}	31.7 MHz
DC bias level	ζ	3.2
PD responsivity	η_{pd}	0.4 A/W
PD physical area	A_{pd}	1 cm^2
Number of subcarriers	K	512
Power decrease factor	σ_P	20 dB
Cell center illuminance from BS	E_v	500 lux
Illuminance from ambient light	$E_{v,ab}$	100 lux
Absolute temperature	T	300 K
Receiver load resistance	R_L	500 Ω

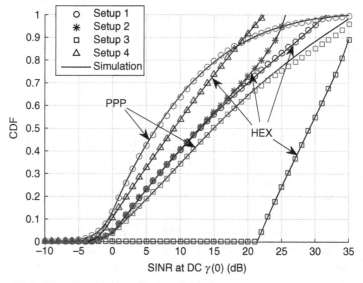

Figure 14.16 CDF of the SINR at DC. Setup 1: $R = 2.5$ m, $\phi_{1/2} = 40°$, $\Delta = 1$, 100% output. Setup 2: same as setup 1 except for $\zeta = 2.5$. Setup 3: $R = 3$ m, $\phi_{1/2} = 50°$, $\Delta = 3$, 100% output. Setup 4: same as setup 1 except for 15% output and 1000 lux ambient light illuminance. Parameters not specified are listed in Table 14.1.

PPP network performs worse than a HEX network. In addition, the ambient light level was assumed to be 100 lux. Therefore, the BSs operate at maximum power to provide sufficient illumination. The highest SINR, of above 30 dB, shows that the noise at the receiver has little effect on the system performance. In setup 2, the DC bias level was modified to 2.5. This results in a higher signal-clipping level. Consequently, the highest SINR in this system is limited by the clipping noise. In setup 3, $R = 3$ m, $\phi_{1/2} = 50°$, and $\Delta = 3$. The other parameters were the same as in setup 1. The high reuse factor leads to a lower level of CCI and the overall SINR level is improved significantly compared with that of setup 1. Therefore, the corresponding SINR is improved compared with the case of setup 1 for both HEX and PPP networks. Setup 4 considered a special case where there was sufficient illumination from ambient light, with an illuminance of 1000 lux. Thus, the BS operates in a dimmed mode with only 15% of its normal output. Owing to the reduced signal power and increased noise level, the overall SINR is decreased (by -3 dB to 22 dB). This demonstrates that the system will work in strong ambient-light conditions, and even in dimmed mode. Furthermore, energy-efficient modulation techniques such as enhanced unipolar OFDM (eU-OFDM) [43, 44] may be used to improve performance further when the lights are dimmed.

14.7 Cell Data Rate and Outage Probability

In this section, the average cell data rate and outage probability are analyzed. Since the information about the per-subcarrier SINR and its statistics is known, different modulation and coding schemes can be assigned to each subcarrier adaptively according to the value of $\gamma(k)$. The average achievable data rate in a LiFi attocell can be computed using:

$$
s = \frac{1}{\Delta} \sum_{k=1}^{K/2-1} \sum_{n=1}^{N} W_{\text{sc}} \varepsilon_n \mathbb{P}[T_n < \gamma(k) < T_{n+1}]
$$

$$
= \frac{F_s}{\Delta K} \sum_{k=1}^{\frac{K}{2}-1} \sum_{n=1}^{N} \varepsilon_n \left(\mathbb{P}[\gamma(k) < T_{n+1}] - \mathbb{P}[\gamma(k) < T_n] \right), \tag{14.68}
$$

where W_{sc} is the per-subcarrier bandwidth, ε_n is the spectral efficiency (bits/symbol) of the nth adaptive modulation and coding (AMC) level, and T_n is the corresponding minimum SINR required to achieve ε_n. In this study, two AMC schemes are considered, which are listed in Table 14.2. AMC scheme 1 is the uncoded QAM modulation [45], achieving a maximum bit error ratio (BER) target of 1×10^{-3}. This scheme is reliable and simple to implement, and has been used in several experimental studies [22–24]. However, this scheme achieves a relatively low spectral efficiency, and the minimum required SINR is as high as 9.8 dB. AMC scheme 2 is used in Long Term Evolution (LTE) systems [46] and is more spectrally efficient, and the lowest acceptable SINR is -6 dB. However, this scheme is more complex to implement.

The outage probability is defined as the probability that the received SINR on all subcarriers are below the minimum SINR required for the AMC scheme. Since $\gamma(1)$ is the highest of the SINRs for all the subcarriers, the outage probability can be defined as follows:

$$
\mathcal{P}_{\text{out}} = \mathbb{P}[\gamma(1) < T_1]. \tag{14.69}
$$

Moving on, the accuracy of the cell data rates was evaluated and the cell data rate/outage probability performance of a LiFi attocell was analyzed. The results included results for systems with a HEX/PPP network model and systems with reuse

Table 14.2 Adaptive modulation and coding.

	AMC scheme 1		AMC scheme 2	
n	T_n (dB)	ε_n (bits/symbol)	T_n (dB)	ε_n (bits/symbol)
0	–	0	–	0
1	9.8	2	−6	0.1523
2	13.4	3	−5	0.2344
3	16.5	4	−3	0.3770
4	19.6	5	−1	0.6016
5	22.5	6	1	0.8770
6	25.5	7	3	1.1758
7	28.4	8	5	1.4766
8	–	–	8	1.9141
9	–	–	9	2.4063
10	–	–	11	2.7305
11	–	–	12	3.3223
12	–	–	14	3.9023
13	–	–	16	4.5234
14	–	–	18	5.1151
15	–	–	20	5.5547

factors of $\Delta = 1$ and $\Delta = 3$. The cell radius R and the modulation bandwidth F_s were considered as the variables of interest. Figure 14.17 shows the cell data rate versus the cell radius R. As shown in Section 14.6.4, a network operating where the BSs operate at full power will not be limited by noise. Therefore, according to the analysis in Section 14.5.3, the emission order m was configured based on (14.36) to achieve better performance. The other system parameters were the same as those listed in Table 14.1 if not otherwise specified. For all of the systems, Monte Carlo simulation results show close agreement with analytical results obtained from (14.68). As expected, the HEX network performs better than the PPP network for the same system parameters. The cell data rate generally decreases with an increase in R. This is because a system with smaller cells has a higher value of m according to (14.36), which results in less CCI to nearby BSs. Firstly, systems using AMC scheme 1 are considered. With the same cell deployment, the system with $\Delta = 1$ always achieves a higher data rate than the system with $\Delta = 3$. In the case of the HEX cell deployment, the $\Delta = 1$ system achieves a 40% to 100% higher data rate than the $\Delta = 3$ system. However, the $\Delta = 1$ system always has a much higher outage probability than the $\Delta = 3$ system, as shown in Figure 14.18. For example, in the case of the HEX cell deployment, the $\Delta = 1$ system has an outage probability of about 30–45%. In contrast, the $\Delta = 3$ system has an outage probability of zero. In Section 14.6, it was demonstrated that with a $\Delta = 1$ system with an appropriate configuration, the minimum SINR can be kept above −5 dB. Therefore, by using AMC scheme 2 in a HEX network, a zero outage probability can be achieved even with $\Delta = 1$. In addition, the cell data rate is improved further by 60 Mbps to 140 Mbps.

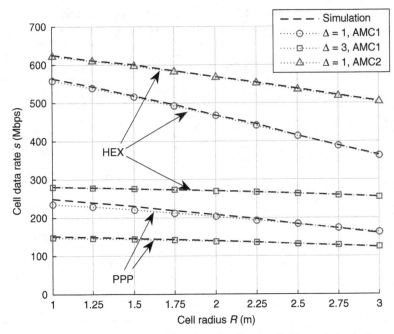

Figure 14.17 Achievable cell data rate as a function of cell radius R. The emission order m was configured based on (14.36). Other system parameters are listed in Table 14.1 if not specified.

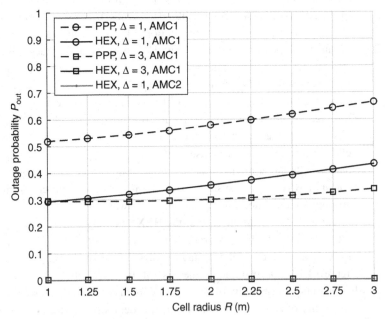

Figure 14.18 Outage probability as a function of cell radius R. The emission order m was configured based on (14.36). Other system parameters are listed in Table 14.1 if not specified.

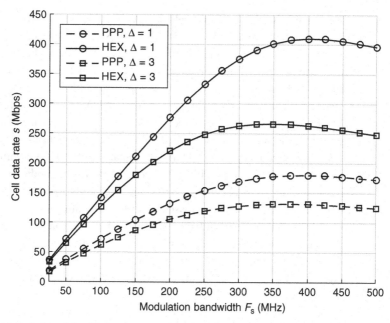

Figure 14.19 Achievable cell data rate as a function of modulation bandwidth F_s with $R = 2.5$ m, $\phi_{1/2} = 40°$ and AMC scheme 1. Other parameters are listed in Table 14.1 if not specified.

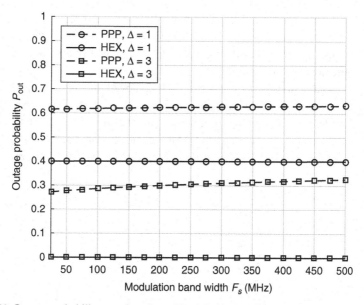

Figure 14.20 Outage probability as a function of modulation bandwidth F_s with $R = 2.5$ m, $\phi_{1/2} = 40°$, and AMC scheme 1. Other parameters are listed in Table 14.1 if specified.

The relationship between the cell data rate/outage probability and the modulation bandwidth was examined, as shown in Figures 14.19 and 14.20. A cell radius of $R = 2.5$ m, a half-power semiangle of $\phi_{1/2} = 40°$, and AMC scheme 1 were used in the system. The other system parameters were the same as those listed in Table 14.1 if not specified. With an increase in the modulation bandwidth F_s, the cell data rate increases as expected. However, when the modulation bandwidth increases further, the channel quality on the high-frequency subcarriers is insufficient owing to the band-limiting effects of the front-end devices. Meanwhile, the total transmission power is spread over a wider frequency band. Thus the signal power on each subcarrier decreases proportionally. In addition, with a further increase in the modulation bandwidth, the SINR of cell edge users becomes less than the threshold for transmission. Consequently, the outage probability also increases as the bandwidth increases, as highlighted in Figure 14.20. When the modulation bandwidth is far beyond the 3-dB bandwidth, too much signal power is "wasted" on subcarriers that are subject to unfavorable channel conditions, compromising the signal quality on those subcarriers which exhibit good channel conditions. As a result, the cell data rate starts to decrease.

14.8   Performance of Finite Networks and Multipath Effects

In the previous sections, infinite networks were analyzed to approximate the performance of practical systems. In this section, the performance of a finite network deployed in a room is compared with the corresponding infinite network with the same system configuration. The MP effects due to reflections from the internal surfaces of the room are now considered.

The finite network considered was deployed in a room of size 20 m × 11 m × 3 m and followed the HEX model, as shown in Figure 14.21. The reflectivity of the ceiling and walls was 0.7, and that of the floor was 0.3. A cell radius of $R = 2.5$ m, a half-power semiangle of $\phi_{1/2} = 40°$, and AMC scheme 2 were used in this system. The remaining parameters were as given in Table 14.1. Firstly, two typical users at the edges of a cell in

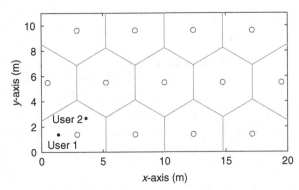

Figure 14.21 Cell deployment of the finite HEX network in a room of size 20 m × 11 m × 3 m.

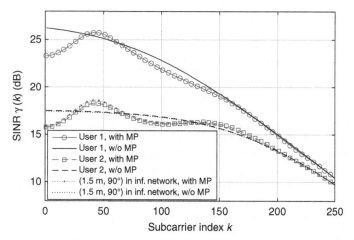

Figure 14.22 SINR on subcarrier k. User 1 and user 2 were located at the bottom left cell in the finite network considered as shown in Figure 14.21.

this finite HEX network were considered, as shown in Figure 14.21. The two users were both 1.5 m away from the cell center and close to one of the edges of the network of hexagon cells. A third user at $(1.5, 40°)$ in an infinite HEX network, denoted as user 3, corresponding to the positions of user 1 and user 2 was also considered for comparison. Figure 14.22 shows the achieved SINR on each subcarrier with and without MP effects. Owing to the low-pass effect of the front-end devices, the achieved SINR decreases with increasing subcarrier index. It can be observed that the SINR without considering MP effects provides a very close estimate of the values determined with consideration of MP effects for each user. Compared with user 2, user 1 is closer to the room edge and further away from the interfering BSs. Consequently, the overall SINR achieved by user 1 is much higher than that achieved by user 2. However, owing to the stronger MP effect, the SINR without the MP effect slightly overestimates the value with the MP effect. In contrast, user 2 is closer to the room center, which makes the situation similar to the case in an infinite network. Therefore, the performances of user 2 and user 3 are very similar.

Next, the performance of this finite network in terms of data rate is considered. Figure 14.23 shows the simulated statistics of the cell data rate. The results both with and without MP effects were simulated and, as shown, these MP effects did not cause any significant variance in the cell data rate performance of the systems. It can be observed that the infinite network offers the worst cell data rate. In contrast, the finite network achieves a slightly improved cell data rate. Furthermore, the cell data rate achieved in the room edge cell is the highest cell data rate because of the lower CCI level. Therefore, it can be concluded that MP effects do not limit the performance of a LiFi attocell system. In addition, a worst-case performance can be obtained by evaluating an infinite network.

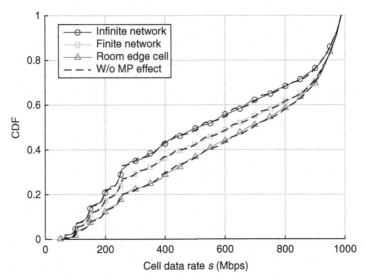

Figure 14.23 Cell data rate statistics. AMC scheme 2 was used in these systems. The finite network considered corresponds to the system shown in Figure 14.21. The room edge cell considered corresponds to the cell in the bottom left cell in Figure 14.21.

14.9   Practical Cell Deployment Scenarios

In a practical network arrangement where the existing lighting infrastructure is used, a BS layout with a regular hexagonal lattice is possible but not the norm. On the other hand, a PPP network is also not entirely practical. This is because having light fixtures deployed in a completely random manner is impractical. In order to demonstrate the significance of the analyzed networks HEX/PPP, the following two scenarios were considered in addition in order to model typical LiFi attocell systems with cell deployments that might be used in practice.

14.9.1   Square Network

The first more practical network model considered was a square lattice cellular model, in which BSs are placed on a square lattice, as shown in Figure 14.24(a). This arrangement is common in indoor lighting networks for several reasons, including design simplicity, good illumination uniformity, and compliance with rectangular-shaped rooms. In this square network, the cell size is controlled by a parameter R_{sq} which is defined as the distance between the two closest BSs. In order to achieve a fair comparison, R_{sq} needs to be consistent with the circular-cell radius R. This requires $R_{sq} = \sqrt{A_{cell}} = \sqrt{\pi R^2} \approx 1.77R$.

14.9.2   Hard-Core Point Process Network

In a PPP network, two BSs can be extremely close to each other, which is unlikely in practice. This is the main drawback of the PPP network model. Therefore, a Matérn

(a)   (b)

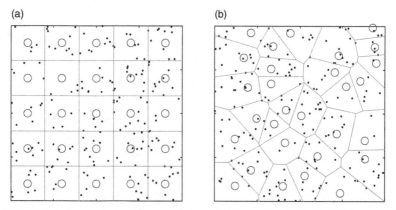

Figure 14.24 Cell deployments used for comparison. (a) Square Network. (b) HCPP network.

type I hard-core point process (HCPP) was used to approximate the network model, as illustrated in Figure 14.24(b). The HCPP model is based on the PPP model with the condition that the shortest distance between any two nodes is greater than a specified threshold, c. To arrive at a set of nodes according to the HCPP model, first a PPP network of density Λ_0 was generated. Then each point was tagged with a random number, and a dependent thinning process was carried out for each marked node as follows: the marked node was retained if there was no other node within a circle centered on the marked node with a radius of c. After the thinning, the density of nodes was reduced. Therefore, to generate an HCPP network with a density of Λ, the initial PPP density Λ_0 had to be [47] $\Lambda_0 = -\ln(1 - \Lambda\pi c^2)/\pi c^2$. In order to have a fair comparison, the choice of Λ_0 also had to be such as to make sure that the average cell area was the same as the area of an equivalent circular cell with a radius of R. Therefore, $\Lambda_0 = -\ln(1 - c^2/R^2)/\pi c^2$.

14.9.3   Performance Comparison

In Figure 14.25, the CDFs of the SINR at the DC level of the systems with different cell deployments are given and compared. An equivalent circular-cell radius of $R = 3$ m and a half-power semiangle of $\phi_{1/2} = 40°$ were used. The remaining system parameters are listed in Table 14.1. Figure 14.25 shows that the SINR distributions of the square network and HCPP network are bounded by the results achieved by the PPP and HEX networks within the SINR region of interest. Similarly to the conclusion drawn in [36], the SINR performance of a PPP and a HEX network can be considered as a lower and an upper bound, respectively, for practical LiFi attocell systems.

14.10   LiFi Attocell Networks Versus Other Small-Cell Networks

In this subsection, the performance of LiFi attocell networks is compared with that achieved by RF femtocell networks and mmWave indoor networks. A LiFi attocell

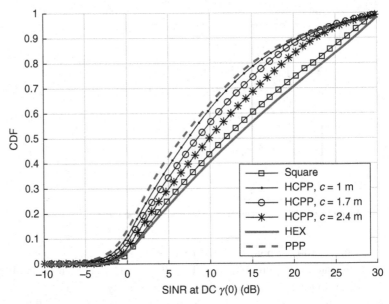

Figure 14.25 Comparison of SINR statistics at DC for different cell deployments. An equivalent circular cell radius of $R = 3$ m and a half-power semiangle of $\phi_{1/2} = 40°$ were used. For the HCPP networks, $c = 1, 1.7, 2.4$ m. Other parameters are listed in Table 14.1.

network achieves a high communication performance owing to its extremely high spatial reuse ($R_{\text{atto}} \in [1,3]$ m). Compared with RF femtocell networks, LiFi attocell networks have a relatively large license-free modulation bandwidth (100 MHz to > 1 GHz) for a given optical spectrum. In contrast, a femtocell has a relatively larger cell size ($R_{\text{femto}} \in [10,40]$ m) and a limited downlink bandwidth of about 10 MHz [48]. The modulation bandwidth of an indoor mmWave system, such as a 60 GHz wireless personal area network, is generally in the range of 500 MHz to > 2 GHz. Therefore, a data rate of up to 7 Gbps can be achieved using the Wireless Gigabit Alliance (WiGig) specification for a single link [49], while the maximum data rate that can be achieved by a single LED source with a 60 MHz device bandwidth has shown to be 3 Gbps [24]. However, owing to the high cost of hardware and to CCI issues, typically only one mmWave access point is available in a room. In contrast, multiple optical BSs can be installed in a room. This enables orders-of-magnitude higher network densities, leading to orders-of-magnitude improvements in data densities. The benefits of a high data density are obvious when a large number of devices in a room need high-speed wireless services such as in Industry 4.0 applications and the Internet of Things (IoT) in general. To demonstrate the high data density achieved by a LiFi attocell system, a metric termed the area data rate is defined as follows:

$$S_{\text{area}} = \frac{S}{A_{\text{cell}}}. \tag{14.70}$$

Figure 14.26 shows the area data rate performance of different systems. The results for the femtocell are extrapolated from [48, 50–52]. As shown, the indoor wireless

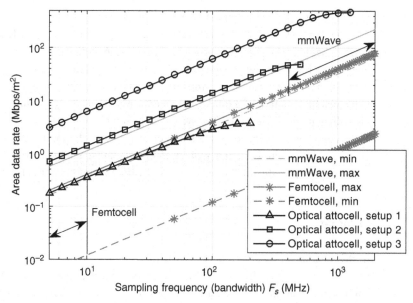

Figure 14.26 Area data rate comparison among LiFi attocell networks, RF femtocell networks, and mmWave networks. LiFi attocell network setup 1: FE1 with $F_{\text{fe}} = 15.2$ MHz, PPP network, $\Delta = 3$, AMC scheme 1, $R = 2.5$ m, and $\phi_{1/2} = 40°$. Setup 2: FE3 with $F_{\text{fe}} = 31.7$ MHz, HEX network, $\Delta = 1$, AMC scheme 2, $R = 2$ m, and $\phi_{1/2} = 30°$. Setup 3: FE4 with $F_{\text{fe}} = 81.5$ MHz, HEX network, $\Delta = 1$, AMC scheme 2, $R = 1$ m, and $\phi_{1/2} = 15°$.

personal area network achieved by the femtocell network is generally in the range of 0.03 to 0.0012 bps/Hz/m². With a bandwidth of 10 MHz, the area data rate achieved by the femtocell networks is in the range of 0.012 to 0.3 Mbps/m². The results for the mmWave systems are extrapolated from [49, 53, 54]. The spectral efficiency achieved by the mmWave systems is generally in the range of 3.24 to 11.25 bps/Hz. Assuming a room of size 10 m × 10 m, with a bandwidth in the range of 400 MHz to 2 GHz, the achievable area data rate is in the range of 13 Mbps/m² to 225 Mbps/m². The estimated minimum and maximum values for these two systems are used the benchmarks here.

Firstly, a low-performance LiFi attocell system using setup 1 is considered. This scenario used FE1 with $F_{\text{fe}} = 15.2$ MHz and a PPP cell deployment with $R = 2.5$ m, $\phi_{1/2} = 40°$, and $\Delta = 3$. It used AMC scheme 1. Figure 14.26 shows that the performance of this system compares closely with that of the femocell network. Note that the maximum achievable data rate of 3.82 Mbps/m², with a bandwidth of 200 MHz, is much higher than that achieved by the femtocell networks, with a bandwidth of 10 MHz. Note also that the performances of systems with different bandwidths are being compared. This is because of the difference in the cost and availability of the two types of frequency bands. For the LiFi system, the frequency band is totally unlicensed and does not cause any interference to a system in an adjacent frequency band. Therefore, there is no requirement for spectrum masks. In contrast, RF spectral resources are scarce, and hence expensive and rigorous spectrum masks are considered. Next, two LiFi attocell systems with moderate (setup 2) and high (setup 3) performance are demonstrated. In setup 2, a HEX cell deployment with $R = 2$ m, $\phi_{1/2} = 30°$, $\Delta = 1$, and AMC scheme 2 was used. In setup 3, a HEX cell deployment with $R = 1$ m, $\Delta = 1$, $\phi_{1/2} = 15°$, and AMC scheme 2 was used. As shown in Figure 14.26, both LiFi attocell systems perform better than the femtocell systems. For the attocell system with setup 2, the maximum achievable area data rate of 49 Mbps/m^2 with a modulation bandwidth of 500 MHz is in a range similar to that of a mmWave system. In particular, a maximum area data rate of 469 Mbps/m^2 is achieved by the system with setup 3 with a modulation bandwidth of 1.26 GHz. This is about two times higher than that of the high-performance mmWave system with a spectral efficiency of 11.25 bps/Hz and a bandwidth of 2 GHz. This result particularly highlights the huge potential of LiFi attocell networks for 5G and beyond.

14.11   Summary

In this chapter, the performance of a LiFi attocell network was analyzed. In order to be able to optimally design a LiFi attocell system, it is important to understand how key network parameters such as the cell size and the network configuration affect the system performance. This is particularly important when piggybacking a LiFi attocell network on an existing lighting infrastructure which leaves little possibility to optimize the network for communication. To this end, an analysis of the SINR distribution and the corresponding data rate assuming different cell deployments was performed. The analysis in this chapter offers an accurate estimation of the downlink performance of a LiFi attocell system that has a large number of parameters. This study provides detailed guidelines for appropriate choices of these parameters. Because of the potential benefits of combining LiFi attocell networks with existing lighting infrastructures and because of other practical constraints, optimized regular hexagonal cell deployments may not always be achievable. Therefore, in this study, several other network topologies such as square and random cell deployments were also considered. In particular, a LiFi attocell network with a PPP cell deployment was considered in order to closely model a random real-world scenario where there are no underlying network-planning considerations. An extensive simulation study confirms that the hexagonal and PPP cell deployments represent the best- and worst-case performance, respectively, of practical attocell deployments. The simulation results also demonstrate that attocell networks deployed in a finite room offer better performance than networks which are horizontally infinite because of the CCI at the room edges being very low. In addition, the simulation results also imply that the multipath effect due to reflections from the internal surfaces of the room is minor relative to the effect of CCI. Because LiFi attocells can be deployed densely in a room, these networks can typically achieve very high data densities. In order to demonstrate this advantage, the downlink performance of LiFi attocell systems was compared with that achieved by RF femtocell networks and indoor mmWave systems in terms of area data rate (achievable data rate per unit area). The results show that LiFi attocell networks generally outperform femtocell networks. In particular, a high-performance LiFi attocell network can achieve an area data rate of 469 Mbps/m^2, which is twice that achieved by a high-performance mmWave system.

The high-performance mmWave system used a spectral efficiency of 11.25 bps/Hz and a bandwidth of 2 GHz in a room of size 10 m × 10 m. The improvement of the system provided by a reduction in cell size is more pronounced for regular hexagonal network deployments than for random PPP network deployments.

References

[1] X. Ge, S. Tu, G. Mao, C. X. Wang, and T. Han, "5G ultra-dense cellular networks," *IEEE Wireless Commun.*, vol. 23, no. 1, pp. 72–79, Feb. 2016

[2] S. Dimitrov and H. Haas, *Principles of LED Light Communications: Towards Networked Li-Fi*, Cambridge University Press, 2015.

[3] H. Haas, "Wireless data from every light bulb," TED, Aug. 2011. Available at www.ted.com/.

[4] S. Arnon, J. Barry, G. Karagiannidis, R. Schober, and M. Uysal, *Advanced Optical Wireless Communication Systems*, Cambridge University Press, 2012.

[5] D. Tsonev, H. Chun, S. Rajbhandari, J. McKendry, S. Videv, E. Gu, M. Haji, S. Watson, A. Kelly, G. Faulkner, M. Dawson, H. Haas, and D. O'Brien, "A 3-Gb/s single-LED OFDM-based wireless VLC link using a gallium nitride μLED," *IEEE Photonics Technol. Lett.*, vol. 26, no. 7, pp. 637–640, Apr. 2014.

[6] S. Dimitrov and H. Haas, "Information rate of OFDM-based optical wireless communication systems with nonlinear distortion," *IEEE J. Lightw. Technol.*, vol. 31, no. 6, pp. 918–929, Mar. 2013.

[7] H. Haas, "High-speed wireless networking using visible light," SPIE Newsroom, 2013.

[8] Z. Wang, D. Tsonev, S. Videv, and H. Haas, "On the design of a solar-panel receiver for optical wireless communications with simultaneous energy harvesting," *IEEE J. Sel. Areas Commun.*, vol. 33, no. 8, pp. 1612–1623, Aug. 2015.

[9] S. Rajagopal, R. Roberts, and S.-K. Lim, "IEEE 802.15.7 visible light communication: Modulation schemes and dimming support," *IEEE Commun. Mag.*, vol. 50, no. 3, pp. 72–82, Mar. 2012.

[10] H. Haas, L. Yin, Y. Wang, and C. Chen, "What is LiFi?" *IEEE J. Lightw. Technol.*, vol. 34, no. 6, pp. 1533–1544, Mar. 2016.

[11] Y. Li, M. Safari, R. Henderson, and H. Haas, "Optical OFDM with single-photon avalanche diode," *IEEE Photonics Technol. Lett.*, vol. 27, no. 9, pp. 943–946, May 2015.

[12] E. Sarbazi, M. Uysal, M. Abdallah, and K. Qaraqe, "Ray tracing based channel modeling for visible light communications," in *Proc. of Signal Processing and Communications Applications Conf. (SIU)*, Apr. 2014.

[13] A. Farid and S. Hranilovic, "Capacity bounds for wireless optical intensity channels with Gaussian noise," *IEEE Trans. Inf. Theory*, vol. 56, no. 12, pp. 6066–6077, Dec. 2010.

[14] B. Rofoee, K. Katsalis, Y. Yan, Y. Shu, T. Korakis, L. Tassiulas, A. Tzanakaki, G. Zervas, and D. Simeonidou, "First demonstration of service-differentiated converged optical sub-wavelength and LTE/WiFi networks over GEAN," in *Proc. of Optical Fiber Communications Conf. and Exhibition (OFC)*, Mar. 2015.

[15] H. Chun, S. Rajbhandari, G. Faulkner, D. Tsonev, E. Xie, J. McKendry, E. Gu, M. Dawson, D. C. O'Brien, and H. Haas, "LED based wavelength division multiplexed 10 Gb/s visible light communications," *IEEE J. Lightw. Technol.*, vol. 34, no. 13, pp. 3047–3052, Jul. 2016.

[16] D. Tsonev, S. Videv, and H. Haas, "Towards a 100 Gb/s visible light wireless access network," *Opt. Express*, vol. 23, no. 2, pp. 1627–1637, Jan. 2015.

[17] C. Chen, D. A. Basnayaka, and H. Haas, "Downlink performance of optical attocell networks," *IEEE J. Lightw. Technol.*, vol. 34, no. 1, pp. 137–156, Jan. 2016.

[18] K. Chandra, R. Venkatesha Prasad, and I. Niemegeers, "An architectural framework for 5G indoor communications," in *Proc. of International Wireless Communications and Mobile Computing Conf. (IWCMC)*, Aug. 2015.

[19] M. B. Rahaim, A. M. Vegni, and T. D. C. Little, "A hybrid radio frequency and broadcast visible light communication system," in *Proc. of IEEE GLOBECOM Workshops (GC Wkshps)*, Dec. 2011.

[20] S. Dimitrov, S. Sinanovic, and H. Haas, "Clipping noise in OFDM-based optical wireless communication systems," *IEEE Trans. Commun.*, vol. 60, no. 4, pp. 1072–1081, Apr. 2012.

[21] H. L. Minh, D. O'Brien, G. Faulkner, L. Zeng, K. Lee, D. Jung, Y. Oh, and E. T. Won, "100-Mb/s NRZ visible light communications using a postequalized white LED," *IEEE Photonics Technol. Lett.*, vol. 21, no. 15, pp. 1063–1065, Aug. 2009.

[22] J. Vucic, C. Kottke, S. Nerreter, K. D. Langer, and J. W. Walewski, "513 Mbit/s visible light communications link based on DMT-modulation of a white LED," *IEEE J. Lightw. Technol.*, vol. 28, no. 24, pp. 3512–3518, Dec. 2010.

[23] A. M. Khalid, G. Cossu, R. Corsini, P. Choudhury, and E. Ciaramella, "1-Gb/s transmission over a phosphorescent white LED by using rate-adaptive discrete multitone modulation," *IEEE Photonics J.*, vol. 4, no. 5, pp. 1465–1473, Oct. 2012.

[24] D. Tsonev, H. Chun, S. Rajbhandari, J. J. D. McKendry, S. Videv, E. Gu, M. Haji, S. Watson, A. E. Kelly, G. Faulkner, M. D. Dawson, H. Haas, and D. O'Brien, "A 3-Gb/s single-LED OFDM-based wireless VLC link using a gallium nitride μLED," *IEEE Photonics Technol. Lett.*, vol. 26, no. 7, pp. 637–640, Apr. 2014.

[25] Z. Chen, D. Tsonev, and H. Haas, "A novel double-source cell configuration for indoor optical attocell networks," in *Proc. of IEEE Global Communications Conf. (GLOBECOM)*, Dec. 2014.

[26] Z. Chen, N. Serafimovski, and H. Haas, "Angle diversity for an indoor cellular visible light communication system," in *Proc. of IEEE Vehicular Technology Conf. (VTC Spring)*, May 2014.

[27] J. M. Kahn and J. R. Barry, "Wireless infrared communications," *Proc. IEEE*, vol. 85, no. 2, pp. 265–298, Feb. 1997.

[28] F. J. López-Hernández, R. Pérez-Jiménez, and A. Santamaría, "Ray-tracing algorithms for fast calculation of the channel impulse response on diffuse IR wireless indoor channels," *Opt. Eng.*, vol. 39, no. 10, pp. 2775–2780, 2000.

[29] V. Jungnickel, V. Pohl, S. Nonnig, and C. von Helmolt, "A physical model of the wireless infrared communication channel," *IEEE J. Sel. Areas Commun.*, vol. 20, no. 3, pp. 631–640, Apr. 2002.

[30] European Standard, "Lighting of indoor work places," EN 12464-1, Jan. 2009.

[31] J. R. Meyer-Arendt, "Radiometry and photometry: Units and conversion factors," *Appl. Opt.*, vol. 7, no. 10, pp. 2081–2084, Oct. 1968.

[32] Integrated System Technologies Ltd, "VESTA 165mm recessed LED downlighter." Available at www.istl.com/vesta.php.

[33] Wellmax LED, "64 W LED panel light." Available at http://wellmaxled.com/portfolio/64-w-led-panel-light/.

[34] B. Ghimire and H. Haas, "Self-organising interference coordination in optical wireless networks," *EURASIP J. Wireless Commun. Netw.*, vol. 2012, p. 131, Apr. 2012.

[35] F. R. Gfeller and U. Bapst, "Wireless in-house data communication via diffuse infrared radiation," *Proc. IEEE*, vol. 67, no. 11, pp. 1474–1486, Nov. 1979.

[36] J. Andrews, F. Baccelli, and R. Ganti, "A tractable approach to coverage and rate in cellular networks," *IEEE Trans. Commun.*, vol. 59, no. 11, pp. 3122–3134, Nov. 2011.

[37] H. Elgala, R. Mesleh, and H. Haas, "Non-linearity effects and predistortion in optical OFDM wireless transmission using LEDs," *Int. J. Ultra Wideband Commun. Syst.*, vol. 1, no. 2, pp. 143–150, 2009.

[38] D. Tsonev, S. Sinanovic, and H. Haas, "Complete modeling of nonlinear distortion in OFDM-based optical wireless communication," *IEEE J. Lightw. Technol.*, vol. 31, no. 18, pp. 3064–3076, Sep. 2013.

[39] B. Almeroth, A. Fehske, G. Fettweis, and E. Zimmermann, "Analytical interference models for the downlink of a cellular mobile network," in *Proc. of IEEE GLOBECOM Workshops (GC Wkshps)*, Dec. 2011.

[40] E. Sousa and J. Silvester, "Optimum transmission ranges in a direct-sequence spread-spectrum multihop packet radio network," *IEEE J. Sel. Areas Commun.*, vol. 8, no. 5, pp. 762–771, Jun. 1990.

[41] J. Bowers and L. Newton, "Expansion of probability density functions as a sum of gamma densities with applications in risk theory," *Trans. Soc. Actuaries*, vol. 18, no. 52, pp. 125–147, 1966.

[42] M. Haenggi, "On distances in uniformly random networks," *IEEE Trans. Inf. Theory*, vol. 51, no. 10, pp. 3584–3586, Oct. 2005.

[43] D. Tsonev, S. Videv, and H. Haas, "Unlocking spectral efficiency in intensity modulation and direct detection systems," *IEEE J. Sel. Areas Commun.*, vol. 33, no. 9, pp. 1758–1770, Sep. 2015.

[44] D. Tsonev and H. Haas, "Avoiding spectral efficiency loss in unipolar OFDM for optical wireless communication," in *Proc. of IEEE International Conf. on Communications (ICC)*, Jun. 2014.

[45] F. Xiong, *Digital Modulation Techniques*, 2nd edn, Artech House, 2006.

[46] H. Burchardt, S. Sinanović, Z. Bharucha, and H. Haas, "Distributed and autonomous resource and power allocation for wireless networks," *IEEE Trans. Commun.*, vol. 61, no. 7, pp. 2758–2771, Aug. 2013.

[47] D. Stoyan, W. S. Kendall, and J. Mecke, *Stochastic Geometry and Its Applications*, 2nd edn, Wiley, 1995.

[48] V. Chandrasekhar, J. Andrews, and A. Gatherer, "Femtocell networks: A survey," *IEEE Commun. Mag.*, vol. 46, no. 9, pp. 59–67, Sep. 2008.

[49] C. Hansen, "WiGiG: Multi-gigabit wireless communications in the 60 GHz band," *IEEE Wireless Commun.*, vol. 18, no. 6, pp. 6–7, Dec. 2011.

[50] P. Chandhar and S. Das, "Area spectral efficiency of co-channel deployed OFDMA femtocell networks," *IEEE Trans. Wireless Commun.*, vol. 13, no. 7, pp. 3524–3538, Jul. 2014.

[51] H.-S. Jo, P. Xia, and J. Andrews, "Downlink femtocell networks: Open or closed?" in *Proc. of IEEE International Conf. on Communications (ICC)*, Jun. 2011.

[52] W. C. Cheung, T. Quek, and M. Kountouris, "Throughput optimization, spectrum allocation, and access control in two-tier femtocell networks," *IEEE J. Sel. Areas Commun.*, vol. 30, no. 3, pp. 561–574, Apr. 2012.

[53] D. Muirhead, M. Imran, and K. Arshad, "Insights and approaches for low-complexity 5G small-cell base-station design for indoor dense networks," *IEEE Access*, vol. 3, pp. 1562–1572, Aug. 2015.

[54] C. Yiu and S. Singh, "Empirical capacity of mmWave WLANs," *IEEE J. Sel. Areas Commun.*, vol. 27, no. 8, pp. 1479–1487, Oct. 2009.

Part III

Network Protocols, Algorithms, and Design

15 Massive MIMO Scheduling Protocols

Giuseppe Caire

In this chapter, we present a general approach to data-oriented downlink scheduling in a wireless network, possibly formed from multiple base stations and users. Following commonly used acronyms, base stations will be denoted by "BS" (base station) and user devices by "UE" (user equipment). Specifically, we consider the case where the BSs have a large number of antennas and serve a given number of downlink data streams using multiuser multiple-input multiple-output (MIMO) spatial multiplexing. When the number of antennas is large and is significantly larger than the number of downlink data streams, such systems are referred to as "massive MIMO." As we shall see, a network operating in the massive MIMO regime has several advantages, not only in terms of achievable spectral efficiency per cell but also in terms of simplified signal processing, rate allocation, and user scheduling. This nontrivial system simplification is due to the fact that the large number of antennas and not-so-large number of simultaneously transmitted data streams has the consequence that the signal-to-interference-plus-noise ratio (SINR) at each UE becomes an almost deterministic quantity that depends only on the distance-dependent path loss and large-scale fading (shadowing) of the propagation channel between the UE and the serving BS, and not on the small-scale multipath fading. Since distance-dependent path loss and shadowing are relatively slowly varying in time and frequency nonselective, in contrast to the time- and frequency-selective small-scale fading, it follows that the scheduling protocol can learn quite accurately the rate at which each user can be served from each BS. Based on this knowledge, a scheduling protocol can decide dynamically which subset of users should be served from which BS. In this chapter, we will see how such a dynamic scheduling policy with given optimality performance guarantees can be systematically designed.

15.1   Introduction

Wireless data traffic has grown dramatically in recent years. Unlike traditional voice-oriented interactive communications, wireless data is typically asymmetric (the downlink traffic is much higher than the uplink traffic) and more delay tolerant. For example, a typical killer application is represented by on-demand video streaming, which is predicted to account for 75% of the total mobile data traffic by 2019 [1]. The streaming process requires that video frames arrive at the receiver within their playback deadlines. However, with a sufficiently large playback buffer, it is possible to smooth out the random fluctuations in the delivery delay such that, as long as the average delivery rate is slightly larger than the playback rate, the playback buffer has a positive drift which allows uninterrupted playback with high probability even in the presence of some delay jitter. Other type of data-oriented traffic, such as file downloads and Internet browsing, is even more delay tolerant.

Another characteristic of data traffic is that it may be delivered at different levels of quality. For example, in *Dynamic Adaptive Streaming over Hypertext Transfer Protocol (HTTP)* (DASH) [2, 3]*, each user (client) monitors the available capacity during a video streaming session and chooses adaptively and dynamically the most appropriate video quality level. Different quality levels can be obtained either by having multiple versions of the same video source encoded at different bit rates available at the video server, or by using scalable video coding and sending an adaptive number of refinement layers [4]. Similar considerations apply for virtually any multimedia information that can be represented at different quality or rate levels.

As a matter of fact, the emphasis on downlink traffic and the possibility of exploiting delay tolerance have been the driving principles of the *Evolution-Data Optimized* (EV-DO) and *High-Speed Downlink Packet Access* (HSDPA) third-generation systems [5]. Also, the fourth-generation of wireless systems, based on *orthogonal frequency division multiple access* (OFDMA), has adopted similar ideas through elaborate rate adaptation and user scheduling of the *orthogonal frequency division multiplexing* (OFDM) transmission resource blocks. However, in order to accommodate the above-mentioned ever-growing traffic demand, it is apparent that the spatial dimension, in addition to the classical time–frequency dimension, must be exploited. Hence, the current trend consists of moving aggressively toward multiuser MIMO (MU-MIMO), where the BSs are equipped with multiple transmit antennas and simultaneously serve multiple downlink data streams on the same resource block, using some form of MU-MIMO precoding.

Motivated by the above facts, this chapter focuses on the case of a network formed from several BSs, each equipped with multiple antennas and supporting a number of downlink data streams using MU-MIMO. We follow the system assumptions of [6], and consider a network operating in the so-called *massive MIMO* regime, where at each BS the number of antennas is significantly larger than the number of downlink data streams. In this regime, the achievable user rate has a particularly simple expression, and is an almost deterministic quantity, despite the presence of small-scale multipath frequency-selective fading. This "channel-hardening" effect can be exploited by the scheduling policy, which can take advantage of knowledge of the "instantaneous" user rates and make its scheduling decisions based only on slowly varying large-scale quantities, such as distance-dependent path loss and large-scale fading (shadowing). Generally speaking, such decisions involve the determination of what level of quality each user should request, from which BS such a request should be handled, and at which *physical layer* (PHY) rate it should be served. Notice that we distinguish between the PHY rate, or "instantaneous" rate, which is the rate at which a given user can be served

---

* This includes industry products such as Microsoft Smooth Streaming and Apple HTTP Live Streaming.

by a given BS in a single scheduling slot, and the *throughput rate*, which is the rate seen at the application layer (e.g., the rate at which fresh data enters the playback buffer in a streaming session). These two rates are conceptually quite different. Remarkably, most of the "communication-theoretic" literature and "system level simulations" on wireless/cellular communication networks has focused on the PHY rate. This is often treated as a random variable, such that people have been traditionally interested in its cumulative distribution function over the user population. In contrast, in this chapter, we focus on the much more relevant *user throughput rate*, i.e., the rate "seen by the application layer." Fundamentally, it is this rate and not the PHY rate which is responsible for the so-called *user quality of experience* (QoE).

The link between the PHY rate and the throughput rate is represented by the scheduling policy, which allocates the transmission resources (time–frequency and spatial downlink data streams) to the users, in order to achieve a certain target performance. Since we are considering a shared channel, the objectives of the different users are in conflict: if one user is given many transmission resources, inevitably some other user is left starved. Therefore, it is *essential* to consider fairness along with network performance maximization. Fairness can be naturally incorporated into the problem by defining a suitable network utility function, and considering the corresponding network utility maximization (NUM) problem.

15.2   Network Model and Problem Formulation

We consider a wireless network with multiple UEs and multiple BSs sharing the same bandwidth. The network is associated to a bipartite graph $\mathcal{G} = (\mathcal{U}, \mathcal{B}, \mathcal{E})$, where $\mathcal{U}$ denotes the set of UEs, $\mathcal{B}$ denotes the set of BSs, and $\mathcal{E}$ contains edges for all pairs (b, u) such that BS b can transmit information to user u. We denote by $\mathcal{N}(u) \subseteq \mathcal{B}$ the neighborhood of user u, i.e., $\mathcal{N}(u) = \{b \in \mathcal{B} : (b, u) \in \mathcal{E}\}$. Similarly, $\mathcal{N}(b) = \{u \in \mathcal{U} : (b, u) \in \mathcal{E}\}$. Each user $u \in \mathcal{U}$ has an active downlink session (e.g., it is streaming a video file). The session is organized into "chunks" (or application-layer packets). For example, we may imagine an HTTP session where each chunk is an HTTP-requested block of data [4]. Let f denote the index of a specific file, and f_u denote the file requested by user u. Since the data chunks may contain encoded multimedia content, we assume that each file f is associated with $N_f \geq 1$ quality levels [3]. The quality–rate profile of a given file f may vary from chunk to chunk. In particular, we let $D_f(m, i)$ and $B_f(m, i)$ denote the quality measure (see, e.g., [7]) and the size (in number of bits), respectively, of the ith chunk of file f at quality level m.

15.2.1   Timescales

It is important to note that the timescale on which chunks are requested and the timescale on which PHY layer transmissions are scheduled differ by one to three orders of magnitude (see Figure 15.1). For instance, in current video-streaming technology [2], a typical video chunk spans a duration between 0.5 and 2 seconds, while the duration

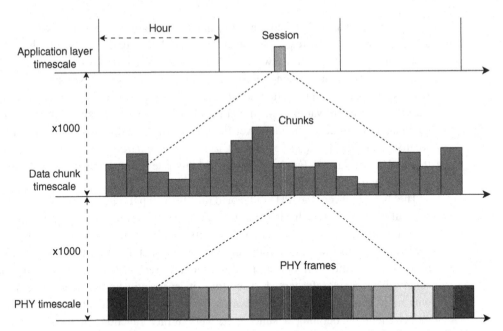

Figure 15.1 Timescale decomposition. The different heights of the chunks indicate that they may have different sizes in terms of bits, although they are requested at a constant rate (e.g., in a video-streaming session). The different shades of the PHY frames indicate that the PHY transmission rate may vary from frame to frame, owing to dynamic rate allocation.

of a PHY frame is of the order of milliseconds [5]. In the following, we define our dynamic scheduling policies to operate on the PHY frame timescale, i.e., they provide a scheduling/resource allocation decision at each PHY frame time $t \in \mathbb{Z}$. However, new chunks are requested at multiples of the chunk time, i.e., at times $t = in$ for $i \in \mathbb{Z}$, where n denotes the number of PHY frames per chunk time, assumed here to be an integer for simplicity. In the rest of the chapter we will use consistently the following notation: the index t denotes the PHY frame *transmission slot*, and the index i denotes the *application layer data chunk*.

15.2.2   Request Queues and Network Utility Maximization

At the beginning of the ith chunk time, each user $u \in \mathcal{U}$ requests a particular quality mode for the ith chunk of its data session. That is, in each slot $t \in \{0, n, 2n, 3n, \ldots\}$, each user $u \in \mathcal{U}$ specifies the quality mode $m_u(t) \in \{1, 2, \ldots, N_{f_u}\}$ for its next chunk. This decision specifies the quality $D_{f_u}(m_u(t), t/n)$ and the number of bits $B_{f_u}(m_u(t), t/n)$ associated with the chunk requested at slot t. As these decisions are made only at times t that are multiples of n, it is convenient to define

$$D_{f_u}(m_u(t), t) = B_{f_u}(m_u(t), t) = 0 \text{ for } t \neq in, \text{ for some } i \in \mathbb{Z}. \qquad (15.1)$$

The bits $B_{f_u}(m_u(t), t)$ are called the *requested bits* of user u at slot t, and are placed in a *request queue* $Q_u(t)$. The request queue evolves over the transmission slots $t \in$

$\{0, 1, 2, \ldots\}$ as

$$Q_u(t+1) = \left[Q_u(t) - \mu_u(t) + B_{f_u}(m_u(t), t) \right]_+ \quad \forall u \in \mathcal{U}, \tag{15.2}$$

where $\mu_u(t)$ is the number of bits downloaded by user u in slot t and where $[\cdot]_+ = \max\{\cdot, 0\}$ denotes the positive-part operator. Note that the request queue in (15.2) can decrease at every transmission slot t as new bits are downloaded, but can only increase in slots t for t equal to an integer multiple of n. Intuitively, $Q_u(t)$ consists of bits associated with all chunks that have been requested by user u but not yet fully received.

The quantity $\mu_u(t)$ indicates the instantaneous aggregate downloading rate of user u in slot t, expressed in bits per slot. This is given by

$$\mu_u(t) = \sum_{b \in \mathcal{N}(u)} \mu_{bu}(t), \tag{15.3}$$

where $\mu_{bu}(t)$ is the rate served by BS b to user u in slot t. The matrix of *instantaneous* rates $[\mu_{bu}(t)]$ depends on the rate allocation policy, and for each t it is selected as a point in the feasible region of the network PHY rate. Specifically, let $\omega(t)$ represent the network topology state in slot t (e.g., the distances between UEs and BSs and their large-scale fading). Assume that $\omega(t)$ takes values in an abstract set Ω, possibly an infinite/uncountable set. For each $\omega \in \Omega$, we define $\mathcal{R}(\omega)$ to be the feasible rate region of the network for state ω. Then, the feasible instantaneous-rate region is $\mathcal{R}(\omega(t))$. For example, the set $\mathcal{R}(\omega)$ may include the constraint that each user can receive a positive rate from at most one BS and/or constrain each BS $b \in \mathcal{B}$ to transmit at most S_b downlink data streams (see [8] for a discussion of various wireless multiple-access scenarios and interference models that fit this general framework). The set $\mathcal{R}(\omega)$ can also handle models that allow simultaneous downloading from multiple BSs (for instance, in a cellular *code division multiple access* (CDMA) system with macrodiversity or in a *Long Term Evolution* (LTE) system when receivers can perform successive interference cancellation), or information-theoretic capacity regions for various network topology models, inclusive of broadcast and interference constraints. In Section 15.5, we provide a simple approximate characterization of $\mathcal{R}(\omega)$ for the relevant wireless scenario where the BSs employ massive MIMO, which is the primary focus of this chapter.

Remark 15.1. Each user u maintains $Q_u(t)$ and updates it according to (15.2) in every transmission slot t. A small amount of bookkeeping is also required by the user to associate the bits $Q_u(t)$ with their appropriate chunks. Specifically, each user maintains a list of the chunks it has requested but not yet fully received, along with the quality modes it requested for each chunk. It can receive new bits in slot t only from a BS that has its requested file, and only if $Q_u(t) > 0$. In order to download these bits, the user fetches (or "pulls") them from the selected BS.   ◊

Since we care about throughput rates or, more generally, long-term average quality performance measures, we introduce a convenient notation for such quantities as follows. Let $X(t)$ denote a random process. We define

$$\overline{X} = \lim_{t \to \infty} \frac{1}{t} \sum_{\tau=0}^{t-1} \mathbb{E}[X(\tau)], \qquad (15.4)$$

and assume that such a limit exists.[†] Let $\phi_u(\cdot)$ be a concave, continuous, nondecreasing function defining the utility of $u \in \mathcal{U}$. The NUM problem that we wish to solve is given by

$$\text{maximize} \sum_{u \in \mathcal{U}} \phi_u(\overline{D}_{f_u}) \qquad (15.5a)$$

$$\text{subject to } \overline{Q}_u < \infty, \quad \forall\, u \in \mathcal{U}, \qquad (15.5b)$$

$$[\mu_{bu}(t)] \in \mathcal{R}(\omega(t)), \quad \forall\, t, \qquad (15.5c)$$

$$m_u(t) \in \{1, 2, \ldots, N_{f_u}\}, \quad \forall\, u \in \mathcal{U}, \forall\, t, \qquad (15.5d)$$

where the requirement of finite $\overline{Q}_u$ corresponds to the *strong stability* condition for all the queues [9]. In short, under this condition "all requested bits will be eventually delivered."

By appropriately choosing the functions $\phi_u(\cdot)$, we can impose some desired notion of fairness. For example, a general class of concave functions suitable for this purpose is given by the α-fairness network utility, defined by [10]

$$\phi_u(x) = \begin{cases} \log x, & \alpha = 1, \\ \dfrac{x^{1-\alpha}}{1-\alpha}, & \alpha > 0, \; \alpha \neq 1. \end{cases} \qquad (15.6)$$

In this case, it is well known that $\alpha = 0$ yields the maximization of the sum quality (with no fairness), $\alpha \to \infty$ yields the maximization of the worst-case quality (max–min fairness), and $\alpha = 1$ yields the maximization of the geometric mean of the quality (proportional fairness).

In order to solve problem (15.5), it is convenient to transform it into an equivalent problem that involves the maximization of a single time average. This transformation is achieved through the use of auxiliary variables $\gamma_u(t)$ and the corresponding virtual queues $\Theta_u(t)$ with buffer evolution:

$$\Theta_u(t+1) = \left[\Theta_u(t) + \gamma_u(t) - D_{f_u}(m_u(t), t)\right]_+ . \qquad (15.7)$$

[†] The existence of the limit in (15.4) is assumed temporarily for ease of exposition of the scheduling problem but is not required for the derivation of the scheduling policy or for the proof of Theorem 15.3.

Consider the transformed problem,

$$\text{maximize} \quad \sum_{u \in \mathcal{U}} \overline{\phi_u(\gamma_u)} \tag{15.8a}$$

$$\text{subject to} \quad \overline{Q}_u < \infty, \quad \forall u \in \mathcal{U}, \tag{15.8b}$$

$$\overline{\gamma}_u \leq \overline{D}_{f_u}, \quad \forall u \in \mathcal{U}, \tag{15.8c}$$

$$D_{f_u}^{\min} \leq \gamma_u(t) \leq D_{f_u}^{\max}, \quad \forall u \in \mathcal{U}, \forall t, \tag{15.8d}$$

$$[\mu_{bu}(t)] \in \mathcal{R}(\omega(t)), \quad \forall t, \tag{15.8e}$$

$$m_u(t) \in \{1, 2, \dots, N_{f_u}\}, \quad \forall u \in \mathcal{U}, \forall t, \tag{15.8f}$$

where $D_{f_u}^{\min}$ and $D_{f_u}^{\max}$ are uniform lower and upper bounds, respectively, on the quality function $D_{f_u}(\cdot, t)$. Notice that the constraints (15.8c) imply the stability of the virtual queues Θ_u, since $\overline{\gamma}_u$ and $\overline{D}_{f_u}$ are the time-averaged arrival rate and the time-averaged service rate for the virtual queue given in (15.7). We have the following lemma.

Lemma 15.1   *Problems (15.5) and (15.8) are equivalent.*

Proof. The proof is well known (see [9, 11], for instance) and is briefly given here for the sake of completeness. Let ϕ_1^{opt} and ϕ_2^{opt} be the optimal solutions of problems (15.5) and (15.8), respectively. Fix $\epsilon > 0$ and let $a^*(t)$ be a policy that satisfies all constraints of the transformed problem (15.8) and achieves a utility not smaller than $\phi_2^{\text{opt}} - \epsilon$. We have

$$\phi_2^{\text{opt}} - \epsilon \leq \sum_{u \in \mathcal{U}} \overline{\phi_u(\gamma_u^*)} \stackrel{(a)}{\leq} \sum_{u \in \mathcal{U}} \phi_u(\overline{\gamma_u^*}) \stackrel{(b)}{\leq} \sum_{u \in \mathcal{U}} \phi_u(\overline{D_{f_u}^*}) \stackrel{(c)}{\leq} \phi_{\text{opt}}^1, \tag{15.9}$$

where (a) follows from Jensen's inequality applied to the concave function $\phi_u(\cdot)$, (b) follows by noticing that the policy $a^*(t)$ satisfies the constraint (15.8c) and $\phi_u(\cdot)$ is nondecreasing, and (c) follows from the fact that since $a^*(t)$ is feasible for problem (15.8) then it also satisfies the constraints of problem (15.5) and therefore is feasible for the latter. As this holds for all $\epsilon > 0$, we conclude that $\phi_2^{\text{opt}} \leq \phi_1^{\text{opt}}$.

Now, let $a'(t)$ be a policy for the original problem (15.5), achieving a utility not smaller than $\phi_1^{\text{opt}} - \epsilon$. Since $a'(t)$ is feasible for (15.5), it also satisfies the constraints (15.8b) and (15.8e) of the transformed problem. Furthermore, we choose $\gamma'(t) = \mathbf{D}'$ for all slots t, where the latter is a vector containing the components $\overline{D}'_{f_u}$ for all $u \in \mathcal{U}$. Such a choice of $\gamma'(t)$ together with the policy $a'(t)$ forms a feasible policy for problem (15.8). Therefore,

$$\phi_1^{\text{opt}} - \epsilon \leq \sum_{u \in \mathcal{U}} \phi_u(\overline{D'_{f_u}}) = \sum_{u \in \mathcal{U}} \overline{\phi_u(\gamma_u')} \leq \phi_2^{\text{opt}}. \tag{15.10}$$

As this holds for all $\epsilon > 0$, we conclude that $\phi_1^{\text{opt}} \leq \phi_2^{\text{opt}}$. Thus, (15.9) and (15.10) imply that $\phi_1^{\text{opt}} = \phi_2^{\text{opt}}$ and, by comparing the constraints, we can immediately conclude that an optimal policy for the transformed problem can be directly turned into an optimal policy for the original problem.

15.3   Dynamic Scheduling Policy

In this section, we provide an algorithm that provably approaches the solution of the NUM problem (15.8) to any degree of accuracy. The distance between the network utility achieved and the optimum is determined by a control parameter, which regulates the trade-off between the gap from optimality and the request queue lengths. The algorithm is based on the well-known Lyapunov optimization framework [9]. The resulting *drift-plus-penalty* (DPP) policy adapts to arbitrarily changing network conditions and in fact is optimal (with respect to the NUM problem) under nonstationary and nonergodic evolution of the underlying network state process [12].

15.3.1   The Drift-Plus-Penalty Expression

Let $\mathbf{Q}(t), \mathbf{\Theta}(t), \boldsymbol{\gamma}(t), \boldsymbol{\mu}(t), \mathbf{B}(t)$, and $\mathbf{D}(t)$ denote column vectors containing the elements $Q_u(t), \Theta_u(t), \gamma_u(t), \mu_u(t), B_{f_u}(m_u(t), t)$, and $D_{f_u}(m_u(t), t)$, respectively, for all $u \in \mathcal{U}$. Let $\mathbf{G}(t) = \left[\mathbf{Q}^\mathsf{T}(t), \mathbf{\Theta}^\mathsf{T}(t) \right]^\mathsf{T}$ be the composite vector of all queue backlogs, and we define the quadratic Lyapunov function $L(\mathbf{G}(t)) = \frac{1}{2} \mathbf{G}^\mathsf{T}(t) \mathbf{G}(t)$. Intuitively, taking actions to push $L(\mathbf{G}(t))$ down tends to maintain the stability of all queues. We define

$$\Delta(\mathbf{G}(t)) \triangleq L(\mathbf{G}(t+1)) - L(\mathbf{G}(t)) \qquad (15.11)$$

as the one-step drift of the Lyapunov function at slot t. The DPP policy described below acquires information about the queue states $\mathbf{G}(t)$, the rate–quality profile $(B_{f_u}(\cdot, t), D_{f_u}(\cdot, t))$ for all users u, and the channel state $\omega(t)$ at every slot t, and chooses control actions $\{m_u(t)\}$, $[\mu_{bu}(t)] \in \mathcal{R}(\omega(t))$, and $\boldsymbol{\gamma}(t)$, subject to $D_{f_u}^{\min} \leq \gamma_u(t) \leq D_{f_u}^{\max}$, in order to minimize a bound on the *drift-plus-penalty*

$$\Delta(\mathbf{G}(t)) - V \sum_{u \in \mathcal{U}} \phi_u(\gamma_u(t)), \qquad (15.12)$$

where V is a nonnegative control parameter that affects the performance of the algorithm. Intuitively, V controls the relative weight with which the control actions of the algorithm in slot t emphasize utility maximization with respect to drift minimization. The bound on (15.12) is given by the following lemma.

Lemma 15.2   *For any choice of the control actions, the drift-plus-penalty expression satisfies*

$$\Delta(\mathbf{G}(t)) - V \sum_{u \in \mathcal{U}} \phi_u(\gamma_u(t)) \leq \mathcal{K} - V \sum_{u \in \mathcal{U}} \phi_u(\gamma_u(t)) + (\mathbf{B}(t) - \boldsymbol{\mu}(t))^\mathsf{T} \mathbf{Q}(t)$$

$$+ (\boldsymbol{\gamma}(t) - \mathbf{D}(t))^\mathsf{T} \mathbf{\Theta}(t). \qquad (15.13)$$

where $\mathcal{K}$ is a uniform upper bound on the term

$$\frac{1}{2} \left[(\mathbf{B}(t) - \boldsymbol{\mu}(t))^\mathsf{T} (\mathbf{B}(t) - \boldsymbol{\mu}(t)) + (\boldsymbol{\gamma}(t) - \mathbf{D}(t))^\mathsf{T} (\boldsymbol{\gamma}(t) - \mathbf{D}(t)) \right]$$

under the realistic assumption that the chunk sizes, the transmission rates, and the video quality are bounded.

Proof. Expanding the quadratic Lyapunov function, we have

$$L(\mathbf{G}(t+1)) - L(\mathbf{G}(t))$$

$$= \frac{1}{2}\left(\mathbf{Q}^{\mathsf{T}}(t+1)\mathbf{Q}(t+1) - \mathbf{Q}^{\mathsf{T}}(t)\mathbf{Q}(t)\right) + \frac{1}{2}\left(\mathbf{\Theta}^{\mathsf{T}}(t+1)\mathbf{\Theta}(t+1) - \mathbf{\Theta}^{\mathsf{T}}(t)\mathbf{\Theta}(t)\right)$$

$$= \frac{1}{2}\left([\mathbf{Q}(t) - \boldsymbol{\mu}(t) + \mathbf{B}(t)]_+^{\mathsf{T}}[\mathbf{Q}(t) - \boldsymbol{\mu}(t) + \mathbf{B}(t)]_+ - \mathbf{Q}^{\mathsf{T}}(t)\mathbf{Q}(t)\right)$$

$$+ \frac{1}{2}\left([\mathbf{\Theta}(t) + \boldsymbol{\gamma}(t) - \mathbf{D}(t)]_+^{\mathsf{T}}[\mathbf{\Theta}(t) + \boldsymbol{\gamma}(t) - \mathbf{D}(t)]_+ - \mathbf{\Theta}^{\mathsf{T}}(t)\mathbf{\Theta}(t)\right), \qquad (15.14)$$

where we have used the queue evolution equations (15.2) and (15.7), and $[\cdot]_+$ is applied componentwise.

Using the fact that for any nonnegative scalar quantities $\Theta, \gamma,$ and D we have the inequality

$$[\Theta + \gamma - D]_+^2 \leq (\Theta + \gamma - D)^2 = \Theta^2 + (\gamma - D)^2 + 2\Theta(\gamma - D), \qquad (15.15)$$

we have

$$L(\mathbf{G}(t+1)) - L(\mathbf{G}(t)) \leq \frac{1}{2}(\mathbf{B}(t) - \boldsymbol{\mu}(t))^{\mathsf{T}}(\mathbf{B}(t) - \boldsymbol{\mu}(t)) + (\mathbf{B}(t) - \boldsymbol{\mu}(t))^{\mathsf{T}}\mathbf{Q}(t)$$

$$+ \frac{1}{2}(\boldsymbol{\gamma}(t) - \mathbf{D}(t))^{\mathsf{T}}(\boldsymbol{\gamma}(t) - \mathbf{D}(t)) + (\boldsymbol{\gamma}(t) - \mathbf{D}(t))^{\mathsf{T}}\mathbf{\Theta}(t).$$

$$(15.16)$$

Under the realistic assumption that the chunk sizes, the transmission rates, and the video quality measures are bounded above by some constants independent of t, the term

$$\frac{1}{2}\left[(\mathbf{B}(t) - \boldsymbol{\mu}(t))^{\mathsf{T}}(\mathbf{B}(t) - \boldsymbol{\mu}(t)) + (\boldsymbol{\gamma}(t) - \mathbf{D}(t))^{\mathsf{T}}(\boldsymbol{\gamma}(t) - \mathbf{D}(t))\right]$$

is bounded above by a constant $\mathcal{K}$. Using this fact and adding the penalty term $-V\sum_{u\in\mathcal{U}}\phi_u(\gamma_u(t))$ on both sides of the inequality (15.16) yields the result.

The nonconstant part in the right-hand side of (15.13) can be rewritten as

$$\left[\mathbf{B}^{\mathsf{T}}(t)\mathbf{Q}(t) - \mathbf{D}^{\mathsf{T}}(t)\mathbf{\Theta}(t)\right] - \left[V\sum_{u\in\mathcal{U}}\phi_u(\gamma_u(t)) - \boldsymbol{\gamma}^{\mathsf{T}}(t)\mathbf{\Theta}(t)\right] - \boldsymbol{\mu}^{\mathsf{T}}(t)\mathbf{Q}(t). \quad (15.17)$$

The resulting control actions are given by the minimization, at transmission slot t, of the expression in (15.17). Notice that the first term of (15.17) depends only on $\{m_u(t)\}$, the second term of (15.17) depends only on $\boldsymbol{\gamma}(t)$, and the third term of (15.17) depends only on $\boldsymbol{\mu}(t)$. Thus, the overall minimization decomposes into three separate subproblems, which can be identified as "pull congestion control," i.e., adaptation of the quality level of the requested chunks, "greedy maximization" of the individual utility functions through the auxiliary variables $\boldsymbol{\gamma}(t)$, and *PHY rate scheduling*. Next, we address the minimization of (15.17), focusing on its (separable) components.

15.3.2   Pull Congestion Control at the UEs

The first term in (15.17) is given by

$$\sum_{u \in \mathcal{U}} \left\{ Q_u(t) B_{f_u}(m_u(t), t) - \Theta_u(t) D_{f_u}(m_u(t), t) \right\}. \tag{15.18}$$

The minimization variables $m_u(t)$ appear in separate terms of the sum and hence can be optimized separately for each user $u \in \mathcal{U}$. Each user observes the queues $Q_u(t), \Theta_u(t)$ and is aware of the rate–quality profile $(B_{f_u}(\cdot, t), D_{f_u}(\cdot, t))$ in slot t. Hence, it can choose the quality level of the requested chunk at every chunk slot i, i.e., at transmission slots $t \in \{in : i \in \mathbb{Z}\}$, as

$$m_u(t) = \arg\min_{m \in \{1, \dots, N_{f_u}\}} \left\{ Q_u(t) B_{f_u}(m, t) - \Theta_u(t) D_{f_u}(m, t) \right\}. \tag{15.19}$$

As defined in (15.1), for all transmission slots t which are not integer multiples of n, there is no chunk requested and therefore $B_{f_u}(m_u(t), t)$ and $D_{f_u}(m_u(t), t)$ are equal to 0.

Qualitatively, the control action in (15.19) chooses an appropriate chunk quality level that balances the desire for high quality (reflected by the term $-\Theta_u(t) D_{f_u}(m, t)$) and the desire for low request queue lengths (reflected by the term $Q_u(t) B_{f_u}(m, t)$). The resulting chunk request is placed in the request queue and, eventually, the bits will be pulled from some BS in $\mathcal{N}(u)$. This policy is reminiscent of the current DASH technology [4], where the client (user) progressively fetches a video file by downloading successive chunks, and makes adaptive decisions about the source encoding quality based on its current knowledge of the congestion of the underlying server–client connection.

15.3.3   Greedy Maximization of the Individual Utilities at the UEs

The second term in (15.17), after a change of sign, is given by

$$\sum_{u \in \mathcal{U}} \left\{ V\phi_u(\gamma_u(t)) - \gamma_u(t)\Theta_u(t) \right\}. \tag{15.20}$$

Again, this is maximized by maximizing separately each term, yielding the simple one-dimensional maximization (solvable by line search for example)

$$\gamma_u(t) = \arg\max_{\gamma \in [D_{f_u}^{\min}, D_{f_u}^{\max}]} \left\{ V\phi_u(\gamma) - \Theta_u(t)\gamma \right\}. \tag{15.21}$$

Notice that, in order to compute (15.19) and (15.21), each user needs to know only *local information* consisting of the locally maintained request queue backlog $Q_u(t)$ and by the local virtual-queue backlog $\Theta_u(t)$.

15.3.4   PHY Rate Scheduling at the BSs

At transmission slot t, the network controller observes the queues $\mathbf{Q}(t)$ and the network state $\omega(t)$, and chooses the feasible instantaneous-rate matrix $[\mu_{bu}(t)] \in \mathcal{R}(\omega(t))$ to maximize the weighted sum rate of the transmission rates achievable in transmission slot t. Namely, the network of BSs must solve the max-weighted sum rate (MWSR) problem

$$\text{maximize} \sum_{b \in \mathcal{B}} \sum_{u \in \mathcal{N}(b)} Q_u(t) \mu_{bu}(t)$$

$$\text{subject to} \quad [\mu_{bu}(t)] \in \mathcal{R}(\omega(t)). \tag{15.22}$$

It can immediately be seen that, after a change of sign, the maximization of the third term in (15.17) yields problem (15.22).

In order to make the MWSR transmission scheduling decision represented by (15.22), the network controller needs the instantaneous global information represented by $\mathbf{Q}(t)$ and $\omega(t)$. Under certain assumptions about the system, the solution to the general MWSR problem lends itself to a simple distributed implementation where each BS b makes its own scheduling decisions using knowledge of the local request queues $\{Q_u(t) : u \in \mathcal{N}(b)\}$. This information can be learned over-the-air at the cost of a small protocol overhead. Notice that this overhead is nothing more than a rate priority request in the form of a recursively computed rate weight, as currently implemented in proportional fairness scheduling [13]. Therefore, it is expected that the implementation of the DPP policy will not be significantly more complicated than the current DASH on top of standard PHY resource allocation schemes. When particularizing the general DPP policy to the massive MIMO multicell network considered in Section 15.5, we will see that the resulting scheme falls into this fortunate class of systems (see also Remark 15.3).

15.4   Policy Performance

In practice, the quality and rate profiles $D_f(m,t)$ and $B_f(m,t)$ of a given session depend on the specific file f and can be considered as arbitrarily varying, and not necessarily stationary and ergodic processes. Also, the network state $\omega(t)$ depends on the propagation path loss between the BSs and UEs, and this in turn depends on the users' motion and large-scale fading effects (e.g., the presence of trees, buildings, and hills), which is also somewhat arbitrary. Therefore, it is meaningful to consider the optimality of the DPP policy for an *arbitrary sample path* of $\{\omega(t), D_f(m,t), B_f(m,t)\}$. Following in the footsteps of [9, 12], we compare the network utility achieved by our DPP policy with that achieved by an optimal oracle policy with T-slot lookahead, i.e., knowledge of the future sample path over an interval of length T slots. Time is split into frames of duration T slots, and we consider F such frames. The static optimization problem over the jth frame is given by

$$\text{maximize} \sum_{u \in \mathcal{U}} \phi_u \left(\frac{1}{T} \sum_{\tau=jT}^{(j+1)T-1} D_{f_u}(m_u(\tau), \tau) \right) \tag{15.23}$$

$$\text{subject to} \frac{1}{T} \sum_{\tau=jT}^{(j+1)T-1} \left[B_{f_u}(m_u(\tau), \tau) - \mu_u(\tau) \right] \leq 0, \quad \forall u \in \mathcal{U}, \tag{15.24}$$

$$[\mu_{bu}(\tau)] \in \mathcal{R}(\omega(\tau)), \quad \forall \tau \in \{jT, \ldots, (j+1)T-1\}, \tag{15.25}$$

$$m_u(\tau) \in \{1, \ldots, N_{f_u}\}, \quad \forall u \in \mathcal{U},$$

$$\forall \tau \in \{jT, \ldots, (j+1)T-1\}, \tag{15.26}$$

and we let ϕ_j^{opt} denote the resulting maximum of the network utility function for frame j. We have the following result (the proof is omitted owing to space limitations, and can be found in [14]).

Theorem 15.3 *The DPP scheduling policy achieves a per-sample path network utility*

$$\sum_{u \in \mathcal{U}} \phi_u(\overline{D}_u) \geq \lim_{F \to \infty} \frac{1}{F} \sum_{j=0}^{F-1} \phi_j^{\text{opt}} - O\left(\frac{1}{V}\right) \tag{15.27}$$

with bounded queue backlogs satisfying

$$\lim_{F \to \infty} \frac{1}{FT} \sum_{\tau=0}^{FT-1} \left(\sum_{u \in \mathcal{U}} Q_u(\tau) + \sum_{u \in \mathcal{U}} \Theta_u(\tau) \right) \leq O(V), \tag{15.28}$$

where $O(1/V)$ indicates a term that vanishes as $1/V$ and $O(V)$ indicates a term that grows linearly with V, as the policy control parameter V grows large.

An immediate corollary of Theorem 15.3 is the following.

Corollary 15.4 *When the evolution of the topology state $\omega(t)$, the rate function $B_f(m, t)$, and the quality function $D_f(m, t)$ are stationary and ergodic, then*

$$\sum_{u \in \mathcal{U}} \phi_u(\overline{D}_u) \geq \phi^{\text{opt}} - O\left(\frac{1}{V}\right), \tag{15.29}$$

where ϕ^{opt} is the optimal value of the NUM problem (15.5) in the stationary ergodic case,[‡] and

$$\sum_{u \in \mathcal{U}} \overline{Q}_u + \sum_{u \in \mathcal{U}} \overline{\Theta}_u \leq O(V). \tag{15.30}$$

In particular, if the network state is an independent and identically distributed (i.i.d.) process, the bounding term in (15.29) is explicitly given by $O(1/V) = K/V$, and the bounding term in (15.30) is explicitly given by $(K + V(\phi_{\max} - \phi_{\min}))/\epsilon$, where $\phi_{\min} = \sum_{u \in \mathcal{U}} \phi_u(D_{f_u}^{\min})$, $\phi_{\max} = \sum_{u \in \mathcal{U}} \phi_u(D_{f_u}^{\max})$, $\epsilon > 0$ is the slack variable corresponding to the constraint (15.24), and the constant K is defined in (15.13).

[‡] Notice that in the stationary and ergodic case the value ϕ^{opt} is generally achieved by an instantaneous policy with perfect knowledge of the state statistics or, equivalently, by a policy with infinite look-ahead, since the state statistics can be learned arbitrarily well from any sample path with probability 1 because of ergodicity.

15.5   Wireless System Model with Massive MU-MIMO Helpers

In this section, we specialize the region of instantaneous service rates $\mathcal{R}(\omega(t))$ to the specific case of a PHY layer comprising a massive MU-MIMO system at each BS. Then, we discuss the solution to the MWSR problem (15.22) for this system. We shall see that, thanks to the channel-hardening effect in high-dimensional MIMO channels, the MWSR problem can be optimally solved by a low-complexity greedy algorithm which can be implemented in a distributed manner, with each BS independently choosing user subsets for MU-MIMO beamforming.

15.5.1   PHY Rates of Massive MIMO BSs

For simplicity of exposition, we present the system under the assumption that the user fading-channel vectors are composed of i.i.d. zero-mean unit-variance complex circularly symmetric coefficients, which are perfectly known at the BS transmitters. In the spirit of massive MIMO [6], we consider per-BS MU-MIMO precoding, i.e., we do not allow base station cooperation. Therefore, the intercell interference at each user receiver is treated as additive noise. Extensions and alternative (more realistic) PHY models are briefly discussed in Remark 15.2 at the end of this section. Each BS b, equipped with a large number of antennas M_b, implements MU-MIMO to serve the users in its neighborhood $\mathcal{N}(b)$. We assume that the number of spatially multiplexed data streams that BS b can transmit is some fixed number $1 \leq S_b \ll M_b$, i.e., following [6, 15, 16], we assume that the number of downlink data streams is limited by the uplink pilot dimension rather than by the number of BS antennas. We assume further that each BS performs linear zero-forcing beamforming (LZFBF) to the set of selected users (referred to in the following as "active users").

The wireless channel is modeled by the well-known and widely accepted OFDM block-fading model, where at each transmission slot t, the channel corresponding to the BS–user link (b,u) in $\mathcal{E}$, on a given OFDM subcarrier $v = 1, \ldots, N$, is given by

$$y_u(t; v) = \sqrt{g_{bu}(t)} \boldsymbol{\xi}_{bu}^{\mathsf{H}}(t; v) \mathbf{V}_b(t; v) \mathbf{x}_b(t; v)$$
$$+ \sum_{b' \neq b} \sqrt{g_{b'u}(t)} \boldsymbol{\xi}_{b'u}^{\mathsf{H}}(t; v) \mathbf{V}_{b'}(t; v) \mathbf{x}_{b'}(t; v) + z_u(t; v), \qquad (15.31)$$

where $\boldsymbol{\xi}_{bu}(t; v)$ is the $M_b \times 1$ column vector of channel coefficients from the antenna array of BS b to the receiving antenna of user u, $g_{bu}(t)$ is the large-scale path loss or shadowing from BS b to UE u (independent of the subcarrier index v, since the path loss is frequency-flat), $\mathbf{V}_b(t; v)$ is the downlink precoding matrix of BS b, and $\mathbf{x}_b(t; v)$ is the corresponding vector of transmitted complex symbols. Also, $z_u(t; v)$ denotes the additive Gaussian noise at the uth UE receiver. Notice that (15.31) takes fully into account the intercell interference from the signals sent by other BSs $b' \neq b$ on the link from BS b to user u.

We use $\mathcal{S}_b(t)$ to denote the subset of users scheduled for transmission by BS b in slot t. Collecting the channel vectors $\boldsymbol{\xi}_{bu}(t; f)$ for $u \in \mathcal{S}_b(t)$ into the columns of an $M_b \times |\mathcal{S}_b(t)|$

channel matrix $\boldsymbol{\Xi}_b(t;\nu)$, the LZFBF precoding matrix of dimension $M \times |\mathcal{S}_b(t)|$ is given by the column-normalized pseudo-inverse

$$\mathbf{V}_b(t;\nu) = \boldsymbol{\Xi}_b(t;\nu)(\boldsymbol{\Xi}_b^{\mathsf{H}}(t;\nu)\boldsymbol{\Xi}_b(t;\nu))^{-1}\boldsymbol{\Lambda}_b^{1/2}(t;\nu), \qquad (15.32)$$

where $\boldsymbol{\Lambda}_b(t;\nu)$ is a diagonal matrix with the uth diagonal element given by

$$\Lambda_u(t;\nu) = \frac{1}{\left[\left(\boldsymbol{\Xi}_b^{\mathsf{H}}(t;\nu)\boldsymbol{\Xi}_b(t;\nu)\right)^{-1}\right]_{uu}} \qquad (15.33)$$

($[\cdot]_{uu}$ denotes the uth diagonal element of the matrix argument). Using the fact that $\boldsymbol{\Xi}_b^{\mathsf{H}}(t;\nu)\mathbf{V}_b(t;\nu) = \boldsymbol{\Lambda}^{1/2}(t;\nu)$, the resulting downlink channel to user $u \in \mathcal{S}_b(t)$ becomes

$$y_u(t;\nu) = \sqrt{g_{bu}(t)\Lambda_u(t;\nu)}x_{bu}(t;\nu)$$
$$+ \sum_{b' \neq b} \sqrt{g_{b'u}(t)}\boldsymbol{\xi}_{b'u}^{\mathsf{H}}(t;\nu)\mathbf{V}_{b'}(t;\nu)\mathbf{x}_{b'}(t;\nu) + z_u(t;\nu). \qquad (15.34)$$

Our goal here is to obtain an accurate yet simple characterization of the feasible instantaneous-rate region $\mathcal{R}(\omega(t))$ for the channel model (15.34), where the network state is defined by the path loss coefficients (reflecting the network topology), i.e., $\omega(t) = [g_{bu}(t)]$. For this purpose, we shall exploit some standard results in large-random-matrix theory (see [17–21]), consider achievable rates under Gaussian random coding and worst-case uncorrelated additive noise plus interference [22], and make a simple application of Jensen's inequality. By dividing the channel coefficient by $\sqrt{M_b}$ and scaling up the BSs' transmit power by M_b we obtain an equivalent channel model, for which the following deterministic approximation holds: as M_b and $|\mathcal{S}_b(t)|$ become large with fixed ratio $|\mathcal{S}_b(t)|/M_b \leq 1$, it is well known that

$$\Lambda_u(t;\nu) \approx \left(1 - \frac{|\mathcal{S}_b(t)| - 1}{M_b}\right), \qquad (15.35)$$

where the approximation error $\Lambda_u(t;\nu) - (1 - |\mathcal{S}_b(t)| - 1/M_b)$ is asymptotically Gaussian with a variance that vanishes as $O(1/M_b^2)$ [17, 18]. By treating the intercell interference as (uncorrelated) additive noise, and assuming a very large number of time–frequency symbols per slot, we can immediately show that the following rate is achievable:

$$R_{bu}(t) = \frac{1}{N}\sum_{\nu=1}^{N}\log\left(1 + \frac{\Lambda_u(t;\nu)g_{bu}(t)M_bP_b/|\mathcal{S}_b(t)|}{1 + \sum_{b' \neq b}g_{b'u}(t)\|\boldsymbol{\xi}_{b'u}^{\mathsf{H}}(t;\nu)\mathbf{V}_{b'}(t;\nu)\|^2 M_{b'}P_{b'}/|\mathcal{S}_{b'}(t)|}\right). \qquad (15.36)$$

Now, using the deterministic approximation (15.35), replacing the arithmetic mean over the subcarrier index with an ensemble average over the fading statistics, and using Jensen's inequality, we find the desired approximate achievable-rate expression as (see [14] for details)

$$R_{bu}(t) \approx \log\left(1 + \frac{M_b - |\mathcal{S}_b(t)| + 1}{|\mathcal{S}_b(t)|}\frac{g_{bu}(t)P_b}{1 + \sum_{b' \neq b}g_{b'u}(t)P_{b'}}\right), \qquad (15.37)$$

where we have used the fact that, by construction of the normalization of the LZFBF precoder in (15.32), we have $\mathbb{E}[\|\xi_{b'u}^H(t;v)\mathbf{V}_{b'}(t;v)\|^2] = 1/M_{b'}\text{tr}(\mathbb{E}[\mathbf{V}_{b'}^H(t;v)\mathbf{V}_{b'}(t;v)]) = |\mathcal{S}_{b'}(t)|/M_{b'}$.

For the sake of notational simplicity, we define the rate (raw) vectors $\mathbf{c}_b(\mathcal{S}_b(t),t) \in \mathbb{R}_+^{|\mathcal{U}|}$ with components

$$\mathbf{c}_{bu}(\mathcal{S}_b(t),t) = \begin{cases} 0, & \text{if } u \notin \mathcal{S}_b(t), \\ T_d R_{bu}(t), & \text{if } u \in \mathcal{S}_b(t), \end{cases} \tag{15.38}$$

where $R_{bu}(t)$ is given by (15.37) and T_d denotes the number of time–frequency data symbols per PHY frame. Notice that for each BS b and each set of scheduled users $\mathcal{S}_b(t) \subseteq \mathcal{N}(b)$, the vector $\mathbf{c}_b(\mathcal{S}_b(t),t)$ yields the number of information bits that the users $u \in \mathcal{S}_b(t)$ can successfully decode from BS b during slot t. Hence, the desired concise expression for $\mathcal{R}(\omega(t))$ is given as follows.

Proposition 15.1 *Feasible instantaneous-rate region. For every t, the region of instantaneous feasible rates $\mathcal{R}(\omega(t))$ is formed by all $|\mathcal{B}| \times |\mathcal{U}|$ rate matrices $[\mu_{bu}(t)]$ whose bth row is $\mathbf{c}_b(\mathcal{S}_b(t),t)$, for some $\mathcal{S}_b(t) \subseteq \mathcal{N}(b)$, for all $b \in \mathcal{B}$.*

We assume that the UEs are "smart" in the sense that they can decode multiple streams in the same transmission slot, i.e., user u, in transmission slot t, can receive $\mu_u(t) = \sum_{b \in \mathcal{N}(u)} \mu_{bu}(t)$ information bits by simultaneously downloading $\mu_{bu}(t)$ bits from BSs $b \in \mathcal{N}(u)$. Notice that each stream is achievable (in an information-theoretic sense) by treating the other streams as Gaussian noise, i.e., we do not make use of multiuser detection schemes (e.g., based on successive interference cancellation) at the user receivers. Therefore, our rate expressions are representative of what can be achieved with today's user device technology. For the sake of comparison, in the simulation results presented in Section 15.6 we also consider a *dumb receiver heuristic* where each user u decodes only the strongest data stream and therefore downloads only $\max_{b \in \mathcal{N}(u)} \mu_{bu}(t)$ information bits. While the dumb receiver heuristic is a degradation of the optimal solution involving advanced receivers, the simulation results in Section 15.6 show that this degradation is quite small. This also implicitly indicates that, in most relevant practical topologies and path loss scenarios, it is unlikely that the same user is scheduled by more than one BS in the same transmission slot, i.e., $\mathcal{S}_b(t) \cap \mathcal{S}_{b'}(t)$ is empty with high probability for $b \neq b'$.

Remark 15.2. Although we have chosen here, for the sake of clarity, to consider the somewhat simplistic case of i.i.d. zero-mean channel vectors and perfect channel state information, we hasten to say that introducing deterministic approximations to the instantaneous rates of users in more involved and realistic cases including BS antenna correlation [23–27] and/or pilot-based channel state information, typically obtained through uplink pilots and *time division duplexing* (TDD) reciprocity [6, 15, 16] and including the effect of pilot contamination, is just a simple exercise. However, since this chapter is focused on cross-layer optimization of multiple data-oriented sessions over a multicell multiuser wireless network employing (massive) MU-MIMO, we have chosen to use the simple model defined above. In fact, while the feasible instantaneous-rate region takes on the same form as that given in Proposition 15.1, the expression for the instantaneous feasible rates may be significantly more complicated. Furthermore, it has been recognized that the problem of pilot contamination manifests itself as the limiting system factor mainly in the regime of a very large number of antennas ($M_b \to \infty$). For a large but finite number of antennas, as considered here, and especially for LZFBF precoding, it is known that standard intercell interference rather than pilot contamination is the dominating limiting factor [15, 16].

15.5.2   Transmission Scheduling with Massive MIMO BSs

We now particularize problem (15.22) to the specific case of a wireless system with massive MIMO BSs. Since the rate vectors at each BS (i.e., the rows $c_b(S_b(t),t)$) can be chosen independently at each BS, (15.22) decouples into separate maximizations for each BS b given by the following discrete optimization problem:

$$\text{maximize} \sum_{u \in \mathcal{N}(b)} Q_u(t)\mu_{bu}(t)$$

$$\text{subject to} \{\mu_{bu}(t)\}_{u \in \mathcal{N}(b)} \in \{c_b(S_b,t) : S_b \subseteq \mathcal{N}(b) : |S_b| \leq S_b\}. \tag{15.39}$$

This corresponds to maximizing, for each BS b, the weighted sum rate over the discrete set of vectors $\{c_b(S_b,t) : S_b \subseteq \mathcal{N}(b), |S_b| \leq S_b\}$, with an exponential number of choices for the subset of active users (the optimization is over all subsets of size up to S_b of the users in $\mathcal{N}(b)$). The key observation that allows us to eliminate this exponential complexity is that when BS b schedules the subset S_b of users for MU-MIMO beamforming, the rate of each user $u \in S_b$ depends only on the cardinality $|S_b|$ and not on the identity of the members of the subset S_b. This is an important consequence of the massive MIMO "deterministic" rate behavior, due to the channel hardening in the presence of a large number of antennas. As a consequence, for a fixed subset size S, the subset $\mathcal{U}^*(S,t)$ of users maximizing the weighted sum rate can be obtained by sorting the users in $\mathcal{N}(b)$ according to the weighted rate

$$Q_u(t) \log \left(1 + \frac{M_b - S + 1}{S} \frac{P_b g_{bu}(t)}{\left(1 + \sum_{b' \neq b} P_{b'u} g_{b'u}(t)\right)} \right)$$

and choosing greedily the best S users. Thus, we have

$$\mathcal{U}_b^*(S,t)$$

$$= \text{arg max-}S \left\{ Q_u(t) \log \left(1 + \frac{M_b - S + 1}{S} \frac{P_b g_{bu}(t)}{1 + \sum_{b' \neq b} P_{b'u} g_{b'u}(t)} \right) : u \in \mathcal{N}(b) \right\},$$

$$\tag{15.40}$$

where arg max-S denotes the operation of choosing the first S elements of a set of real numbers sorted in decreasing order.

This *sort & greedy selection* procedure is repeated for every subset size, yielding all the subsets $\{\mathcal{U}^*(S,t) : S = 1,\ldots,\min\{S_b, |\mathcal{N}(b)|\}\}$. Then, from these subsets, the subset $\mathcal{S}_b^*(t)$ which has the maximum weighted sum rate is chosen, yielding the optimal solution to (15.39).

A typical sorting algorithm has complexity $O(|\mathcal{N}(b)|\log(|\mathcal{N}(b)|))$ and since the sorting procedure is repeated for every subset size, the algorithm has complexity $O(|\mathcal{N}(b)|^2\log(|\mathcal{N}(b)|))$, which improves upon existing user-scheduling algorithms [28] for the MIMO broadcast channel.

Remark 15.3. Notice that, unlike the case for conventional cellular systems, we do not assign a fixed set of users to each BS. In contrast, the BS–user association is dynamic, and results from transmission-scheduling decisions according to the selection of the served user subset $\mathcal{S}_b^*(t)$ from the set (15.40). For a practical implementation of the PHY rate-scheduling algorithm, each BS b first needs to learn locally the request queue lengths Q_u of the users in its neighborhood $\mathcal{N}(b)$. Furthermore, the $\mu_{bu}(t)$ information bits transmitted by BS b to user u need to correspond to the chunks at the head of the request queue Q_u, at the quality level requested by the user in some previous chunk slot according to the pull scheme (15.19). Thus, each user u must also broadcast the metadata (chunk number and quality level) of the chunks at the head of the queue along with Q_u to the BSs in $\mathcal{N}(b)$. Explicit accounting for this control information is neglected here for brevity, but it should be noticed that this is not very different from what is already implemented today in HTTP/DASH.

15.6   Numerical Experiments

In this section, we present a few numerical examples illustrating the behavior of the DPP scheduling policy applied to a multicell massive MIMO scenario. We considered an 80 m × 80 m region with five BSs (indicated by o's) as shown in Figure 15.2. Approximately 500 users (indicated by *'s) were placed according to a nonhomogeneous Poisson point process with a higher density in a central region of size 80/3 m × 80/3 m, as shown in Figure 15.2.

Each BS had M antennas and served user sets of size up to S, with a transmission power of 40 dBm. The path loss from a BS to a user was given by $1/(1 + (d/40)^{3.5})$, with d representing the BS–user distance (assuming a torus wrap-around model to avoid boundary effects). We assumed a PHY frame duration of 10 ms and a total system bandwidth of 18 MHz as specified in the LTE 4G standard. With one OFDM resource block (7 × 12 channel symbols) spanning 0.5 ms in time and 180 kHz in bandwidth (corresponding to 12 adjacent subcarriers, each with 15 kHz bandwidth), each transmission slot spanned $T_d = 84 \times 100 \times 20$ channel symbols.

We considered on-demand video-streaming sessions. Each video file consisted of a long sequence of chunks, each of duration 0.5 s and with a frame rate of 30 frames per second. We considered a specific video sequence formed from 800 chunks, constructed using several standard video clips from the database in [29]. The chunks were encoded in different quality modes, with the quality index measured using the structural similarity (SSIM) index defined in [30]. The chunks from 1 to 200 were

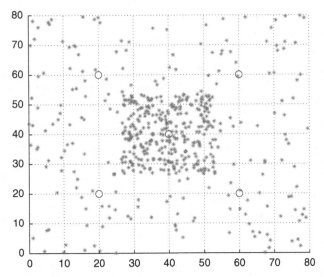

Figure 15.2 Simulation layout [14]. Reproduced with permission from the IEEE.

encoded in eight quality modes with an average bit rate of 631 kbps. Chunks 201–400 were encoded in four quality modes at an average bit rate of 3908 kbps. Similarly, chunks 401–600 and 601–800 were encoded into four and eight quality modes with average bit rates of 6679 kbps and 556 kbps, respectively. In the simulation, each user started its streaming session of 1000 chunks from some arbitrary position in this reference video sequence and successively requested 1000 chunks by cycling through the sequence. The scheduling policy used the utility function $\phi_u(\cdot) = \log(\cdot) \; \forall \, u \in \mathcal{U}$ to impose proportional fairness.

We studied the performance of our algorithm with $M = 40$ antennas per BS and a maximum active-user subset size $S = 10$, for different values of the policy control parameter V, and observed that the QoE metric (average network utility) did not increase significantly for V larger than 2×10^{14}. Hence, we used this value for the rest of the simulations presented here.

The QoE improvement due to the use of MU-MIMO with respect to a legacy single-user MIMO (SU-MIMO) system was evidenced by the cumulative distribution function (CDF) of the QoE metrics over the user population for three different cases: (1) MU-MIMO with $M = 40$ antennas and maximum active user subset size $S = 10$; (2) MU-MIMO with $M = 20$ antennas and maximum active user subset size $S = 5$; and (3) SU-MIMO with $M = 10$ antennas and $S = 1$. From Figure 15.3, we can observe that there is significant improvement in video-streaming performance in terms of average video quality, average delay, and percentage of time spent in buffering mode when MU-MIMO is employed in the PHY layer in comparison with SU-MIMO. This clearly indicates that upgrading current SU-MIMO systems to massive MU-MIMO is a promising approach not only in terms of instantaneous PHY rate but also in terms of "application layer" QoE metrics incorporating a fairness criterion.

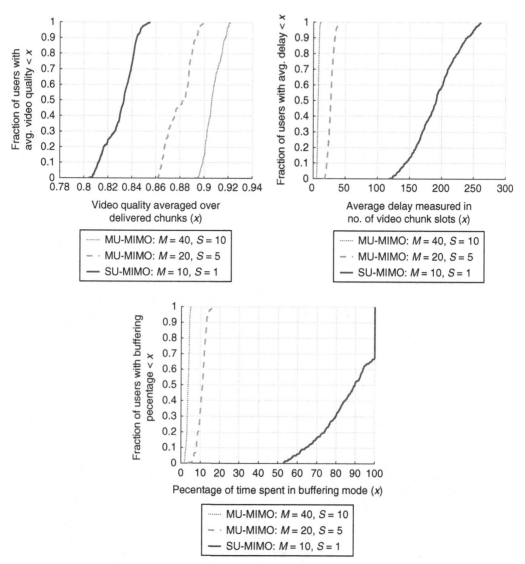

Figure 15.3 Video-streaming QoE improvement with MU-MIMO over SU-MIMO [14]. Reproduced with permission from the IEEE.

We have also studied the benefits of using a cross-layer approach in comparison with a baseline scheme representative of present technology. We considered a case where each BS employed SU-MIMO with $M = 10$ antennas. For the baseline scheme, every user first fixed its association with the unique BS that provided the maximum received signal strength (RSSI) $P_b g_{bu}$ and then used the same control decision (15.19) to choose the quality levels for the chunks that arrived in the request queue every video chunk slot. Furthermore, we assumed that the BSs *locally* employed proportional fairness scheduling [13], which under the massive MIMO deterministic rate approximation reduces to *equal-airtime* scheduling. In brief, each BS b scheduled the users associated

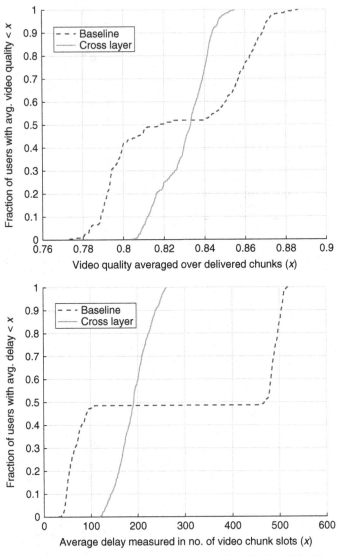

Figure 15.4 Performance comparison of a cross-layer approach with a baseline scheme [14]. Reproduced with permission from the IEEE.

with it through the max-RSSI scheme in a round-robin fashion across the transmission slots *independent of the request queue lengths at the users*. This baseline scheme is representative of current practical systems where the decisions in different layers are independent and there is no interaction between the upper and lower layers. We have plotted the CDFs over the user population of the average video quality and the average delay in the reception of chunks in Figure 15.4. We can observe that the cross-layer scheme treats the users in a fair manner, while the baseline scheme favors some users at the expense of other users in the system. Since the topology in Figure 15.2 has a central region with a higher user density (hotspot) and the user–BS association is max-RSSI based in the baseline scheme, most of the users in the hotspot become associated with the BS in the center of the topology. As a consequence, this BS becomes *overloaded*, with many users associated with it. Moreover, since the BS employs the round-robin scheduling policy in the baseline scheme, each user in the hotspot is scheduled only on a small fraction of the transmission slots, resulting in poor video quality and delay performance. However, the users outside the hotspot are associated with the lightly loaded BSs and therefore experience better video-streaming QoE. This explains the skewed nature of the CDFs for the baseline scheme in Figure 15.4.

On the other hand, with the cross-layer scheme, the user–BS association is dynamic and changes during the course of the control policy depending on the transmission-scheduling decisions, which in turn depend on the achievable rates of the users and their dynamically changing request queue lengths. Such dynamic user–BS association implicitly balances the load across all the BSs in the system, leading to fairness in QoE performance across the user population.

15.7   Conclusion

In this chapter, we have developed a systematic framework for scheduling in a multicell massive MIMO wireless network comprising multiple users with data-oriented downlink sessions. Each session is formed from a long sequence of data chunks. In general, such data may represent a multimedia source, which can be represented at different levels of the quality–rate trade-off. The framework is based on defining a network utility function that captures the users' individual QoEs and imposes some desired notion of fairness between different sessions. The corresponding NUM problem is solved through a dynamic policy derived from the Lyapunov DPP approach in a system for efficient streaming of video content in a network of BSs capable of implementing the advanced physical layer technique of massive MU-MIMO. We have formulated an NUM problem to maximize the global network utility function and solve it using the Lyapunov optimization technique to derive a DPP dynamic policy. The proposed policy decomposes into decentralized congestion control decisions, greedy maximization of the individual utility functions at the UEs, and global PHY rate-scheduling decisions at the BSs, obtained by solving an MWSR problem. For the special case of massive MIMO networks, we have devised a low-complexity greedy user selection scheme to solve the MWSR problem optimally. This optimal low-complexity decentralized solution is possible thanks to the channel-hardening effect in massive MIMO, which provides an additional nontrivial advantage at the system architecture level. Overall, the proposed DPP scheduling policy is amenable to easy implementation based only on local information and some metadata describing the quality–rate profile of the requested files.

References

[1] Cisco, "Cisco visual networking index: Global mobile data traffic forecast update, 2015–2020," Tech. Rep., Feb. 2016. Available at http://goo.gl/1XYhqY.

[2] DASH Industry Forum, "MPEG DASH standard." Available at http://mpeg.chiariglione.org/standards/mpeg-dash.

[3] A. Begen, T. Akgul, and M. Baugher, "Watching video over the web: Part 1: Streaming protocols," *IEEE Internet Comput.,* vol. 15, no. 2, pp. 54–63, Mar. 2011.

[4] Y. Sánchez, T. Schierl, C. Hellge, T. Wiegand, D. Hong, D. De Vleeschauwer, W. Van Leekwijck, and Y. Lelouedec, "iDASH: Improved dynamic adaptive streaming over http using scalable video coding," in *Proc. of ACM Multimedia Systems Conf. (MMSys),* Feb. 2011.

[5] A. F. Molisch, *Wireless Communications,* 2nd edn, Wiley, 2011.

[6] T. L. Marzetta, "Noncooperative cellular wireless with unlimited numbers of base station antennas," *IEEE Trans. Wireless Commun.,* vol. 9, no. 11, pp. 3590–3600, Nov. 2010.

[7] Z. Wang, A. C. Bovik, H. R. Sheikh, and E. P. Simoncelli, "Image quality assessment: From error visibility to structural similarity," *IEEE Trans. Image Process.,* vol. 13, no. 4, pp. 600–612, Apr. 2004.

[8] M. J. Neely, "Wireless peer-to-peer scheduling in mobile networks," in *Proc. of Conf. on Information Sciences and Systems (CISS),* Mar. 2012.

[9] M. Neely, "Stochastic network optimization with application to communication and queueing systems," *Synth. Lect. Commun. Netw.,* vol. 3, no. 1, pp. 1–211, 2010.

[10] J. Mo and J. Walrand, "Fair end-to-end window-based congestion control," *IEEE/ACM Trans. Netw.,* vol. 8, no. 5, pp. 556–567, Oct. 2000.

[11] D. Bethanabhotla, G. Caire, and M. J. Neely, "Adaptive video streaming for wireless networks with multiple users and helpers," *IEEE Trans. Commun.,* vol. 63, no. 1, pp. 268–285, Jan. 2015.

[12] M. Neely, "Universal scheduling for networks with arbitrary traffic, channels, and mobility," in *Proc. of IEEE Conf. on Decision and Control (CDC),* Dec. 2010.

[13] A. Eryilmaz and R. Srikant, "Fair resource allocation in wireless networks using queue-length-based scheduling and congestion control," in *Proc. of IEEE International. Conf. on Computer Communications (INFOCOM),* Mar. 2005.

[14] D. Bethanabhotla, G. Caire, and M. Neely, "WiFlix: Adaptive video streaming in massive MU-MIMO wireless networks," *IEEE Trans. Wireless Commun.,* vol. 15, no. 6, pp. 4088–4103, Jun. 2016.

[15] H. Huh, G. Caire, H. Papadopoulos, and S. Ramprashad, "Achieving massive MIMO spectral efficiency with a not-so-large number of antennas," *IEEE Trans. Wireless Commun.,* vol. 11, no. 9, pp. 3226–3239, Sep. 2012.

[16] J. Hoydis, S. Ten Brink, and M. Debbah, "Massive MIMO in the UL/DL of cellular networks: How many antennas do we need?" *IEEE J. Sel. Areas Commun.,* vol. 31, no. 2, pp. 160–171, Feb. 2013.

[17] A. Tulino and S. Verdú, *Random Matrix Theory and Wireless Communications,* NOW Publishers, 2004.

[18] R. Couillet and M. Debbah, *Random Matrix Methods for Wireless Communications,* Cambridge University Press, 2011.

[19] D. Bethanabhotla, O. Y. Bursalioglu, H. C. Papadopoulos, and G. Caire, "Optimal user-cell association for massive MIMO wireless networks," *IEEE Trans. Wireless Commun.*, vol. 15, no. 3, pp. 1835–1850, Mar. 2016.

[20] H. Huh, A. M. Tulino, and G. Caire, "Network MIMO with linear zero-forcing beam-forming: Large system analysis, impact of channel estimation, and reduced-complexity scheduling," *IEEE Trans. Inf. Theory*, vol. 58, no. 5, pp. 2911–2934, May 2012.

[21] Y. Lim, C. Chae, and G. Caire, "Performance analysis of massive MIMO for cell-boundary users," *IEEE Trans. Wireless Commun.*, vol. 14, no. 12, pp. 6827–6842, Dec. 2015.

[22] B. Hassibi and B. M. Hochwald, "How much training is needed in multiple-antenna wireless links?" *IEEE Trans. Inf. Theory*, vol. 49, no. 4, pp. 951–963, Apr. 2003.

[23] A. Adhikary, J. Nam, J.-Y. Ahn, and G. Caire, "Joint spatial division and multiplexing the large-scale array regime," *IEEE Trans. Inf. Theory*, vol. 59, no. 10, pp. 6441–6463, Oct. 2013.

[24] J. Nam, A. Adhikary, J.-Y. Ahn, and G. Caire, "Joint spatial division and multiplexing: Opportunistic beamforming, user grouping and simplified downlink scheduling," *IEEE J. Sel. Top. Signal Process.*, vol. 8, no. 5, pp. 876–890, Oct. 2014.

[25] A. Adhikary, E. Al Safadi, M. K. Samimi, R. Wang, G. Caire, T. S. Rappaport, and A. F. Molisch, "Joint spatial division and multiplexing for mm-Wave channels," *IEEE J. Sel. Areas Commun.*, vol. 32, no. 6, pp. 1239–1255, Jun. 2014.

[26] H. Yin, D. Gesbert, M. Filippou, and Y. Liu, "A coordinated approach to channel estimation in large-scale multiple-antenna systems," *IEEE J. Sel. Areas Commun.*, vol. 31, no. 2, pp. 264–273, Feb. 2013.

[27] S. Wagner, R. Couillet, M. Debbah, and D. Slock, "Large system analysis of linear precoding in correlated MISO broadcast channels under limited feedback," *IEEE Trans. Inf. Theory*, vol. 58, no. 7, pp. 4509–4537, Jul. 2012.

[28] T. Yoo and A. Goldsmith, "On the optimality of multiantenna broadcast scheduling using zero-forcing beamforming," *IEEE J. Sel. Areas Commun.*, vol. 24, no. 3, pp. 528–541, Mar. 2006.

[29] xiph.org, "Xiph.org video test media." Availabel at http://media.xiph.org/video/derf/.

[30] Z. Wang, A. C. Bovik, H. R. Sheikh, and E. P. Simoncelli, "The SSIM index for image quality assessment." Available at http://goo.gl/ngR0UL.

16 Mobile Data Offloading for Heterogeneous Wireless Networks

Man Hon Cheung, Haoran Yu, and Jianwei Huang

16.1 Introduction

According to Cisco's forecast [1], mobile data traffic will grow to 30.6 exabytes per month by 2020, which amounts to a nearly eight-fold increase between 2015 and 2020 globally. Such a huge amount of traffic is putting increasing pressure on the cellular network operators. On the other hand, traditional network expansion methods, such as acquiring more spectrum and upgrading to more advanced communication technologies such as Long Term Evolution (LTE)-Advanced, are often costly and time-consuming. An efficient way to increase the network capacity in a cost-effective and timely manner is to use complementary technologies, such as Wi-Fi or small cells, to offload the traffic originally targeted toward the cellular network. In fact, Cisco showed that offloaded mobile data traffic exceeded cellular traffic for the first time in 2015, where 51% of the total mobile data traffic was offloaded to the fixed network through Wi-Fi or femtocell networks [1]. Owing to the popularity of Wi-Fi usage and deployment, we will focus our attention on the offloading of mobile data through Wi-Fi networks for the rest of this chapter.

In general, there are two main approaches to the initiation of Wi-Fi offloading, namely user-initiated and operator-initiated offloading. In the early days of implementation, when Wi-Fi networks were not tightly integrated with cellular networks, *user-initiated* offloading was the common choice, where the mobile users needed to manually select the network that they intended to use. However, in such a cellular and Wi-Fi *coexistence* scenario, the cellular operators usually lose their visibility of the users' activities and thus cannot provide a guaranteed quality of experience (QoE) to users.

In contrast, with the ongoing standardization efforts that we will discuss in the next section, cellular and Wi-Fi networks are becoming increasingly tightly coupled together, so that the performance of the Wi-Fi network is usually within the mobile operator's control. This enables *operator-initiated* offloading, where the connection manager in a mobile device connects with the mobile operator's server and retrieves the mobile operator's policy to initiate the offloading procedure. In other words, the mobile operators have more control over the users' network selections and thus their

QoE under the operator-initiated offloading. Overall, the benefits in this cellular and Wi-Fi *integration* scenario are as follows:

- *Seamless authentication.* Currently, a user usually needs to input their username and password to be authenticated to a Wi-Fi network. However, with new technologies such as Hotspot 2.0, a mobile device can be connected to a Wi-Fi network automatically based on a roaming agreement between the mobile operator and the Wi-Fi owner without any user intervention, which improves the user's QoE.
- *Automatic network selection.* With cellular and Wi-Fi coexistence, a user may need to go through a list of available hotspots and choose to connect to a particular network based on limited information about those hotspots. However, with cellular and Wi-Fi integration, the network discovery and selection can be done automatically. Moreover, a mobile operator can provide a better QoE by selecting a more suitable network for a user based on the mobile operator's policy, the network performance, and the user's data plan.
- *Policy integration.* Besides the integration of network access between the cellular and Wi-Fi networks, the policy and charging functions can also be integrated. Thus, more consistent services can be offered, and more advanced tiering and charging options can be supported [2].
- *Seamless roaming.* Since a common authentication and network selection platform is used in the integration model, more seamless roaming among mobile operators can be achieved.

In the following section, we will describe the current standardization efforts that aim to facilitate the implementation of mobile data offloading.

16.2   Current Standardization Efforts

In this section, we discuss a few closely related standardization efforts related to mobile data offloading, namely Access Network Discovery and Selection Function (ANDSF), Hotspot 2.0, and Next Generation Hotspot (NGH). We also introduce advancements in the management of radio resources between cellular and Wi-Fi networks in the 3rd Generation Partnership Project (3GPP) standard.

16.2.1   Access Network Discovery and Selection Function (ANDSF)

ANDSF is a primary enabler of intelligent network selection for communication between the cellular and Wi-Fi networks specified in the 3GPP standard. In this architecture, an ANDSF server provides users with *network information* and *operator-defined policies*, which are rules that are used to prioritize and restrict access to networks. Some of the network information and operator policies specified in ANDSF are as follows [3, 4]:

Table 16.1 Comparisons between today's hotspot and Hotspot 2.0.

	Hotspot today	Hotspot 2.0
Network selection	Manual	Automatic (IEEE 802.11u)
User authentication	Manual	Automatic (IEEE 801.1x)
Over-the-air encryption	No	Yes (IEEE 802.11i)

- *User location.* A user may need to report their current location to the ANDSF server.
- *Discovery information.* Once the ANDSF server knows the user's current location, it can provide *discovery information* to the user about the access networks that are available in the user's vicinity.
- *Intersystem mobility policy (ISMP).* ISMP consists of a number of prioritized rules, which list the priorities of the networks that a user can assess at a particular location and time, and these are sent by the ANDSF server to the users. ISMP allows a user to route Internet Protocol (IP) traffic over a *single* radio interface (e.g., either a cellular network or a Wi-Fi network) at a time.
- *Intersystem routing policy (ISRP).* Similarly to ISMP, ISRP is an operator-defined policy sent by the ANDSF server to users for network selection. However, ISRP allows a user to route IP traffic over *multiple* radio interfaces (e.g., both cellular and Wi-Fi networks) simultaneously.

In addition, ANDSF can operate in both a push and a pull mode, where the ANDSF server can initiate a push to distribute its information to the target users in the former case, while the users query the ANDSF server to pull the desired information in the latter case [4].

16.2.2   Hotspot 2.0

Hotspot 2.0 is a working group in the Wi-Fi Alliance (WFA), and its certification program is called *Passpoint*. Its target is to make Wi-Fi hotspot usage and roaming as easy as is the case in a cellular network [4]. It is based primarily on the IEEE 802.11u, 802.11i, and 802.1x standards [5]. A brief comparison between the today's hotspot and Hotspot 2.0 is given in Table 16.1.

In particular, for the IEEE 802.11u standard related to network selection, its target is to allow a device to obtain more information about an access point (AP) *before* deciding to connect with it. During the network discovery process, without this standard, the information that the users obtain is the service set identifier (SSID) of the AP and some basic security information. In contrast, the IEEE 802.11u standard enables the discovery of additional hotspot information by users before joining the network. Such information includes the mobile operator's name, the roaming partners accessible through the hotspot, pricing, Internet speed, the credential type (e.g., subscriber identification module (SIM) or username/password), the Extensible Authentication Protocol (EAP) types supported for authentication, and the IP address type availability (e.g., IPv4 or IPv6).

As a brief comparison between ANDSF and Hotspot 2.0, they can be regarded as both partners and competitors with each other [4]. More specifically, in Passpoint Release 1, Hotspot 2.0 can provide useful network information to ANDSF, so the two technologies are complementary to each other. However, in Passpoint Release 2, Hotspot 2.0 starts to provide network-selection-related instructions to users, which may compete with ANDSF.

16.2.3   Next Generation Hotspot (NGH)

Closely related to the Hotspot 2.0 efforts of the WFA, NGH is an initiative driven by the Wireless Broadband Alliance (WBA) that also aims to promote seamless and secure Wi-Fi network access. While Hotspot 2.0 focuses mainly on the interoperability of the *equipment*, NGH focuses more on the partnership among network and service *operators*. In particular, it has published the global Wi-Fi roaming specification WISPr 2.0, which is a standard for Wi-Fi roaming among operators [6].

16.2.4   Radio Resource Management

Today's cellular and Wi-Fi networks are usually loosely coupled, and there is often no active cooperation between them. The Wi-Fi networks are beyond the control of the mobile operators, who lose visibility of their users whenever those users decide to switch to Wi-Fi. As a result, users may experience a loss of IP session continuity during network switching, which leads to a QoE degradation [4].

In the near future, it is anticipated that cellular and Wi-Fi networks will become more tightly coupled, although the cooperation will still be relatively *loose* and *static*. In the *network-assisted* radio resource management (RRM) in 3GPP Release 12 [7], the mobile operator may provide network information or network selection parameters (e.g., through the ANDSF framework) to the user, who is responsible for their own network selection. Alternatively, in the *network-controlled* RRM in 3GPP Release 13 [7], the mobile operator makes network selection decisions based on Wi-Fi measurements from users.

Going forward, cellular and Wi-Fi networks will have *tighter* and more *dynamic* cooperation, which will result in a single unified radio network. For example, the LTE–WLAN aggregation (LWA) in 3GPP Release 13 specifies the joint packet-level scheduling of the LTE and Wi-Fi networks, where there is a shared transmission queue across the LTE and Wi-Fi schedulers [7]. This increasing level of cellular and Wi-Fi RRM cooperation in the 3GPP standard is summarized in Table 16.2.

16.2.5   Design Considerations in Data Offloading Algorithms

Although the aforementioned standardization efforts outline the architecture for automatic and seamless network selection, they do not specify the detailed offloading

Table 16.2 Increasing level of cooperation in RRM between cellular and Wi-Fi networks.

Cooperation level	Radio resource management	Standard
No cooperation	User controlled	No
Loose cooperation	Network assisted/controlled	3GPP R12/R13
Tight cooperation	Joint packet-level scheduling between cellular and Wi-Fi	3GPP R13

policies, which are supposed to be proprietary operational decisions of the mobile operators. In fact, the design of an efficient offloading policy needs to consider a number of aspects:

- *Delay.* A cellular network usually has better coverage than a Wi-Fi network. In the case of ubiquitous coverage in the cellular network, a user can assess the network at any location all the time without any delay. On the other hand, since the Wi-Fi deployment is usually location dependent, a user may experience network access delay, which depends on their mobility.
- *Data rate.* The comparison of data rates between cellular and Wi-Fi networks depends heavily on the specific communication standards. In fact, the communication technologies in both the cellular standard (e.g., LTE-Advanced) and the Wi-Fi standard (e.g., IEEE 802.11ac) can deliver a peak data rate on the order of gigabits per second. However, the Wi-Fi standard usually has an advantage in the data rate due to the shorter communication distance between the device and the Wi-Fi AP.
- *Energy.* The energy consumption of a user's device for using a Wi-Fi network is usually lower than that for a cellular network, probably owing to the shorter communication distance.
- *Cost.* The cost of using a Wi-Fi network is usually lower than that for a cellular network. At some specific locations, such as the user's home, shopping malls, or cafes, free Wi-Fi is often available.
- *Scalability.* In a typical data offloading scenario, a mobile operator may need to coordinate the network selection for hundreds of users within the coverage of some cellular base stations. Thus, an efficient data offloading scheme needs to scale well with increasing user population.

In the following two sections, we will describe two data offloading algorithms to address these issues. In Section 16.3, we propose an *optimal* delay-aware algorithm that considers that *cost and delay* trade-off of a *single* user. In Section 16.4, we design a *close-to-optimal* and *scalable* algorithm that trade-offs the *energy and delay* of *multiple* users.

16.3   DAWN: Delay-Aware Wi-Fi Offloading and Network Selection

16.3.1   System Model

As shown in Figure 16.1, we consider a mobile user (MU) moving within the coverage of a cellular network, such that a cellular connection is always available to the MU. Occasionally, the MU may be able to access Wi-Fi APs at some locations (e.g., in a coffee shop or in a shopping mall). In other words, the Wi-Fi connection is *location dependent* and may not be available to the MU at all times. The MU is running a file transfer application which requires the transfer of K bits within T time slots. In other words, the file transfer application is *delay-tolerant* with a deadline T. For example, an MU on the road may want to send an email with a large attachment of size 20 Mbyte through a smartphone in the next 10 minutes. The MU moves within a set $\mathcal{L} = \{1,\ldots,L\}$ of possible locations, following a Markovian mobility model that can be derived based on the past mobility pattern of the MU [8].

We consider the *usage-based pricing* used by mobile operators, where the usage price of the cellular network is often higher than that of the Wi-Fi network. It should be noted that this pricing scheme is general, and includes free Wi-Fi as a special case. When making offloading decisions, the MU needs to take into account the payments for different network types and its QoE requirement in terms of file transfer completion. First, the MU has an incentive to offload as much data traffic to the Wi-Fi network

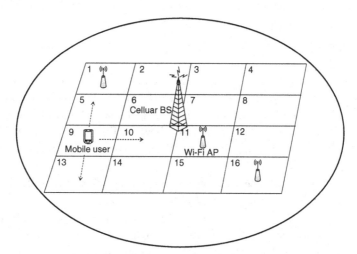

Figure 16.1 An example of the network setting, where the MU is moving within a set of $\mathcal{L} = \{1,\ldots,16\}$ locations. The MU is always in the coverage of a cellular BS, but Wi-Fi is only available at three locations, i.e., $\mathcal{L}^{(1)} = \{1,11,16\}$. The rest of the locations do not have Wi-Fi, i.e., $\mathcal{L}^{(0)} = \mathcal{L}\backslash\mathcal{L}^{(1)}$. We assume that the MU is sending a file of size K bits and that the transfer needs to be completed by a deadline T. Given the mobility pattern of the MU, it the MU aims to decide whether it should remain idle ($a = 0$), use the cellular network ($a = 1$), or use the Wi-Fi network ($a = 2$) if available, in each time slot to reduce its payment under usage-based pricing, while taking into account the potential penalty if the deadline constraint is violated.

as possible, so as to reduce its payment. This means that the MU prefers to defer the transmission until a Wi-Fi hotspot is available. On the other hand, the MU must also consider whether it can complete the file transfer by the deadline. For example, if the time remaining before the deadline is short, then a deferred transmission may violate the deadline if the MU does not have enough opportunity to transmit through Wi-Fi in the near future. In this case, instead, the MU should start the file transfer using the ubiquitous cellular connection as soon as possible to reduce the latency. To sum up, an efficient delay-aware Wi-Fi offloading scheme needs to achieve a good *trade-off* between the total data usage payment and the MU's QoE requirement.

As the Wi-Fi offloading problem involves decision making in multiple time slots before the deadline, we formulate it as a finite-horizon sequential decision problem in the following section. We aim to find the MU's optimal transmission policy, which minimizes the MU's data usage payment while taking into account the deadline of the file transfer application. By defining the total *cost* as the total payment and a *penalty* for not finishing the file transfer by the deadline, we can derive the optimal transmission policy through dynamic programming.

16.3.2   Problem Formulation

In this section, we formulate the delay-aware Wi-Fi offloading problem for a *single* MU as a *finite-horizon sequential* decision problem [9]. Without loss of generality, we normalize the length of a time slot to one. The MU needs to choose an action (to be explained later) in each *decision epoch*

$$t \in \mathcal{T} = \{1, \ldots, T\}. \tag{16.1}$$

The system *state* is defined as $s = (k, l)$. The state element $k \in \mathcal{K} \subseteq [0, K]$ represents the *remaining* size (in bits) of a file to be transferred. The state element $l \in \mathcal{L} = \{1, \ldots, L\}$ is the location index, where L is the total number of possible locations that the MU may reach within the T time slots. As shown in Figure 16.1, we let $\mathcal{L}^{(0)} \subseteq \mathcal{L}$ and $\mathcal{L}^{(1)} \subseteq \mathcal{L}$ be the sets of locations where Wi-Fi is not and is available, respectively, such that $\mathcal{L}^{(0)} = \mathcal{L} \backslash \mathcal{L}^{(1)}$.

The *action* a specifies the transmission decision of the MU in each decision epoch. Specifically, we have $a \in \mathcal{A} = \{0, 1, 2\}$, where $a = 0$ means that the MU chooses to remain idle, $a = 1$ means that the MU transmits through the cellular network, and $a = 2$ represents that the MU transmits through Wi-Fi. Notice that actions $a = 0$ and $a = 1$ are always available to the MU at all locations. Action $a = 2$, however, is only available at a location $l \in \mathcal{L}^{(1)}$. Thus, the available choice of action a depends on the state element l, so $a \in \mathcal{A}^{(l)} \subseteq \mathcal{A}$, where $\mathcal{A}^{(l)}$ is the set of available transmission actions at location l:

$$\mathcal{A}^{(l)} = \begin{cases} \{0, 1, 2\}, & \text{if } l \in \mathcal{L}^{(1)}, \\ \{0, 1\}, & \text{if } l \in \mathcal{L}^{(0)}. \end{cases} \tag{16.2}$$

We adopt the commonly used usage-based pricing, where the payment made by an MU is directly proportional to its data usage. Let $p(l, a)$ be the price per unit of usage for choosing action $a \in \mathcal{A}^{(l)}$ at location l, where $p(l,0) = 0, \forall l \in \mathcal{L}$ for the idle action. It should be noted that we consider a general location and network-dependent pricing, which includes the commonly used location-independent pricing as a special case. Let $\mu(l,a)$ be the estimated throughput of the user at location l with action $a \in \mathcal{A}^{(l)}$, where $\mu(l,0) = 0, \forall l \in \mathcal{L}$ when the MU remains idle (i.e., when $a = 0$). The *payment* of the MU in state s with action $a \in \mathcal{A}^{(l)}$ at time slot $t \in \mathcal{T}$ is

$$c_t(s,a) = c_t(k,l,a) = \min\{k, \mu(l,a)\} p(l,a), \tag{16.3}$$

which is equal to the data usage payment in the time slot.

After the deadline has passed, we define the *penalty* for not being able to finish the file transfer in state s as

$$\hat{c}_{T+1}(s) = \hat{c}_{T+1}(k,l) = h(k), \tag{16.4}$$

where $h(k) \geq 0$ is a nondecreasing function with $h(0) = 0$. The subscript $T+1$ means that we compute the penalty at the beginning of the time slot $T+1$ (immediately after the deadline). In fact, the MU chooses $h(k)$ according to the QoE requirement of its application.

The *state transition probability* $p(s'|s,a) = p\big((k',l')|(k,l),a\big)$ is the probability that the system will go into state $s' = (k',l')$ in the next time slot if action a is taken in state $s = (k,l)$. Since the movement of the MU from location l to location l' is independent of the file size k and transmission action a, we have

$$p(s'|s,a) = p\big((k',l')|(k,l),a\big) = p(l'|l)\, p\big(k'|(k,l),a\big), \tag{16.5}$$

where

$$p\big(k'|(k,l),a\big) = \begin{cases} 1, & \text{if } k' = [k - \mu(l,a)]^+ \text{ and } a \in \mathcal{A}^{(l)}, \\ 0, & \text{otherwise}, \end{cases} \tag{16.6}$$

and $[x]^+ = \max\{0,x\}$. Here, $p(l'|l)$ is the probability that the MU will move from location l to location l', and it is estimated based on the past mobility pattern of the MU [8].

Let $\delta_t : \mathcal{K} \times \mathcal{L} \to \mathcal{A}$ be a function that specifies the transmission decision of the MU in state $s = (k,l)$ and time slot t. We define a *policy* $\pi = (\delta_t(k,l), \forall k \in \mathcal{K}, l \in \mathcal{L}, t \in \mathcal{T})$ as the set of decision rules for states and time slots. We denote by $s_t^\pi = (k_t^\pi, l_t^\pi)$ the state at time slot t if policy π is used, and we let Π be the feasible set of π. The MU aims to find an optimal policy π^* that minimizes the sum of the expected total payment from $t = 1$ to $t = T$ and the penalty at $t = T+1$ as follows:

$$\text{minimize}_{\pi \in \Pi} \quad E_{s_1}^\pi \left[\sum_{t=1}^{T} c_t\big(s_t^\pi, \delta_t(s_t^\pi)\big) + \hat{c}_{T+1}(s_{T+1}^\pi) \right]. \tag{16.7}$$

Here, $E_{s_1}^\pi$ denotes the expectation with respect to the probability distribution of the MU mobility model and the policy π for an initial state $s_1 = (K,l_1)$, where l_1 is the location of the MU at $t = 1$.

16.3.3   General DAWN Algorithm

In this section, we solve problem (16.7) *optimally* using a *finite-horizon* dynamic programming model for the general penalty function, network usage price, and cellular/Wi-Fi data rate. We propose a general DAWN algorithm that computes the optimal policy.

Let $v_t(s)$ be the minimum expected total cost for the MU from time slot t to $T+1$, given that the system is in state s immediately before the decision at time slot t. The *optimality equation* [9, p. 83] relating the minimum expected total cost in different states for $t \in \mathcal{T}$ is given by

$$v_t(s) = v_t(k,l) = \min_{a \in \mathcal{A}^{(l)}} \{\psi_t(k,l,a)\}, \tag{16.8}$$

where for $k \in \mathcal{K}$, $l \in \mathcal{L}$, and $a \in \mathcal{A}^{(l)}$ we have

$$\psi_t(k,l,a) = c_t(k,l,a) + \sum_{l' \in \mathcal{L}} \sum_{k' \in \mathcal{K}} p\big((k',l') \,|\, (k,l),a\big) v_{t+1}(k',l') \tag{16.9}$$

$$= \min\{k, \mu(l,a)\} p(l,a) + \sum_{l' \in \mathcal{L}} p(l' \,|\, l) v_{t+1}\big([k-\mu(l,a)]^+, l'\big). \tag{16.10}$$

The first and second terms on the right-hand side of (16.9) are the *immediate cost* and the *expected future cost*, respectively, in the remaining time slots for choosing action a. The derivation of (16.10) from (16.9) follows directly from (16.3), (16.5), and (16.6). For $t = T+1$, we set the boundary condition as

$$v_{T+1}(s) = \hat{c}_{T+1}(k,l) = h(k), \quad \forall k \in \mathcal{K}, l \in \mathcal{L}. \tag{16.11}$$

With the optimality equation, we are ready to propose the general DAWN algorithm, shown in Algorithm 16.1. The algorithm consists of two phases, namely a planning phase and a transmission and Wi-Fi offloading phase. Let $\sigma > 0$ be the granularity of the discrete state element k in the algorithm (such as 1 Mbit). First, in the planning phase, based on the optimality equation (16.8) and the boundary condition (16.11), we obtain the *optimal policy* π^* that solves problem (16.7) using *backward induction* [9, p. 92]. Specifically, we first set $v_{T+1}(k,l)$ based on the boundary condition (line 2) of Algorithm 16.1. Then, we obtain the values of $\delta_t^*(k,l)$ and $v_t(k,l)$ by updating them recursively backward from time slot $t = T$ to time slot $t = 1$ (lines 3 to 16). Algorithm 16.1 has a computational complexity of $\mathcal{O}(KLT/\sigma)$ [10].

Theorem 16.1   *The policy $\pi^* = (\delta_t^*(k,l), \forall k \in \mathcal{K}, l \in \mathcal{L}, t \in \mathcal{T})$, where*

$$\delta_t^*(k,l) = \arg\min_{a \in \mathcal{A}^{(l)}} \{\psi_t(k,l,a)\}, \tag{16.12}$$

is the optimal solution of problem (16.7).

Proof. Using the principle of optimality [11, p. 18], we can show that π^* is the optimal solution of problem (16.7).

Notice that the optimal policy π^* is a *contingency plan* that contains information about the optimal transmission decision for *all* possible states (k,l) in any time slot

---

Algorithm 16.1 General delay-aware Wi-Fi offloading and network selection (DAWN) algorithm.

1: Planning phase:
2: Set $v_{T+1}(k,l), \forall k \in \mathcal{K}, \forall l \in \mathcal{L}$ using (16.11)
3: Set $t := T$
4: **while** $t \geq 1$
5:   **for** $l \in \mathcal{L}$
6:     Set $k := 0$
7:     **while** $k \leq K$
8:       Calculate $\psi_t(k,l,a), \forall a \in \mathcal{A}^{(l)}$ using (16.10)
9:       Set $\delta_t^*(k,l) := \arg\min_{a \in \mathcal{A}^{(l)}}\{\psi_t(k,l,a)\}$
10:      Set $v_t(k,l) := \psi_t\big(k,l,\delta_t^*(k,l)\big)$
11:      Set $k := k + \sigma$
12:     **end while**
13:   **end for**
14:   Set $t := t - 1$
15: **end while**
16: Output the optimal policy π^* for the transmission and Wi-Fi offloading phase
17: Transmission and Wi-Fi offloading phase:
18: Set $t := 1$ and $k := K$
19: **while** $t \leq T$ **and** $k > 0$
20:   Determine the location index l from GPS
21:   Set action $a := \delta_t^*(k,l)$ based on the optimal policy π^*
22:   **If** $a > 0$
23:     Send $\mu(l,1)$ bits to the cellular network if $a = 1$
       or offload $\mu(l,2)$ bits to the Wi-Fi network if $a = 2$
24:     Set $k := [k - \mu(l,a)]^+$
25:   **end if**
26:   Set $t := t + 1$
27: **end while**

$t \in \mathcal{T}$, and the system computes it *offline* before the file transfer begins in the second phase. In the second phase, the MU first determines the location index l in each time slot based on location information obtained by use of the Global Positioning System (GPS) (line 20). Then, the MU carries out transmission decisions based on the optimal policy π^* through checking a table (lines 21–25), and updates the state element k accordingly (line 24). As the complexity of Algorithm 16.1 is high in general, we will next design an approximation algorithm with a lower computational complexity.

16.3.4　Threshold Policy

In this section, we establish sufficient conditions under which the optimal policy has a *threshold* structure in the remaining file size k and time t. When these conditions are not satisfied, the threshold policy can serve as the basis for the design of a low-complexity approximation algorithm.

Specifically, we consider the following assumptions.

Assumption 16.2 *(a) The penalty function h(k) is convex and nondecreasing in k. (b) Wi-Fi is free to the MU (i.e., $p(l,2) = 0, \forall l \in \mathcal{L}^{(1)}$). (c) The cellular price is location independent (i.e., $p(l,1) = p(l',1), \forall l,l' \in \mathcal{L}, l \neq l'$). (d) The cellular and Wi-Fi data rates are location independent (but these two rates are different in general). That is, $\mu_1 = \mu(l,1), \forall l \in \mathcal{L}$ and $\mu_2 = \mu(l,2), \forall l \in \mathcal{L}^{(1)}$. (e) We can approximate $\min\{k, \mu(l,1)\}$ in (16.3) by $\mu(l,1)$ for action $a = 1$.*

Under Assumption 16.2, we can characterize the optimality of the threshold policy in both of the dimensions k and t as follows.

Theorem 16.3 *Under Assumption 16.2, the optimal policy $\pi^* = (\delta_t^*(k,l), \forall k \in \mathcal{K}, l \in \mathcal{L}, t \in \mathcal{T})$ has a threshold structure in both k and t as follows:*
For a location $l \in \mathcal{L}^{(0)}$ without Wi-Fi, we have

$$\delta_t^*(k,l) = \begin{cases} 1 \ (cellular), & if \ k \geq k^*(l,t), \\ 0 \ (idle), & otherwise, \end{cases} \forall t \in \mathcal{T}, \qquad (16.13)$$

and

$$\delta_t^*(k,l) = \begin{cases} 1 \ (cellular), & if \ t \geq t^*(k,l), \\ 0 \ (idle), & otherwise, \end{cases} \forall k \in \mathcal{K}, \qquad (16.14)$$

where $k^*(l,t)$ and $t^*(k,l)$ are location- and time-dependent thresholds in the dimensions k and t, respectively.
For a location $l \in \mathcal{L}^{(1)}$ with Wi-Fi, if the data rate of Wi-Fi is lower than that of cellular (i.e., $\mu_2 \leq \mu_1$), we have

$$\delta_t^*(k,l) = \begin{cases} 1 \ (cellular), & if \ k \geq k^*(l,t), \\ 2 \ (Wi\text{-}Fi), & otherwise, \end{cases} \forall t \in \mathcal{T}, \qquad (16.15)$$

and

$$\delta_t^*(k,l) = \begin{cases} 1 \ (cellular), & if \ t \geq t^*(k,l), \\ 2 \ (Wi\text{-}Fi), & otherwise, \end{cases} \forall k \in \mathcal{K}. \qquad (16.16)$$

Otherwise (i.e., $\mu_1 < \mu_2$), we have

$$\delta_t^*(k,l) = 2 \ (Wi\text{-}Fi), \forall k \in \mathcal{K}, t \in \mathcal{T}. \qquad (16.17)$$

Theorem 16.3 states that when k is above a threshold (i.e., there are many bits waiting to be transmitted) or when t is above a threshold (i.e., the deadline is close), the MU should use the cellular network immediately to avoid the penalty (if Wi-Fi is not available or Wi-Fi is not fast enough). The proof of Theorem 16.3 is given in [12].

With Theorem 16.3, we can modify Algorithm 16.1 and propose a monotone DAWN algorithm with a much lower complexity. The details can be found in [12].

16.3.5   Performance Evaluation

In this section, we evaluate the performance of the general and monotone DAWN schemes by comparing them with three benchmark schemes (the no-offloading, on-the-spot offloading (OTSO) [13], and Wiffler [14] schemes). In our simulations, the MU moved within $L = 16$ possible locations in a 4×4 grid (similar to that in Figure 16.1). To generate the trajectory of the MU, we considered the state transition probabilities $p(l'|l)$, where we assumed that the probability that the MU stay at the same location between two consecutive time slots was $p(l|l) = 0.1, \forall l \in \mathcal{L}$. Moreover, it was equally likely for the MU to move to any one of the neighboring locations. As an example, in Figure 16.1, at location 7, the probability that the MU would move to one of the locations 3, 6, 8, or 11 would be equal to $(1-0.1)/4 = 0.225$. As another example, at location 1, the probability that the MU would move to one of the neighboring locations 2 and 5 would be equal to $(1 - 0.1)/2 = 0.45$. Unless specified otherwise, we assumed that the cellular data rate $\mu(l,1), \forall l \in \mathcal{L}$ and the Wi-Fi data rate $\mu(l,2), \forall l \in \mathcal{L}^{(1)}$ were truncated (in the range $[0,\infty)$) normally distributed random variables with means μ_c and μ_w, respectively, and with standard deviations equal to 5 Mbps. We assumed that the cellular usage price $p(l,1), \forall l \in \mathcal{L}$ was US\$6/Gbyte, while the Wi-Fi was free, such that $p(l,2) = 0, \forall l \in \mathcal{L}^{(1)}$. The probability that a Wi-Fi connection was available at a particular location was 0.5. The length of a time slot Δt was equal to 10 s. We assumed that the MU was transferring a file (e.g., a movie), where the deadline for the file transfer was D minutes (so $T = 60D/\Delta t$). We set the file size granularity σ to 10 Mbit. For the delay violation penalty, we used the convex function

$$h(k) = bk^2, \quad \forall k \in \mathcal{K}, \tag{16.18}$$

where $b \geq 0$ is a constant. In our simulations, we took $b = 1$.

We compared the performance of the five schemes under stringent and nonstringent deadline requirements. First, we considered a large file size $K = 750$ Mbyte, which made it challenging to complete the transmission when the deadline was short. The mean cellular and Wi-Fi data rates were $\mu_c = 90$ Mbps and $\mu_w = 20$ Mbps, respectively, which are reasonable parameters for a 4G LTE-A cellular system [15] and a congested Wi-Fi network [16]. In Figure 16.2, we plot the probability of completing the file transfer against the deadline D. As D increases, it is more likely that the file transfer will finish before the deadline, so the probability of completing the file transfer in the five schemes increases. Moreover, we observe that the general DAWN and no-offloading schemes achieve the highest probability of completing the file transfer, and the monotone DAWN scheme achieves a slightly lower probability. On the other hand, we observe that the OTSO [13] and Wiffler [14] schemes are not able to complete the file transfer around 38% of the time when $D = 2$ mins. The reason is that these two schemes always offload the traffic to the Wi-Fi network whenever Wi-Fi is available. However, they ignore the QoE requirement of the application in terms of the stringent deadline. When the cellular data rate is higher than the Wi-Fi data rate, the MU may prefer to use the cellular network to increase the chance of file transfer completion despite the higher payment.

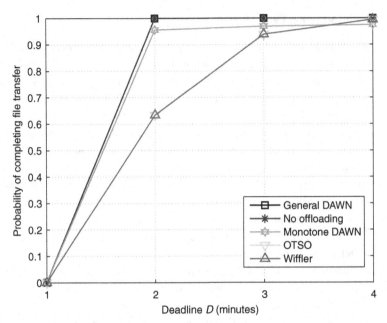

Figure 16.2 Probability of completing file transfer versus deadline D for $p(l|l) = 0.1, \forall l \in \mathcal{L}$, $K = 750$ Mbyte, $\mu_c = 90$ Mbps, and $\mu_w = 20$ Mbps.

Next, we considered a case with a nonstringent deadline requirement due to a smaller file size $K = 92.5$ Mbyte, where all the schemes had a very high probability of completing the file transfer in this setup. In Figure 16.3, we plot the total payment against the deadline D under the five schemes. For the no-offloading scheme, since it always uses the more expensive cellular network, the payment is highest and is independent of D. For the OTSO scheme, the payment is lower than for the no-offloading scheme, because it uses free Wi-Fi networks whenever they are available. However, the OTSO scheme is not aware of the deadline, so it often incurs a significant penalty for violating the deadline. In contrast, the general DAWN, monotone DAWN, and Wiffler schemes are deadline-aware, in that they evaluate the chance of file transfer completion by the deadline. As D increases, these three schemes use the Wi-Fi network more often to complete the file transfer, so the total payment decreases. In Figure 16.3, we observe that the monotone DAWN scheme achieves the same lowest payment as the general DAWN scheme.

16.4   Data Offloading Considering Energy–Delay Trade-off

In the previous section, we designed the optimal data offloading policy for a single mobile user. In order to address the data offloading problem in the case of a large-scale system, where there are multiple cellular and Wi-Fi networks and multiple users, we introduce a low-complexity close-to-optimal data offloading algorithm in this section.

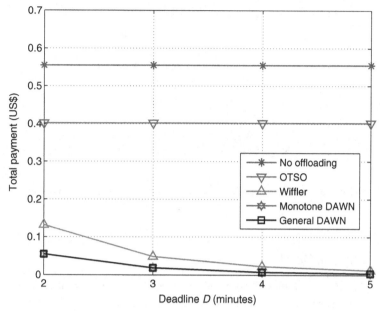

Figure 16.3 Total data usage payment of the user versus deadline D for $p(l\,|\,l) = 0.1, \forall l \in \mathcal{L}$, $K = 92.5$ Mbyte, $\mu_c = 90$ Mbps, and $\mu_w = 20$ Mbps.

16.4.1   Background on Energy-Aware Data Offloading

The explosive growth of global mobile data traffic has led to an increase in the energy consumption of communication networks. Based on the study in [17], the information and communications technology sector constituted about 2% of global CO_2 emissions. Furthermore, the energy consumption of communication networks accounts for a significant proportion of the operational expenditure of mobile operators [18]. An efficient approach to reducing this energy consumption is to offload mobile traffic onto Wi-Fi networks, which have less energy consumption than a cellular network owing to their shorter communication distances.

We focus on the design of an energy-aware *network selection* and *resource allocation* algorithm. First, the mobile operator reduces the energy consumption of the system by selecting networks with a low energy consumption (e.g., Wi-Fi networks) for mobile users. Second, within the cellular network, the mobile operator reduces the transmission power while maintaining system throughput by allocating subchannels and power to cellular users with good channel conditions.

There are two major challenges in our problem. First, we are considering a stochastic system, where users' locations, channel conditions, and traffic demands change over time. This requires the mobile operator to design an algorithm that dynamically selects networks and allocates resources for users based on limited information about the future. Second, the mobile operator needs to reduce the total power consumption while providing delay guarantees to all users. This requires the mobile operator to keep a good balance between power consumption and fairness among users.

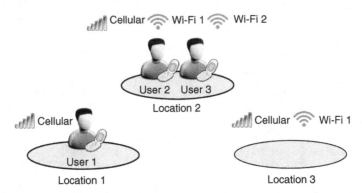

Figure 16.4 An example of the system model, where users 1, 2, and 3 are moving within the set of locations $S = \{1,2,3\}$. The cellular network covers all locations. Each location is covered by a set of Wi-Fi networks, for example, $\mathcal{N}_1 = \emptyset$, $\mathcal{N}_2 = \{1,2\}$, $\mathcal{N}_3 = \{1\}$.

16.4.2   System Model

We consider the downlink transmission in a slotted system, indexed by $t \in \{0,1,\ldots\}$. We focus on the monopoly case, where a single mobile operator serves users with its own cellular and Wi-Fi networks. We consider an orthogonal frequency division multiplexing (OFDM) system for the cellular network [19]. We introduce the following notation:

- $\mathcal{I} \triangleq \{1,2,\ldots,I\}$: set of users;
- $\mathcal{N} \triangleq \{1,2,\ldots,N\}$: set of Wi-Fi networks;
- $\mathcal{S} \triangleq \{1,2,\ldots,S\}$: set of locations;
- $\mathcal{M} \triangleq \{1,2,\ldots,M\}$: set of subchannels.

We assume that the cellular base station covers all S locations, and we use $\mathcal{N}_s \subseteq \mathcal{N}$ to denote the set of available Wi-Fi networks at location $s \in \mathcal{S}$. We illustrate the system model with an example in Figure 16.4.

We group T time slots into a *frame*, and define the kth frame ($k \in \mathbb{N}$) as the time interval that contains a set $\mathcal{T}_k \triangleq \{kT, kT+1, \ldots, kT+T-1\}$ of time slots. We consider the users' random locations, channel conditions, and traffic demands, and assume that the users' locations change every frame while the users' channel conditions and traffic demands change every time slot. The reason is that users' locations usually change much less frequently than the other two kinds of randomness. We define $S_i(kT) \in \mathcal{S}$ as user i's location during the kth frame. We use $H_{im}(t)$ to denote the channel condition for user i on subchannel m at time slot t, and we use $A_i(t) \in [0, A_{\max}]$ to denote the arrival rate of traffic for user i at time slot t, where the constant $A_{\max}$ is the upper bound of the traffic arrival rate. We define $S(kT) \triangleq (S_i(kT), \forall i \in \mathcal{I})$, $H(t) \triangleq (H_{im}(t), \forall i \in \mathcal{I}, m \in \mathcal{M})$, and $A(t) \triangleq (A_i(t), i \in \mathcal{I})$.

16.4.2.1   Network Selection

The mobile operator aims to reduce the total power consumption through network selection, subchannel allocation, and power allocation. We assume that a network selection decision is made every frame instead of every time slot. This is because frequent switching among different networks interrupts data delivery and incurs a nonnegligible cost (e.g., in the form of energy consumption, quality-of-service degradation, and delays). At time slot $t = kT$, i.e., the beginning of the kth frame, the mobile operator determines the network selection for the kth frame. We denote the network selection by $\alpha(kT) = (\alpha_i(kT), \forall i \in \mathcal{I})$, where $\alpha_i(kT)$ indicates the network that user i is connected to during the kth frame. Since the availabilities of Wi-Fi networks are location dependent, we have the following constraint on $\alpha(kT)$:

$$\alpha_i(kT) \in \mathcal{N}_{S_i(kT)} \cup \{0\}, \quad \forall i \in \mathcal{I}, k = 0, 1, \ldots, \tag{16.19}$$

where the selection $\alpha_i(kT) = 0$ indicates that user i is connected to the cellular network.

16.4.2.2 Subchannel Allocation

At each time slot t, the mobile operator determines the subchannel and power allocation. We denote the subchannel allocation by $x(t) = (x_{im}(t), \forall i \in \mathcal{I}, m \in \mathcal{M})$, where the variable $x_{im}(t) \in \{0, 1\}$ for all i and m: if subchannel m is allocated to user i, $x_{im}(t) = 1$; otherwise, $x_{im}(t) = 0$. We assume that each subchannel can be allocated to one user at most:

$$\sum_{i=1}^{I} x_{im}(t) \leq 1, \quad \forall m \in \mathcal{M}. \tag{16.20}$$

Since the mobile operator can only allocate subchannels to those users who are connected to the cellular network, we have the following constraint on $x(t)$:

$$\alpha_i(t_T) x_{im}(t) = 0, \quad \forall i \in \mathcal{I}, m \in \mathcal{M}, t \geq 0. \tag{16.21}$$

Here, $t_T \triangleq \lfloor t/T \rfloor T$ is the beginning of the frame that time slot t belongs to, and the network selection $\alpha_i(t_T)$ indicates user i's associated network during the frame.

16.4.2.3 Power Allocation

We denote the power allocation by $p(t) = (p_{im}(t), \forall i \in \mathcal{I}, m \in \mathcal{M})$, where the variable $p_{im}(t) \geq 0$ denotes the power allocated to user i on subchannel m. We have the following power budget constraint:

$$\sum_{m=1}^{M} \sum_{i=1}^{I} p_{im}(t) \leq P_{\max}^C, \quad \forall t \geq 0, \tag{16.22}$$

where $P_{\max}^C > 0$ is the power budget for the cellular network. Similarly to (16.21), the mobile operator can only allocate power to those users who are connected to the cellular network. We have the following constraint on $p(t)$:

$$\alpha_i(t_T) p_{im}(t) = 0, \quad \forall i \in \mathcal{I}, m \in \mathcal{M}, t \geq 0. \tag{16.23}$$

Based on the network selection $\alpha(t_T)$, subchannel allocation $x(t)$, power allocation $p(t)$, and channel condition $H(t)$, user i's transmission rate at time slot t is given by a function $r_i(\alpha(t_T), x(t), p(t), H(t))$. The mobile operator's total power consumption at time slot t is given by a function $P(\alpha(t_T), p(t))$. We assume that both functions are bounded. In other words, there are positive constants r_{max} and P_{max} such that $r_i(\alpha(t_T), x(t), p(t), H(t)) \in [0, r_{max}]$ and $P(\alpha(t_T), p(t)) \in [0, P_{max}]$ for all $i \in \mathcal{I}$ and $\alpha(t_T), x(t), p(t)$ satisfying (16.19), (16.20), (16.21), (16.22), and (16.23). Our results apply to a general transmission rate function $r_i(\alpha(t_T), x(t), p(t), H(t))$ and power consumption function $P(\alpha(t_T), p(t))$. A detailed example of the transmission rate function and power function can be found in [20].

16.4.3   Problem Formulation

We assume that each user has a data queue, the length of which denotes the amount of unserved traffic. Let $Q(t) = (Q_i(t), \forall i \in \mathcal{I})$ be the queue length vector, where $Q_i(t)$ is user i's queue length at time slot t. We assume that all queues are initially empty, i.e.,

$$Q_i(0) = 0, \ \forall i \in \mathcal{I}. \tag{16.24}$$

The queue length evolves according to the traffic arrival rate and transmission rate as

$$Q_i(t+1) = \left[Q_i(t) - r_i(\alpha(t_T), x(t), p(t), H(t))\right]^+ + A_i(t), \forall i \in \mathcal{I}, t \geq 0. \tag{16.25}$$

Here, $[x]^+ = \max\{x, 0\}$: this occurs because the actual number of served packets cannot exceed the current queue size.

The objective of the mobile operator is to design an online network selection and resource allocation algorithm that minimizes the expected time-average power consumption, while keeping the network stable. This can be formulated as the following optimization problem:

$$\text{minimize} \quad \overline{P} \triangleq \limsup_{K \to \infty} \frac{1}{KT} \sum_{t=0}^{KT-1} \mathbb{E}\{P(\alpha(t_T)), p(t)\}$$

$$\text{subject to} \quad \overline{Q_i} \triangleq \limsup_{K \to \infty} \frac{1}{KT} \sum_{t=0}^{KT-1} \mathbb{E}\{Q_i(t)\} < \infty, \ \forall i \in \mathcal{I}, \tag{16.26}$$

$$\text{constraints } (16.19), (16.20), (16.21), (16.22), (16.23),$$

$$\text{variables} \quad \alpha(t_T), x(t), p(t), \ \forall t \geq 0.$$

Here, $\overline{Q_i}$ is user i's time-average queue length, and the constraint $\overline{Q_i} < \infty$ for all $i \in \mathcal{I}$ ensures the stability of the network. According to Little's law, $\overline{Q_i}$ is proportional to user i's time-average traffic delay. We will show that our algorithm guarantees an upper bound for $\overline{Q_i}$ and thus achieves a bounded traffic delay.

16.4.4   Energy-Aware Network Selection and Resource Allocation (ENSRA) Algorithm

$$\text{minimize} \quad V \sum_{\tau=kT}^{kT+T-1} P\big(\boldsymbol{\alpha}\,(kT),\boldsymbol{p}\,(\tau)\big)$$

$$-\sum_{i=1}^{I} Q_i\,(kT) \sum_{\tau=kT}^{kT+T-1} r_i\big(\boldsymbol{\alpha}\,(kT),\boldsymbol{x}\,(\tau),\boldsymbol{p}\,(\tau),\boldsymbol{H}\,(\tau)\big) \qquad (16.27)$$

$$\text{subject to} \quad \text{constraints } (16.19),(16.20),(16.21),(16.22),(16.23),$$

$$\text{variables} \quad \boldsymbol{\alpha}\,(kT),\boldsymbol{x}\,(\tau),\boldsymbol{p}\,(\tau),\forall \tau \in \mathcal{T}_k.$$

We assume that the mobile operator has complete information about the channel conditions in the current frame, i.e., at time slot $t = kT$ (the beginning of the kth frame), the mobile operator has information about $\boldsymbol{H}\,(\tau)$ for all $\tau \in \mathcal{T}_k$. We introduce our energy-aware network selection and resource allocation (ENSRA) algorithm in Algorithm 16.2. Basically, at the beginning of each frame, the mobile operator solves problem (16.27) to determine the network selection and resource allocation for the whole frame; at the end of each time slot, the mobile operator updates the queue length vector $\boldsymbol{Q}\,(t)$. The intuition behind problem (16.27) can be understood as follows:

- If user i's queue length $Q_i(kT)$ is small, the mobile operator will focus more on the term $V \sum_{\tau=kT}^{kT+T-1} P\big(\boldsymbol{\alpha}\,(kT),\boldsymbol{p}\,(\tau)\big)$ to minimize the objective function in problem (16.27). This implies that the mobile operator will wait for good channels or low-power-cost Wi-Fi networks to serve user i. Since $Q_i(kT)$ is small, suspending user i's traffic in several time slots does not greatly increase the average queue length. According to Little's law, this also does not cause much delay.
- If $Q_i\,(kT)$ is large, the mobile operator will focus more on the term $-Q_i(kT) \cdot \sum_{\tau=kT}^{kT+T-1} r_i\big(\boldsymbol{\alpha}\,(kT),\boldsymbol{x}\,(\tau),\boldsymbol{p}\,(\tau),\boldsymbol{H}\,(\tau)\big)$. This implies that there exists a big "pressure" to push the mobile operator to serve user i immediately, even when the user has a poor channel condition or the power needed for serving this user is high. As a result, user i's queue length is reduced and the mobile operator avoids a severe traffic delay.

Algorithm 16.2 Energy-aware network selection and resource allocation (ENSRA) algorithm

---

1: Set $t = 0$ and $\boldsymbol{Q}\,(0) = \boldsymbol{0}$;
2: **while** $t < t_{end}$ **do** //t_{end} *denotes the number of running time slots for ENSRA.*
3:     **if** $\bmod\,(t,T) = 0$
4:       Set $k = t/T$ and solve problem (16.27) to determine $\boldsymbol{\alpha}\,(kT),\boldsymbol{x}\,(\tau),\boldsymbol{p}\,(\tau),\forall \tau \in \mathcal{T}_k$;
5:     **end if**
6:     Update $\boldsymbol{Q}\,(t+1)$, according to (16.25);
7:     $t \leftarrow t+1$.
8: **end while**

In summary, by adjusting the control parameter $V > 0$, the mobile operator can achieve a good trade-off between the power consumption and the traffic delay.

16.4.5   Performance Analysis of ENSRA

For ease of exposition, we analyze the performance of ENSRA by assuming that the randomness in the system is independent and identically distributed (i.i.d.). Note that with the technique developed in [21], we can obtain similar results under Markovian randomness.

We define the capacity region Λ as the closure of the set of arrival vectors that can be stably supported, considering all network selection and resource allocation algorithms. We assume that the mean traffic arrival is strictly interior to Λ, i.e., there exists an $\eta > 0$ such that

$$\mathbb{E}\{A(t)\} + \eta \cdot \mathbf{1} \in \Lambda. \tag{16.28}$$

We use P_{av}^* to denote the optimal expected time-average power consumption for problem (16.26). The performance of ENSRA is described in the following theorem.

Theorem 16.4   *ENSRA achieves*

$$P_{av}^{ENSRA} \triangleq \limsup_{K \to \infty} \frac{1}{KT} \sum_{t=0}^{KT-1} \mathbb{E}\{P(\alpha(t_T), p(t))\} \leq P_{av}^* + \frac{B}{V}, \tag{16.29}$$

$$Q_{av}^{ENSRA} \triangleq \limsup_{K \to \infty} \frac{1}{KT} \sum_{i=1}^{I} \sum_{t=0}^{KT-1} \mathbb{E}\{Q_i(t)\} \leq \frac{B + VP_{max}}{\eta} + \frac{T-1}{2} I A_{max}, \tag{16.30}$$

where $B = \frac{1}{2} TI \left(A_{max}^2 + r_{max}^2 \right)$.

Here, P_{av}^{ENSRA} and Q_{av}^{ENSRA} are the expected time-average power consumption and expected time-average queue length, respectively, under ENSRA. Notice that the inequality (16.30) ensures $\overline{Q}_i < \infty$ for all $i \in \mathcal{I}$. Theorem 16.4 implies that, by increasing the parameter V, the mobile operator can push the power consumption arbitrarily close to the optimal value, i.e., P_{av}^*, but at the expense of an increase in the average traffic delay.

16.4.6   Performance Evaluation

We simulated the problem with $I = 10$ users, one cellular network, $N = 10$ Wi-Fi networks, and $S = 100$ locations. We set the time slot length to 10 ms, and the frame length to be 1 s, i.e., $T = 100$. We assumed that the cellular network covered all locations, and the Wi-Fi networks were randomly distributed spatially with a coverage of one to four connected locations. We chose the mean traffic arrival rate per user to be 2 Mbps, and ran each experiment in MATLAB for 5000 frames.

In Figure 16.5, we plot the average power consumption P_{av}^{ENSRA} and the average traffic delay per user of ENSRA against the control parameter V. From Little's law, we can

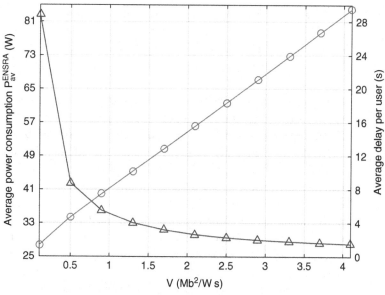

Figure 16.5 Power-delay trade-off of ENSRA.

compute the average traffic delay per user as the ratio between the average queue length per user and the mean traffic arrival rate per user. We observe that as V increases, P_{av}^{ENSRA} decreases. According to (16.29), the upper bound of P_{av}^{ENSRA} decreases with an increase in V, which is consistent with our observation here. Furthermore, as V increases, the average delay per user increases, which is consistent with the result in (16.30). Therefore, Figure 16.5 illustrates the power–delay trade-off of ENSRA.

16.5   Open Problems

There are still some open problems in the area of mobile data offloading for heterogeneous networks.

- *Learning and prediction.* First, it would be interesting to incorporate learning and prediction of the network information into the design of a data offloading algorithm. For example, the study in [20] assumed that when making the network selection and resource allocation decisions, the mobile operator had complete information about the users' mobilities, channel conditions, and traffic demands for the next few minutes. It was shown that by incorporating such future information into the offloading-algorithm design, the mobile operator could significantly improve the performance on the power–delay trade-off. While [20] investigated the fundamental benefit of having future network information, an open problem is to study the approach of learning and the predicting future information and incorporating this into a data offloading algorithm.

- *Heterogeneous QoE requirements.* Second, it is critical to study data offloading algorithms for the situation where different users have heterogeneous QoE requirements. For example, video streaming requires a minimum service rate to guarantee fluency. Moreover, it requires consecutive service and cannot tolerate frequent interruptions. These features prevent a mobile operator from offloading video traffic onto Wi-Fi networks with large capacity fluctuations. In contrast, traffic such as file downloading usually has milder QoE requirements compared with video traffic. Therefore, the mobile operator can fully utilize the network resources by serving traffic with different QoE requirements using networks with different characteristics.
- *Design of Distributed algorithms.* Third, most current data offloading algorithms require the mobile operator to have full information about the system when making network selection and resource allocation decisions for any single user. Furthermore, these offloading algorithms are usually operated in a centralized manner. In order to reduce the computational complexity and avoid frequent information exchange, it is important to consider distributed algorithms that perform data offloading based only on local information. Moreover, it would also be interesting to characterize the performance gaps between distributed and centralized offloading algorithms.

16.6 Conclusion

In this chapter, we introduced mobile data offloading as a cost-effective strategy to address the cellular-network congestion problem. We discussed the current standardization efforts and the challenges related to data offloading. We presented two data offloading algorithms, namely DAWN and ENSRA, where DAWN is an optimal delay-aware algorithm that considers the cost and delay trade-off for a single user, while ENSRA is a close-to-optimal and scalable algorithm that trades off the energy and delay of multiple users. Finally, we concluded this chapter by discussing several open problems in this area.

Acknowledgment

This work was supported by the General Research Funds (Project Number CUHK 14202814) established by the University Grant Committee of the Hong Kong Special Administrative Region, China.

References

[1] Cisco Systems, "Cisco visual networking index: Global mobile data traffic forecast update, 2015–2020," White Paper, Feb. 2016.
[2] M. Paolini, "Wi-Fi and cellular integration: From Wi-Fi offload to HetNets," White Paper, 2014.

[3] Alcatel-Lucent and British Telecommunications, "Wi-Fi roaming: Building on ANDSF and Hotspot2.0," White Paper, 2012.

[4] 4G Americas, "Integration of cellular and Wi-Fi networks," White Paper, Sep. 2013.

[5] Ruckus Wireless, "How interworking works: A detailed look at 802.11u and Hotspot 2.0 mechanism," White Paper, 2013.

[6] Wireless Broadband Alliance, "WISPr 2.0," Specification, 2010.

[7] 4G Americas, "LTE aggregation and unlicensed spectrum," White Paper, Nov. 2015.

[8] A. J. Nicholson and B. D. Noble, "BreadCrumbs: Forecasting mobile connectivity," in *Proc. of ACM International Conf. on Mobile Computing and Networking (MobiCom)*, Sep. 2008.

[9] M. L. Puterman, *Markov Decision Processes: Discrete Stochastic Dynamic Programming.* Wiley, 2005.

[10] M. H. Ngo and V. Krishnamurthy, "Optimality of threshold policies for transmission scheduling in correlated fading channels," *IEEE Trans. Commun.*, vol. 57, no. 8, pp. 2474–2483, Aug. 2009.

[11] D. P. Bertsekas, *Dynamic Programming and Optimal Control,* Volume 1, 3rd edn. Athena Scientific, 2005.

[12] M. H. Cheung and J. Huang, "DAWN: Delay-aware Wi-Fi offloading and network selection," *IEEE J. Sel. Areas Commun.*, vol. 33, no. 6, pp. 1214–1223, Jun. 2015.

[13] K. Lee, I. Rhee, J. Lee, S. Chong, and Y. Yi, "Mobile data offloading: How much can WiFi deliver?" in *Proc. of ACM CoNEXT*, Nov. 2010.

[14] A. Balasubramanian, R. Mahajan, and A. Venkataramani, "Augmenting mobile 3G using WiFi," in *Proc. of ACM International Conf. on Mobile Systems, Applications, and Services (MobiSys)*, Jun. 2010.

[15] Wikipedia, "4G." Available at http://en.wikipedia.org/wiki/4G

[16] "IEEE, IEEE 802.11." Available at http://standards.ieee.org/getieee802/download/802.11-2007.pdf.

[17] A. Fehske, G. Fettweis, J. Malmodin, and G. Biczók, "The global footprint of mobile communications: The ecological and economic perspective," *IEEE Commun. Mag.*, vol. 49, no. 8, pp. 55–62, Aug. 2011.

[18] E. Oh, B. Krishnamachari, X. Liu, and Z. Niu, "Toward dynamic energy-efficient operation of cellular network infrastructure," *IEEE Commun. Mag.*, vol. 49, no. 6, pp. 56–61, Jun. 2011.

[19] J. Huang, V. G. Subramanian, R. Agrawal, and R. A. Berry, "Downlink scheduling and resource allocation for OFDM systems," *IEEE Trans. Wireless Commun.*, vol. 8, no. 1, pp. 288–296, Jan. 2009.

[20] H. Yu, M. H. Cheung, L. Huang, and J. Huang, "Power–delay tradeoff with predictive scheduling in integrated cellular and Wi-Fi networks," *IEEE J. Sel. Areas Commun.*, vol. 34, no. 4, pp. 735–742, Apr. 2016.

[21] L. Huang and M. J. Neely, "Max-weight achieves the exact $[O(1/V), O(V)]$ utility–delay tradeoff under Markov dynamics," arXiv preprint arXiv:1008.0200, 2010.

17   Cellular 5G Access for Massive Internet of Things

Germán Corrales Madueño, Nuno Pratas, Čedomir Stefanović, and Petar Popovski

17.1   Introduction to the Internet of Things (IoT)

The IoT refers to the paradigm of physical and virtual "things" that communicate and collaborate over the Internet, with or without human intervention. The spectra of things that may be connected within the IoT ranges from complex machines, such as aircraft and cars, to everyday appliances, such as consumer refrigerators, and very simple devices such as humidity sensors. The emphasis of the IoT is on services, which represent the primary driver for interconnecting things. Examples of IoT services include micro-climate monitoring of homes, asset tracking during transportation, and, on a larger scale, controlling the power consumption of all the refrigerators in a country depending on the load. Current and forecast market evaluations (such as Cisco's forecast of a \$14.4 trillion global IoT market by 2022 [1]) show that the IoT has a huge revenue potential, to be shared between operators, service providers, hardware vendors, and testing-solutions vendors. Thus, it is not surprising that the IoT is currently one of the hottest topics in the telecommunications world, endorsed by both industry and academia.

A term closely related, but not identical to IoT is *machine-to-machine* (M2M) *communications*, or, in the Third Generation Partnership Project (3GPP) terminology, *machine-type-communications* (MTC). M2M communications refer to the concept in which machines (i.e., standalone devices) communicate with a remote server without human intervention. "M2M can be considered as the plumbing of IoT" [2] or, more formally stated, M2M communications are the key enabler of IoT services. A natural question that arises is how well the existing networking solutions and technologies can serve as the basis for M2M communications and, more broadly, IoT services and, when they cannot support them, how to design other, suitable connectivity solutions. These questions have in recent years instigated a significant body of research and development by industry, standardization bodies, and academia. The general conclusion is that the existing technologies, in their present form, cannot efficiently support M2M communications. The reason is that existing communication systems, particularly in the wireless domain, are designed to efficiently support human-type communications (HTC), such as web browsing, voice calls, and video streaming, where high data rates are essential but the volume of users that simultaneously require service is far beyond the expected number of interconnected devices. In contrast, M2M communications are characterized by the transmission and reception of small data packets from a potentially massive number of devices. This fundamentally changes the goal of the design of the communication system from increasing link capacity to the implementation of low-overhead signaling protocols and handling a large number of simultaneous connections. In summary, the identified deficiencies of the existing communication technologies have led to development of new, typically proprietary solutions [3, 4] or to enhancing the current technologies [5], and these are competing to become the communication protocol of choice for M2M communications and, therefore, the IoT.

Cellular technologies are seen as viable candidates in this race, owing to their maturity and worldwide availability and the use of reserved spectrum. Nevertheless, the main obstacle to the efficient support of M2M communications in cellular networks is that such networks are not designed to support a large number of simultaneous connection attempts. Specifically, owing to the sporadic nature of data transmission in M2M communications, where a device only occasionally transmits a short report, very often a device has to first establish a connection with an access point, i.e., base station, using a connection establishment procedure. This procedure involves a considerable amount of signaling exchange and starts in a slotted ALOHA-based manner, involving several bottlenecks when a massive number of devices are active simultaneously [6]. In such cases, the connection establishment will fail for most of the devices, resulting in outages, i.e., service deprivation, which makes the cellular access mechanisms for M2M/IoT communications extremely important. Furthermore, many IoT devices will be placed inside buildings and basements, involving significant a decrease in signal power due to penetration losses. This calls for a technology that supports extended coverage. Some IoT devices are expected to operate for more than 10 years without charging or replacing the battery, thereby requiring ultra-low power consumption. Finally, in addition to supporting massive IoT, the next generation of cellular technologies should also support another class of IoT devices that offer *mission-critical communication*. These are characterized by requirements for ultra-high reliability and low latency. Although massive M2M is usually treated separately from mission-critical M2M communication, in this chapter we will focus mainly on massive M2M, occasionally referring to the technology issues for mission-critical M2M communications.

17.2   IoT Traffic Patterns in Network Access

The IoT is characterized by a wide application range, for example monitoring and control of household appliances, home security, elderly care, e-health, smart metering, monitoring and control of smart grids, industrial processes, and traffic and automotive applications. All of these applications share common features that impact access to the cellular network [7]:

- Small data transmissions, with a packet/report size below 1000 bytes.

- Uplink-dominated transmissions.
- Typically, delay-tolerant transmissions for a regular/periodic mode of reporting.
- Service priority and low delays for an alarm mode of reporting.
- Low-mobility devices, which are in most cases sensors or controllers deployed at a specific position.
- Low-complexity and low-power transceivers with limited processing capabilities and lack of software updates, in order to achieve extremely low cost.
- Strict requirements on energy consumption, as it is assumed that the device battery is not rechargeable, while the operational time should be more than 10 years.
- A potentially very large number of devices in a cell.

Moreover, there are two dominant traffic patterns that can be observed in all potential IoT applications: *asynchronous* and *synchronous* reporting. In asynchronous reporting, the report arrivals are not correlated across different devices, i.e., the devices report in an independent and random manner. For example, asynchronous reporting arises in the case where earthquake sensors periodically inform a remote server that they are operational. Although the cause of the reporting is the same, the likelihood of multiple devices reporting at the same time is almost negligible. On the other hand, in synchronous reporting, there is a strong correlation in the traffic patterns across devices such that a considerable number of devices become active simultaneously or nearly simultaneously, typically owing to an external event that triggers these devices. For example, synchronous reporting arises in the case where earthquake sensors are triggered by an actual earthquake, when a multitude of devices will try report the detected seismic activity to the remote server almost simultaneously.

Obviously, the asynchronous/synchronous nature of traffic patterns is strongly related to the reporting mode of the IoT devices. Specifically, there are three main reporting modes, depicted in Figure 17.1: (i) *periodic* reporting, (ii) *on-demand* reporting, and (iii) *alarm* reporting, where the first two modes give rise to the traffic patterns of asynchronous reporting and the last mode to synchronous reporting.

Periodic reporting occurs when reports are sent to or exchanged with the remote server under standard operation conditions in some monitoring application. Despite the

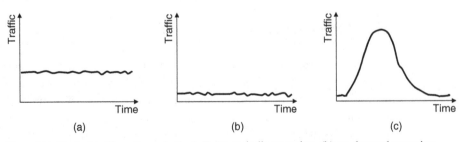

Figure 17.1 Typical traffic patterns in the IoT: (a) periodic reporting, (b) on-demand reporting, and (c) alarm reporting.

Table 17.1 Possible cellular scenario with IoT applications [8, 10]. Overhead due to IP packets not included.

IoT application	RI min	Payload (bytes)	Number of devices
Smart meters	1, 5, 15, > 60	100–1000	20 911
Home security systems	10	20	4 647
Elderly sensor devices	1	128	465
Credit machines in groceries	2	24	108
Credit machines in shops	30	24	1 650
Roadway signs	0.5	1	4 444
Traffic lights	1	1	540
Traffic sensors	1	1	540

name "periodic,"* periodic reporting is commonly modeled as a uniform or Poisson process [8] in terms of the distribution of the times of report arrivals over the reporting devices. In the latter case, it assumed that the intensity of the process is $\lambda_p = 1/RI$, where RI is the reporting interval of the device in seconds [9]. The expected number of report arrivals $\Lambda_p(\tau)$ for a single device during an interval of τ is

$$E[\Lambda_p(\tau)] = \lambda_p \tau, \tag{17.1}$$

while the probability of k report arrivals during τ is

$$P[\Lambda_p(\tau) = k] = \frac{(\lambda_p \tau)^k}{k!} e^{-\lambda_p \tau}. \tag{17.2}$$

The aggregate of the traffic arrivals over devices is also Poisson distributed, with an (aggregate) intensity of $N\lambda_p$, where N is the number of reporting devices in a cell; it is assumed that the traffic arrivals are independent and identically distributed (i.i.d.) processes over the reporting devices involved in the same application. The expected number of aggregate arrivals during τ is simply $N\lambda_p \tau$. The number of reporting devices in an IoT application per cell varies from source to source, spanning a range from several thousand [10] to up to 300 000 per cell in some use cases foreseen for fifth generation (5G) systems in coming years [11]. This number also depends on the type of cell being considered, i.e., suburban or urban. Table 17.1 shows a potential scenario in terms of the expected number of devices (both commercial and in-home) and their payloads in a sub-urban cell of radius 1500 m with three sectors [8, 10]. The main parameters considered are the average message size in bytes, the average message transaction rate per second (i.e., the intensity), and the traffic distribution. Smart meters represent a special case, with reporting intervals ranging from 1 minute to several hours.

On-demand reporting occurs when the end user or the application server requests information from the device, for example, an energy consumption report demanded by

---

* In fact, it would be more suitable to use the term "regular reporting," since it is implicitly assumed that the devices involved in the same application are configured or initialized such that their reporting periods are somewhat randomized in time. If this is not the case, then multiple devices may end up using the same time instants for reporting, thereby making the reporting pattern synchronous and thus increase the chances of congesting the network.

a customer in order to learn about the amount of energy consumed. The distribution of traffic arrivals for on-demand reporting can also be modeled as a Poisson process [6], with an intensity λ_o that is much lower than λ_p. Thus, the total arrival intensity of asynchronous traffic per device is given by $\lambda_a = \lambda_p + \lambda_o$, while the total arrival intensity aggregated over the reporting devices is simply $N\lambda_a$. Finally, it should be noted that this type of reporting is generally considered as delay tolerant [6, 12], as it can be delayed for a certain amount of time without causing service disruption. Taking a smart metering example, if an energy consumption report cannot be delivered at the time of generation, it can be delayed until the next scheduled reporting interval.

Finally, alarm reporting corresponds to messages generated owing to observation of an alarm condition. A common example of alarm reporting occurs when there is a power outage event in the electrical grid, where potentially thousands of monitoring devices will be affected and will try to report the outage simultaneously or near simultaneously. The delay constraints of alarm reporting are more strict than for periodic or on-demand reporting where the exact value of the tolerable delay depends on the particular service/application. In another example, in smart power metering the tolerable delay for the reporting of an outage is below 0.5 s, owing to the depletion of the integrated battery (also known as last-gasp reporting) [10]. In [13], 3GPP proposed a model for highly correlated traffic arrivals, i.e., synchronous traffic, where the arrival time interval follows a Beta distribution,

$$p(t) = \frac{t^{\alpha-1}(T-t)^{\beta-1}}{T^{\alpha+\beta-1}\mathrm{B}(\alpha,\beta)} \text{ for } 0 \leq t \leq T, \tag{17.3}$$

where $\alpha > 0$ and $\beta > 0$ are shape parameters, $\mathrm{B}(\alpha,\beta)$ is the Beta function [13], and T is the activation period, i.e., the time period from the activation of the first station until the last device is activated. The values of the parameters α, β, and T suggested in [13] are $\alpha = 3$, $\beta = 4$, and $T = 10$ s. The number of devices triggered during an interval τ is given by

$$N_A = N \int_0^\tau p(t)\,dt, \tag{17.4}$$

where N is the total number of devices.

The assumption that alarm traffic can be modeled via a Beta distribution seems to be commonly accepted; however [13] does not provide any specific reasoning for choosing this particular distribution or for choosing the particular values of the parameters α, β, and T. Some insights are provided in [6], where a simulation of the physical propagation of an alarm event, specifically, an earthquake, was carried out, and the results obtained were compared with the results of the model proposed in [14]. The primary wave of an earthquake propagates at different speeds depending on a number of factors, such as the type of ground; in the simulated model it was assumed that the speed was $v = 4$ km/s [13]. The simulation assumed a circular cell with a radius $r = 1500$ m and that the number of monitoring devices was $N = 10\,000$. The epicenter of the earthquake could be located anywhere in the cell. Furthermore, it was expected that the likelihood of a device being triggered by an event would decrease with distance. This was modeled via

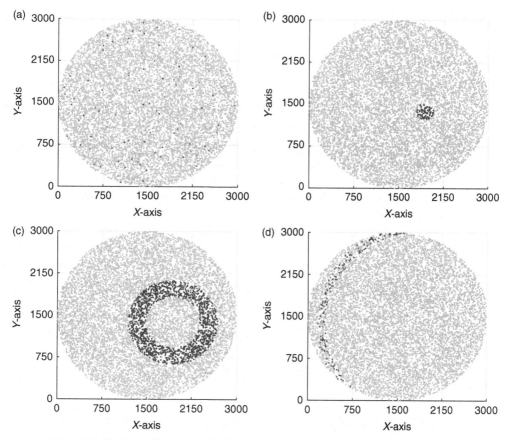

Figure 17.2 Typical traffic patterns in the IoT: (a) periodic reporting, (b) alarm event starts, (c) alarm event propagates, and (d) alarm event ends.

a function $\theta(t)$ that indicates the probability that a device will report an alarm event in a time instant t:

$$\theta(t) = \Psi(d) \cdot \delta\left(t - t_0 - \frac{d}{v}\right), \tag{17.5}$$

where $\Psi(d)$ is the spatial correlation factor, $\delta(t - t_0 - d/v)$ is a delta pulse that describes the propagation of the event that triggered the alarm reporting, t_0 is the moment of occurrence of the alarm event, d is the distance of the device from the epicenter of the alarm event, and v is the propagation speed of the alarm event. The spatial correlation factor $\Psi(d)$ expresses the probability that a device is affected by an alarm event that took place at a distance d. There are multiple models for $\Psi(d)$, here we focus on two simple examples. The first model corresponds to the situation where all devices in the cell are affected by the event:

$$\Psi(d) = 1. \tag{17.6}$$

In the second model, we consider a square root function

$$\Psi^{Sq}(d) = \begin{cases} \sqrt{d_{max}^2 - d^2}, & 0 \le d \le d_{max}, \\ 0, & \text{otherwise,} \end{cases} \tag{17.7}$$

where d_{max} is the maximum distance from the epicenter to which the event can reach.

An instance of the propagation of an alarm event is presented in Figure 17.2. At the time instant $t = 0$, the normal operation regime can be observed in Figure 17.2(a), i.e., periodic reporting, where only a few devices are active in the cell. At the time instant $t = 0.1$ s, the first devices triggered by the alarm event can be seen in Figure 17.2(b). At the time instant $t = 0.2$ s, the alarm triggers the devices shown in Figure 17.2(c), and at the time instant $t = 0.3$ s, the devices shown in Figure 17.2(d). Interestingly, it was shown that the expected distribution of the interarrival time between activations could indeed be modeled by a Beta distributed [6]. However, while the results confirm the choice of a Beta distribution to model the interarrival time for synchronous reporting, they also show that the choice of the shape parameters and the activation period depends on the alarm event that one wishes to model. More specifically, in the case of an earthquake, the activation period is in fact one order of magnitude shorter than the value proposed in [13] having a tremendous impact on the system performance. In other words, the parameters of the Beta distribution have to be carefully tuned to the properties of the phenomenon that describes the spreading of the alarm event, and there is no one-size-fits-all solution as suggested by [13].

17.3   The Features of Cellular Access That Are Suitable for the IoT

One of the major obstacles to the proliferation of efficient cellular access for the IoT stems from the deficiencies of the cellular access mechanism, which is currently designed to enable connection establishment from a low number of accessing devices, each with a relatively high requirement for data rate, as described in Section 17.1. Enhancement of the access scheme for IoT traffic has been the focus of both the research community [15, 16] and standardization efforts [17–19]. However, there is a common understanding that the IoT traffic requirements call for a more radical redesign of cellular access.

Taking into account the traffic patterns and communication characteristics of machine-type devices, an adequate access protocol should have the following, inter-related features:

1. A capability to provide extended coverage, i.e., accommodating stationary devices at locations with poor link quality and low-power transceivers. This is a rather new situation compared with classic mobile communications, where the mobility of terminals leads to time diversity and hence a device is unlikely to experience a permanently bad connection.
2. A capability to discriminate between/adapt operation to different types of IoT services/applications. This is particularly important at the level of alarm versus

regular reporting, as these require differentiated treatment from the very beginning of the service.

3. Graceful performance degradation in terms of reliability and delay when the load, i.e., the number of accessing devices, increases above the anticipated operating range. In this context, it is of paramount importance that the access protocol does not collapse under the "weight" of any load, causing service deprivation. Moreover, in the case of an overload, the system should be able to prioritize its resources in order to get the most critical information and discard the correlated information. For example, if many sensors are reporting the same alarm, the system should be able to detect this and reorganize the access in such a way as to avoid excessive collisions that could hinder its performance.

4. Low, ideally negligible, signaling overhead in comparison with data payloads.

5. Low power consumption on the device side.

6. Provisions for secure transmission, which can be challenging owing to the fact that the packets are short and the processing capability of the machine-type devices is modest.

Feature 1 is partially supported by today's cellular access protocols that use power-ramping procedure when the device reattempts an access [20]. However, this is insufficient to address the challenges of indoor coverage, since the total amount of power transmitted by a base station is tightly regulated and machine-type devices are typically low-power. Feature 2 is only partially supported by the access class barring and extended access class barring mechanisms proposed as enhancements of the Long Term Evolution (LTE) access protocol [17, 18]. These mechanisms are only able to block delay-tolerant traffic/devices and there is still no priority mechanism in place that can react promptly to instantaneous service requirements. Furthermore, although a major direction of the standardization efforts [17] and research work [16] is related to provision of the behavior required by feature 3, the ultimate result is that the point of collapse of the cellular access mechanisms is pushed to higher load values. In particular, the initial phase of the cellular access protocol is based on traditional slotted ALOHA in all instances and releases of the standard, and suffers from low efficiency and rather slow adaptation to the varying load. Moreover, the amount of overhead of the connection establishment procedure in terms of exchange of control information typically significantly exceeds the amount of data exchanged, requiring a fundamental rethinking of the cellular access mechanism in the light of M2M communications. Finally, features 5 and 6 have recently come under consideration in emerging technologies such as those explained in Section 17.8.

17.4   Overview of Cellular Access Protocols

A cellular device can be in one of two main states: *idle*, where the device is disconnected from the network, and *connected*, where the device has a connection established with the network. While in the connected state, a device can directly request system resources to send or receive data to or from the network; in the idle state, the device needs to establish a connection prior to data transfer, which is done via an access protocol. The current cellular access is connection oriented, i.e., prior to data transmission, the devices exchange a significant amount of signaling, which changes the device from an idle state to a connected state, see Section 17.4.4 for a detailed example. IoT traffic is characterized by the devices spending a majority of their time unconnected from the network; therefore, the ideal access protocol for the IoT is one where the data transmission can be performed with minimal signaling exchanges. This access is termed *connectionless*, and Figure 17.3(a) and (b) depict the two possible types of connectionless access protocols. Additionally, when the IoT traffic arrives periodically, an alternative way of serving this traffic could be to keep the devices in a connected state to the network and assign periodic resources. We denote this access scheme as *connection-oriented* and depict it in Figure 17.3(c).

We note that although current cellular networks use a common access protocol for all traffic types [16], it is foreseen that future-generation networks will have the capability to choose from a pool of access protocols the one suitable for serving a specific application.

17.4.1   One-Stage Access

In the one-stage access protocol, sometimes called one-shot access, each device attempts access by contending directly using its data packets. Specifically, there are prespecified, typically periodically recurring uplink time–frequency resources (i.e., slots) that are reserved for contention, and the devices send their data packets in a randomly selected slot. In addition to the payload, the data packets also include additional signaling, identifying the device and containing information about the security context. The simplest realization of this protocol is the slotted ALOHA scheme, as depicted in Figure 17.4. In this scheme, a packet that is the only one being transmitted in a slot, depicted by the dark packets in Figure 17.4, has the highest chance of being successfully received; this chance also depends on the noise. On the other hand, the chance of the successful reception of packets experiencing collision with other packets in the same slot is much lower and depends on the potential of the capture effect due to power imbalance of the colliding packets. Successfully received packets are

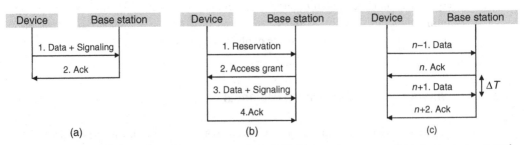

Figure 17.3 Overview of access protocols for IoT: (a) one-stage access, (b) two-stage access, and (c) periodic reporting.

Figure 17.4 One-stage access protocol.

acknowledged via message 2 in the downlink, while unsuccessful ones are not, and the corresponding devices reattempt contention after a related timer expires.

Assuming Poisson arrivals with an average arrival rate λ, the probability of a device transmitting successfully, also termed the access reliability and denoted by P_s, is given by

$$P_s = \frac{\lambda e^{-\lambda}}{1 - e^{-\lambda}}, \tag{17.8}$$

which decreases rapidly with increasing λ. A common technique to improve the access reliability is to allow devices to retransmit up to K times, which, although providing moderate performance gains, induces extra latency and device energy expenditure due to the extra retransmissions. In conclusion, this simple realization of the one-stage access scheme does not scale with increasing load, where the main performance bottleneck is due to the occurrence of collisions.

More recently, several one-stage access schemes have been put forward that can take advantage of the capabilities of iterative receivers to decode the collided transmissions of multiple devices [21].

17.4.2 Two-Stage Access

In the two-stage access protocol, the contention takes place in the first stage. It is performed by the transmission of a reservation request in message 1, instead of the actual data packets. The contention process is the same as in the one-stage access, but the main advantage is that in the case of collision there is the potential to spend fewer network resources. This becomes increasingly evident with an increase in the data payload and the associated signaling, which is also the reason why all current cellular protocols adopt this approach, thereby being optimized for higher data payloads.

The reservation request can take one of two forms: (i) an informative request, where the device informs the network about the kind of transmission it requires, and (ii) a non-informative request, where the device signals that it needs to connect to the network, but does not provide any other a priori information. Depending on the reservation request in place, the network can provide customized access in the remaining access stages. The main advantage of the non-informative request types is that they can be realized using signals for which detection is very robust, such as the case of the Zadoff–Chu sequences used in LTE for the random access preambles [22].

The remaining message exchanges correspond to the actual data exchanges and any necessary signaling. Note that in current cellular systems, where the access is connection oriented, the actual access protocol includes several more stages.

Also, M2M-optimized schemes have recently been proposed which can optimize cellular access in the direction of a connectionless two-stage access, for example the signature-based access [23]. Section 17.6 provides a description of a possible realization of a two-stage access protocol.

17.4.3   Periodic Reporting

In the periodic access protocols, it is assumed that the device is in the connected state and that the network knows the average time interval between two arrivals. The network can then reserve a set of resources that recur periodically over time and are at the device's disposal. This type of access differs from the current scheduled access mode in cellular systems, such as LTE, in that each device has a set of dedicated resources. These resources are not necessarily used (e.g., if there is no arrival), but when they are used, the device performs a one-stage access without contention. The main drawback of this type of access is that by itself it is not resource or efficient/robust with respect to the uncertainty imposed by the report arrivals.

17.4.4   Case Study: LTE Connection Establishment

In order to understand better the need to reconsider the fundamental design of cellular access protocols, we present a summarized study of the access protocol used in LTE-Advanced (LTE-A) and the associated bottlenecks. As previously mentioned, the LTE-A access protocol is a multi-stage access protocol.

The access reservation protocol (ARP) in LTE consists of the exchange of four medium access control (MAC) messages between the accessing device and the base station, as shown in Figure 17.5. Message 1 is a random access preamble that is sent in the physical random access channel (PRACH), a channel defined in the uplink consisting of periodically appearing time–frequency resources denoted as random

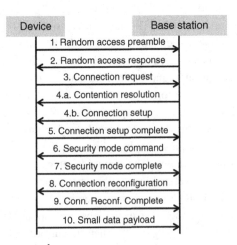

Figure 17.5 LTE-A access protocol.

access opportunities (RAOs). The preambles are randomly chosen from a set of 64 orthogonal preambles. However, typically, the eNodeB can only detect which preambles have been activated, and not whether multiple activations (collisions) have occurred [24]. Message 2 is sent for all detected preambles. If a preamble is activated by multiple devices, each of the devices will interpret message 2 as sent exclusively to it and send message 3 in the same uplink resources as the rest of the devices that activated the same preamble. Consequently, their messages 3 will collide, lowering the chances of the successful reception of any of them. If the collision of messages 3 is detected, the base station will not respond with message 4 and the ARP will be restarted after a related timer expires.

Following the ARP a series of messages are exchanged between the IoT device and the eNodeB to authenticate the device to the network and to establish the security and ciphering. Finally, after 10 messages the device is ready to send the data packet to the network and then go back to sleep if no more data is available in the buffer until the next scheduled transmission. In other words, a total of 10+ different messages are exchanged in order to fully establish the connection and start the data payload transfer.

To illustrate the cost of the connection establishment, we consider smart metering, which can be seen as a showcase for the IoT. We consider the upcoming generation of smart meters, which are expected to report more frequently and with larger payloads [25]. More specifically, we investigate the performance of LTE for smart meters reporting every 10 s with payloads of 100 and 1000 bytes. We consider a typical PRACH configuration with one RAO every 10 ms. The main limitation in the uplink is the collisions stemming from the random nature of the ARP, as well as the limited capacity of the data channel. In the downlink, the main limitation on massive access is the control channel. Specifically, every downlink message sent to a device requires explicit signaling in the control channel indicating where its data can be found. The amount of resources in the downlink control channel is typically limited to three messages per subframe in a 1.4 MHz LTE system; new releases of LTE have introduced the concept of enhanced control channels, which partially removes these limitations, but at the cost of reducing the amount of resources available for data in the downlink. Finally, we assume that both the control information and the data used the lowest modulation scheme in LTE, which is to be expected in devices with poor link quality [26].

The evaluation was performed in terms of blocked devices and the aggregated number of arrivals at the system. A device is blocked if it has not been served before the maximum number of transmission attempts in the PRACH, which is set to 10, has been reached. To illustrate the impact of connection establishment, we considered two different cases with respect to where the data transmission actually starts. In the first case, data transmission starts immediately after the ARP (i.e., after message 4), which is referred to as *lightweight-signaling* access and corresponds to an extreme case of signaling overhead reduction, beyond what has been proposed by 3GPP [27, 28]. The second case corresponds to the normal LTE connection establishment, where data transmission starts after the ARP and the additional signaling. As shown in Figure 17.6, two main effects can be observed: (1) the impact of the payload size and

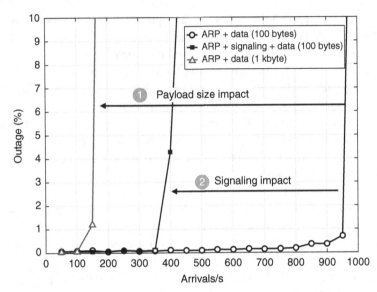

Figure 17.6 Comparison of outage for only ARP and data transmission (ARP + data) and full message exchange (ARP + signaling + data).

(2) the impact of the additional messages exchanged after the ARP procedure, such as the establishment of the security context. As expected, larger payloads require more resources, representing a considerable burden for the limited bandwidth of a 1.4 MHz system. More surprising is the impact of the additional signaling, which consumes a considerable amount of resources, decreasing the number of supported devices by a factor of three. In this case it can be shown that the main limitation is the capacity of the control channel in the downlink [25]. These results highlight the importance of simplified signaling for the IoT, which, although it has been considered by 3GPP, has not yet been standardized. Finally, it can be seen that up to 1000 devices reporting 100 bytes every 10 s can be supported in LTE, or approximately 150 devices if larger payloads are used.

In the following sections, we address the main use cases identified for the IoT that affect cellular access procedures: (i) ultra-reliability, (ii) massive access, and (iii) extended coverage and ultra-low power consumption. The use cases (i) and (ii) were originally described in [11], while use case (iii) is a matter of research and discussion in both the industry and the standardization bodies for the emerging technologies, such as LTE for M2M (LTE-M) and narrowband IoT (NB-IoT).

17.5   Improving the Performance of One-Stage Access for 5G Systems

The main bottleneck of one-stage access and, in general, of the contention-based stage of cellular access mechanisms is the waste of resources due to collisions of multiple packets/connection requests sent by users. However, recently it has been shown that the

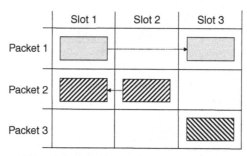

Figure 17.7 Illustration of successive interference cancellation: packet 2 is recovered in the singleton slot 2, and its replica is removed from the collision slot 2. Slot 2 now becomes a singleton slot, enabling recovery of packet 1, whose replica is removed from the collision slot 3. Slot 3 becomes a singleton slot and packet 3 is recovered from it.

use of successive interference cancellation can "unlock" the collisions and substantially improve the performance of the access protocol [29]. The main idea is based on the observation that a device, in general, when contending for access has to send multiple replicas of its packets until a replica occurs in a singleton slot. When a replica is successfully received, the base station can learn in which (collision) slots the other replicas of the same packet have occurred and, in principle, remove (i.e., cancel) those replicas. Cancellation of replicas may, in turn, resolve some of the collisions and enable recovery of new packets, giving rise to a new succession of interference cancellations and packet recoveries, as depicted in Figure 17.7.

In essence, the application of successive interference cancellation is analogous to iterative belief propagation erasure decoding, motivating application of the theory and tools of codes-on-graphs for the design of advanced random access schemes, i.e., *coded* random access [30, 31]. It has been shown that coded random access schemes asymptotically achieve a throughput of 1 for the collision channel model [32]. This is equal to the performance of scheduled access and far exceeds the well-known bound on the asymptotic throughput of slotted ALOHA, which is equal to 0.37. In more refined models of wireless channels which include the capture effect, the throughput of coded random access can actually exceed 1 [33], i.e., more than a single packet can be recovered per slot. The price to pay for the increase in throughput performance, increased reliability, and potentially lower latency is the increased complexity of the receiver operation. Specifically, the base station has to buffer the signals observed in the collision slots and to perform interference cancellation, which may involve complex physical layer processing.

17.6   Reliable Two-Stage Access for 5G Systems

When the number of IoT devices that attempt to connect simultaneously becomes very large, the network becomes unable to offer reliable service. This is a situation of *overload* and the traditional approach in cellular systems to cope with it is to enact a backoff mechanism, which spreads the traffic arrivals over time. The goal of this

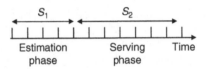

Figure 17.8 Proposed access frame, consisting of S_1 slots for the estimation phase and S_2 slots for the serving phase.

procedure is to change the incoming IoT traffic from "Beta distributed" to "uniformly distributed," or in other words to change from the profile depicted in Figure 17.1(c) to that in Figure 17.1(a). The backoff mechanism works as follows. Upon transmission of an access request to the base station, the device starts a predefined timer; if no response has been received before the timer expires, the backoff mechanism kicks in, forcing the device to wait for a random time before a new access request is sent to the base station. However, as the number of contending devices is unknown, the result depends strongly on the backoff value used. If the backoff value is not large enough, multiple collisions will take place again. On the other hand, the larger the value used, the larger the delay experienced by a device that is requesting service, potentially causing the device to miss the deadline for its message. This solution does not provide reliable service, since it blindly spreads the requests over time, not knowing the number of accessing devices nor taking into account the needs of the different IoT applications using the system.

Our proposal for reliable access in 5G in the presence of synchronous traffic arrivals, for example alarm events, is to split the first stage of the access mechanism, i.e., the contention phase, into two substages or phases: (1) an estimation phase and (2) a serving phase. We illustrate the scheme in Figure 17.8.

In the estimation phase, S_1 random access slots are periodically provided every T slots, and each device randomly chooses $s \in \{1, 2, \ldots, S_1\}$ slots in which to transmit a random access request, where each slot has a different probability of being chosen. Based on the ternary outcome for each slot, i.e., idle, singleton, or collision, the number of contending devices can be determined. This concept was originally proposed in [34] for the LTE system, with the peculiarity that in LTE the outcome was only binary, i.e., either idle or active. Once the number of contending devices is determined, the appropriate length of the serving phase can be determined and the devices are informed of the number of serving slots, denoted by S_2, where we note that $S_2 \gg S_1$. However, owing to the periodicity of the mechanism, a maximum of S access slots within T is possible, where $S_1 + S_2 \leq S$. Thus, after the estimation phase, the base station imposes an access barring parameter to limit the number of stations accessing the serving phase. It should be noted that in the serving phase the normal access mechanism takes place, followed by the second access stage, i.e., the data transmission stage. Given T and S, the maximum number of devices for a given reliability can be determined [34]. As shown in [34], the proposed mechanism can be extended to support multiple priority classes. In this case, one initial estimation round takes place, where the number of accessing devices of each type is determined and then the serving phase takes place, serving first the devices belonging to the highest priority type.

17.7   Reliable Periodic Reporting Access for 5G Systems

In this section we describe an efficient and reliable access mechanism, which exploits the delay tolerance of asynchronous IoT traffic. Recall from the previous sections that asynchronous traffic in the IoT typically refers to the normal operation of IoT devices, such as a smart meter periodically reporting energy consumption. If a given report cannot be delivered, it can still be delivered during the next scheduled transmission time. One might ask why we do not simply provide periodically recurring resources to each IoT device; however, we note that this approach only works if the number of packets or the packet size is deterministic and if there are no packet errors. If the number of packets accumulated between two reporting instances is random and the probability of a packet error is not zero, then the amount of transmission resources required per device in each transmission period is random.

Consider a generic system, where the granularity of resources, or in other words the minimum set of data resources allocatable to a device, is referred to as a resource element (RE). The total number of REs in the system per unit of time (i.e., slot), is denoted by Y, where Y_{IoT} REs are reserved for IoT service. Each device, as mentioned earlier, can send no, one, or multiple reports within a reporting interval (RI). The time elapsed between two consecutive RIs is denoted by T_{RI}.

The idea is a system with a pool of periodically recurring resources that are reserved for IoT asynchronous operation and are shared for uplink transmission by all IoT devices subscribed to the service. The period is selected such that if a report is transmitted successfully with the upcoming resource pool, then the reporting deadline is met. We assume that the uplink resources are split into two pools, one reserved for IoT asynchronous service and the other used for other services. Within the pool of resources for IoT, we can distinguish two different parts as depicted in Figure 17.9, referred to as the preallocated and the common pool, and where the minimum chunk of resources allocatable to a device is referred to as an RE.

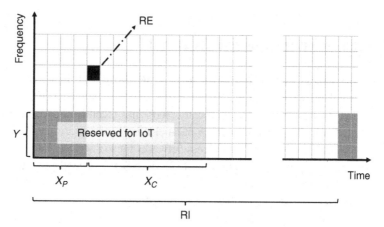

Figure 17.9 Illustration of the proposed method, where a set of resources are reserved for IoT service, while the rest are kept for other purposes.

Consider a system with N reporting devices, where each device is preallocated the required amount of REs from the preallocated pool, sufficient to transmit a single report and an additional message to indicate if there are more reports to be sent, referred to as excess reports. The common pool is used to allocate resources for the excess reports, as well as for all the retransmissions of reports/packets that were erroneously received. The lengths of the IoT resource pool, preallocated pool, and common pool, expressed as numbers of slots, are denoted by X, X_P, and X_C, respectively (see Figure 17.9), such that

$$X = X_P + X_C = \gamma N + X_C, \tag{17.9}$$

where $\gamma \leq 1$ denotes the fraction of REs per slot required to accommodate a report transmission, and where the value of X_C should be chosen such that a report is served with the required reliability. Finally, we note that the duration of X has a direct impact on the delay. As shown in [35], the probability of report failure, i.e., the probability that the report has not been successfully delivered after all attempted (re)transmissions, if the common pools consists of X_C slots which can accommodate C transmissions (i.e., C is the capacity of the common pool in number of transmissions), is given by

$$P[\Phi] \leq Q\left(\frac{C-\mu}{\sigma}\right)(1 - p_e^L) + p_e^L, \tag{17.10}$$

where $\mu = E[R]$, $\sigma = \sqrt{\sigma^2[R]}$, where R denotes the number of transmissions required by all devices, L is the maximum number of packet retransmissions, p_e is the target packet error probability for the data transmissions; and $Q(\cdot)$ is the Q-function. Therefore, although the numbers of reports, excess reports, and failures are random, we can determine the amount of resources required to provide a given reliability R, i.e., $R = 1 - P[\Phi]$.

In [35] it was shown that this scheme can be used to drastically reduce the amount of resources that needs to be reserved in an LTE-A system for the IoT in order to achieve the same reliability.

17.8   Emerging Technologies for the IoT

In parallel with the efforts toward 5G systems, there are several other 3GPP standardization initiatives in which the goal is to tailor the current cellular networks to IoT applications [36]. Specifically, 3GPP has recently concentrated its standardization efforts in this regard on three parallel tracks, which are (i) LTE-M, focusing on the modification of LTE radio access networks for IoT services and targeted at devices with reduced air interface capabilities [37]; (ii) NB-IoT, recently adopted for standardization, which targets low-cost narrowband devices with reduced functionalities [38] and (iii) Extended Coverage Global System for Mobile Communications (EC-GSM), addressing the same needs as NB-IoT, but using GSM as a basis [39]. The objectives of these activities can be summarized as follows [7]: improved indoor coverage (15–20 dB compared with current cellular systems) and outdoor coverage up to 15 km, support for massive numbers of low-data-rate devices with modest device complexity, improved power efficiency to ensure longer battery life, reduced access latency, and efficient coexistence with legacy cellular systems. In the following, we provide a brief overview of the expected features of each of these systems.

17.8.1   LTE-M: LTE for Machines

The motivation for LTE-M comes from the need to support simpler devices than the UE types defined currently, while still being able to take advantage of the existing LTE capabilities and network support. The changes in comparison with the LTE system will take place at both the device and the network infrastructure levels, where the most important change is the reduction in bandwidth from 20 to 1.4 MHz in both the downlink and the uplink [37]. The consequence of this change is that the control signals (e.g., synchronization or broadcast of system block information), which are currently spread over the 20 MHz band, will be altered to support the coexistence of both LTE-M UEs and the standard, more capable, UEs. Another feature of this new UE category is a reduced power consumption, achieved by reduction of the complexity of the transceiver chain, such as support for uplink and downlink rates of 1 Mbps, half-duplex operation, use of a single antenna, a reduced operation bandwidth of 1.4 MHz, and reduction of the allowed maximum transmission power from 23 to 20 dBm. Furthermore, there is a requirement to increase the cellular coverage of these LTE-M UEs by providing up to 15 dB extra in the cellular link budget. Several techniques have been proposed to improve coverage, including the use of network-controlled network frequency hopping, improved hybrid automatic repeat request (HARQ), and the use of repetition coding in the physical control channels.

We outline here the most important features of connection establishment in LTE-M, required to be completed before the device can transmit its data. From the network perspective, this entails a transition of the device from the unconnected state (idle) to the connected state (connected), as explained previously. This transition implies a substantial number of signaling exchanges. Therefore, in cases where the interval between user transmissions is not long enough to warrant disconnection of the device from the network (i.e., a return to the idle state), the device can enter a dormant state (DRX) and wake up periodically to listen to the network for pending downlink transmissions and to perform uplink transmissions. Discontinuous transmission and dormant periods of this kind, denoted respectively as *DRX operation* and *DRX cycles*, are already present in standard LTE. LTE-M supports larger DRX cycles in order to reduce signaling further and optimize battery life.

17.8.2   Narrowband IoT (NB-IoT): A 3GPP Approach to Low-Cost IoT

NB-IoT pertains to a clean-slate design of an access network dedicated to serving a massive number of low-throughput, delay-tolerant, and ultra-low-cost devices. NB-IoT can be seen as an evolution of LTE-M with respect to the optimization of the trade-off between device costs and capabilities, as well as a substitute for legacy General Packet

Radio Service (GPRS) to serve low-rate IoT applications. The main technical features foreseen are: (i) a reduced bandwidth of 180 kHz in the downlink and uplink; (ii) a maximum device transmission power of 23 dBm; and (iii) a link budget increased by 20 dB compared with commercially available legacy GPRS, specifically to improve the coverage of indoor IoT devices. This coverage enhancement can be achieved by power boosting of the data and control signals, message repetition, and relaxed performance requirements, for example by allowing a longer signal acquisition time and a higher error rate.

An important enabler for this coverage enhancement is the introduction of multiple coverage classes, which allows the network to adapt to a device's coverage impairments. The coverage classes foreseen include (i) a normal coverage class, similar to legacy GPRS coverage; (ii) an extended coverage class, corresponding to about 10 dB improvement relative to GPRS; and (iii) an extreme coverage class, corresponding to 20 dB improvement relative to GPRS.

The different coverage classes correspond to operation with different modulation orders, coding rates, and repetition factors, among other things, in order to match each device's data rate to its available link budget. This allows devices with better coverage conditions to operate at higher data rates and with lower latency than devices that have poor coverage. The main benefit of this multicoverage class scheme is that it allows NB-IoT to be designed to meet the throughput and latency requirements of devices requiring extreme coverage extension, while allowing devices with normal or extended coverage conditions to improve their performance. This is achieved by segregating the devices into different physical layer resources according to the required coverage class. The main enabler is the control channel time interval, which scales according to the coverage class.

17.8.3   Extended Coverage GSM (EC-GSM): Evolution of GSM for the IoT

The idea of the EC-GSM standardization track is to moderately change legacy GSM/GPRS in order to achieve extended coverage, while allowing coexistence with existing GSM deployments. Under normal coverage conditions, the same physical layer speeds as today can be achieved, and legacy devices are supported. When a device is out of coverage in a legacy network, the extended-coverage features are obtained via blind repetitions of messages. Finally, it should be noted that like NB-IoT, EC-GSM also features a reduced level of signaling traffic, obtained through new, simplified control messages.

17.9   Conclusion

By design, current cellular systems are not well equipped to efficiently and reliably support IoT traffic. The expected volume of low-rate connections in IoT scenarios, on the order of tens of thousands, demands rethinking and redesigning of the current cellular access mechanisms. Moreover, the difference in nature of the two typical categories of IoT applications, i.e., periodic and alarm reporting, imposes additional burdens on the cellular access process if it is to provide the required service and be a conclusive solution for IoT access. In this chapter we have outlined the main features that cellular access has to offer for that purpose, and provided some guidelines on how to address the shortcomings of the current cellular mechanisms. However, it is probable that no one-size-fits-all solution can be implemented in future cellular standards. Instead, a vertical approach in which differentiation among IoT services/traffic types takes place at the radio network access level and each traffic type/service has its "own" optimized access protocol may have to be adopted.

References

[1] Cisco, "Embracing the Internet of Everything to capture your share of $14.4 trillion," 2015 White Paper. Available at www.cisco.com/c/dam/en_us/about/ac79/docs/innov/IoE_Economy.pdf.

[2] "IoT and M2M," 2015. Available at http://iotandm2m.blogspot.dk/2015/06/iot-m2m.html.

[3] SigFox, "Global cellular connectivity for the Internet of Things," 2015. Available at www.sigfox.com/en/.

[4] LoRa Alliance, "Wide area networks for the Internet of Things," 2015. Available at www.lora-alliance.org/.

[5] 3GPP, "Overview of 3GPP release 12," Tech. Rep., 2014.

[6] G. C. Madueño, C. Stefanovic, and P. Popovski, "Reliable and efficient access for alarm-initiated and regular M2M traffic in IEEE 802.11 ah systems," *IEEE Internet Things J.*, vol. 3, no. 5, pp. 673–682, Oct. 2016.

[7] 3GPP, "Study on machine-type communications (MTC) and other mobile data applications communications enhancements," TR 22.368, Dec. 2013.

[8] G. C. Madueño, C. Stefanovic, and P. Popovski, "Reengineering GSM/GPRS towards a dedicated network for massive smart metering," in *Proc. of IEEE International Conf. on Smart Grid Communications (SmartGridComm)*, Nov. 2014.

[9] G. C. Madueño, C. Stefanović, and P. Popovski, "How many smart meters can be deployed in a GSM cell?" in *Proc. of IEEE Internatonal Conf. on Communications Workshops (ICC)*, Jun. 2013.

[10] IEEE, "IEEE 802.16p machine to machine (M2M) evaluation methodology document (EMD)," IEEE 802.16 Broadband Wireless Access Working Group (802.16p), EMD 11/0005, 2011.

[11] FP-7 METIS, "Requirements and general design principles for new air interface," Deliverable D2.1, 2013.

[12] M. Laner, P. Svoboda, N. Nikaein, and M. Rupp, "Traffic models for machine type communications," in *Proc. of the Tenth International Symposium on Wireless Communication Systems (ISWCS)*, Aug. 2013.

[13] 3GPP, "Study on RAN improvements for machine-type communications," TR 37.868 V11.0, Aug. 2010.

[14] G. Andrews, R. Askey, and R. Roy, *Special Functions*, 1st edn, Cambridge University Press, 2000.

[15] J.-P. Cheng, C. Han Lee, and T.-M. Lin, "Prioritized random access with dynamic access barring for RAN overload in 3GPP LTE-A networks," in *Proc. of IEEE GLOBECOM Workshops (GC Wkshps)*, Dec. 2011.

[16] A. Laya, L. Alonso, and J. Alonso-Zarate, "Is the random access channel of LTE and LTE-A suitable for M2M communications? A survey of alternatives," *IEEE Commun. Surv. Tutor.*, vol. 16, no. 1, pp. 4–16, First Quarter 2014.

[17] 3GPP, "Access barring for delay tolerant access in LTE," TR R2-113013, May 2011.

[18] 3GPP, "MTC simulation results with specific solutions," TR R2-104662, Aug. 2010.

[19] 3GPP, "Backoff enhancements for RAN overload control," TR R2-112863, May 2011.

[20] 3GPP, "RACH congestion evaluation and potential solutions," TR R2-102824, May 2011.

[21] E. Paolini, C. Stefanovic, G. Liva, and P. Popovski, "Coded random access: Applying codes on graphs to design random access protocols," *IEEE Commun. Mag.*, vol. 53, no. 6, pp. 144–150, Jun. 2015.

[22] 3GPP, "Medium access control (MAC) protocol specification," TR 36.321, Tech. Rep., 2014.

[23] N. K. Pratas, C. Stefanovic, G. C. Madueño, and P. Popovski, "Random access for machine-type communication based on bloom filtering." Available at http://arxiv.org/abs/1511.04930, Nov. 2015.

[24] H. Thomsen, N. Pratas, C. Stefanovic, and P. Popovski, "Analysis of the LTE access reservation protocol for real-time traffic," *IEEE Commun. Lett.*, vol. 17, no. 8, pp. 1616–1619, Aug. 2013.

[25] G. C. Madueño, J. J. Nielsen, D. M. Kim, N. K. Pratas, C. Stefanovic, and P. Popovski, "Assessment of LTE wireless access for monitoring of energy distribution in the smart grid," *IEEE J. Sel. Areas Commun.*, vol. 34, no. 3, pp. 675–688, Mar. 2016.

[26] J. J. Nielsen, G. C. Madueño, N. K. Pratas, R. B. Sørensen, C. Stefanovic, and P. Popovski, "What can wireless cellular technologies do about the upcoming smart metering traffic?" *IEEE Commun. Mag.*, vol. 53, no. 9, pp. 41–47, Sep. 2015.

[27] 3GPP, "Study on RAN improvements for machine-type communications, rel. 11," TR 37.868, Tech. Rep., Sep. 2011.

[28] 3GPP, "Study on enhancements to machine-type communications (MTC) and other mobile data applications; Radio access network (RAN) aspects, rel. 12," TR 37.869, Tech. Rep., Sep. 2013.

[29] E. Casini, R. D. Gaudenzi, and O. del Rio Herrero, "Contention resolution diversity slotted ALOHA (CRDSA): An enhanced random access scheme for satellite access packet networks," *IEEE Trans. Wireless Commun.*, vol. 6, no. 4, pp. 1408–1419, Apr. 2007.

[30] G. Liva, "Graph-based analysis and optimization of contention resolution diversity slotted ALOHA," *IEEE Trans. Commun.*, vol. 59, no. 2, pp. 477–487, Feb. 2011.

[31] E. Paolini, C. Stefanovic, G. Liva, and P. Popovski, "Coded random access: How coding theory helps to build random access protocols," *IEEE Commun. Mag.*, vol. 53, no. 6, pp. 144–150, Jun. 2015.

[32] E. Paolini, G. Liva, and M. Chiani, "Coded slotted ALOHA: A graph-based method for uncoordinated multiple access," *IEEE Trans. Inf. Theory*, vol. 61, no. 12, pp. 6815–6832, Dec. 2015.

[33] C. Stefanovic, M. Momoda, and P. Popovski, "Exploiting capture effect in frameless ALOHA for massive wireless random access," in *Proc. of IEEE Wireless Communications and Networking Conf. (WCNC)*, May 2014.

[34] G. C. Madueño, N. K. Pratas, C. Stefanovic, and P. Popovski, "Massive M2M access with reliability guarantees in LTE systems," in *Proc. of IEEE International Conf. on Communications (ICC)*, Jun. 2015.

[35] G. C. Madueño, C. Stefanovic, and P. Popovski, "Reliable reporting for massive M2M communications with periodic resource pooling," *IEEE Wireless Commun. Lett.*, vol. 3, no. 4, pp. 429–432, Aug. 2014.

[36] F. Boccardi, R. Heath, A. Lozano, T. Marzetta, and P. Popovski, "Five disruptive technology directions for 5G," *IEEE Commun. Mag.*, vol. 52, no. 2, pp. 74–80, Feb. 2014.

[37] 3GPP, "Further LTE physical layer enhancements for MTC," Tech. Rep. RP-151186, May 2015.

[38] 3GPP, "Narrowband IOT," Tech. Rep. RP-151621, 2015.

[39] 3GPP, "EC-GSM concept description," Tech. Rep. GP-150132, 2015.

18 Medium Access Control, Resource Management, and Congestion Control for M2M Systems

Shao-Yu Lien and Hsiang Hsu

18.1 Introduction

The current mobile networks provide seamless and reliable streaming (voice/video) services to an increasing number of mobile users. From Global System for Mobile Communications (GSM) General Packet Radio Service (GPRS) and Universal Mobile Telecommunication System (UMTS) to Long Term Evolution (LTE) and LTE-Advanced (LTE-A), the transmission data rates have increased tremendously. Relying on the deployment of a heterogeneous network (HetNet) [1–5] involving macrocells, small cells (femtocells and picocells), and/or relay nodes, ubiquitous support for basic multimedia and Internet browsing applications has been tractable. In addition to mobile networks, wireless local area networks (WLANs) such as IEEE 802.11a/b/g/e/n/ac/ax also support moderate distance data exchange with best effort services or a certain level of quality of service (QoS). As a result, it seems that the needs of human-to-human (H2H) communication applications can be satisfied using the existing network architectures and technologies. However, to substantially facilitate human daily activities, providing only basic voice/video and Internet access services may be insufficient.

Recently, a novel communication paradigm known as machine-to-machine (M2M) communications has had a significant impact on the development of the next generation of wireless applications. Achieving "full automation" and "everything-to-everything" (X2X) connection facilitated by M2M communications has been regarded as two urgent and ultimate targets not only in the industry but also in the context of economics, social communities/activities, transportation, agriculture, and energy allocation [6]. "Full automation" implies a significant enhancement of human beings' sensory and processing capabilities, which embraces unmanned or remotely controlled vehicles/robots/offices/factories/stores, augmented/virtual/kinetic reality, and immersive sensory experiences. On the other hand, X2X connection implies that diverse entities, including human and machines, will be able to form new types of communities in addition to those based on H2H communication, such as social networks using human-to-machine (H2M) and M2M communication [7–12]. The applications include intelligent transportation systems (ITSs) [13], volunteer information networks (in which each entity in the network shares information with other entities to obtain a global view of the environment and the community) [14], the Internet of Things (IoT) [15–17], and smart buildings/cities/grids [18–20], to name but a few.

To support these emerging applications, boosting the transmission data rate is only one of the diverse requirements for providing some of the necessary services. In addition, the performance in terms of end-to-end transmission latency [6], energy efficiency, reliability, scalability, cost efficiency, and stability must be fundamentally enhanced. For example, ultra-low latency and reliable data exchange are particularly required for unmanned or remotely controlled vehicles/robots, augmented/virtual/kinetic reality, immersive sensory experiences, and ITSs. On the other hand, as a large number of sensors may be involved in the IoT and smart buildings/cities/grids, scalability, cost efficiency, and energy efficiency are a primary concern. Unfortunately, the state-of-the-art radio access schemes (e.g., UMTS/LTE/LTE-A and IEEE 802.11a/b/g/n/ac/ax) designed solely to optimize data rates may face an unprecedented challenge in meeting all the requirements for the two targets above. As a result, envisioning effective designs is of crucial importance. To satisfy the diverse application requirements of M2M communications, novel designs of system architectures, medium access control (MAC) schemes, and resource management techniques are needed.

18.2   Architectures for M2M Communications

18.2.1   WLAN Architecture for M2M Communications

In past decades, developments of WLANs such as IEEE 802.11 primarily emphasized data rate enhancement. For example, the peak data rate of IEEE 802.11b/g/n is 150 Mbps, and the peak data rate of IEEE 802.11n can be 600 Mbps using four antennas. IEEE 802.11ac can achieve a peak data rate of 3.46 Gbps using eight antennas. A similar development trend has also occurred in short-range communication networks such as next-generation Bluetooth. However, if they are to be deployed in the 2.4 GHz and 5 GHz unlicensed bands, the communication ranges of these systems are limited. In addition, the power consumption and the cost of the devices may also be unaffordable in M2M scenarios. The above concerns have consequently driven the development of a new WLAN particularly designed for M2M/IoT communications, known as IEEE 802.11ah [21–25].

In 2013, the Wi-Fi Alliance launched a system called "HaLow" based on IEEE 802.11ah. "HaLow" is named from a combination of "ah" and "low" for "low-power." The target of IEEE 802.11ah and thus HaLow is that it will be deployed in frequency bands under 1 GHz, to achieve a data rate of more than 100 kbps even if the communication range exceeds 1 km, as illustrated in Figure 18.1. IEEE 802.11ah is designed to have a very low power consumption, which makes this system competitive with Bluetooth and ZigBee. Layer 1 of IEEE 802.11ah partially adopts the physical layer functions of IEEE 802.11ac, including 256 quadrature amplitude modulation (QAM), beamforming, and multiuser multiple-input multiple-output (MIMO). However, the

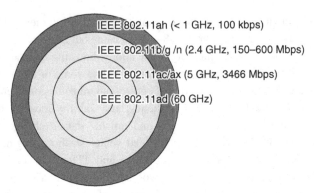

Figure 18.1 The deployment frequency bands of IEEE 802.11ah for M2M applications are less than 1 GHz to achieve a data rate of 100 kbps.

bandwidth of IEEE 802.11ah is very limited as compared with 802.11ac (only 1, 2, 4, 8, 16 MHz are supported).

18.2.2   Cellular Radio Access Network for M2M Communications

To support mobility, ubiquitous wireless services, QoS guarantees, and robust connections, data exchange among machines may rely on the existing cellular infrastructure of mobile networks. For this purpose, the standards for cellular-based M2M communications were launched with the Third Generation Partnership Project (3GPP) Release 11 for machine-type communications (MTC) in 2012. The goal of Release 11 MTC was to utilize the 3GPP infrastructure, which provides (wired) connections between all stations. In Release 11 (i.e., LTE-A), these stations can be Evolved Universal Terrestrial Radio Access (E-UTRA) node Bs (eNB) in macrocells or picocells, relay nodes (RNs), or home eNBs (HeNBs) in small cells. These stations thus provide ubiquitous wireless access in both outdoor and indoor environments. By attaching to these stations, higher-layer connections between all MTC devices can be provided. To fully leverage the 3GPP infrastructure, in Release 11, the Uu interface (i.e., the air interface between an eNB/HeNB and a user equipment (UE)) is reused as the air interface to connect an MTC device with the network, as shown in Figure 18.2. As a result, an MTC device is regarded as a special type of UE. Although this paradigm quickly boosted the development of Release 11 MTC, new challenges have consequently emerged.

Since 2011, 3GPP has been conducting a series of studies to investigate the characteristics of MTC traffic. The results show that the characteristics of traffic to support MTC could be greatly different from those of conventional traffic flowing through UMTS/LTE-A. The differences between MTC traffic and human communication traffic include *infrequent transmissions* (MTC devices may send or receive data at a low duty cycle), *small data transmissions* (MTC devices may only send/receive a small amount of data), *time control* (MTC devices can send or receive data only during an access grant time interval and the network must reject access requests, sending/receiving data,

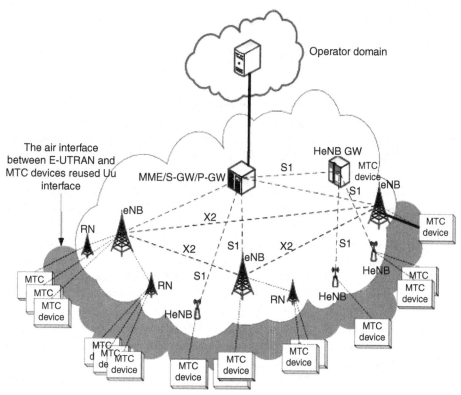

Figure 18.2 In 3GPP Release 11 MTC, the Uu interface (between a UE and an eNB) is reused as the interface between an MTC device and an eNB. In Release 12 MTC, the Uu interface is further simplified. In this figure, X2 is the interface between the eNBs, while S1 is the interface between an eNB/HeNB and a mobility management entity (MME), serving gateway (S-GW), packet data network gateway (P-GW), or HeNB gateway (HeNB GW).

and signaling of MTC devices within a forbidden time interval), and *group-based MTC* (for certain management or resource allocation purposes, the system must provide a mechanism to associate one MTC device with one or more MTC groups).

To further support the above new traffic characteristics, the enhanced MTC (eMTC) standard was launched in Release 12 (see [26]). In eMTC, since an MTC device may transmit/receive very small amount of data, link adaptation schemes such as adaptive modulation and coding schemes, beamforming, and spatial multiplexing to support high data rates are not utilized, to simplify the design of MTC devices. In addition, to support infrequent transmissions, the cell search and synchronization procedures for MTC devices are also simplified.

18.2.3 Heterogeneous Cloud Radio Access Network for M2M Communications

The heterogeneous cloud radio access network (H-CRAN) architecture is suitable for supporting M2M communications which require high mobility and link reliability, such as ITS-based M2M applications. The concept of H-CRAN originates from the hierarchical network architecture of UMTS, in which each radio network controller (RNC) coordinates a number of NodeBs [27]. In UMTS, radio resource management (RRM) is conducted by each RNC, while NodeBs only perform physical signal transmissions/receptions. Although RRM was subsequently implemented so that it could be performed by an eNB in LTE/LTE-A/Evolved Packet Core (EPC), this concept opens up the design of integrating the radio resource optimization in individual eNBs into a joint optimization. Coordinated multipoint (CoMP) transmissions/receptions are thus a practical paradigm for joint resource scheduling/optimization among multiple eNBs [28–31]. In [32], I et al. proposed a novel evolution of CoMP to separate the radio resource optimization and front-end/baseband signal processing, in which the resource optimization is efficiently executed via cloud computing technology. Based on this design, the baseband unit (BBU) and the radio head do not have to be collocated within an eNB. Instead, a number of BBUs and remote radio heads (RRHs) can be separated from an eNB and massively deployed. Using fiber-optic cables to connect eNBs and BBUs/RRHs, the coverage of eNBs can be extended. This architecture is the cloud radio access network (CRAN) [33–36]. Subsequently, Peng et al. [37] and Lei et al. [38] introduced the concept of H-CRAN. Using the wired or wireless interfaces provided (i.e., S1, X2, and Un), not only BBUs/RRHs but also multiple HeNBs, RNs, and eNBs are able to exchange information for joint resource scheduling and allocation [6], as depicted in Figure 18.3.

By fully exploiting the technical merits of HetNets to densely deploy different kinds of cells with different coverage, potential coverage holes can be effectively eliminated to adapt to all kinds of environment with H-CRAN. Furthermore, via cloud computing, a universal resource scheduling/allocation optimization can be achieved to mitigate potential (intercell or intracell) interference. Therefore, in the H-CRAN architecture, high-mobility machines are able to improve their QoS. However, if it is deliver enhanced wireless service experiences, the cost of H-CRAN can be high. Owing to the large number of machines involved in M2M communications, the complexity of resource optimization in H-CRAN may be a growing and inevitable issue, even with the facilitation provided by cloud computing technology. Second, along with the growing number of machines, the amount of traffic may increase by more than a thousandfold. In the H-CRAN architecture, all the traffic relies on the infrastructure for data transmission among machines. Such a star-like topology centered on the cloud for traffic flows may eventually be a burden on both the fronthaul and the backhaul links. In H-CRAN, three facts are generally ignored:

1. Traffic may be exchanged socially or locally among a group of machines. It is assumed that each packet from each machine may be delivered to any other machine in the world in H-CRAN. However, this assumption may not be generally true, as more and more social applications only require data exchange in close physical proximity.

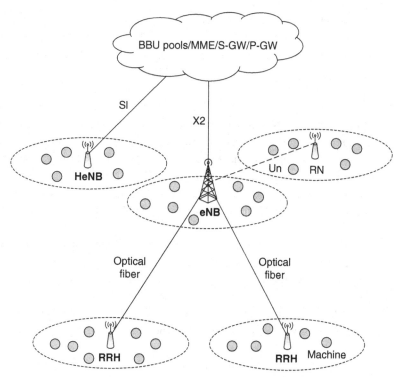

Figure 18.3 In H-CRAN, using the wired or wireless interfaces provided (i.e., S1, X2, and Un), not only BBUs/RRHs but also multiple HeNBs, RNs, and eNBs are able to exchange information for joint resource scheduling and allocation. In this figure, a Un is the interface between an eNB and an RN.

2. Each machine may exchange information with machines within its social network more frequently than with other machines. Such a social network may be a set of servers and machines.

3. The downlink traffic to different machines or the uplink traffic from different machines may have strong correlations. For example, a group of densely deployed sensors measuring a common physical quantity may obtain similar results, and thus the uplink data to the cloud may also be correlated.

As a result, the technical merits of H-CRAN also cause these engineering challenges to limit the performance of H-CRAN. This predicament thus motivates us to revisit the fog computing network (FogNet) architecture for M2M communications.

18.2.4   FogNet Architecture for M2M Communications

The concept of FogNet emerged from fog computing technology [39]. In contrast to cloud computing, fog computing facilitates processing and computing capabilities at edge entities, where not all information for performance optimization needs to be delivered to the cloud. Instead, those tasks (and the corresponding information for optimization) that cannot be processed well by edge entities are handled by the cloud. Fog computing therefore may significantly alleviate the computing burdens in cloud computing and thereby achieve scalability (which is especially important for M2M communications involving a large number of machines). Aryafar *et al.* applied the concept of fog computing to form a new type of network architecture known as FogNet [40]. Under FogNet, a machine that has social messages to be exchanged with other machines does not rely on traffic relaying via the cloud, and machines in close physical proximity are able to share messages locally. This design leads to the concept of socially aware traffic management, where the aim is to significantly decrease the amount of traffic to be supported by the cloud. Recently, radio access technologies such as device-to-device (D2D) communications [41–46] using small cells as smart data/traffic routing gateways, Bluetooth, and ZigBee have been successful examples where the use of FogNet is possible. Subsequently, Peng *et al.* took the correlation among traffic to and from different edge entities into consideration further [47]. When a group of machines have highly correlated traffic to be delivered to the cloud, each entity does not have to upload traffic individually. Instead, the common part of the traffic is delivered once by a single machine. This concept is also known as the *swarm architecture* for M2M networks. On the other hand, when the cloud has highly correlated traffic to be forwarded to multiple machines, it does not forward the traffic individually to each machine. Instead, the cloud selects only one machine to forward the traffic, and then the selected machine autonomously shares the traffic with other machines. Consequently, the amount of traffic supported by both the fronthaul and the backhaul links can be reduced.

Although the FogNet architecture has considerable technical virtues that allow one to potentially tackle the issues of complexity, scalability, and heavy traffic burdens in H-CRAN, new challenges are also created. First, although data can be shared socially among machines, there is no guarantee that all machines that require this service will be able to successfully receive the data. Therefore, reliability of data delivery becomes a primary concern. Second, mobility management and service continuity may not be sufficiently supported in FogNet. Third, owing to the lack of adequate resource coordination among machines, interference may drastically impact the performance of FogNet.

18.3   MAC Design for M2M Communications

The major features of M2M communications lie in (i) the very large number of machines, and (ii) infrequent transmissions of a large number of small amounts of data. Thus, an effective MAC is of crucial importance to eliminate interference and congestion among machines. Without an effective design, machines may suffer continuous collision when a large number of machines are involved. MAC design can be divided into two categories: scheduling-based MAC and random-access-based MAC.

In this section, various MAC schemes for different M2M architectures to support the above features of M2M communications are presented.

18.3.1   Grouping-Based M2M MAC in H-CRAN

In M2M communications, some applications (e.g., reports of measured data from meters in smart grids and navigation signal transmissions in ITSs) require strict timing constraints. Therefore, in M2M communications, providing hard QoS guarantees is an important requirement. Although a lot of schemes and methods have been proposed for providing QoS guarantees in H2H communications, these schemes can be difficult to apply directly to M2M communications. The reason is two-fold. First, the packet arrival periods in M2M communications can range from 10 ms to several minutes or hours owing to the infrequent-transmission feature. Such enormously diverse QoS requirements cause difficulties in the design of radio resource allocation algorithms for M2M communications for providing guaranteed jitter performance, as the jitter (defined by the difference between the time of two successive packet departures and the time of two successive packet arrivals) fully captures the timing performance of periodic traffic. Secondly, unlike the case of multimedia traffic, typically with bursty packet arrivals, each machine may only occupy a small amount of radio resource in each transmission owing to the small amounts of data transmitted. Furthermore, a large number of machines may need to be supported by H-CRAN. As a consequence, the complexity of radio resource allocation for M2M communications can be high.

To support (hard) QoS guarantees in M2M communications, a scheduling-based MAC may provide better performance owing to its capability to optimize resource allocation. However, the major concern for a scheduling-based MAC originates from complexity and scalability, which becomes extremely severe when it is required to support a large number of machines. The key for this to be applied to M2M communications is to reduce the complexity. This idea thus motivates a grouping-based M2M MAC. This grouping-based M2M MAC is a scheduling-based MAC. The main idea is to group machines with similar QoS requirements and traffic characteristics into a cluster. Then, the network schedules radio resources for each cluster, instead of allocating resources to individual machines, as shown in Figure 18.4. Within a cluster, all machines are associated with similar QoS requirements and traffic characteristics. Thus, resource sharing within a cluster may not be a challenging issue. In this case, the number of entities involved in the optimization is substantially reduced, and therefore the complexity of scheduling optimization can be substantially reduced.

To implement a grouping-based M2M MAC, the network may need knowledge regarding the QoS requirements and traffic characteristics of all machines. This knowledge is usually available in the H-CRAN architecture, while it may not be available in the FogNet and WLAN M2M architectures. For a cellular RAN architecture (i.e., MTC in Release 11/12), this knowledge may be available within a small number of eNBs supporting CoMP. As a result, a grouping-based M2M MAC is favorable for H-CRAN and can be partially supported by cellular CRAN.

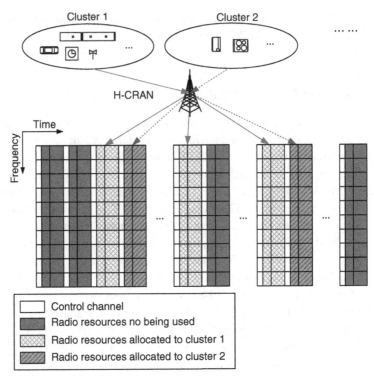

Figure 18.4 In a grouping-based M2M MAC, machines with similar QoS requirements and traffic arrival patterns are grouped into a cluster. Radio resources are then allocated on a cluster basis.

18.3.2   Access Class Barring Based M2M MAC in FogNet/WLAN

MACs based on access class barring (ACB) were originally designed for access control in the random access channel (RACH) of LTE/LTE-A [48, 49]. An ACB-based MAC provides different levels of access latency performance to support access requests from users in different priority classes. In LTE/LTE-A ACB, there are 16 access classes (ACs): AC 0−9 represent normal UEs, AC 10 represents an emergency call, and AC 11−15 represent specific high-priority services. In LTE/LTE-A ACB, an eNB broadcasts access probabilities q and AC barring times for UEs corresponding to ACs 0−9. The UE randomly draws a value p, $0 \leq p \leq 1$. If $p < q$, then the UE proceeds to the random access procedure; otherwise, the UE is barred for the duration of the AC barring time. In this case, the channel access probability of each UE is basically under the control of q, announced by the eNB. When the channel is severely congested, the eNB may announce a small value of q to prevent a large number of UEs trying to access the channel.

The ACB-based MAC is regarded as an effective random access MAC for the RACH of MTC devices. Figure 18.5 shows the performance in terms of access delay using ACB, where 30 000 MTC devices were involved in channel contention. In this simulation, a Beta distribution was used as the traffic arrival model for each MTC device [48], and the AC barring time was 4 s. Figure 18.5 shows that when the barring

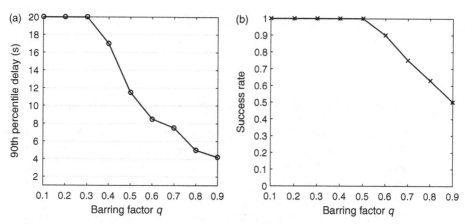

Figure 18.5 (a) Access delay of MTC devices for different barring parameters q. (b) Success rate of MTC devices for different barring parameters q.

parameter q is less than 0.3, the 90th percentile delay exceeds 20 s. In other words, only when the number of MTC devices exceeds 30 000 does the access delay become unacceptable (i.e., exceeding 20 s). These results demonstrate the effectiveness of ACB for dealing with massive accesses by MTC devices.

Because of this performance, ACB is not limited to access schemes for the RACH of LTE/LTE-A. ACB can be extended to a general M2M MAC scheme to prevent channel starvation, especially in FogNet and LAN M2M architectures. However, if the number of machines continues to increase, a very small q may be inevitable, which may lead to unacceptably large latency during contention. As a result, severe congestion may occur on a channel, and how to alleviate such channel congestion may be a crucial issue. In the next section, congestion control schemes for an ACB-based M2M MAC will be discussed further.

18.3.3 Random-Backoff-Based M2M MAC

In a random-access-based MAC such as carrier sense multiple access (CSMA), (slotted) ALOHA or ACB, devices may suffer from continuous collisions when the number of devices is extremely large. Random backoff is a possible collision resolution scheme. In a random-backoff-based M2M MAC, when a collision occurs, each machine randomly selects a value that specifies the size of a backoff window. This backoff value is stored in a counter, and this counter counts down when a particular event occurs (e.g., at an idle slot or at each slot time). The machine can access the channel when its counter reaches zero. It is expected that a large backoff window size can alleviate the problem collisions and thus facilitate the collision resolution. However, although a backoff-based MAC can provide performance improvements for a low congestion level, it cannot resolve a high congestion level when the channel is overloaded. A MAC scheme similar to the backoff-based MAC is dynamic resource allocation. In this scheme, the network dynamically allocates additional resources (in the time, frequency, or code domain) for random access based on the congestion levels and overall traffic load [50]. Although the dynamic allocation of random access resources can be effective in most cases in supporting M2M communications, the performance improvement is limited by the availability of additional resources.

18.3.4   Harmonized M2M MAC for Low-Power/Low-Complexity Machines

It is well known that a scheduling-based MAC may provide better QoS guarantees, while a random-access-based MAC may have the engineering merit of simplicity in implementation. However, the design of an effective MAC scheme generally does not take channel conditions into consideration. This may be reasonable in designs for H2H communications, in which a device is capable of implementing link adaptation technologies. Nevertheless, for M2M communications, low power and low complexity may be critical design requirements in the implementation of the machines. Under this constraint, a machine may not be able to increase the transmission power to enhance the received signal-to-noise ratio at the receiver side. A machine may also not be able to combat severe channel fading through sophisticated channel estimation/equalization. In addition, for those M2M applications which require long-distance communication, the channel quality may have a crucial impact on the design.

In the literature, the integration of a scheduling-based MAC and a random-access-based MAC has been widely discussed to strike an engineering trade-off between QoS and simplicity. A harmonized MAC with scheduling and random access may provide benefits in terms of throughput enhancement for M2M communications that lack the capability to combat channel fading. If a scheduling-based MAC is adopted for M2M communications, when a machine requires a radio resource to upload/download data from/to the network, it needs to send a request to the network. If this resource request is granted by the network, then the network may allocate a radio resource to the machine. However, when the network allocates a radio resource to a machine, there are two obstacles obstructing the attempt of the network and machine to combat channel fading on this allocated resource:

1. On the network side, although the network is able to perform channel estimation to avoid allocating a radio resource suffering from a channel fading condition, such channel estimation typically requires the machine to provide and transmit reference signals to the network. However, transmitting reference signals may be a heavy burden for a low-power machine. As a result, the performance of channel estimation on the network side may be very limited.
2. Without effective channel estimation, a channel fading condition can be eliminated on the machine side via channel equalization. However, as mentioned above, channel equalization may not be affordable for a low-complexity machine. As a consequence, when the network allocates a radio resource to a machine, both the network and the machine are uncertain whether fading conditions may occur on this allocated resource. If channel fading unfortunately occurs on the allocated resource, then transmissions/receptions on this radio resource may not be successful. In other words, if the probability of occurrence of a fading channel on a radio resource is g, then the

probability that a machine can successfully utilize this radio resource is $1 - g$, which is completely subject to g.

To improve the probability that a machine will successfully utilize an allocated radio resource, one possible scheme is that the network allocates two radio resources to a machine which needs only one radio resource. In this case, the probability that the machine can successfully utilize at least one radio resource is $1 - g^2$, which is larger than $1 - g$. Nevertheless, the average resource utilization of each of these two radio resources is $\left(1 - g^2\right)/2$, which may be less than $1 - g$. As a result, if the network allocates more radio resources to a machine needing only one radio resource, then the probability that a machine can successfully utilize at least one radio resource can be significantly boosted. In this case, these radio resources allocated to a machine can be regarded as a resource pool. However, the average resource utilization of each resource may decrease drastically, which is especially unfavorable for M2M applications.

To enhance the average utilization of each resource, we observe that different machines at different geographic locations may suffer from different levels of channel fading conditions with the same resource. The network may thus allow other machines to share the resource pool. In other words, the pool of radio resources is allocated to more than one machine, and each machine can randomly select a radio resource from within the pool to perform transmission. In this case, if multiple machines select different resources from each other, then this pool of radio resources can be fully utilized. Nevertheless, if some machines select the same radio resource, then collisions may occur on that radio resource and the average utilization of each resource may decrease. We note, particularly, that a pool of radio resources randomly shared by multiple machines is exactly the concept of a random-access-based MAC.

The above discussions reveal a trade-off between a scheduling-based MAC and a random-access-based MAC for M2M communications. This trade-off depends on the probability of deep channel fading. As a result, it is difficult to know whether a scheduling-based MAC or a random-access-based MAC is able to provide better average utilization of each radio resource. In this case, there needs to be a harmonization between these two sorts of MAC scheme. In other words, if g is smaller than a certain value, then a scheduling-based MAC should be adopted; otherwise, a random-access-based MAC should be adopted, to optimize the average utilization of each radio resource for all values of q. Such a scheme where the radio access switches between a scheduling-based MAC and a random-access-based MAC is known as a harmonized MAC. To derive the switching condition g analytically, we need to analyze the average of each radio resource when a random-access-based MAC is adopted.

Given a resource pool composed of M radio resources indexed by $m = 1,\ldots,M$, to be shared by N machines indexed by $n = 1,\ldots,N$, let

$$\mathbf{I}_{m,i} = \begin{cases} 1, & i\text{th machine selects } m\text{th resource,} \\ 0, & \text{otherwise} \end{cases}$$

be an indicator function. The probability that the ith machine has one radio resource that allows it to transmit data without suffering deep channel fading is

$$\Pr\left\{\sum_{m=1}^{M}\mathbf{I}_{m,i}=1\right\}=1-g^{M},$$

while the probability that the ith machine has no radio resource that allows it to transmit data without suffering deep channel fading is

$$\Pr\left\{\sum_{m=1}^{M}\mathbf{I}_{m,i}=0\right\}=g^{M}.$$

The mean of $\sum_{m=1}^{M}\mathbf{I}_{m,i}$ is $\mathrm{E}\left[\sum_{m=1}^{M}\mathbf{I}_{m,i}\right]=\sum_{m=1}^{M}\mathrm{E}\left[\mathbf{I}_{m,i}\right]=M\mathrm{E}\left[\mathbf{I}_{m,i}\right]=1-q^{M}$. Thus, $E\left[\mathbf{I}_{m,i}\right]=\Pr\left\{\mathbf{I}_{m,i}=1\right\}=\left(1-g^{M}\right)/M$ can be obtained. For each machine, the probability that the mth radio resource is selected and no collision occurs on this radio resource is given by

$$\Pr\left\{\sum_{m=1}^{M}\mathbf{I}_{m,i}\right\}=N\left(\frac{1-g^{M}}{M}\right)\left(1-\frac{1-g^{M}}{M}\right)^{N-1}.$$

Assuming that the probability of deep channel fading on every radio resource is independently and identically distributed (i.i.d.), the average utilization of each radio resource can be expressed as

$$\rho=N\left(\frac{1-g^{M}}{M}\right)\left(1-\frac{1-g^{M}}{M}\right)^{N-1}.$$

The optimum number of machines N^* to share the pool of radio resources can be found from

$$N^{*}=\arg\max_{N}\left\{N\left(\frac{1-g^{M}}{M}\right)\left(1-\frac{1-g^{M}}{M}\right)^{N-1}\right\}$$

and N^* can be obtained as

$$N^{*}=\begin{cases}\dfrac{M}{1-g^{M}}, & \text{if }(1-g^{M})/M\text{ is an integer,}\\[2mm]\left\lfloor\dfrac{M}{1-g^{M}}\right\rfloor, & \text{otherwise,}\end{cases}$$

which leads to the optimum average utilization of each radio resource as follows:

$$\rho^{*}=\begin{cases}\left(1-\dfrac{1-g^{M}}{M}\right)^{(1-g^{M})/M-1}, & \text{if }(1-g^{M})/M\text{ is an integer,}\\[3mm]\left\lfloor\dfrac{M}{1-g^{M}}\right\rfloor\left(\dfrac{1-g^{M}}{M}\right)\left(1-\dfrac{1-g^{M}}{M}\right)^{\lfloor M/(1-g^{M})\rfloor-1}, & \text{otherwise.}\end{cases}$$

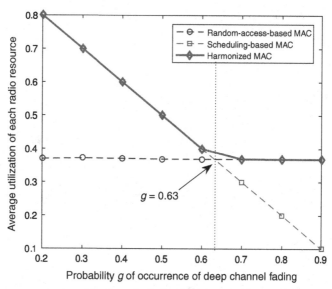

Figure 18.6 Average resource utilization of a harmonized M2M MAC compared with two other schemes.

On the other hand, when a scheduling-based MAC is adopted, the average utilization of each radio resource is $1 - g$. Consequently, a random-access-based MAC should be adopted if

$$\left\lfloor \frac{M}{1 - g^M} \right\rfloor \left(\frac{1 - g^M}{M} \right) \left(1 - \frac{1 - g^M}{M} \right)^{\lfloor M/(1 - g^M) \rfloor - 1} \geq 1 - g.$$

Since

$$\left\lfloor \frac{M}{1 - g^M} \right\rfloor \left(\frac{1 - g^M}{M} \right) \left(1 - \frac{1 - g^M}{M} \right)^{\lfloor M/(1 - g^M) \rfloor - 1} \approx \left(1 - \frac{1 - g^M}{M} \right)^{M/(1 - g^M) - 1},$$

the above inequality can be rewritten as

$$\left(1 - \frac{1 - g^M}{M} \right)^{M/(1 - g^M) - 1} \geq 1 - g \Rightarrow g \geq 1 - \left(1 - \frac{1}{M} \right)^{M-1}.$$

When radio resources are abundant, we may use the approximation

$$\lim_{M \to \infty} \left(1 - \frac{1}{M} \right)^{M-1} \approx 0.63.$$

From the above analysis, it can be seen that a harmonized M2M MAC should switch to a random-access-based MAC if $q \geq 0.63$; otherwise, it should switch to a scheduling-based MAC. Such a harmonized M2M MAC is illustrated in Figure 18.6. When the probability of channel fading is less than 0.63, a scheduling-based MAC provides better performance. In contrast, if the probability of channel fading is larger than 0.63, then a random-access-based MAC can fully exploit "statistical multiplexing" arising from the random access of multiple machines. These results demonstrate that

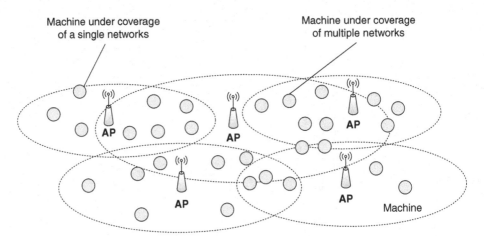

Figure 18.7 Each machine may be located within overlapping coverage areas of multiple M2M networks.

a harmonized M2M MAC is a promising scheme for low-power, low-complexity machines.

18.4   Congestion Control and Low-Complexity/Low-Throughput Massive M2M Communications

Owing to the feature of a large number of machines, traffic congestion may be an inevitable issue. In this section, a promising mechanism to alleviate traffic congestion in a Layer 2 ACB-based M2M MAC is consequently discussed.

18.4.1   Congestion Control in ACB-Based M2M MAC

As mentioned above, an ACB-based M2M MAC is effective for reducing the access latency when a large number of machines are involved in channel contention. However, when the number of machines continue to increase, a very small q may be inevitable, which may lead to unacceptably large latency during contention. As a result, severe congestion may occur on a channel. To alleviate access congestion on a channel further, we observe that, in practical deployments of M2M networks, each machine may be located in overlapping coverage areas of multiple M2M networks, as shown in Figure 18.7. Each M2M network may support a different number of machines and thus suffer from different levels of congestion. If a machine suffering from severe congestion can possibly access an M2M network with a lower congestion level, then the access delay (and thus the congestion) experienced by the machine can be reduced. The main idea is to achieve balancing of the access load from machines between networks.

In an ACB-based M2M MAC, the congestion control of a network can be controlled via deciding on an appropriate ACB parameter q. If the network is highly congested, then q may be set to a small value. On the other hand, if the network does not suffer from congestion, q may be set to a large value. For a machine located in overlapping areas of multiple networks, a simple strategy for a machine is to choose a network announcing a large q for access. However, such a simple strategy may not be optimal for achieving the optimum access balance. When each machine always selects the network with the largest q, then all machines may access that network simultaneously, and this may lead to severe congestion on that network. As a result, the optimum network selection strategy for each machine may be one in which the probability of selecting a network is proportional to the value of q announced by the network. In other words, if a network announces a larger g, then each machine may select that network with a higher probability.

With our knowledge regarding the network selection strategy of machines, we know that the value of q announced by each access point (AP) of a network cannot be decided individually by each AP. Instead, all values of q announced by the APs must be decided jointly. Specifically, suppose that there are $|N_m|$ machines attempting to access the mth AP. In a conventional ACB-based M2M MAC, the ACB parameter q_m of the mth AP should be $q_m = 1/|N_m|$, such that the expected number of machine access is $q_m |N_m| = 1$. However, given the optimum network selection strategy of the machines, if $|N_m|$ is large, then $q_m < 1/|N_m|$ to force machines to access other networks. On the other hand, if $|N_m|$ is small, then $q_m > 1/|N_m|$ to attract more machines to access that network. As a consequence, the joint optimization of all q values can be formulated as an optimization problem.

18.4.2   Massive MTC and Low-Complexity/Low-Throughput IoT Communications

In 2015, the International Telecommunication Union Recommendation Sector (ITU-R) launched its vision of International Mobile Telecommunications for 2020 and beyond, termed IMT-2020. Along with the trend of "full automation" and "X2X connection," the targets of IMT-2020 and beyond include supporting very low-latency and high-reliability human-centric and machine-centric communications, supporting high user densities, maintaining high quality at high mobility, enhanced multimedia services, the IoT, convergence of applications, and ultra-accurate positioning applications. Among these targets, the use cases of ultra-reliable/low-latency M2M communications and massive MTC (mMTC) are receiving significant attention. The use cases of ultra-reliable/low-latency M2M communications include wireless control of industrial manufacturing and production processes, remote medical surgery, distributed automation in smart grids, and transportation safety. The use cases of mMTC include sensors and detectors. To support these use cases, relying solely on MTC/eMTC may not be sufficient. As mentioned above, 3GPP MTC/eMTC reuses the Uu interface to support general use cases of M2M applications. Although MTC/eMTC may support connection reliability and mobility management, this paradigm may lead to high deployment and implementation costs, which are particularly unfavorable for mMTC. These concerns are thus driving 3GPP to develop an alternative system supporting an ultra-low-complexity and low-throughput IoT [51]. LTE and LTE-A (and thus MTC/eMTC) are regarded as wideband solutions for M2M. Besides, MTC/eMTC, two narrowband solutions for low-complexity mMTC have been developed: Extended Coverage GSM (EC-GSM) and narrowband IoT (NB-IoT).

Owing to the engineering merits of simplicity, GSM may be feasible for low-complexity machines. In addition, since GSM is an open-loop system (i.e., feedback channels for sophisticated link adaptation schemes such as hybrid automatic repeat requests, beamforming, MIMO, dynamic modulation, and coding schemes are avoided), it is suitable for low-throughput and low-power M2M applications. To be deployed in the 800–900 MHz frequency bands (close to the frequency bands of HaLow), EC-GSM has an extended communication range, and the evolution from GSM to EC-GSM can be undertaken without significant system impacts.

In October 2015, 3GPP agreed to study the integration of EC-GSM and NB-IoT into a single NB-IoT solution, and the first release of NB-IoT standardization was included in 3GPP Releases 13 and 14. NB-IoT is also known as the clean-slate technology, and in Releases 13 and 14, three different operation modes are supported:

1. Standard-alone operation utilizes the spectrum currently being used by GSM and by Enhanced Data Rates for GSM Evolution (EDGE) Radio Access Networks (GERAN) systems as a replacement for one or more GSM carriers.
2. Guard-band operation utilizes the unused resource blocks within an LTE carrier's guard band.
3. In-band operation utilizes resource blocks within a normal LTE carrier.

In particular, the following communication schemes are supported. The bandwidth for both uplink and downlink transmissions is 180 kHz. For downlink transmissions, orthogonal frequency division multiple access (OFDMA) is adopted, and two numerology options are being considered for inclusion: 15 kHz subcarrier spacing (with a normal or extended cyclic prefix) and 3.75 kHz subcarrier spacing. For uplink transmissions, two options are being considered: frequency division multiple access (FDMA) with Gaussian minimum-shift keying (GMSK) modulation (see Section 7.3 in [51]) and single-carrier FDMA (SC-FDMA) with quadrature phase-shift keying (QPSK) as a baseline modulation. In both downlink and uplink transmissions, single-tone (where one single 15 kHz or 3.75 kHz subcarrier is utilized as a carrier) and multitone (multiple 15 kHz or 3.75 kHz subcarriers are utilized as a carrier) schemes may be used. Table 18.1 summarizes the differences between MTC/eMTC, EC-GSM, and NB-IoT.

Similarly to the development of MTC/eMTC and EC-GSM, to support low-complexity, low-power, and low-throughput machines, a number of features supported by LTE/LTE-A are no longer supported by NB-IoT, including dual connectivity, carrier aggregation, real-time services, handover, relay, measurement reports, D2D, WLAN interworking, etc. Compared with LTE/LTE-A, since the bandwidth of NB-IoT is greatly reduced, the duty cycle of NB-IoT is extended. In LTE/LTE-A, the length of a transmission time interval is 1 ms, which is also the basic time unit for radio resource scheduling. For single-tone transmissions with 15 kHz subcarrier spacing, one schedulable time unit is 8 ms. For single-tone transmissions with 3.75 kHz subcarrier spacing, one schedulable time unit is 32 ms. For three-tone transmissions, one

Table 18.1 MTC/eMTC, EC-GSM, and NB-IoT.

	MTC/eMTC (Rel-11/12)	NB-IoT (Rel-13)	EC-GSM (Rel-13)
Range	$\leq$ 11 km	$\leq$ 15 km	$\leq$ 15 km
Maximum coupling loss	156 dB	164 dB	164 dB
Spectrum	Licensed (700–900 MHz)	Licensed (700–900 MHz)	Licensed (800–900 MHz)
Bandwidth	1.4 MHz or shared	200 kHz (180 kHz + 20 kHz guard band)	2.4 MHz or shared
Data rate	$\leq$ 1 Mbps	$\leq$ 150 kbps	10 kbps
Battery life	~ 10 years	~ 10 years	~ 10 years
Availability	2015	2016	2016

schedulable time unit is 4 ms. For six-tone transmissions, one schedulable time unit is 2 ms. In Layer 2, the duty cycle for broadcasting network information is also extended as compared with LTE/LTE-A. For example, in LTE/LTE-A, the master information block (MIB) is broadcast via the broadcast channel every 40 ms. In NB-IoT, the MIB is broadcast every 640 ms. In LTE/LTE-A, the system supports 20 system information blocks (SIBs) carrying different sorts of network information, while in NB-IoT, only seven SIBs are used.

For link adaptation, stop-and-wait-based HARQ is supported in NB-IoT in both the uplink and the downlink. In the uplink, asynchronous adaptive HARQ is supported (i.e., there is no fixed time period between a data transmission and the corresponding feedback acknowledgment (ACK) or negative ACK (NACK) messages). In the downlink, ACKs/NACKs in response to uplink (re)transmissions are sent on the downlink control channel. This design is very different from that in LTE/LTE-A, in which ACKs/NACKs are sent on the physical HARQ indication channel (PHICH). In other words, in NB-IoT, PHICH is avoided. The main idea of the above design is to simplify the implementation of low-complexity, low-power, and low-throughput machines. As the required bandwidth of each NB-IoT carrier is limited to 200 kHz, it is expected that the design will accommodate more machines to support massive deployment.

18.5   Conclusion

In this chapter, MAC and congestion control designs for some promising M2M architectures to support the next generation of M2M applications were discussed. These discussions demonstrate a top-down design approach, originating from M2M applications, to M2M networks and specific MAC and resource management designs. Considering that M2M applications may dominate wireless services in the coming decades, the insights provided in this chapter may help provide an engineering foundation for practice in 5G mobile networks.

References

[1] Y. S. Soh, T. Q. S. Quek, M. Kountouris, and H. Shin, "Energy efficient heterogeneous cellular networks," *IEEE J. Sel. Areas Commun.*, vol. 31, no. 5, pp. 840–850, May 2013.

[2] Q. Li, R. Q. Hu, Y. Qian, and G. Wu, "Intracell cooperation and resource allocation in a heterogeneous network with relays," *IEEE Trans. Veh. Technol.*, vol. 62, no. 4, pp. 1770–1784, May 2013.

[3] A. Damnjanovic, J. Montojo, Y. Wei, T. Ji, T. Luo, M. Vajapeyam, T. Yoo, O. Song, and D. Malladi, "A survey on 3GPP heterogeneous networks," *IEEE Wireless Commun.*, vol. 18, no. 3, pp. 10–21, Jun. 2011.

[4] M. Wildemeersch, T. Q. S. Quek, M. Kountouris, A. Rabbachin, and C. H. Slump, "Successive interference cancellation in heterogeneous networks," *IEEE Trans. Commun.*, vol. 62, no. 12, pp. 4440–4453, Dec. 2014.

[5] Y. L. Lee, T. C. Chuah, J. Loo, and A. Vinel, "Recent advances in radio resource management for heterogeneous LTE/LTE-A networks," *IEEE Commun. Surv. Tutor.*, vol. 16, no. 4, pp. 2142–2180, Fourth Quarter 2014.

[6] S. Y. Lien, S. C. Hung, K. C. Chen, and Y. C. Liang, "Ultra-low latency ubiquitous connections in heterogeneous cloud radio networks," *IEEE Wireless Commun.*, vol. 22, no. 3, pp. 22–31, Jun. 2015.

[7] K. C. Chen and S. Y. Lien, "Machine-to-machine communications: Technologies and challenges," *Ad Hoc Netw.*, vol. 18, pp. 3–23, Jul. 2014.

[8] S. Y. Lien, K. C. Chen, and Y. Lin, "Toward ubiquitous massive accesses in 3GPP machine-to-machine communications," *IEEE Commun. Mag.*, vol. 49, no. 4, pp. 66–74, Apr. 2011.

[9] A. Rajandekar and B. Sikdar, "A survey of MAC layer issues and protocols for machine-to-machine communications," *IEEE Internet Things J.*, vol. 2, no. 2, pp. 175–186, Apr. 2015.

[10] H. L. Fu, P. Lin, H. Yue, G. M. Huang, and C. P. Lee, "Group mobility management for large-scale machine-to-machine mobile networking," *IEEE Trans. Veh. Technol.*, vol. 63, no. 3, pp. 1296–1305, Mar. 2014.

[11] X. Xiong, K. Zheng, R. Xu, W. Xiang, and P. Chatzimisios, "Low power wide area machine-to-machine networks: Key techniques and prototype," *IEEE Commun. Mag.*, vol. 53, no. 9, pp. 64–71, Sep. 2015.

[12] I. Stojmenovic, "Machine-to-machine communications with in-network data aggregation, processing, and actuation for large-scale cyber-physical systems," *IEEE Internet Things J.*, vol. 1, no. 2, pp. 122–128, Apr. 2014.

[13] G. Dimitrakopoulos and P. Demestichas, "Intelligent transportation systems," *IEEE Veh. Technol. Mag.*, vol. 5, no. 1, pp. 77–84, Mar. 2010.

[14] C. Chakrabarti and S. Roy, "Adapting mobility of observers for quick reputation assignment in a sparse post-disaster communication network," in *Proc. of IEEE Applications and Innovations in Mobile Computing (AIMoC)*, Feb. 2015.

[15] R. Pozza, M. Nati, S. Georgoulas, K. Moessner, and A. Gluhak, "Neighbor discovery for opportunistic networking in Internet of things scenarios: A survey," *IEEE Access*, vol. 3, pp. 1101–1131, 2015.

[16] M. R. Palattella, N. Accettura, X. Vilajosana, T. Watteyne, L. A. Grieco, G. Boggia, and M. Dohler, "Standardized protocol stack for the Internet of (important) things," *IEEE Commun. Surv. Tutor.*, vol. 15, no. 3, pp. 1389–1406, Third Quarter 2013.

[17] C. Perera, A. Zaslavsky, P. Christen, and D. Georgakopoulos, "Context aware computing for the Internet of things: A survey," *IEEE Commun. Surv. Tutor.*, vol. 16, no. 1, pp. 414–454, First Quarter 2014.

[18] K. Katsaros, W. Chai, N. Wang, G. Pavlou, H. Bontius, and M. Paolone, "Information-centric networking for machine-to-machine data delivery: A case study in smart grid applications," *IEEE Netw.*, vol. 28, no. 3, pp. 58–64, May–Jun. 2014.

[19] Z. M. Fadlullah, M. M. Fouda, N. Kato, A. Takeuchi, N. Iwasaki, and Y. Nozaki, "Toward intelligent machine-to-machine communications in smart grid," *IEEE Commun. Mag.*, vol. 49, no. 4, pp. 60–65, Apr. 2011.

[20] A. Zanella, N. Bui, A. Castellani, L. Vangelista, and M. Zorzi, "Internet of things for smart cities," *IEEE Internet Things J.*, vol. 1, no. 1, pp. 22–32, Feb. 2014.

[21] S. Aust, R. V. Prasad, and I. G. M. M. Niemegeers, "Outdoor long-range WLANs: A lesson for IEEE 802.11ah," *IEEE Commun. Surv. Tutor.*, vol. 17, no. 3, pp. 1761–1775, Third Quarter 2015.

[22] T. Adame, A. Bel, B. Bellalta, J. Barcelo, and M. Oliver, "IEEE 802.11ah: The WiFi approach for M2M communications," *IEEE Wireless Commun.*, vol. 21, no. 6, pp. 144–152, Dec. 2014.

[23] C. W. Park, D. Hwang, and T. J. Lee, "Enhancement of IEEE 802.11ah MAC for M2M communications," *IEEE Commun. Lett.*, vol. 18, no. 7, pp. 1151–1154, Jul. 2014.

[24] M. Park, "IEEE 802.11ah: Sub-1-GHz license-exempt operation for the Internet of things," *IEEE Commun. Mag.*, vol. 53, no. 9, pp. 145–151, Sep. 2015.

[25] V. Jones and H. Sampath, "Emerging technologies for WLAN," *IEEE Commun. Mag.*, vol. 53, no. 3, pp. 141–149, Mar. 2015.

[26] 3GPP, "Study on provision of low-cost machine-type communications (MTC) user equipments (UEs) based on LTE," TR 36.888 v12.0.0, Jun. 2013.

[27] A. Samukic, "UMTS universal mobile telecommunications system: Development of standards for the third generation," *IEEE Trans. Veh. Technol.*, vol. 47, no. 4, pp. 1099–1104, Nov. 1998.

[28] Q. Cui, H. Wang, P. Hu, X. Tao, P. Zhang, J. Hamalainen, and L. Xia, "Evolution of limited-feedback CoMP systems from 4G to 5G: CoMP features and limited-feedback approaches," *IEEE Veh. Technol. Mag.*, vol. 9, no. 3, pp. 94–103, Sep. 2014.

[29] S. Sun, Q. Gao, Y. Peng, Y. Wang, and L. Song, "Interference management through CoMP in 3GPP LTE-Advanced networks," *IEEE Wireless Commun.*, vol. 20, no. 1, pp. 59–66, Feb. 2013.

[30] O. Onireti, F. Heliot, and M. A. Imran, "On the energy efficiency–spectral efficiency trade-off in the uplink of CoMP system," *IEEE Trans. Wireless Commun.*, vol. 11, no. 2, pp. 556–561, Feb. 2012.

[31] V. S. Annapureddy, A. El Gamal, and V. V. Veeravalli, "Degrees of freedom of interference channels with CoMP transmission and reception," *IEEE Trans. Inf. Theory*, vol. 58, no. 9, pp. 5740–5760, Sep. 2012.

[32] C. L. I, J. Huang, R. Duan, C. Cui, J. X. Jiang, and L. Li, "Recent progress on C-RAN centralization and cloudification," *IEEE Access*, vol. 2, pp. 1030–1039, 2014.

[33] M. Peng, S. Yan, and H. V. Poor, "Ergodic capacity analysis of remote radio head associations in cloud radio access networks," *IEEE Wireless Commun. Lett.*, vol. 3, no. 4, pp. 365–368, Aug. 2014.

[34] F.A. Khan, H. He, J. Xue, and T. Ratnarajah, "Performance analysis of cloud radio access networks with distributed multiple antenna remote radio heads," *IEEE Trans. Signal Process.*, vol. 63, no. 18, pp. 4784–4799, Sep. 2015.

[35] S. N. Hong and J. Kim, "Joint coding and stochastic data transmission for uplink cloud radio access networks," *IEEE Commun. Lett.*, vol. 18, no. 9, pp. 1619–1622, Sep. 2014.

[36] S. H. Park, O. Simeone, O. Sahin, and S. Shamai Shitz, "Fronthaul compression for cloud radio access networks: Signal processing advances inspired by network information theory," *IEEE Signal Process. Mag.*, vol. 31, no. 6, pp. 69–79, Nov. 2014.

[37] M. Peng, K. Zhang, J. Jiang, J. Wang, and W. Wang, "Energy-efficient resource assignment and power allocation in heterogeneous cloud radio access networks," *IEEE Trans. Veh. Technol.*, vol. 64, no. 11, pp. 5275–5287, Nov. 2015.

[38] L. Lei, Z. Zhong, K. Zheng, J. Chen, and H. Meng, "Challenges on wireless heterogeneous networks for mobile cloud computing," *IEEE Wireless Commun.*, vol. 20, no. 3, pp. 34–44, Jun. 2013.

[39] F. Bonomi, R. Milito, J. Zhu, and S. Addepalli, "Fog computing and its role in the Internet of things," in *Proc. of ACM Mobile Cloud Computing Workshop (MCC)*, Aug. 2012.

[40] E. Aryafar, A. Keshavarz-Haddad, M. Wang, and M. Chiang, "RAT selection games in HetNets," in *Proc. of IEEE International Conf. on Computer Communications (INFOCOM)*, Apr. 2013.

[41] G. Fodor, S. Parkvall, S. Sorrentino, P. Wallentin, Q. Lu, and N. Brahmi, "Device-to-device communications for national security and public safety," *IEEE Access*, vol. 2, pp. 1510–1520, 2014.

[42] Y. Li, D. Jin, J. Yuan, and Z. Han, "Coalitional games for resource allocation in the device-to-device uplink underlaying cellular networks," *IEEE Trans. Wireless Commun.*, vol. 13, no. 7, pp. 3965–3977, Jul. 2014.

[43] L. Lei, Z. Zhong, C. Lin, and X. Shen, "Operator controlled device-to-device communications in LTE-Advanced networks," *IEEE Wireless Commun.*, vol. 19, no. 3, pp. 96–104, Jun. 2012.

[44] A. T. Gamage, H. Liang, R. Zhang, and X. Shen, "Device-to-device communication underlaying converged heterogeneous networks," *IEEE Wireless Commun.*, vol. 21, no. 6, pp. 98–107, Dec. 2014.

[45] Q. Ye, M. Al-Shalash, C. Caramanis, and J. G. Andrews, "Distributed resource allocation in device-to-device enhanced cellular networks," *IEEE Trans. Commun.*, vol. 63, no. 2, pp. 441–454, Feb. 2015.

[46] S. Y. Lien, C. C. Chien, F. M. Tseng, and T. C. Ho, "3GPP device-to-device communications for beyond 4G cellular networks," *IEEE Commun. Mag.*, vol. 54, no. 3, pp. 29–35, Mar. 2016.

[47] M. Peng, S. Yan, K. Zhang, and C. Wang, "Fog computing based radio access networks: Issues and challenges," *IEEE Netw.*, vol. 30, no. 4, pp. 45–53, Jul. 2016.

[48] 3GPP, "Study on RAN improvements for machine-type communications," TR 37.868 V11.0.0, Sep. 2011.

[49] R. G. Cheng, J. Chen, D. W. Chen, and C. H. Wei, "Modeling and analysis of an extended access barring algorithm for machine-type communications in LTE-A networks," *IEEE Trans. Wireless Commun.*, vol. 14, no. 6, pp. 2956–2968, Jun. 2015.

[50] G. Y. Lin, S. R. Chang, and H. Y. Wei, "Estimation and adaptation for bursty LTE random access," *IEEE Trans. Veh. Technol.*, vol. 65, no. 4, pp. 2560–2577, Apr. 2016.

[51] 3GPP, "Cellular system support for ultra-low complexity and low throughput Internet of Things (CIoT)," TR 45.820 v13.1.0, Nov. 2015.

19 Energy-Harvesting Based D2D Communication in Heterogeneous Networks

Howard H. Yang, Jemin Lee, and Tony Q. S. Quek

19.1 Introduction

Direct communication between user equipment (UE) – termed device-to-device (D2D) communication – is envisioned as an intriguing solution to meet the growing demand for local wireless service in fifth generation (5G) networks. Taking advantage of physical proximity, D2D communication is blazing the trail for a flexible infrastructure and boasts the potential benefits of high spectral efficiency, low power consumption, and reduced end-to-end latency. Meanwhile, the heterogeneous network has been emerging as another promising technology for 5G, where by overlaying macrocells with a large number of small-cell access points (APs), it can provide higher coverage and throughput. The idea of using D2D communication to perform mobile relaying in a heterogeneous network is attractive, since together with the better link quality provided by the heterogeneous network in the first hop, D2D communication is able to provide flexible relay selection and enhanced link quality in the second hop, and an overall throughput improvement is therefore foreseeable. However, a problem of fairness arises as the UE relay (UER) needs to consume power to forward information to other UEs. One way to address this issue is to use energy harvesting (EH) technology, which enables devices to harvest energy from their surrounding environments. By adopting EH techniques at each UE, devices can harvest energy from the surrounding environment and use only the harvested energy for relaying, thus preventing power loss from their own battery. In this chapter, we try to coalesce EH technology, D2D communication, and the heterogeneous network into one called the D2D-communication-provided EH heterogeneous network (D2D-EHHN), and investigate the effect of different network parameters as well as provide design insights.

D2D communication has been proposed as a new way to enhance network performance by allowing UEs to communicate directly with their corresponding destinations instead of using a base station (BS) or AP [1–3]. To realize the potential advantages of D2D communication, efforts also need to be made to address the challenges that abound, including peer discovery, mode selection, and interference management in shared networks. In response, various solutions have been proposed. In particular, for resource management, methods to enhance the network throughput include allocating optimal proportions of time [4, 5] or spectrum [6] to activate D2D communication, and joint spectrum scheduling and power control [7]. With regard to mode selection, solutions range from simple distance-based solutions [6] to more sophisticated ones that consider channel state information (CSI) [8], and even a complicated model that involves a system equation [9]. As a result, D2D communication can extend cellular coverage further and facilitate new types of wireless peer-to-peer service by using UEs as a relay [10, 11].

On the other hand, the heterogeneous network is deemed an essential way to combat the continued surge in mobile data traffic, where by offloading traffic from macrocells to smaller but denser cells, the distance between transmitter and receiver can be reduced and a significant enhancement in spectral efficiency can be attained [12–14]. Efforts have been made by wireless researchers to improve the performance of heterogeneous networks: the methods vary from cognitive sensing [15] and sleep mode management [16] for reducing energy consumption to load balancing [17], full-duplex [18], and deploying relays [19] with the purpose of improving network throughput. Using relays, especially UERs, in a heterogeneous network has the advantage of flexible relay selection, and thus is able to effectively reduce path loss and improve throughput [20]. However, the power consumption incurred in the relay nodes becomes a crucial issue, especially when the network has a heavy traffic profile. One possible solution comes from the direction of the newly emerging EH technology [21].

With high expectations of advancing progress in the EH technique, the use of harvested energy to support data transmission has become an attractive research field, [22]. However, with the many new possibilities it brings along, EH technology also poses lots of challenges. To exploit the benefits of energy harvesting, communication protocols need to be redesigned to adapt to the randomness in the external power source [23, 24]. The most common approach is to adjust the duration of energy harvesting and data transmission [25], different strategies have been proposed for both simple scenarios such as point-to-point communication [26] and more complicated situations such as two-hop transmission with relays [27]. In the context of communication networks, possible EH schemes are based on harvesting energy from the received wireless signal [28, 29], or deploying an external energy supplier termed a "power beacon" to provide wireless energy [30]. Further studies have also investigated the performance of an EH-enabled ad hoc network [31] and a fixed relay network with an external power source [32].

To probe these ideas and obtain insights into a heterogeneous network with EH-enabled D2D communication, it is of critical importance to develop a framework that takes into account the presence of harvested energy and the randomness in the locations of D2D UEs. In the following, we borrow tools from stochastic geometry to conduct a study of the D2D-EHHN. Stochastic geometry has recently been used for the analysis and performance evaluation of wireless networks, where, by modeling the locations of BSs and UEs as Poisson point processes (PPPs), tractable analytical expressions for the distribution of the signal-to-interference ratio (SIR) can be obtained to help optimize the network performance [33, 34]. Using stochastic geometry, we develop an analytical framework which considers the effect of EH efficiency, load, and random topology on the network outage probability. In particular, we apply stochastic geometry to model the randomness in the locations of APs and UEs, and model the status of the harvested energy at each UE using a Markov chain. We then derive the density of the UERs, which are the UEs with harvested energy larger than a certain threshold. After deriving the optimal UER locations in terms of outage probability, we propose a transmission mode selection strategy to determine the choice of transmission mode between the AP mode, in which APs transmit directly to UEs, and the D2D mode, which allows APs to transmit their data in two hops via a UER. Based on the mode selection strategy, we derive the outage probability of the D2D-EHHN in closed form. Results from the analysis enable us to understand the impact of network parameters and provide useful guidelines for system design.

19.2   Energy Harvesting Heterogeneous Network

Let us consider the downlink of a K-tier heterogeneous network, which consists of APs that are capable of performing wireless power transfer and UEs that can harvest energy from their nearby APs. In this network, if a UE has an amount of harvested energy exceeding a certain threshold, it is able to act as a UER and help APs forward information in a two-hop transmission. We model the location of the APs in the ith tier as a homogeneous PPP Φ_i with spatial density λ_i, and the location of the UEs an independent PPP Φ_u with spatial density λ_u. We assume that each AP in tier i transmits with power P_i, and each UE has a transmit power of P_u. The channels between pairs of antennas in this network are modeled as independent, narrowband, and affected by two attenuation components, namely small-scale Rayleigh fading and large-scale path loss. We assume that each UE associates with the AP that provides the largest received biased power. Specifically, a UE located at $\mathbf{x}$ associates with an AP $\mathbf{z}^*$ in tier i^*, given by

$$i^* = \arg \max_{1 \leq i \leq K} \max_{\mathbf{z} \in \Phi_i} P_i B_i \|\mathbf{x} - \mathbf{z}\|^{-\alpha}, \tag{19.1}$$

$$\mathbf{z}^* = \arg \max_{\mathbf{z} \in \Phi_{i^*}} \|\mathbf{x} - \mathbf{z}\|^{-\alpha}, \tag{19.2}$$

where B_i is the value of the offloading bias for tier i, and α is the path loss exponent. As UEs do not always communicate with an AP for their own data, we assume that each UE independently receives data for itself with probability p_{rc}.

In light of its high spectral utilization, we consider co-channel deployment, i.e., the APs in all tiers share the same frequency band for transmission. Furthermore, orthogonal access is assumed in this network, where the cellular spectrum is divided into M_c subchannels and each UE communicates with an AP through one of the subchannels. In this sense, the load ρ_i on tier i can be defined as the fraction of APs using the same spectrum at any random time slot, and expressed as follows [35]:

$$\rho_i = \frac{p_{rc} \lambda_u A_i}{\lambda_i M_c}, \tag{19.3}$$

where A_i is the probability that a UE associates with an AP in tier i, given by

$$A_i = \frac{\lambda_i (P_i B_i)^{2/\alpha}}{\sum_{j=1}^{K} \lambda_j (P_j B_j)^{2/\alpha}}. \tag{19.4}$$

We assume further that the D2D communication uses an orthogonal spectrum for APs to avoid severe interference that might degrade the quality of relaying, and the D2D spectrum is similarly divided into M_d subchannels, with UERs selecting one of the subchannels for transmission each time.

19.2.1   Energy Harvesting Region

In order to harvest radio frequency (RF) energy, the UEs are equipped with a power conversion circuit that is able to extract direct current (DC) power from the received electromagnetic wave. Such devices have a sensitivity constraint in reality which requires the input power to exceed a predesigned threshold to activate the EH circuit. In this regard, the power received by a UE from a nearby AP cannot be small, which means that the UE can only harvest energy from a certain region around the AP. We call the region in which a UE can harvest energy the *energy harvesting region* (EHR). In particular, the EHR in tier i consists of circles of radius R_{h_i} centered on APs, where R_{h_i} is given by

$$R_{h_i} = \left(\zeta \frac{P_i}{C} \right)^{\nu}, \tag{19.5}$$

where $\zeta \in (0, 1]$ is the energy harvesting efficiency, C is the energy harvesting threshold for activating the EH circuit, and ν is the path loss exponent for the EH transmission. It should be noted that the transfer distance in the EHR is generally short, which results in line-of-sight power transfer links. Hence, ν is not necessarily the same as the path loss exponent for the data link [30]. As data transmission and energy harvesting cannot happen simultaneously within one time slot, only those UEs in the EHR which have no data to receive can harvest energy from APs. The probability that a UE is located within the EHR can be derived using the void probability in the Poisson process, given by

$$p_{eh} = 1 - e^{-\sum_{i=1}^{K} \lambda_i \pi R_{h_i}^2}. \tag{19.6}$$

When a UE does not need to receive its own data and has an amount of harvested energy exceeding $P_u = NC$, where N is the factor by which the transmit power P_u exceeds C, the UE can help another UE by acting as a UER for forwarding received data from an AP to another UE in a decode-and-forward (DF) manner. Since D2D communication has taken place between the UER and the intended UE, we denote the two-hop transmission as the D2D mode and term the direct transmission without the UER the AP mode. Depending on the location of the UE and UER, either the D2D mode or the AP mode may outperform the other in a typical transmission. It is thus important to develop a mode selection strategy that can adjust to different network conditions and choose the better mode. In the following, after characterizing the distribution of UERs,

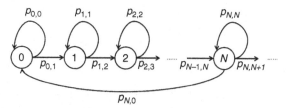

Figure 19.1 State transition diagram of the harvested-energy levels. The state i denotes the amount iC of harvested energy at a UE.

we provide an effective mode selection method which allows us to determine the most appropriate transmission scheme for each UE in this network.

19.2.2   Energy Harvesting Process and UE Relay Distribution

We consider a highly dynamic network where the UEs are moving fast and changing their position at every time slot. Furthermore, we make the assumption that each UE has infinite battery capacity to simplify our notation. In this regard, we can model the amount of harvested energy at each UE using an infinite-state Markov chain, shown in Figure 19.1. When a UE does not harvest enough energy, i.e., the harvested energy is smaller than P_u, which corresponds to states $i = 0, 1, \ldots, N-1$ in Figure 19.1, the UE cannot act as a UER and there are no transitions back to the previous states. When the UE has enough harvested energy, i.e., the UE is in a state $i \geq N$, the UE can support other UE as a UER by using P_u, and its harvested-energy state transits to state $i - N$ after the relaying.

We denote by p_{ij} the transition probability from state i to state j, and give details of this transition probability in the following. In states $i \leq N-1$, the forward transition from state i to state $i+1$ happens only when the UE is located inside the EHR and has no data to be received. Hence,

$$p_{i,i+1} = \left(1 - e^{-\sum_{k=1}^{K} \lambda_k \pi R_{h_k}^2}\right)(1 - p_{rc}), \quad \forall\, i \leq N-1, \tag{19.7}$$

where p_{rc} is the probability that the UE is receiving its own data. Similarly, the transition from state i back to state i happens when the UE either is located outside the EHR or is located inside the EHR while receiving its own data, resulting in

$$p_{i,i} = e^{-\sum_{k=1}^{K} \lambda_k \pi R_{h_k}^2} + \left(1 - e^{-\sum_{k=1}^{K} \lambda_k \pi R_{h_k}^2}\right)p_{rc}, \quad \forall\, i \leq N-1. \tag{19.8}$$

The transition probability changes when $i \geq N$, as the UE has harvested enough energy and is able to relay information at this time. In this case, the transition from state i to state $i+1$ happens when the UE, if it is located inside the EHR, has no data to receive and has not received a request from another UE to perform D2D communication. If we use p_{r_k} to denote the probability that a UE in a state $i \geq N$ has a request from another UE in tier k, the probability of a UER being requested by an arbitrary UE can then be calculated as

$$p_r = 1 - \prod_{k=1}^{K} (1 - p_{r_k}),$$ (19.9)

and the corresponding forward transition probability $p_{i,i+1}$ is given by

$$p_{i,i+1} = \left(1 - e^{-\sum_{k=1}^{K} \lambda_k \pi R_{h_k}^2}\right)(1 - p_{rc})(1 - p_r), \ \forall \ i \geq N.$$ (19.10)

The state of harvested energy in a typical UE remains unchanged if one of the following scenarios happens: (1) the UE is receiving its own data or (2) the UE, located outside the EHR, is neither receiving its own data nor being used by another UE as a UER. In these cases, the transition probability $p_{i,i}$ is given by

$$p_{i,i} = p_{rc} + (1 - p_{rc})(1 - p_r)e^{-\sum_{k=1}^{K} \lambda_k \pi R_{h_k}^2}, \ \forall \ i \geq N.$$ (19.11)

If a UE in state $i \geq N$ is not receiving its own data and is being requested by others to perform relaying, its harvested energy then transits from state i to state $i - N$. Hence, the transition probability $p_{i,i-N}$ is given by

$$p_{i,i-N} = (1 - p_{rc})p_r.$$ (19.12)

With all the transition probabilities derived, we are able to construct the transition matrix

$$\mathbf{P} = [p_{ij}], \quad i,j = 1,2,\ldots.$$ (19.13)

Let $\mathbf{v} = (v_1, v_2, \ldots)$ denote the steady-state probability vector of this Markov chain. We can then solve for each element using

$$\mathbf{v}\mathbf{P} = \mathbf{v}.$$ (19.14)

The probability p_a of a UE being in a state $i \geq N$ is then given by [36]

$$p_a = \sum_{i=N}^{\infty} v_i$$

$$= \frac{\left(1 - e^{-\sum_{i=1}^{K} \lambda_i \pi R_{h_i}^2}\right) p_r^{-1}}{1 - e^{-\sum_{i=1}^{K} \lambda_i \pi R_{h_i}^2} + N}.$$ (19.15)

Since the events of having enough harvested energy and receiving data are independent for each UE, UERs can be identified from among the UEs by removing those which have not yet harvested enough energy or are receiving their own data. In this respect, the location of the UERs Φ_{UER} is a point process thinned from Φ_u, and the thinning property of the PPP indicates that the distribution of UERs follows a homogeneous PPP in the network with a spatial density $\lambda_{\text{UER}} = \lambda_u p_a (1 - p_{rc})$.

19.2.3   Transmission Mode Selection and Outage Probability

The performance of D2D communication depends on which UER is used, and thus the D2D mode may not always have better performance than the AP mode. There is be no need for a UE to use the D2D mode if that performs worse than the AP mode. Hence, the mode selection, i.e., determining when to switch from AP mode to D2D mode and vice versa, is of crucial importance for the D2D-EHHN. Furthermore, in the D2D mode, the UEs need to compare the CSI or the locations of all UERs distributed throughout the network to choose the one that will achieve the best performance. However, it is generally not possible to obtain this information from all distributed UERs, owing to the constraint on feedback overhead, and a UE may therefore only be able to obtain information about a limited number of UERs. In this respect, to enhance the D2D communication performance, it is important to determine which UERs we need to consider and how to select the best UER from among that limited number of UERs.

In the following, we define a *feasible UER region* (FUR) for mode selection and propose an efficient UER selection scheme. We focus on the performance metric of the outage probability, which is defined as the probability that the received SIR γ at a typical UE fails to reach an SIR threshold θ, given by

$$P_o = \Pr(\gamma \leq \theta).$$

The outage probability represents the portion of UEs in the network that are unable to attain the target SIR, which should be kept as small as possible.

We start with the analysis of the outage probability in the AP mode. If we denote the distance between an AP in the ith tier and its intended UE by x, the outage probability can be determined by classical stochastic geometry as [35]

$$P_{o,i}^{AP}(x) = 1 - \exp\left\{-\pi \sum_{j=1}^{K} \rho_j \lambda_j \left(\frac{P_j}{P_i}\right)^{2/\alpha} \mathcal{Z}\left(\theta, \alpha, \frac{B_j}{B_i}\right) x^2\right\}, \qquad (19.16)$$

where θ is the SIR threshold and $\mathcal{Z}(\theta, \alpha, B)$ is given by

$$\mathcal{Z}(\theta, \alpha, B) = \frac{2\theta B^{2/\alpha - 1}}{\alpha - 2} {}_2F_1\left(1, 1 - \frac{2}{\alpha}; 2 - \frac{2}{\alpha}, -\frac{\theta}{B}\right), \qquad (19.17)$$

with ${}_2F_1(a,b;c,x)$ being the hypergeometric function.

Next, we consider the outage probability for a two-hop transmission. If we denote the distance between the AP and the UER as d_1 and the distance between the UER and the intended UE as d_2, the outage probability in the D2D mode can be represented as [36]

$$P_{o,i}^{D2D}(d_1, d_2) = 1 - \exp\left\{-\xi(\theta_d, \alpha)\left(\sum_{j=1}^{K} \rho_j \lambda_j \left(\frac{P_j}{P_i}\right)^{2/\alpha} d_1^2 + \frac{p_r \lambda_{UER}}{M_d} d_2^2\right)\right\}, \qquad (19.18)$$

where θ_d is the SIR threshold for the D2D mode,* and $\xi(\theta,\alpha)$ is given by

$$\xi(\theta,\alpha) = \frac{2\pi^2}{\alpha}\csc\left(\frac{2\pi}{\alpha}\right)\theta^{2/\alpha}. \qquad (19.19)$$

It can easily be seen from (19.18) that the relay location plays an important role in the performance in terms of outage probability, as varying the position of the UER changes d_1 and d_2 simultaneously, thus leading to significant changes in the outage probability. Fortunately, between an AP at (x_{a_i},y_{a_i}) and a UE located at (x_u,y_u), there exists an optimal UE location that minimizes the outage probability, given by [37]

$$(x_{0_i},y_{0_i}) = \left(\frac{x_{a_i}+\Lambda_{n_i}x_u}{1+\Lambda_{n_i}}, \frac{y_{a_i}+\Lambda_{n_i}y_u}{1+\Lambda_{n_i}}\right), \qquad (19.20)$$

where Λ_{n_i} is a scaling parameter, given by

$$\Lambda_{n_i} = \frac{p_r P_i^{2/\alpha}\lambda_{\text{UER}}}{M_d\sum_{j=1}^{K}\rho_j\lambda_j P_j^{2/\alpha}}. \qquad (19.21)$$

We denote the distance from the optimal relay location to the UER by D_i; then, using the cosine law, the outage probability in D2D mode can be represented in another form as [36]

$$P_{0,i}^{\text{D2D}}(x,D_i) = 1 - \exp\left\{-\frac{\xi(\theta_d,\alpha)}{P_i^{2/\alpha}}\left(\sum_{j=1}^{K}\rho_j\lambda_j P_j^{2/\alpha}\right)\left(\frac{\Lambda_{n_i}x^2}{1+\Lambda_{n_i}}+(1+\Lambda_{n_i})D_i^2\right)\right\}. \qquad (19.22)$$

To make a meaningful decision about the mode selection, we define the FUR as a region such that UERs located inside it can guarantee that the intended UE will achieve a smaller outage probability in D2D mode than in AP mode. Specifically, for a D2D-mode UE in the ith tier, the FUR is a region such that UERs located inside it satisfy $P_{0,i}^{\text{D2D}}(x,D_i) \leq P_{0,i}^{\text{AP}}(x)$, which is equivalent to a circle centered on the optimal UER location (x_{0_i},y_{0_i}) with radius $\kappa_i x$, where κ_i is given by

$$\kappa_i = \sqrt{\left[\left(\frac{\varphi(\theta,\alpha,B_i)}{\xi(\theta_d,\alpha)\left(\sum_{j=1}^{K}\rho_j\lambda_j P_j^{\frac{2}{\alpha}}\right)} - \frac{\Lambda_{n_i}}{1+\Lambda_{n_i}}\right)\frac{1}{1+\Lambda_{n_i}}\right]^+}, \qquad (19.23)$$

with $[x]^+ = \max\{x,0\}$ and

$$\varphi(\theta,\alpha,B_i) = \pi\sum_{j=1}^{K}\rho_j\lambda_j P_j^{2/\alpha}\mathcal{Z}\left(\theta,\alpha,\frac{B_j}{B_i}\right). \qquad (19.24)$$

---

* Note that θ_d is generally larger than the threshold θ in the AP mode, since one degree of freedom has been lost in the two-hop transmission, and a higher SIR is required to maintain the achievable rate.

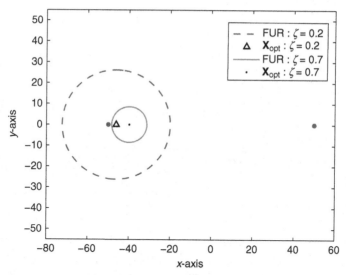

Figure 19.2 An example of FURs for different EH efficiencies ζ, where the AP is located at $(-50, 0)$ and the destination UE is located at $(50, 0)$.

Figure 19.2 shows an example of FURs; it can be seen that the EH efficiency directly affects the FUR in terms of both the optimal relay location and the radius of the FUR circle. In order to adjust the aggregated interference in the network and optimize the achievable network performance, we propose a transmission mode selection scheme for UEs in this network using the following steps [36]:

1. The UE associates with an AP that provides the largest biased received power.
2. The UE checks whether there is any UER in its FUR:

- If there is no UER in its FUR, the UE communicates directly with the AP in the AP mode.
- Otherwise, it communicates in D2D mode, and from among UERs in the FUR, it uses the nearest UER to the optimal UER location for its communication.

Note that this mode selection strategy is both efficient and effective, as the UEs can choose their transmission mode by using only information about the location of UERs in the FUR, which significantly reduces the computational burden. As a consequence, closed-form expressions for the outage probability in the AP and D2D modes can be derived as follows [36]:

$$P_{\mathrm{o}}^{\mathrm{AP}} = 1 - \sum_{i=1}^{K} \frac{\lambda_i \pi \, (P_i B_i)^{2/\alpha}}{\varphi\left(\theta, \alpha, B_i\right) (P_i B_i)^{2/\alpha} + \sum_{k=1}^{K} \lambda_k \pi \, (P_k B_k)^{2/\alpha}} \tag{19.25}$$

and

$$P_{o}^{D2D}$$

$$= 1 - \sum_{i=1}^{K} \left\{ \frac{\beta_i(1 + \Lambda_{n_i})\lambda_i \pi}{\xi\left(\theta_d, \alpha\right)\Lambda_{n_i}\sum_{j=1}^{K}\rho_j\lambda_j\left(\frac{P_j}{P_i}\right)^{2/\alpha} + (1 + \Lambda_{ni})\sum_{k=1}^{K}\lambda_k\pi\,(P_kB_k/P_iB_i)^{2/\alpha}} \right.$$

$$\left. + \frac{(1 - \beta_i)\lambda_i\pi}{\pi\sum_{j=1}^{K}\rho_j\lambda_j\left(P_j/P_i\right)^{2/\alpha}\mathcal{Z}\left(\theta_i, \alpha, B_j/B_i\right) + \lambda_{UER}\kappa_i^2\pi + \sum_{k=1}^{K}\lambda_k\pi\,(P_kB_k/P_iB_i)^{\frac{2}{\alpha}}} \right\},$$

(19.26)

where β_i is given by

$$\beta_i = \frac{\lambda_{UER}\pi P_i^{2/\alpha}}{\lambda_{UER}\pi P_i^{2/\alpha} + \sum_{j=1}^{K}\rho_j\lambda_j P_j^{2/\alpha}(1 + \Lambda_{n_i})\xi\left(\theta_d, \alpha\right)}.$$

(19.27)

Equations (19.25) and (19.26) quantify how all the key features of a D2D-EHHN, i.e., EH efficiency, load, and interference, affect the outage probability of the network. Several numerical results based on (19.25) and (19.26) will be presented in the next section to give more practical insights into the design of a D2D-EHHN.

19.3   Numerical Analysis and Discussion

In this section, we evaluate the performance of a two-tier D2D-EHHN by comparing it with that of a two-tier heterogeneous cellular network (HCN) without EH-D2D communication. Unless specified differently, we used $\lambda_u = 10^{-4}$, $p_{rc} = 0.3$, $p_{r_1} = 0.08$, $p_{r_2} = 0.2$, $\nu = 2$, $\alpha = 3.8$, $\theta = 1$, $\theta_d = 3$, $P_1 = 10$ W, $P_2 = 1$ W, $P_u = 10$ mW, $C = 1$ mW, $M_c = 60$, and $M_d = 30$.

In Figure 19.3, the outage probability P_o of the D2D-EHHN is plotted as a function of the EH coefficient ζ, for different tier 1 AP densities λ_1. We observe that an optimal EH factor ζ exists, and its value is smaller for a large density of APs; for example, for $\lambda_1 = \lambda_2$ in Figure 19.3. This is due to the fact that for small ζ, the UER density is small and the D2D link interference is small, and thus using EH-D2D communication can decrease the outage probability. For large ζ, the UER density is high, which makes the D2D link interference large and increases P_o. Hence, having a large EH factor does not always improve the outage performance in a D2D-EHHN. The results also show that for a sparse network with a small AP density, for example for $\lambda_1 = 0.1\lambda_2$ in Figure 19.3, having a large EH factor provides a smaller outage probability for the D2D-EHHN, while for a dense network with a large AP density, it is better to have a smaller EH factor to avoid a large amount of interference that can degrade the D2D transmission link quality.

Figure 19.4 depicts P_o as a function of λ_2 for different values of ζ. It can be seen that the difference between the outage probabilities of the D2D-EHHN and of the HCN without EH-D2D communication is more significant for a sparse network than for a

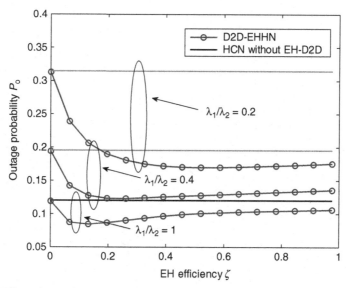

Figure 19.3 Effect of EH efficiency on outage probability, where $B_1 = B_2$ and $\lambda_2 = 10^{-6}$.

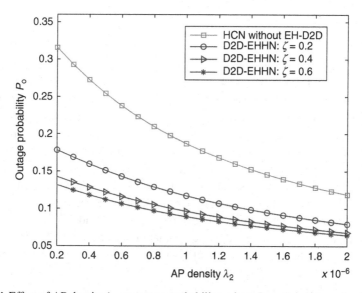

Figure 19.4 Effect of AP density λ_2 on outage probability, where $B_1 = B_2$ and $\lambda_2 = 5 \times 10^{-7}$.

dense network. This observation tells us that using EH-D2D communication is more desirable in a sparse network. Note also that to achieve the same outage probability, the D2D-EHHN requires a smaller λ_2 than does the HCN without EH-D2D. For instance, to obtain an outage probability of 15%, the HCN without EH-D2D requires $\lambda_2 = 2 \times 10^{-6}$, while the D2D-EHHN with $\zeta = 0.4$ needs only $\lambda_2 = 2 \times 10^{-7}$. Consequently, by using EH-D2D communication, we can reduce the energy consumption of the network by turning off some APs while maintaining the same outage probability.

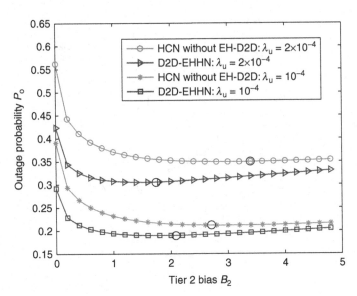

Figure 19.5 Effect of offloading bias on outage probability of different networks, where $B_1 = 1$. The optimal bias B_2 is marked by circles.

Figure 19.5 compares the effects of the offloading bias on the outage probability of the D2D-EHHN and of the HCN without EH-D2D. From Figure 19.5, we can see first that in the absence of EH-D2D communication, the outage probability of the heterogeneous cellular network becomes larger, demonstrating the advantage of using EH-D2D communication. We can also observe that an optimal bias B_2 that minimizes the outage probability exists. This is due to the fact that as B_2 increases, more UEs are associated with APs in tier 2, and the load on tier 1 becomes smaller, and thus the outage probability decreases with increasing B_2 for small B_2. As B_2 continues to increase, more UEs become associated with tier 2, which increases the tier 2 load while the effect of the load decrement in tier 1 becomes negligible, and this results in an increasing outage probability. Note also that the optimal bias B_2 for the D2D-EHHN is smaller than that for the HCN without EH-D2D. This can be attributed to the fact that the average link distance from a UE to an associated AP in tier 1 is larger than that for an AP in tier 2, which results in the FUR for tier 1 being larger than that for tier 2. This indicates that a UE associated with tier 1 has a higher chance of selecting the EH-D2D communication and having a better link quality; thus, a smaller B_2 is desired. Figure 19.5 also shows the impact of the UE density λ_u on the outage probability. As λ_u increases more UEs receive information and the interference in the whole network increases, and hence the outage probabilities of both tiers increase. However, in the D2D-EHHN, the UER density increases with λ_u as well, which provides more UERs for D2D communication. As a result, we observe that for a larger λ_u, the difference in outage probability between the D2D-EHHN and the HCN without EH-D2D is also larger, which indicates that the D2D-EHHN can handle more load.

19.4   Conclusion

In this chapter, we have established the foundations for the concept of the D2D-EHHN, which takes account of the energy harvesting parameter, the AP density, the UE density, and the random network topology. After characterizing the UER density in an energy harvesting context, we proposed an efficient, low-complexity mode selection strategy, including a UER selection method, and also derived the outage probability of a D2D-EHHN. By quantifying the outage probability, we showed how the energy harvesting parameters and the AP densities affect the performance of a D2D-EHHN. Specifically, we showed that an optimal EH efficiency exists, due to the trade-off between the increasing UER density and interference in the D2D communication link, and its value depends on the AP density in the network. Furthermore, the outage probability of a D2D-EHHN is smaller than that of a heterogeneous network without EH-D2D communications. Hence, by exploiting EHD2D communication, energy can be saved by turning off some APs while maintaining a certain level of expected outage probability. We also showed that the optimal offloading bias needs to be redesigned when EH-D2D is available in a heterogeneous network. The D2D-EHHN represents a significant advance toward energy efficiency in 5G networks and opens up several issues for future research, including the effect of outdated UE location information on the UER selection in a D2D-EHHN, introducing multiple antennas at the AP, and the application of full-duplex transmission in a D2D-EHHN.

References

[1] K. Doppler, M. Rinne, C. Wijting, C. Ribeiro, and K. Hugl, "Device-to-device communication as an underlay to LTE-Advanced networks," *IEEE Commun. Mag.*, vol. 47, no. 12, pp. 42–49, Dec. 2009.

[2] G. Fodor, E. Dahlman, G. Mildh, S. Parkvall, N. Reider, G. Miklós, and Z. Turányi, "Design aspects of network assisted device-to-device communications," *IEEE Commun. Mag.*, vol. 50, no. 3, pp. 170–177, Mar. 2012.

[3] X. Lin, J. G. Andrews, A. Ghosh, and R. Ratasuk, "An overview on 3GPP device-to-device proximity services," *IEEE Commun. Mag.*, vol. 52, no. 4, pp. 40–48, Apr. 2014.

[4] Q. Ye, M. Al-Shalash, C. Caramanis, and J. G. Andrews, "Resource optimization in device-to-device cellular systems using time–frequency hopping," *IEEE Trans. Wireless Commun.*, vol. 13, no. 10, pp. 5467–5480, Oct. 2014.

[5] Q. Ye, M. Al-Shalash, C. Caramanis, and J. G. Andrews, "Distributed resource allocation in device-to-device enhanced cellular networks," *IEEE Trans. Commun.*, vol. 63, no. 2, pp. 441–454, Feb. 2015.

[6] X. Lin, J. G. Andrews, and A. Ghosh, "Spectrum sharing for device-to-device communication in cellular networks," *IEEE Trans. Wireless Commun.*, vol. 13, no. 12, pp. 6727–6740, Dec. 2014.

[7] C.-H. Yu, K. Doppler, C. Ribeiro, and O. Tirkkonen, "Resource sharing optimization for device-to-device communication underlaying cellular networks," *IEEE Trans. Commun.*, vol. 10, no. 8, pp. 2752–2763, Aug. 2011.

[8] K. Doppler, C.-H. Yu, C. B. Ribeiro, and P. Janis, "Mode selection for device-to-device communication underlaying an LTE-Advanced network," in *Proc. of IEEE Wireless Communications and Networking Conf. (WCNC)*, Apr. 2010.

[9] S. Hakola, T. Chen, J. Lehtomaki, and T. Koskela, "Device-to-device (D2D) communication in cellular network-performance analysis of optimum and practical communication mode selection," in *Proc. of IEEE Wireless Communications and Networking Conf. (WCNC)*, Apr. 2010.

[10] S. Burleigh, A. Hooke, L. Torgerson, K. Fall, V. Cerf, B. Durst, K. Scott, and H. Weiss, "Delay-tolerant networking: An approach to interplanetary Internet," *IEEE Commun. Mag.*, vol. 41, no. 6, pp. 128–136, Jun. 2003.

[11] H. Nishiyama, M. Ito, and N. Kato, "Relay-by-smartphone: Realizing multihop device-to-device communications," *IEEE Commun. Mag.*, vol. 52, no. 4, pp. 56–65, Apr. 2014.

[12] T. Q. S. Quek, G. de la Roche, I. Güvenç, and M. Kountouris, *Small Cell Networks: Deployment, PHY Techniques, and Resource Management*, Cambridge University Press, 2013.

[13] W. C. Cheung, T. Q. S. Quek, and M. Kountouris, "Throughput optimization, spectrum allocation, and access control in two-tier femtocell networks," *IEEE J. Sel. Areas Commun.*, vol. 30, no. 3, pp. 561–574, Apr. 2012.

[14] H. S. Dhillon, R. K. Ganti, F. Baccelli, and J. G. Andrews, "Modeling and analysis of K-tier downlink heterogeneous cellular networks," *IEEE J. Sel. Areas Commun.*, vol. 30, no. 3, pp. 550–560, Apr. 2012.

[15] M. Wildemeersch, T. Q. S. Quek, C. Slump, and A. Rabbachin, "Cognitive small cell networks: Energy efficiency and trade-offs," *IEEE Trans. Wireless Commun.*, vol. 61, no. 9, pp. 4016–4029, Sep. 2013.

[16] Y. S. Soh, T. Q. S. Quek, M. Kountouris, and H. Shin, "Energy efficient heterogeneous cellular networks," *IEEE J. Sel. Areas Commun.*, vol. 31, no. 5, pp. 840–850, Apr. 2013.

[17] Q. Ye, B. Rong, Y. Chen, M. Al-Shalash, C. Caramanis, and J. G. Andrews, "User association for load balancing in heterogeneous cellular networks," *IEEE Trans. Wireless Commun.*, vol. 12, no. 6, pp. 2706–2716, Jun. 2013.

[18] J. Lee and T. Q. S. Quek, "Hybrid full-/half-duplex system analysis in heterogeneous wireless networks," *IEEE Trans. Wireless Commun.*, vol. 14, no. 5, pp. 2883–2895, May 2015.

[19] M. Peng, Y. Liu, D. Wei, W. Wang, and H.-H. Chen, "Hierarchical cooperative relay based heterogeneous networks," *IEEE Wireless Commun.*, vol. 18, no. 3, pp. 48–56, Jun. 2011.

[20] J. Bang, J. Lee, S. Kim, and D. Hong, "An efficient relay selection strategy for random cognitive relay networks," *IEEE Trans. Wireless Commun.*, vol. 14, no. 3, pp. 1555–1566, Mar. 2015.

[21] I. Krikidis, S. Timotheou, S. Nikolaou, G. Zheng, D. W. K. Ng, and R. Schober, "Simultaneous wireless information and power transfer in modern communication systems," *IEEE Commun. Mag.*, vol. 52, no. 11, pp. 104–110, Nov. 2014.

[22] A. Kurs, A. Karalis, R. Moffatt, J. D. Joannopoulos, P. Fisher, and M. Soljačić, "Wireless power transfer via strongly coupled magnetic resonances," *Science*, vol. 317, no. 5834, pp. 83–86, Jul. 2007.

[23] O. Ozel, K. Tutuncuoglu, J. Yang, S. Ulukus, and A. Yener, "Transmission with energy harvesting nodes in fading wireless channels: Optimal policies," *IEEE J. Sel. Areas Commun.*, vol. 29, no. 8, pp. 1732–1743, Sep. 2011.

[24] K. Tutuncuoglu and A. Yener, "Optimum transmission policies for battery limited energy harvesting nodes," *IEEE Trans. Wireless Commun.*, vol. 11, no. 3, pp. 1180–1189, Mar. 2012.

[25] A. A. Nasir, X. Zhou, S. Durrani, and R. A. Kennedy, "Relaying protocols for wireless energy harvesting and information processing," *IEEE Trans. Commun.*, vol. 12, no. 7, pp. 3622–3636, Jul. 2013.

[26] I. Krikidis, S. Timotheou, and S. Sasaki, "RF energy transfer for cooperative networks: Data relaying or energy harvesting?" *IEEE Commun. Lett.*, vol. 16, no. 11, pp. 1772–1775, Nov. 2012.

[27] Y. Luo, J. Zhang, and K. B. Letaief, "Throughput maximization for two-hop energy harvesting communication systems," in *Proc. of IEEE International Conf. on Communications (ICC)*, Jun. 2013.

[28] M. Maso, S. Lakshminarayana, T. Q. S. Quek, and V. H. Poor, "A composite approach to self-sustainable transmission: Rethinking OFDM," *IEEE Trans. Commun.*, vol. 62, no. 11, pp. 3904–3917, Nov. 2014.

[29] M. Maso, C. F. Liu, C. H. Lee, T. Q. S. Quek, and L. S. Cardoso, "Energy-recycling full-duplex radios for next-generation networks," *IEEE J. Sel. Areas Commun.*, vol. 33, no. 12, pp. 2948–2962, Dec. 2015.

[30] K. Huang and V. Lau, "Enabling wireless power transfer in cellular networks: Architecture, modeling and deployment," *IEEE Trans. Wireless Commun.*, vol. 13, no. 2, pp. 4788–4799, Feb. 2014.

[31] K. Huang, "Spatial throughput of mobile ad hoc networks with energy harvesting," *IEEE Trans. Inf. Theory*, vol. 59, no. 11, pp. 7597–7612, Nov. 2013.

[32] I. Krikidis, "Simultaneous information and energy transfer in large-scale networks with/without relaying," *IEEE Trans. Wireless Commun.*, vol. 62, no. 3, pp. 900–912, Mar. 2014.

[33] J. G. Andrews, F. Baccelli, and R. K. Ganti, "A tractable approach to coverage and rate in cellular networks," *IEEE Trans. Commun.*, vol. 59, no. 11, pp. 3122–3134, Nov. 2011.

[34] B. Blaszczyszyn, M. K. Karray, and H. P. Keeler, "Using Poisson processes to model lattice cellular networks," in *Proc. of IEEE International Conf. on Computer Communications (INFOCOM)*, Apr. 2013.

[35] H. S. Dhillon, R. K. Ganti, and J. G. Andrews, "Load-aware modeling and analysis of heterogeneous cellular networks," *IEEE Trans. Wireless Commun.*, vol. 12, no. 4, pp. 1666–1677, Apr. 2013.

[36] H. H. Yang, J. Lee, and T. Q. S. Quek, "Heterogeneous cellular network with energy harvesting-based D2D communication," *IEEE Trans. Wireless Commun.*, vol. 15, no. 2, pp. 1406–1419, Feb. 2016.

[37] H. H. Yang, J. Lee, and T. Q. S. Quek, "Opportunistic D2D communication in energy harvesting heterogeneous cellular network," in *Proc. of IEEE International Workshop on Signal Processing, Advances in Wireless Communications*, Jun. 2015.

20 LTE-Unlicensed: Overview and Distributed Coexistence Design

Yunan Gu, Lin X. Cai, Lingyang Song, and Zhu Han

With more users, more mobile phones and tablets, more connections to homes and offices, and richer content sharing over wireless communication networks, the industry is facing an exponential increase in mobile broadband traffic in the frequency spectrum from 700 MHz to 2.6 GHz. To meet this demanding challenge, an intuitive idea is to add more licensed spectrum, which would ensure predictable performance in terms of mobility. However, for various reasons, it is possible that sufficient additional licensed spectrum will not be available in the near future. On the other hand, the amount of unlicensed spectrum already assigned or planned to be assigned is comparable or even more than the licensed spectrum. By taking full advantage of the unlicensed spectrum, the burden on the licensed spectrum can be relieved. Unlicensed spectrum has already been utilized in multiple technologies, such as Wi-Fi and Bluetooth, and now there is a new member of the unlicensed family at 5 GHz for mobile broadband. By extending the benefits of Long Term Evolution-Advanced (LTE-A) to the unlicensed spectrum, we can boost the capacity, while maintaining seamless mobility and predictable performance.

Although it has great potential, LTE-Unlicensed is still in its infancy and faces some major challenges. Only by careful design can the LTE-Unlicensed technique truly benefit us with tremendous advantages. Thus, in this chapter, we give a comprehensive introduction to the benefits and design principles of LTE-Unlicensed. In addition, two specific implementation cases are provided for illustration purposes. This chapter is organized as follows. In Section 20.1, the motivation for proposing the LTE-Unlicensed technique will be stated. Then the design challenges in and potential solutions for LTE-Unlicensed will be discussed in Section 20.2. Two distributed resource allocation applications utilizing matching-based approaches will be introduced in Section 20.3. Finally, conclusions are drawn in Section 20.4.

20.1 Motivations

Currently, technologies such as IEEE 802.11 (Wi-Fi), 802.15.1 (Bluetooth), and 802.15.4 (ZigBee) are implemented in the 2.4 GHz ISM (Industry, Scientific, and Medical) and 5 GHz U-NII (Unlicensed National Information Infrastructure) bands, more commonly referred to as the unlicensed bands. Some operators have deployed a large number of Wi-Fi access points (APs) to offload cellular traffic to the unlicensed spectrum. Wi-Fi offloading strikes a trade-off between capacity and performance.

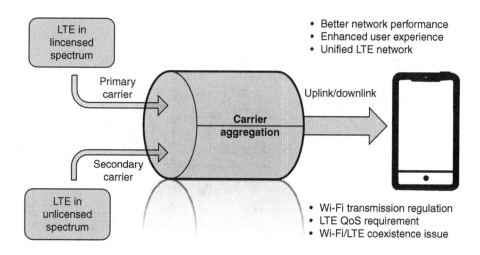

Figure 20.1 LTE-Unlicensed carrier aggregation.

However, such efforts are not achieving expectations in terms of reducing cost and improving performance, owing to the extra cost of the backhaul and core network in addition to the existing cellular infrastructure, and to the lack of good coordination between cellular and Wi-Fi systems [1].

Carrier aggregation (CA), an LTE innovation, has enabled operators to make the best use of the unlicensed bands, as shown in Figure 20.1. By aggregating the unlicensed bands with an LTE anchor in the licensed spectrum, better coverage and higher spectral efficiency can be achieved compared with Wi-Fi offloading, while seamless flow of data across licensed and unlicensed spectrum in a single core network is also guaranteed. Generally, there are two approaches in terms of integrating LTE into the unlicensed spectrum, as shown in Figure 20.2: (1) LTE-Wi-Fi link aggregation (LWA), to leverage Wi-Fi networks using both the 2.4 GHz and the 5 GHz bands, and (2) LTE-Unlicensed carrier aggregation, also referred to as LTE-Unlicensed, where LTE operates in the unlicensed 5 GHz spectrum [2]. The link aggregation needs support from both device side and network side. On the device side, the aggregation is implemented at the modem level. On the network side, it can be implemented between either collocated or separate Wi-Fi and LTE-Wi-Fi APs. The eNBs (Evolved Node B) control the amount of traffic scheduled to LTE and Wi-Fi, and ensure proper load balancing between the LTE and Wi-Fi links. The link aggregation is defined in the Third Generation Partnership Project (3GPP) Release 13 document [3], and its benefits have been demonstrated by Qualcomm [4]. However, compared with LWA, LTE-Unlicensed CA offers the tightest possible aggregation. It comes in multiple flavors: LTE-U [5], Licensed Assisted Access (LAA) [3], and MuLTEfire* [6], which depend on the specific deployment scenario

---

* MuLTEfire does not fall into the scope of this chapter, but is still a good choice of traffic offloading scheme.

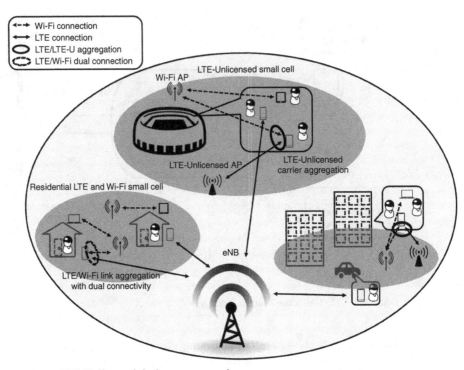

Figure 20.2 LTE-Unlicensed deployment scenario.

and region. LTE-U, based on 3GPP Release 10/11/12, targets early deployment by mobile operators in the United States, Korea, and India and markets without a Listen Before Talk (LBT) requirement. For example, the US deployment of LTE-Unlicensed in the U-NII radio bands ranges from 5.15 to 5.85 GHz, as regulated by the Federal Communications Commission (FCC) [7]. Defined in 3GPP Release 13, LAA targets deployments in Europe, Japan, and beyond, and in regions with specific access procedures with LBT. The performance of LTE-U and LAA, and how they can coexist in a fair manner with Wi-Fi were demonstrated at Mobile World Congress (MWC) 2015. MuLTEfire broadens the LTE ecosystem to new deployment opportunities by operating solely in unlicensed spectrum without a licensed anchor channel.

For both LTE-U and LAA, the signaling and control messages are sent through a reliable licensed anchor, and the unlicensed link is used only for data. Both options use a single, unified core network that provides cost efficiency and simplicity of management to operators, while offering seamless service continuity and a better broadband experience to users. Owing to the low-power restrictions imposed by regulations on unlicensed transmission, the transmission range of the unlicensed bands is relatively small. Therefore, small-cell deployment is more suitable for LTE-Unlicensed, as shown in Figure 20.2. For both indoor and outdoor small-cell deployment, LTE-Unlicensed can provide its greatest benefits, as listed in the following subsections.

20.1.1   Better Network Performance

By aggregating unlicensed carriers with licensed frequency division duplex (FDD) or time division duplex (TDD) carriers in small cells, LTE-Unlicensed augments existing LTE mobile networks to provide enhanced mobile broadband performance. By using the LTE technology, which supports coordinated, synchronized scheduling of resources, and has an efficient radio link with features such as scaling to lower data rates, handling larger delay spreads, hybrid automatic repeat request (HARQ), and so on, LTE-Unlicensed can provide two or more times capacity in dense network deployments. By trading off part of the capacity increase, the network coverage can be improved.

20.1.2   Enhanced User Experience

Owing to its integration into the LTE system, the unlicensed secondary-carrier function can be enabled or disabled seamlessly by the LTE control without users noticing. Thus, mobile users can experience an enhanced data rate and seamless transmission. In addition, for those applications that have strict quality-of-service (QoS) requirements and important signaling traffic, data are routed through the licensed spectrum, thus ensuring high reliability.

20.1.3   Unified LTE Network Architecture

For cellular operators that already have an LTE network, LTE-Unlicensed can be bundled seamlessly with an LTE carrier in licensed spectrum using Supplemental Downlink (SDL)[†] or both downlink (DL) and uplink (UL) CA. The licensed and unlicensed carriers can be managed conveniently as one unified network, and data traffic can be routed flexibly through both carriers.

20.1.4   Fair Coexistence with Wi-Fi

The criterion for good coexistence is that the impact of interferences from LTE-Unlicensed on a given neighboring Wi-Fi node is no more than the impact of interference from other Wi-Fi nodes in the same setup. LTE, as a coordinated system, uses sensing-based channel access and can effectively achieve interference coordination with existing Wi-Fi users. If coexistence restrictions are imposed on the LTE transmission, LTE-Unlicensed users may be a better neighbor to Wi-Fi than Wi-Fi itself [2].

20.2   Coexistence Issues in LTE-Unlicensed

Owing to the low-power and high-frequency transmission requirements imposed by the regulations for unlicensed spectrum, small-cell deployment becomes a perfect

---

† In SDL, the unlicensed spectrum is used for downlink traffic only.

implementation scenario for LTE-Unlicensed. One urgent need for LTE-Unlicensed small cells is to guarantee fair coexistence between newly joined cellular users (CUs) and existing unlicensed users (UUs). For Wi-Fi transmission, which is collision avoidance based, a UU may back off to an LTE-Unlicensed user (operating in the same channel) if its interference level is above the energy detection threshold (e.g., −62 dBm over 20 MHz) [8]. Thus, without proper coexistence mechanisms, LTE-Unlicensed transmissions could cause considerable interference to Wi-Fi transmissions. On the other hand, interference from co-channel Wi-Fi users may also degrade the performance of LTE-Unlicensed devices, leading to a failure to meet the QoS requirements for cellular transmission. In addition, with limited unlicensed bands, LTE-Unlicensed users (among multiple cellular network operators (CNOs)) need to compete among themselves for available unlicensed channels. Thus, there may exist interoperator interference. To summarize, such unplanned and unmanaged deployment can result in excessive interference to both Wi-Fi and LTE-Unlicensed users. Therefore, it is critical to design a coexistence mechanism to avoid such co-channel interference and guarantee the harmonious coexistence of Wi-Fi and LTE systems.

The performance of LTE-Unlicensed has been evaluated in some studies. For example, the work in [9] presented a system performance analysis of LTE and wireless local area networks (WLANs) sharing the unlicensed resource using a simple fractional bandwidth-sharing mechanism. Simulation results showed that co-existence had a negative impact on WLAN system performance if there were no restrictions on LTE transmission, but the severity of the impact could be controlled by restricting LTE activity. Similar evaluations were performed in [10]. The results showed a Wi-Fi user performance degradation of about 70% to 100% without intersystem coordination. In [11], five deployment cases of LTE and Wi-Fi coexistence were evaluated and compared. The results indicated that with coexistence mechanisms, the LTE throughput could be considerably increased; at the same time, an LTE-Unlicensed small cell could be as good a neighbor to a Wi-Fi network as Wi-Fi itself and sometimes even better.

There have been some potential solutions proposed to eliminate coexistence interference between LTE and Wi-Fi. For example, in [12], channel selection, carrier-sensing adaptive transmission (CSAT), and opportunistic SDL were considered. The channel selection mechanism enables small cells to choose an available channel based on carrier sensing. The sensing is performed in both the initial power-up stage and later, periodically, in the SDL operation stage. Whenever interference is detected in the operating channel and there is another, better channel available, the SDL transmission is switched to the new channel using procedures specified in LTE Release 10/11. Therefore, the interference between the small cell and neighboring Wi-Fi devices and other LTE-U small cells can be mitigated. The CSAT mechanism is used when no idle channel is available, which means the small cell may need to share the channel with Wi-Fi. The basic idea is time sharing based on long-term carrier sensing of co-channel Wi-Fi, which is a similar concept to ones in the LBT and CSMA (carrier sense multiple access) techniques. In CSAT, the small cell senses the medium for a longer duration than in LBT and CSMA (tens of milliseconds to 200 m), and, according to the observed activity in the medium, the algorithm turns off the LTE transmission proportionally. The duty cycle of LTE-U transmission is configurable and adaptable. CSAT ensures that LTE-U nodes can share the channel in a fair manner with neighboring Wi-Fi APs even in dense deployments. In opportunistic SDL, the SDL transmission is done opportunistically based on the traffic demand. When either the DL traffic in the small cell exceeds a certain threshold or there are active users within the coverage area of the unlicensed band, the SDL carrier can be turned on for offloading. When the traffic load can be managed by the primary carrier alone, or there are no users within the coverage area of the unlicensed band, the SDL carrier is turned off. This mitigates the interference from continuous LTE-U transmission in the unlicensed channel, which further reduces the interference to neighboring Wi-Fi APs.

In markets where there is no LBT regulation, LTE-U can be deployed based on the Release 10/11/12 CA protocols. This allows the fast-time-to-market launch of small cell systems that can aggregate unlicensed carrier as an SDL carrier with a primary carrier in a licensed band. New radio frequency band support is needed in both small cells and UEs for operation. Moreover, carefully designed mechanisms must be used to ensure harmonious coexistence with Wi-Fi. Without modifying the Release 10/11 LTE physical layer and medium access control (PHY/MAC) standards, the above-mentioned three mechanisms can be used directly for this purpose. In Europe and Japan, where there is an additional unlicensed-band regulation, requiring LBT, the design of coexistence is different from LTE-U, as specified in LAA based on Release 13. The sharing concept is similar to that in LTE-U, but, owing to the LBT requirement, some modifications to the LTE air interface standards are required [2, 12]. In LAA, whenever a device wants to transmit, it senses the channel every 20 ms, which is called the clear channel assessment (CCA) period. When the detected energy level is lower than the threshold (i.e., the channel is considered free), then the equipment can reserve the channel for transmission, and finally transmit for a duration equal to the channel occupancy time (1–10 ms). If the channel is busy, the equipment waits for a specific amount of time, based on a randomized counter (per LBT regulations). After that, if the equipment wishes to continue its transmission, it has to repeat the CCA process. In addition, for the UL transmission, some other modifications are required to meet the regulations.

In addition to the above-mentioned mechanisms, some other approaches can also be considered. Transmit power control (TPC) was proposed in [13] to enable LTE/Wi-Fi coexistence. By measuring the interference at the LTE eNBs, LTE users estimate the presence and proximity of Wi-Fi users, and thus adjusts their transmission power to avoid excessively strong interference to Wi-Fi. Alternatively, the blank subframe method has also been proposed to prevent LTE and Wi-Fi accessing the channel at the same time [14], which is similar to the idea of the LTE technique of the almost blank subframe (ABS), proposed in Release 10. By silencing some of the subframes in LTE UL/DL transmission, Wi-Fi users can reuse the blank subframe of LTE to increase throughput.

20.3   Distributed Resource Allocation Applications of LTE-Unlicensed

In the previous sections, potential solutions to the LTE-Unlicensed coexistence issue have been discussed. However, the use of either TPC, the blank subfram method, CSAT, or opportunistic SDL affects the transmission quality/throughput of the LTE users, to a greater or lesser extent; thus, an intuitive way to ensure fair coexistence with Wi-Fi while simultaneously allowing quality LTE transmission is channel selection. Specifically, channel selection should be deployed in both LTE and Wi-Fi systems. In other words, restrictions should be imposed on both LTE and Wi-Fi users when resource sharing is established between them. For example, if an LTE user causes too much interference to an existing co-channel Wi-Fi user, that LTE user should not be allowed to transmit on this channel. Similarly, if a co-channel Wi-Fi user is causing too much interference to an LTE user such that the LTE user's signal-to-noise ratio becomes lower than its QoS requirement, then the LTE user should give up sharing that unlicensed band. To summarize, the issue of coexistence between Wi-Fi and LTE systems can be treated as a resource (unlicensed band) allocation problem between the CUs and UUs (each associated with an unlicensed resource) under certain interference restrictions. Thus, in this section, we treat LTE-Unlicensed coexistence design as a constrained resource allocation problem, and develop corresponding solutions.

20.3.1   Matching Theory Framework

The existing solutions of resource allocation problems in wireless communications can be generally classified into centralized optimization and distributed game approaches. Despite the optimality of centralized optimization, the network density, the requirement for global information, and the network dynamics may result in extremely high computational overhead and complexity, which makes it less efficient. Game theory can be used as an alternative approach to solve the problem in a distributed manner. The resource allocation problem can be modeled as interactions between players under certain rules. For example in [15], a coordinated hierarchical game was proposed for modeling multioperator spectrum sharing in LTE-Unlicensed, where a Kalai–Smorodinsky bargaining game among operators and a Stackelberg game between operators and users were modeled. However, game theory also has its shortcomings in the sense that each player requires knowledge of the other players' actions in many best response mechanisms, which limits the possibility of a distributed implementation. In addition, many game solutions, such as the Nash equilibrium, apply to a one-sided notion of stability, which may not be applicable to problems between distinct sets of players. Matching theory, as a mathematical framework that attempts to describe the formation of mutually beneficial relationships, can overcome some limitations of game theory and of the centralized optimization approach [16]. The advantages of matching theory include suitable models for various communication issues, preference interpretations for system constraints, and efficient algorithms for desired objectives.

So far, the matching theory framework has been applied to many research topics in wireless communications, such as device-to-device communication [17], physical layer security [18], content caching [19], and small-cell user association [20].

To give a general idea of how matching theory works, we take the classical matching model of *stable marriage* [21] as an example. Assume a set of men and a set of women; each man and woman is called a matching agent. The *preference list* for each agent is an ordered list based on the preferences over the other set of agents of who he/she finds acceptable. A matching consists of (man, woman) pairs. A basic requirement, the *stability* concept, refers to the condition that, in a matching, there exists no (man, woman) pair who both have an incentive to leave their current partners and form a new marriage with each other. A stable matching can be achieved by using the Gale–Shapley (GS) algorithm, which is widely deployed and has been customized to generate stable matchings in many other models. The GS algorithm is composed of operations of proposing and accepting/rejecting, and terminates when no further proposals are made. Stability is essential in both economic applications and applications to wireless communication systems. It generally refers to the situation where no player pairs/groups (e.g., CU and UU pairs in LTE-Unlicensed) have an incentive to violate the current assignment under the table for their own benefit. Thus, most matching algorithms are designed to respect the stability principle.

In the following sections, i.e., Sections 20.3.2 and 20.3.3, two distributed matching-based implementations of LTE-Unlicensed will be developed: (1) the student–project allocation (SPA) model and (2) the hospital resident (HR) model. These two applications tackle the static and dynamic resource allocation cases, respectively, in LTE-Unlicensed, as shown in Figure 20.3.

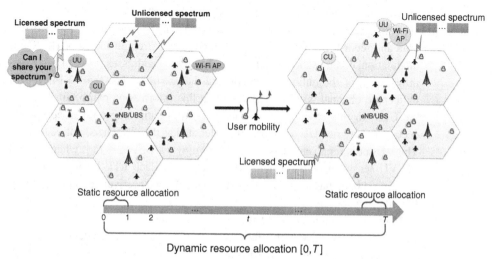

Figure 20.3 LTE-Unlicensed applications: static and dynamic scenarios.

20.3.2   Static Resource Allocation: Student–Project Allocation Matching

As mentioned in Section 20.3.1, the resource allocation problem in LTE-Unlicensed with coexistence constraints can be modeled by suitable matching games. With the objective of optimizing the social welfare (including both the CUs' and the UUs' throughput), as well as the interference restrictions imposed on the CUs and the UUs, in this example, we model the interactions between the CUs and UUs using an SPA game [22]. The coexistence issues are interpreted in the SPA matching model as the interactions between students, projects, and lecturers. The SPA-(S,P) algorithm can be utilized to find a stable resource allocation under the conventional matching assumption. However, multiple co-channel CUs bring an external effect into the matching, and thus the interchannel cooperation (ICC) procedure is introduced. Finally, by using SPA-(S,P) in combination with ICC, a stable matching is returned.

20.3.2.1   Modeling of Coexistence Constraints Using SPA

Consider a cellular network consisting of CUs $\mathcal{CU} = \{cu_1, \ldots, cu_i, \ldots, cu_N\}$ subscribed to one CNO. Each CU is served by its local eNB with an allocated licensed band. However, owing to the time-varying traffic, some transmission requests cannot be satisfied by the allocated licensed bands $\mathcal{W}^C = \{w_1^C, \ldots, w_i^C, \ldots, w_N^C\}$. Thus, the CUs seek to share the unlicensed spectrum using CA to enhance the transmission. A set of LTE-Unlicensed base stations (UBSs) $\mathcal{UBS} = \{ubs_1, \ldots, ubs_i, \ldots, ubs_B\}$ are deployed to help CUs access the unlicensed spectrum $\mathcal{W} = \{w_1, \ldots, w_i, \ldots, w_K\}$. To access a clean unlicensed channel, CUs need to have the channel sensing phase before joining any unlicensed channel. During the channel sensing, CUs detect the transmission energy on the targeted unlicensed channel and decide this channel is clean if the received energy is smaller than a certain threshold. The CUs then communicate with its local UBSs through control signal exchanges using the pre-assigned licensed bands. On the other hand, to model the interference incurred at UUs from the sharing CUs, the Wi-Fi medium utilization (MU) estimation should be performed by the Wi-Fi APs through network listening to decode the preamble of any Wi-Fi packet detected during this time and record its corresponding received signal strength indicator (RSSI), duration, modulation, coding scheme and source/destination addresses. With the above estimated information of the unlicensed bands and the existing UUs, the Wi-Fi APs will share unlicensed channel with the UBSs so that this information can be further shared with the CUs. In this case, for simplification, we assume one unlicensed spectrum only accommodates one UU, and this UU is denoted as the virtualized unlicensed user (VU). The VU set is represented as $\mathcal{VU} = \{vu_j^k \mid ubs_j \in \mathcal{UBS}, w_k \in \mathcal{W}\}$, where vu_j^k represents the VU sensed by ubs_j on unlicensed band w_k. Each LTE-Unlicensed BS and each unlicensed band has a quota, denoted by Q and q, respectively regarding the number of CUs they can accept.

To model the interaction between CUs, VUs, and UBSs, we use the SPA model [23]. In this model, each lecturer offers a variety of projects, and students seek to undertake a project from the lecturers. Each student has preferences over the projects that he/she finds acceptable, whilst each lecturer has preferences over the (project, student) pairs. Both projects and lecturers are limited by quotas. We assume LTE-Unlicensed BSs, unlicensed bands (i.e., VUs), and CUs to be lecturers, projects, and students, respectively. UBSs offer available unlicensed bands to CUs, and CUs propose to these unlicensed bands. The UBSS make decisions based on the revenue from both the CUs and the unlicensed bands. A fair coexistence of the LTE and Wi-Fi systems can be achieved by finding a stable matching between the students and projects under the coexistence constraints.

The coexistence issues in LTE-Unlicensed can generally be classified into two categories: (1) interference between co-channel CUs and VUs, and (2) interference between multiple co-channel CUs. In this model, the coexistence issue (2) is avoided by deploying the time division multiple access (TDMA) method between multiple co-channel CUs. The coexistence issue (1) is satisfied by setting up the preference lists of both the CUs and the UUs in the following way. Each CU lists the group of unlicensed bands which satisfy the minimum required signal-to-interference-plus-noise ratio (SINR) Γ_{min}^{cu}. Similarly, each LTE-Unlicensed UBS checks the interference caused by the CUs to the VUs, which has to satisfy the maximum requirement δ_{max}. We assume that the preference of CU cu_i for vu_j^k is based on cu_i's achievable transmission rate (without considering the effect of other CUs). On the other hand, the preference of UBS ubs_k for the user–band pair (cu_i, vu_j^k) is based on the sum of cu_i and vu_j^k's achievable transmission rates (without considering the effect of other CUs). We denote the CUs' and UBSs' preference lists by $\mathcal{PL}^{cu}$ and $\mathcal{PL}^{ubs}$, respectively.

20.3.2.2   SPA Solution

The SPA-(S,P) algorithm can be applied to find a stable matching between CUs and VUs. It is a generalization of the GS algorithm. The motion of stability here implies robustness to deviations that can benefit both CUs and VUs. The formal definition of stability in our resource allocation problem is given in Definition 20.1. The basic rule of the SPA-(S,P) algorithm is that the students (i.e., CUs) propose to projects (i.e., VUs) and the lecturers (i.e., LTE-Unlicensed BSs) make decisions whether to accept/reject the student and project (i.e., (CU, VU)) pairs, based on the preference lists and on the quota. The procedure terminates when no further proposing is done. A detailed implementation of SPA-(S,P) is presented in Algorithm 20.1.

Definition 20.1 *Stability. A matching $\mathcal{M}$ is said to be stable if there is no blocking pair (BP). A pair (cu_i, vu_j^k) is defined as a BP if both of the following conditions are satisfied:*

(1) For cu_i: either cu_i is unmatched in $\mathcal{M}$, or cu_i prefers vu_j^k to $\mathcal{M}(cu_i)$.

(2) For vu_j^k: either vu_j^k is undersubscribed and one of the following three conditions is satisfied:

   (a) $\mathcal{M}(cu_i) \in vu^k$, and ubs_k prefers (cu_i, vu_j^k) to $(cu_i, \mathcal{M}(cu_i))$;
   (b) $\mathcal{M}(cu_i) \notin vu^k$ and ubs_k is undersubscribed;
   (c) $\mathcal{M}(cu_i) \notin vu^k$ and ubs_k is full and ubs_k prefers (cu_i, vu_j^k) to its worst pair (cu_w, vu_w^k); or vu_j^k is full and ubs_k prefers (cu_i, vu_j^k) to the pair (cu_w, vu_j^k),

---

Algorithm 20.1 SPA-(S,P) algorithm

Input: $\mathcal{CU}, \mathcal{UBS}, \mathcal{W}, q, Q, \mathcal{PL}^{\text{cu}}, \mathcal{PL}^{\text{ubs}}, \mathcal{M} = \emptyset$;
Output: Matching $\mathcal{M}$.

1: **while** some CU cu_i is free and has a nonempty list **do**
2:   **for all** $cu_i \in \mathcal{CU}$ **do**
3:     cu_i proposes to the first VU vu_j^k in $\mathcal{PL}_i^{\text{cu}}$, and remove vu_j^k from the list;
4:     $\mathcal{M} \leftarrow \mathcal{M} \cup (cu_i, vu_j^k)$;
5:   **end for**
6:   **for all** vu_j^k, $ubs_k \in \mathcal{UBS}, w_j \in \mathcal{W}$ **do**
7:     **while** vu_j^k is oversubscribed **do**
8:       Find the worst pair (cu_w, vu_j^k) assigned to vu_j^k in ubs_k's list;
9:       $\mathcal{M} \leftarrow \mathcal{M}/(cu_w, vu_j^k)$;
10:     **end while**
11:   **end for**
12:   **for all** $ubs_k \in \mathcal{UBS}$ **do**
13:     **while** ubs_k is oversubscribed **do**
14:       Find the worst pair (cu_w, vu_j^k) in ubs_k's list;
15:       $\mathcal{M} \leftarrow \mathcal{M}/(cu_w, vu_j^k)$;
16:     **end while**
17:   **end for**
18: **end while**
19: Terminate with a matching $\mathcal{M}$.

---

where cu_w is the worst CU in $\mathcal{M}(vu_j^k)$ and either of the following two conditions is satisfied:

(a) $\mathcal{M}(cu_i) \notin vu^k$;
(b) $\mathcal{M}(cu_i) \in vu^k$ and ubs_k prefers (cu_i, vu_j^k) to $(cu_i, \mathcal{M}(cu_i))$.

In the above, $\mathcal{M}(cu_i)$ represents the partner VU of cu_i in the matching $\mathcal{M}$.

We can then apply the SPA-(S,P) algorithm, as illustrated in Algorithm 20.1, to find an efficient matching between CUs and subbands. The basic rule of the SPA-(S,P) algorithm is to follow the principle of the classical GS algorithm [21]. The final matching is reached by a sequence of interactions between the CUs and UBSs. It has been proven [24] that every instance of SPA-(S,P) admits a stable matching, and an algorithm to find a student-oriented stable matching was given in [24].

Note here that a stable matching is guaranteed under the condition of canonical matching, which implies that the preference of any player depends solely on local information about the other type of players. In other words, the preferences do not depend on the choices or actions of other players. However, this assumption is not exactly true, as CUs' performances are infact affected by the other CUs' choices. Thus, the resulting matching is not necessarily stable, and this calls for further actions to reach stability.

20.3.2.3   ICC Procedure

Owing to the interdependence of the preferences of CUs and subbands (i.e., they are influenced by the existing matching), the matching yielded by the SPA-(S,P) algorithm is not necessarily stable. Since the conventional assumption that the preferences of a player do not depend on the choices of other players cannot be directly applied to solve our resource allocation problem, we call the matching framework with such interdependence matching games with externality [25]. Thus, cooperation between UBSs to transform the existing matching into a stable one is needed for matchings with an external effect. Note that, since UBSs can only operate on the channels allocated to CUs (not VUs), the externality effect can only be dealt with by making changes to the CUs. In other words, only CUs have an incentive to change partners. Thus, we seek to find a new "stability" among the CUs. The new "stability," different from that in Definition 20.1, relies on equilibrium among all CUs (i.e., there is no CU that has an incentive to make any changes). This one-sided "stability" is called "Pareto optimality" in matching theory [23]. The definition of "Pareto optimal" is as follows.

Definition 20.2 *Pareto optimal. A matching is said to be Pareto optimal if there is no other matching in which some player (i.e., CU) is better off, whilst no player (i.e., CU) is worse off.*

Accordingly, we define a new BP for one-sided matching problems in Definition 20.3.

Definition 20.3 *BP in a one-sided matching. A CU pair (cu_i, cu_j) is defined as a BP if both cu_i and cu_j are better off after exchanging their partners.*

The ICC strategy to find a Pareto optimal matching is illustrated in Algorithm 20.2. Note that during each iteration of ICC, the matching result is updated to $\mathcal{M}_l, 0 \leq l \leq s$, where $\mathcal{M}_0$ is the initial matching and $\mathcal{M}_s$ is the final stable matching. For each $\mathcal{M}_l$, we define $vu_{j1}^{k1} = \mathcal{M}_l(cu_{i1})$, and $vu_{j2}^{k2} = \mathcal{M}_l(cu_{i2})$, and the set of unstable CU–CU pairs is denoted by $\mathcal{BP}_l$. We represent the utility of cu_i as $U(cu_i)$, which is the transmission rate, and we have $\Delta U(cu_i) = U(cu_i)' - U(cu_i)$, where $U(cu_i)'$ is the utility after exchanging partners with another CU. The optimal BP is defined as follows:

$$(cu_{i1}^*, cu_{i2}^*) = \arg\max_{(cu_{i1}, cu_{i2})} \sum_{cu_{i1} \in \mathcal{M}_l(vu_{j1}^{k1})} \Delta U(cu_{i1}) + \sum_{cu_{i2} \in \mathcal{M}_l(vu_{j2}^{k2})} \Delta U(cu_{i2}), \qquad (20.1)$$

where the CU pair (cu_{i1}, cu_{i2}) is allowed to exchange partners.

The basic idea can be described as follows: first of all, search for all "unstable" CU–CU pairs (which have an incentive for exchange) regarding the current matching; then, check whether an exchange between such a pair is allowed (i.e., it is beneficial to the relevant CUs); after that, find the allowed pair which provides the greatest throughput improvement, switch their partners, and update the current matching; then keep searching for "unstable" CU–CU pairs until a trade-in-free environment is reached. The convergence of this search process is guaranteed by the irreversibility of each switch. Finally, ICC terminates with a stable matching, and simultaneously improves the system throughput.

Algorithm 20.2 Interchannel cooperation strategy

Input: Existing matching $\mathcal{M}_0$; related preference lists $\mathcal{PL}^{cu}$;

Output: Stable matching $\mathcal{M}_s$.

1:　Set $l = 0$, i.e., $\mathcal{M}_l = \mathcal{M}_0$;
2:　**while** $\mathcal{M}_l$ is not Pareto optimal **do**
3:　　　Search the set of "unstable" CU–CU pairs $\mathcal{BP}_l$ based on $\mathcal{PL}^{cu}$;
4:　　　**for all** $(cu_{i1}, cu_{i2}) \in \mathcal{BP}_l$ **do**
5:　　　　**if** $\exists cu \in \mathcal{M}_l(vu_{j1}^{k1}) \cup \mathcal{M}_l(vu_{j2}^{k2})$, $\Delta U(cu) < 0$ **then**
6:　　　　　(cu_{i1}, cu_{i2}) are not allowed to exchange partners;
7:　　　　**else**
8:　　　　　(cu_{i1}, cu_{i2}) are allowed to exchange partners;
9:　　　　**end if**
10:　　**end for**
11:　　Find the optimal BP (cu_{i1}^*, cu_{i2}^*);
12:　　cu_{i1}^* and cu_{i2}^* switch partners;
13:　　$\mathcal{M}_l \leftarrow \mathcal{M}_l / \{(cu_{i1}^*, \mathcal{M}_l(cu_{i1}^*)), (cu_{i2}^*, \mathcal{M}_l(cu_{i2}^*))\}$;
14:　　$\mathcal{M}_l \leftarrow \mathcal{M}_l \cup \{(cu_{i1}^*, \mathcal{M}_l(cu_{i2}^*)), (cu_{i2}^*, \mathcal{M}_l(cu_{i1}^*))\}$;
15:　　Update $\mathcal{PL}^{cu}$ based on $\mathcal{M}_l$;
16:　　$l = l + 1$;
17:　**end while**
18:　$\mathcal{M}_s = \mathcal{M}_l$.

20.3.2.4　Simulation Analysis

In the simulation that we performed, we assumed a circular network with a radius of $R = 800$ m, consisting of $N \in [0, 300]$ CUs, $B = 3$ UBSs, and $K = 20$ unlicensed subbands. The capacities of each subband and UBS were set to 15 and 100, respectively. The bandwidth of each subband was set to 5 MHz. The SINR requirement for the CUs was a uniform random distribution within $(20, 30)$ dB, while the maximum interference for VUs was -90 dBm. For the propagation gain, we set the path loss constant to 10^{-2} and the path loss exponent to 4, the multipath fading gain followed an exponential distribution with unit mean, and the shadowing gain followed a log-normal distribution with 4 dB deviation [26].

The performance of the proposed ICC strategy after SPA-(S,P) is evaluated in Figure 20.4. Inspired by [27], which adopted a random partner exchange (two-sided matching), we have used an ICC-random (ICC-R) strategy as a benchmark. The difference between ICC and ICC-R is that in ICC, the CU pair which has the most throughput improvement is chosen as the next exchange pair each time, while in ICC-R a random pair (allowed to exchange) is chosen each time. By removing the external effect using either ICC and ICC-R, the system's throughput is further improved (compared with SPA-(S,P)). What is more, ICC outperforms ICC-R, which demonstrates the effectiveness of ICC. During the simulation, we also evaluated the performance degradation of the VUs, which was on average 0.2%. This indicates

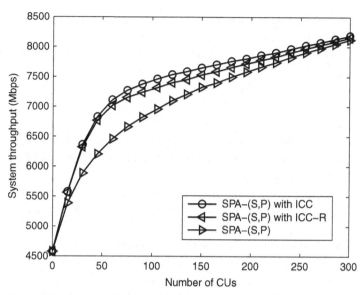

Figure 20.4 System throughput evaluation by a comparison of SPA-(S,P) with ICC, SPA-(S,P) with ICC-R, and SPA-(S,P).

that our proposed matching approach effectively improves system throughput while protecting the Wi-Fi performance well.

20.3.3   Dynamic Resource Allocation: Random Path to Matching Stability

In the previous example, we discussed the static coexistence issue in LTE-Unlicensed. In this example, we present a dynamic resource allocation framework for LTE-Unlicensed with user mobility. To characterize the user mobility and analyze the impact of user mobility on the performance of the proposed mechanisms, we divide the simulation period $[0, T]$ into time slots of identical duration ΔT, which can be set according to specific applications. The slot duration ΔT can be small enough such that during each time slot $(t, t+1), t \in \{1, \ldots, t, \ldots, T\}$, the user distribution and channel conditions can be treated as static. In other words, the dynamic resource allocations can be treated as sequential static resource allocations over continuous time. Intuitively, we can solve the sequential problems independently for each time slot. However, if we take advantage of the network relations between adjacent times, the sequential allocation problem can be solved in a time-related manner.

20.3.3.1   Modeling of Coexistence Constraints Using HR Model

Consider the static system model adopted in Section 20.3.2 in continuous time with user mobility. In the static model, the set of CUs $\mathcal{CU} = \{cu_1, \ldots, cu_i, \ldots, cu_N\}$ are assumed to have the option to choose from the available UBSs to assist their transmission. This increases the complexity of resource allocation compared with assigning a fixed UBS to each CU. Since complexity is a major concern in dynamic allocation, we lower the complexity by simply assuming that each CU is served by its local UBS (i.e., the closest UBS) for UL/DL transmissions. Owing to user mobility and time-varying traffic, some CUs' transmission requests cannot be served in the current allocated licensed bands. Thus, these CUs search for nearby UUs, and seek to share and aggregate their unlicensed spectrum. The set of UUs $\mathcal{UU} = \{uu_1, \ldots, uu_j, \ldots, uu_K\}$ are distributed in the cellular network with a certain distribution, and each UU is allocated a specific unlicensed subband, denoted by $\mathcal{W} = \{w_1, \ldots, w_j, \ldots, w_K\}$, from its local unlicensed AP in the set $\mathcal{AP} = \{ap_1, \ldots, ap_j, \ldots, ap_K\}$. CUs that are allocated unlicensed subbands can aggregate these with their preassigned licensed subbands $\mathcal{W}^C = \{w_1^C, \ldots, w_i^C, \ldots, w_N^C\}$ to enhance transmission.

To model the interactions between CUs and UUs, we adopt the HR matching model [23]. Since, without the selection of UBSs, the matching involves only two type of agents, CUs and UUs, thus the SPA model can be replaced by the HR model. In an instance of the HR problem, which contains a set of residents $\mathcal{R} = \{r_1, \ldots, r_{n1}\}$ and a set of hospitals $\mathcal{H} = \{h_1, \ldots, h_{n2}\}$, each resident seeks to be accepted by a hospital, while each hospital h_j can recruit multiple residents within its quota q. A matching is a product of two set of players R and H, formed under certain rules. The definition of stability in an HR problem is given in Definition 20.4.

Definition 20.4   *Stability. Let I be an instance of HR and $\mathcal{M}$ be a matching in I. A pair (r_i, h_j) blocks $\mathcal{M}$, or is a blocking pair (BP) of $\mathcal{M}$, if the following conditions are satisfied relative to $\mathcal{M}$:*
(1) r_i is unassigned or prefers h_j to $\mathcal{M}(r_i)$;
(2) h_j is undersubscribed or prefers r_i to at least one member of $\mathcal{M}(h_j)$.
$\mathcal{M}$ is said to be stable if it admits no blocking pair.

We model the UUs as hospitals, and the CUs as residents. Again, the two coexistence constraints that we discussed in Section 20.3.2.1 must be satisfied at each time slot. The second coexistence issue, inter-CU interference, can be avoided by adopting the TDMA mechanism for co-channel CUs. The first issue, interference between co-channel CUs and UUs, can be dealt with by setting up preference lists. Each CU lists the group of acceptable UUs which satisfy its minimum required SINR. Similarly, each UU lists the acceptable CUs whose interference is lower than its maximum allowable amount. The preference of a CU $cu_i, cu_i \in \mathcal{CU}$, over its acceptable UUs $uu_j, uu_j \in \mathcal{UU}$, at time t, denoted by $\mathcal{PL}_{i,j}^{CU}(t)$, is based on cu_i's achievable transmission rate if it is sharing spectrum with uu_j. Although multiple CUs may share the same UU's resource, there is no way in which CUs can know other CUs/UUs' decisions. Thus, the preference of cu_i for uu_j at time t is simply assumed to be cu_i's achievable rate if only cu_i itself is sharing uu_j's subband. On the other hand, the preferences of uu_j for its acceptable CUs cu_i at time t, denoted by $\mathcal{PL}_{j,i}^{UU}(t)$, are based on uu_j's achievable transmission rate when sharing spectrum with cu_i.

20.3.3.2   Time-Independent Implementation
Generally, a stable matching for an HR instance can be achieved by using the resident-oriented Gale–Shapley (RGS) algorithm. The RGS algorithm is the many-to-one

---

Algorithm 20.3 Resident-oriented GS (RGS) algorithm

Input: $\mathcal{CU}, \mathcal{UU}, q, \mathcal{PL}^{CU}, \mathcal{PL}^{UU}$

Output: Stable matching $\mathcal{M}$

   Initialization;

   Construct the set of unmatched CUs as $\mathcal{CU}_{un}$, and set $\mathcal{CU}_{un} = \mathcal{CU}$, $\mathcal{M} = \emptyset$;

   while $\mathcal{CU}_{un} \neq \emptyset$ and $\mathcal{PL}^{CU} \neq \emptyset$ **do**

      CUs propose to UUs;

      for all $cu_i \in \mathcal{CU}_{un}$ **do**

         Propose to the first UU uu_j in its preference list $\mathcal{PL}^{CU}$, and set $\mathcal{M} \leftarrow \mathcal{M} \cup \{cu_i, uu_j\}$;

         Remove uu_j from cu_i's preference list;

      end for

      UUs make decisions;

      for all $uu_j \in \mathcal{UU}$ **do**

         if uu_j is not oversubscribed **then**

            uu_j keeps all of the proposed CUs;

            Remove such CUs from $\mathcal{CU}_{un}$;

         else

            uu_j keeps the most preferred q CUs, and rejects the rest;

            Remove such q CUs from $\mathcal{CU}_{un}$;

            Add the rejected CUs to the $\mathcal{CU}_{un}$ (if not included previously), and set $\mathcal{M} \leftarrow \mathcal{M}/\{cu_i, uu_j\}$, where cu_i is rejected by uu_j;

         end if

      end for

   end while

   End of algorithm;

---

generalization of the classical GS algorithm for the stable marriage problem. A stable matching is always guaranteed to be obtained by using the RGS algorithm [23].

The RGS algorithm consists of sequential actions of proposing and accepting/rejecting. Each iteration starts with residents proposing to their favorite hospitals on their current preference lists. After the proposing operations, the hospitals that have been proposed to are removed from the residents' preference lists. Then the hospitals decide whether to accept or reject the proposals they have received so far, based on their preference lists over the residents and on their capacities. If the cumulative number of proposals exceeds their capacity, they choose to increase the number of residents to the capacity that they favor most, and reject the rest. This proposing and accepting/rejecting iteration runs for as many rounds as needed until all residents are matched or all residents' preferences are empty. Implementation details of the RGS algorithm can be found in Algorithm 20.3. Again, owing to the interdependence of the CUs' preferences on each others' choices, an external effect still exists. Thus, in each time slot after using the RGS algorithm, we need to run the ICC procedure to remove externality.

20.3.3.3　Time-Related Implementation

Owing to user mobility, a stable matching in the previous time slot may not be stable in the current time interval. Although we can use RGS–ICC repeatedly to obtain stable solutions, this is in fact solving a new HR problem each time. Consider the case where only one CU moves, and the rest of the network stays unchanged. Thus, only a small number of related users' preferences are changed. Under such a small network variation, instead of redoing the whole matching, we can utilize the network relations between the current and previous time slots to transform a previously unstable matching into a stable one again. Thus, we propose an adaptive matching approach: the random path to stability (RPTS). The basic idea of the RPTS mechanism is to use divorce and remarriage operations to transform a random matching into a stable matching. Based on the previous matching $M(t-1)$ at time $t-1$ and the updated preference lists $\mathcal{PL}^{cu}(t)$, $\mathcal{PL}^{uu}(t)$ at time t, the RPTS algorithm provides a stable matching $M(t)$ at time t.

The RPTS algorithm was originally proposed to solve the dynamic stability problem in the many-to-many matching case. We have modified the RPTS algorithm to fit our many-to-one matching case by assuming the capacity of one side to be 1.

Next, we consider an example of many-to-many matching, namely, the worker–firm (WF) model [28], to illustrate the RPTS algorithm.

In a labor market, where firms seek to hire one or more workers, and workers seek jobs in one or more firms, each firm has preferences over subsets of workers and each worker has preferences over subsets of firms. Let F be the set of firms and W be the set of workers. Let $FW = F \cup W$ be the set of all agents. For each agent $i \in FW$, let P_i denote the set of potential partners for i. Each agent has preferences over 2^{P_i} sets, which are subsets of that agent's potential partners. For each agent $i \in FW$, and a potential partner set $S \subseteq P_i$, the choice of i in S is the preferred partner set of i from among the partners in S. We denote this by $CH_i(S)$. For each matching μ and each agent i, we say that i blocks μ individually if $\mu(i) \neq CH_i(\mu(i))$, where $\mu(i)$ represents i's partners in the matching μ. A matching μ is individually stable if no agent blocks μ. For each matching μ and each pair $(f, w) \in (F \times W)/\mu$, we say that (f, w) is a BP in μ if $w \in CH_f(\mu(f) \cup w)$ and $f \in CH_w(\mu(w) \cup f)$. A matching is pairwise stable if μ is blocked neither individually nor in pairs.

In the general domain of strict preferences, a pairwise-stable matching may not exist. We need restrictions on preferences to guarantee the existence of such a matching. For each agent i, that agent's preference relation is substitutable if for all $S, S' \subseteq P_i$ with $S \subseteq S'$, we have $CH_i(S') \cap S \subseteq CH_i(S)$. In other words, worker w's desire for firm f_i does not depend on any other firm f_j. Another requirement is responsive preferences. For each agent i, her preference relation is responsive with quota q if (1) for all $j, k \in P_i$, and all $S \subseteq P_i/j, k$ with $|S| \leq q$, we have $j \cup S \succeq_i k \cup S \Leftrightarrow j \succeq_i k$; (2) for all $j \in P_i$; and all $S \subseteq P_i/j$ with $|S| \leq q$, we have $j \cup S \succeq_i S \Leftrightarrow j \succeq_i \emptyset$, and (3) for all $S \subseteq P_i$ with $|S| > q$, we have $\emptyset \succeq_i S$. This means that the agent's rankings of two partners are independent of the agent's other partners, unless that agent exceeds their quota, and any set of partners exceeding that quota is preferred less being unmatched. Apparently, both CUs and UUs satisfy the conditions of both responsive and substitutable preferences, since any user's

---

Algorithm 20.4 RPTS algorithm

Input: An individually stable matching $\mathcal{M}_0$
Output: A stable matching $\mathcal{M}$
Initialization: $\mathcal{M} = \mathcal{M}_0, I = \emptyset$;

1: **while** $I \neq FW$ **do**
2:      $\forall \bar{l} \in FW/I$;
3:      $\bar{I} = I \cup \bar{l}$;
4:      **if** $\bar{l}$ is involved in a BP of $\bar{I}$ **then**
5:          Match $\bar{l}$ with its most preferred blocking partner in $\bar{I}$;
6:          **while** Exists any firm-pointed BP (f, w) in I **do**
7:              Match firm f with its most preferred blocking workers in $\bar{I}$;
8:          **end while**
9:          **while** Exists any BP (f, w) in I **do**
10:              Match worker w with its most preferred blocking firms in $\bar{I}$;
11:          **end while**
12:      **end if**
13: **end while**
14: $\mathcal{M}$ is pairwise stable.

---

preferences for another user do not depend on any other users. A conclusion about pairwise stability in the case of many-to-many matching is stated in Theorem 20.5.

Theorem 20.5   *Assume that every agent on one side of the market has substitutable preferences, and every agent on the other side has responsive preferences with a quota. Let μ_0 be a pairwise-unstable matching. There exists a finite sequence of matchings $\mu_0, \ldots, \mu_t$, where μ_i is pairwise stable and, for each $1 \leq i \leq t$, M_i is obtained from M_{i-1} by satisfying a blocking individual or blocking pair.*

The RPTS algorithm in the many-to-many matching case is stated in Algorithm 20.4. It starts with an individually stable matching. If the initial matching is not individually stable, then individual agents are satisfied one at a time to obtain an individually stable matching. During the matching, we use a set I to represent an internally stable set, which is a subset of FW. I is internally stable under μ if no pair or individual in I blocks μ. The RPTS algorithm starts by adding any agent $\bar{i}$ outside I to I obtain $\bar{I}$, and by satisfying i with its most preferred partner that if form a BP with. The algorithm then checks if there is any firm-pointed blocking pair (FPBP) (f, w) in $\bar{I}$, and satisfies the firm f until there is no FPBP in $\bar{I}$. A pair (f, w) is called an FPBP if the involved f blocks not only the current matching but also the initial matching. Then, after all the FPBPs have been satisfied, the algorithm checks if there still exists any BP in $\bar{I}$. If yes, then it satisfies the worker w involved in the BP until there is no BP in $\bar{I}$. It then updates I with $\bar{I}$, checks if $I = FW$, and so on and so forth. For a more detailed explanation of Algorithm 20.4, see [28].

By identifying the CUs as the workers and the UUs as the firms, as well as setting all workers' capacities to 1, we can customize the RPTS algorithm to transform the previous matching into a "stable" one again. The ICC procedure still needs to be used after RPTS to eliminate externality iteratively to ensure dynamic stability.

20.3.3.4   Simulation Analysis

In the simulation that we performed, we set up seven adjacent cells, each with radius $R = 1$ km. $N = 70$ CUs and $M = 20$ UUs were initially randomly distributed in the network. The capacity of each UU was set to $q = 3$. The bandwidth of each unlicensed band was set within $(2,4)$ MHz. The SINR requirement for each CU followed a uniform distribution within $(20,30)$ dB, while the maximum interference for UUs was -90 dBm. To cope with the number of CUs, we assumed that each base station assigned its 20 MHz licensed spectrum equally among its associated CUs. For the propagation gain, we set the path loss constant to 10^{-2} and the path loss exponent to 4 the multipath fading gain followed an exponential distribution with unit mean, and the shadowing gain followed a log-normal distribution with 4 dB deviation. We set the slot duration ΔT to 1 s, and the total simulation time was 90 s. We used the random waypoint (RWP) mobility model, where the stop time was set to 0.5 s for all CUs. The maximum velocity in the RWP model was set to 40 m/s.

Figures 20.5 and 20.6 evaluate the time-dynamic performance of the RGS and RPTS algorithms with the RWP mobility model, with respect to the computational complexity and the throughput, respectively. Apparently, the RPTS algorithm achieves a much lower complexity than the RGS algorithm, as shown in Figure 20.5. The complexity of RPTS is only about 11% of RGS on average during the 90 s simulation period, except at time slot 1, when RPTS transforms an empty matching into a stable one, and thus requires more interactions than in any later time slots. In addition, even at time slot

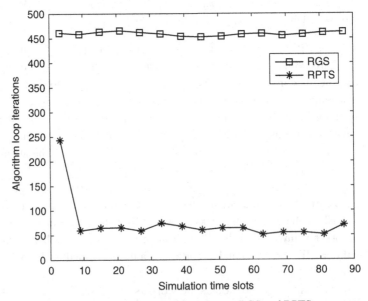

Figure 20.5 Comparison computational complexity between RGS and RPTS.

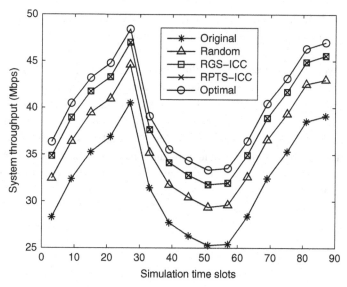

Figure 20.6 Evaluation system throughput: comparison of the Original, Random, RGS-ICC, RPTS-ICC, and Optimal methods.

1, RPTS still achieves better performance than RGS. As both results are stable, which means that the numbers of final partnerships (pairs) are the same, this implies that more connections break during the implementation of RGS than of RPTS. This results from the careful design of RPTS with respect to the order of BP/FPBP selection, which has to be satisfied each time. In other words, RPTS, as an adaptive matching mechanism which takes advantage of the time relations, tries to make as few changes as possible to the network to restabilize the matching. For the system throughput performance, we compared the RGS–ICC and RPTS–ICC methods with three other methods; "Random," "Original," and "Optimal." In the Random method, the unlicensed spectrum (i.e., UUs) was allocated randomly to the CUs, while in the Original method no spectrum sharing happened. As shown in Figure 20.6, both RGS-ICC and RPTS-ICC achieve similar performance, while outperforming the Random and Original methods. In addition, both RGS-ICC and RPTS-ICC are close to optimal, as compared with the Optimal solution. Although it is not shown in the figures, we also evaluated the UUs' throughput before and after the matching, and this showed an average performance decrease by 1.89%, which is in fact trivial.

20.4   Conclusion

LTE-Unlicensed provides flexible and efficient performance enhancement of the conventional LTE system, while relying on existing LTE architectures to guarantee access authority, security, mobility, and QoS management. As the major design challenge, the intersystem interference imposed by LTE/Wi-Fi coexistence needs to be treated carefully. The available techniques, such as channel selection, CSAT, and transmission power control, can be selected to provide effective interference avoidance/elimination solutions according to the specific system requirements. While meeting the minimum system requirements, system performance measures such as overall throughput and stability, can be optimized using either centralized or distributed approaches. As a distributed framework, matching theory has been demonstrated to offer stability guarantees, low complexity, fast implementation, and close-to-optimal performance with respect to the system throughput. Furthermore, the throughput degradation of Wi-Fi is trivial. Not limited to static resource allocation, matching theory also provides efficient dynamic matching mechanisms. The dynamic matchings obtained make use of the structural relations between time-adjacent matchings and maintain as many connections as possible, which reduces the setup cost of new connections in a real-world implementation.

References

[1] Huawei, "U-LTE: Unlicensed spectrum utilization of LTE," White Paper, Sep. 2014.

[2] Qualcomm, "Making the best use of unlicensed spectrum for 1000x," White Paper, 2014.

[3] 3GPP, "LTE in unlicensed spectrum," Jun. 2014. Available at www.3gpp.org/news-events/3gpp-news/1603-lte-in-unlicensed.

[4] Qualcomm, "LTE Wi-Fi link aggregation," Mar. 2015. Available at www.qualcomm.com/videos/lte-wi-fi-link-aggregation.

[5] Verizon, "LTE-U Forum." Available at www.lteuforum.org/.

[6] Qualcomm, "Introducing MuLTEfire: LTE-like performance with Wi-Fi-like simplicity," Jun. 2015. Available at www.qualcomm.com/news/onq/2015/06/11/introducing-multefire-lte-performance-wi-fi-simplicity.

[7] LTE-U Forum, "LTE-U SDL Coexistence Specifications," Jun. 2015.

[8] Alcatel-Lucent, Ericsson, Qualcomm Technologies Inc., Samsung Electronics, and Verizon, "LTE-U technical report: coexistence study for LTE-U SDL v1.0," Tech. Rep., Feb. 2015.

[9] T. Nihtila, V. Tykhomyrov, O. Alanen, M. A. Uusitalo, A. Sorri, M. Moisio, S. Iraji, R. Ratasuk, and N. Mangalvedhe, "System performance of LTE and IEEE 802.11 Coexisting on a shared frequency band," in *Proc. of IEEE Wireless Communications and Networking Conf. (WCNC)*, Apr. 2013.

[10] A. M. Cavalcante, E. Almeida, R. D. Vieira, F. Chaves, R. Paiva, F. Abinader, S. Choudhury, E. Tuomaala, and K. Doppler, "Performance evaluation of LTE and Wi-Fi coexistence in unlicensed bands," in *Proc. of IEEE Vehicular Technology Conf. (VTC)*, Jun. 2013.

[11] R. Zhang, M. Wang, L. X. Cai, Z. Zheng, X. Shen, and L. L. Xie, "LTE-Unlicensed: The future of spectrum aggregation for cellular networks," *IEEE Wireless Commun.*, vol. 22, no. 3, pp. 150–159, Jun. 2015.

[12] Qualcomm, "Qualcomm research LTE in unlicensed spectrum: Harmonious coexistence with Wi-Fi," White Paper, Jun. 2014.

[13] F. S. Chaves, E. P. L. Almeida, R. D. Vieira, A. M. Cavalcante, F. M. Abinader, S. Choudhury, and K. Doppler, "LTE UL power control for the improvement of LTE/Wi-Fi coexistence," in *Proc. of IEEE Vehicular Technology Conf. (VTC)*, Sep. 2013.

[14] E. Almeida, A. M. Cavalcante, R. C. D. Paiva, F. S. Chaves, F. M. Abinader, R. D. Vieira, S. Choudhury, E. Tuomaala, and K. Doppler, "Enabling LTE/WiFi coexistence by LTE

blank subframe allocation," in *Proc. of IEEE International Conf. on Communications (ICC)*, Jun. 2013.

[15] H. Zhang, Y. Xiao, L. X. Cai, D. Niyato, L. Song, and Z. Han, "A hierarchical game approach for multi-operator spectrum sharing in LTE unlicensed," in *Proc. of IEEE Global Communications Conf. (GLOBECOM)*, Dec. 2015.

[16] Y. Gu, W. Saad, M. Bennis, M. Debbah, and Z. Han, "Matching theory for future wireless networks: Fundamentals and applications," *IEEE Commun. Mag.*, vol. 53, no. 5, pp. 52–59, May 2015.

[17] Y. Gu, Y. Zhang, M. Pan, and Z. Han, "Matching and cheating in device to device communications underlying cellular networks," *IEEE J. Sel. Areas Commun.*, vol. 33, no. 10, pp. 2156–2166, Oct. 2015.

[18] S. Bayat, R. H. Y. Louie, Z. Han, Y. Li, and B. Vucetic, "Distributed stable matching algorithm for physical layer security with multiple source-destination pairs and jammer nodes," in *Proc. of IEEE Wireless Communications and Networking Conf. (WCNC)*, Apr. 2012.

[19] Y. Gu, Y. Zhang, M. Pan, and Z. Han, "Student admission matching based content-cache allocation," in *Proc. of IEEE Wireless Communications and Networking Conf. (WCNC)*, Mar. 2015.

[20] W. Saad, Z. Han, R. Zheng, M. Debbah, and H. V. Poor, "A college admissions game for uplink user association in wireless small cell networks," in *Proc. of IEEE INFOCOM*, Apr. 2014.

[21] D. Gale and L. S. Shapley, "College admissions and the stability of marriage," *Am. Math. Monthly*, vol. 69, no. 1, pp. 9–15, Jan. 1962.

[22] Y. Gu, Y. Zhang, L. X. Cai, M. Pan, L. Song, and Z. Han, "Exploiting student–project allocation matching for spectrum sharing in LTE-unlicensed," in *Proc. of IEEE Global Communications Conf. (GLOBECOM)*, Dec. 2015.

[23] D. F. Manlove, *Algorithmics of Matching under Preferences*, World Scientific, 2013.

[24] A. H. A. El-Atta and M. I. Moussa, "Student project allocation with preference lists over (student, project) pairs," in *Proc. of International. Conf. on Computer and Electrical Engineering (ICCEE)*, Dec. 2009.

[25] A. Roth and M. A. O. Sotomayor, *Two-sided Matching: A Study in Game-Theoretic Modeling and Analysis*, Cambridge University Press, 1992.

[26] A. Goldsmith, *Wireless Communications*, Cambridge University Press, 2004.

[27] F. Pantisano, M. Bennis, W. Saad, S. Valentin, and M. Debbah, "Matching with externalities for context-aware user–cell association in small cell networks," in *Proc. of IEEE Global Communications Conf. (GLOBECOM)*, Dec. 2013.

[28] F. Kojima and M. U. Ünver, "Random paths to pairwise stability in many-to-many matching problems: A study on market equilibration," *Int. J. Game Theory*, vol. 36, no. 3, pp. 473–488, 2006.

21 Scheduling for Millimeter Wave Networks

Lin X. Cai, Lin Cai, Xuemin Shen, and Jon W. Mark

21.1 Introduction

The spectrum between 30 and 300 GHz is referred to as the millimeter wave (mmWave) band because the wavelengths for these frequencies are in the range from about one to ten millimeters. The Federal Communications Commission (FCC) has allocated the 57–64 GHz mmWave band for general unlicensed use, opening the door to supporting high data rate wireless applications over the 7 GHz unlicensed band. Given the spectrum deficiency and network densification of cellular systems, how to use the mmWave band to support various machine/human-to-machine/human communications is critically important for fifth generation (5G) cellular systems.

Millimeter wave can be applied to both outdoor and indoor wireless communications. mmWave together with massive multiple-input multiple-output (MIMO) is a promising candidate for 5G outdoor transmission, as discussed in Chapter 15. For indoor uses, mmWave communication has many salient features, listed below, and it is highly desirable for 5G femtocell communications. This chapter focuses on the indoor femtocell scenario.

First, mmWave can achieve very high data rates (up to multi-Gbps), so it can enable many killer applications such as high-definition and interactive streaming services, and the Internet of Things. These applications require not only a high data rate but also stringent quality-of-service (QoS) requirements in terms of delay, jitter, and loss. Second, mmWave can coexist well with other wireless communication systems, such as the existing cellular systems, Wi-Fi (IEEE 802.11), and ultra-wideband (UWB) systems, because of the large frequency difference. Third, oxygen absorption has its peak at 60 GHz, so the transmission and interference ranges of mmWave communication are small, which allows very dense deployment of mmWave-based femtocells. In addition, the fact that the mmWave signal degrades significantly when passing through walls and over distance is helpful for ensuring security of the content.

The special channel characteristics and features of mmWave communication pose new challenges regarding how to coordinate mmWave transmissions to achieve high spatial reuse and guarantee the QoS. In the following, given the unique characteristics of mmWave communications and of the appropriate multiplexing technologies and network architectures for mmWave-based femtocells, we discuss the key opportunities and challenges in resource management of mmWave-based wireless networks, and introduce an appropriate scheduling solution to explore the spatial multiplexing gain in mmWave networks.

21.2 Background

21.2.1 Multiplexing Technologies for mmWave Networks

The main characteristics of mmWave communication are short wavelength and high frequency, large bandwidth, high interaction with atmospheric constituents, and sensitivity to blockage. For mmWave communication at very high data rates (and thus very small symbol durations), intersymbol interference (ISI) due to time dispersion in multipath propagation becomes significant. Orthogonal frequency division multiplexing (OFDM) is a good candidate, since the signals are relatively robust against ISI owing to the reduced symbol rate in each of the subcarriers.

OFDM can be combined with a multiple access scheme such as time division multiple access (TDMA) or code division multiple access (CDMA) for effective multiple access control. OFDM–TDMA is straightforward: different users share the wireless medium in different time slots. Several combinations of OFDM and CDMA have been discussed in [1]. For radio frequency oscillators in the mmWave spectrum, it is very difficult to maintain a low level of phase noise, which affects the signal during frequency conversion operations, and results in a higher bit error rate in effective communications. Different multiple access techniques, including OFDM/TDMA, direct sequence (DS)-CDMA, multicarrier (MC)-CDMA, and MC-DS-CDMA, have different sensitivities to phase noise. According to [2], MC-DS-CDMA is the most robust against phase noise and multiple access interference. MC-DS-CDMA is considered as a promising multiplex access technology for mmWave networks and is therefore considered in the scheduling design presented in this chapter, although this scheduling solution is general enough to be applicable to other types of multiplex access technologies.

21.2.2 Directional Antennas

Because of the unique characteristics of 60 GHz mmWave communication, i.e., the small wavelength and the high path loss due to severe oxygen absorption and atmospheric attenuation, it is practical and highly desirable to use a directional antenna to achieve a much higher antenna gain to enlarge the transmission range, by radiating transmission energy in the desired direction only [3]. There are two types of directional antenna [4]: conventional sectored/switched antenna arrays and adaptive antenna arrays. A sectored antenna array has a number of fixed beams that provide full coverage in azimuth. An adaptive antenna array is able to automatically adapt its radiation pattern by using a beamforming technique that intelligently puts a main beam in the direction of the desired signal and nulls in the directions of the interference and noise. Since the size of the antennas used in mmWave communication can be very small, it is feasible to deploy multiple antenna elements in a device to achieve directivity. In a mmWave network with directional antennas, directivity and high path loss can result in more efficient spectrum reuse and a substantial improvement in network throughput. In addition, directional antennas are more energy efficient.

In networking research, a popular model for a directional antenna is the flat-top model [5, 6]: the antenna gain is constant within the beamwidth and zero outside the beamwidth. Therefore, for a beam with beamwidth θ, the antenna gain of the mainlobe is $2\pi/\theta$, and that of the sidelobe is 0.

In practice, signal radiation outside the mainlobe still exists. Assuming all devices in a network to be in a two-dimensional plane, we use a sector-plus-circle model and define the antenna gains of the mainlobe and sidelobe as $\eta 2\pi/\theta$ and $(1-\eta)2\pi/(2\pi-\theta)$, respectively, where η is the radiation efficiency of the antenna.

21.2.3   Network Architecture

We focus on the indoor environment for mmWave networks, where they can provide high-speed wireless services for many bandwidth-hungry multimedia applications. Since mmWave signals cannot penetrate walls, we consider devices randomly distributed in a room, where an access point (AP) and multiple devices form a femtocell, the basic network element. The AP collects global information about the femtocell and makes scheduling decisions. Data transmissions in the femtocell are based on a time-slotted superframe structure, similar to that in wireless personal area network standards [7]. Since most devices use a directional antenna in mmWave femtocells, a centralized AP is very useful for device/neighbor discovery. The AP broadcasts beacons periodically in all directions which allow other devices to synchronize and determine their locations. All devices send channel time requests and their locations to the AP, which schedules peer-to-peer communications accordingly. How to ensure efficient and fair resource allocation to all users is the main objective of the scheduling design here.

21.3   Exclusive Regions

In wireless scheduling design, as wireless signals can propagate in some area and mutually interfere with each other, we need to limit the mutual interference to ensure transmission reliability, while we aim to allow multiple flows to be transmitted concurrently to maximize spatial reuse. To serve this purpose, we introduce the concept of the exclusive region (ER). Each flow has an ER around the receiver, and the senders of all flows transmitting concurrently should be outside the ERs of other flows. The key issue is how to determine the appropriate ER.

Let P_R denote the received signal power, R denote the channel capacity (or the achievable data rate with an efficient transceiver design), N_0 denote the one-sided spectral density of white Gaussian noise, W denote the channel bandwidth in Hertz, and I denote the total interference power. According to Shannon's theory, $R = W\log_2(\text{SINR}+1)$, where SINR can be approximated by $P_R/(N_0W+I)$.

Consider a network with N flows, $\{f_i \mid i \in 1,2,\dots,N\}$, requesting transmission times in a superframe with N time slots. The distance between the transmitter and receiver of the ith flow is d_i, and the distance between the transmitter of the jth flow and the receiver of the ith flow is $d_{j,i}$. The average transmission power and receiving power of flow f_i are denoted by $P_T(i)$ and $P_R(i)$, respectively. Let $G_T(i)$ and $G_R(i)$ be the antenna gains of the transmitter and receiver, respectively. Taking account of signal dispersion with distance, the average received signal power can be modeled as

$$P_R(i) = k_1 G_T(i) G_R(i) d_i^{-\alpha} P_T(i), \tag{21.1}$$

where $k_1 \propto (\lambda/4\pi)^2$ is a constant coefficient dependent on the wavelength λ, and α is the path loss exponent, which is dependent on the propagation environment and usually takes a value between 2 and 6 [8].* We assume that $G_T(i)$, $G_R(i)$, and α are constant, and all devices use the same transmission power.

In broadband communication systems, adaptive modulation and coding can be applied to adjust the sending rate according to the received SINR in a discrete manner. Given the large bandwidth of mmWave, we assume that the data rate is a continuous variable and simplify the data rate estimation by applying Shannon's theorem. If only one flow is allowed to transmit at a time, i.e., flows are transmitted in a round-robin TDMA fashion, the average data rate of the ith flow during the N slots, R_i, is given by

$$R_i = \frac{k_2 W}{N} \log_2\left(\frac{P_R(i)}{N_0 W} + 1\right), \tag{21.2}$$

where k_2 is a coefficient related to the efficiency of the transceiver design.

If all flows can be transmitted simultaneously in all slots, i.e., the flows are transmitted in a CDMA fashion, the achievable data rate R_i' of the ith flow is given by

$$R_i' = k_2 W \log_2\left(\frac{k_1 G_T(i) G_R(i) P_T(i) d_i^{-\alpha}}{N_0 W + \sum_{i \neq j} I_{j,i}} + 1\right), \tag{21.3}$$

where $I_{j,i}$ is the interference power between the transmitter of the jth flow and the receiver of the ith flow. We assume that the cross-correlation between any two concurrent transmissions is constant, i.e., $G_{j,i} = G_0$, $\forall j \neq i$. The interference power is $I_{j,i} = k_1 G_0 G_T(j) G_R(i) P_T(j) d_{j,i}^{-\alpha}$.

To compare R and R', we consider two cases separately. First, if SINR < 1, the achieved data rate can be approximated as

$$k_2 W \log_2(\text{SINR}+1) \approx k_2 W \times \text{SINR} \log_2 e. \tag{21.4}$$

With this approximation, from (21.2) and (21.3), a sufficient condition to ensure that $R_i' \geq R_i$ is $I_{j,i} \leq N_0$, $\forall j \neq i$, i.e., the average interference level from any other flow should

---

* Note that wireless channels may suffer from fast fading, which will affect the instantaneous received signal and interference power. Here, to minimize complexity and overhead of the channel measurement, the scheduler uses the path loss model to represent the average channel rate to simplify the scheduling process.

be less than the background noise.[†] Thus, if we allow flows with a mutual interference less than the background noise to transmit simultaneously, the throughput of each flow can be higher than that for round-robin TDMA transmissions.

Second, if SINR ≥ 1, the approximation in (21.4) may not hold. Nevertheless, the sufficient condition above can still ensure that $R_i \leq R_i'$. This is because $\log_2(x/N + 1) \geq (1/N)\log_2(x+1)$, $\forall x \geq 1$, $N \geq 1$. If $I_{j,i} \leq N_0 W$, then $R_i'/R_i \geq (1/N)\log_2(\text{SNR} + 1)/\log_2(\text{SNR}/N + 1) \geq 1$. Thus, the above sufficient condition is still applicable.

We assume that the noise power spectrum is constant. To ensure that the interference power is less than the noise, an interferer must be at least $r(i)$ away from the receiver of the ith flow, where $r(i)$ is given by

$$r(i) = \left(\frac{k_1 G_0 G_T(j) G_R(i) P_T(j)}{N_0 W} \right)^{1/\alpha}. \tag{21.5}$$

The ERs are determined by the types of transmitting and receiving antennas, i.e., omni-directional or directional. In the following, we consider four cases in a two-dimensional plane, and the results obtained can also be extended to three-dimensional space.

21.3.1   Case 1: Omni-antenna to Omni-antenna

In this case, both the transmitters and the receivers use omni-antennas, so that $G_T(i) = G_R(i) = 1, \forall i \in 1, 2, \ldots, N$. The interference between flows j and i is $I_{j,i} = k_1 G_0 P_T(j) d_{j,i}^{-\alpha}$. All transmitters use the same power P for transmission. To ensure that the interference from each interferer is less than the noise, all interfering sources must be at least r_0 away from the receiver of the ith flow ($d_{j,i} \geq r_0$), where r_0 is given by

$$r_0 = \left(\frac{k_1 G_0 P}{N_0 W} \right)^{1/\alpha}. \tag{21.6}$$

Therefore, the ER is a circle centered on the receiver, with radius r_0, as shown in Figure 21.1(a).

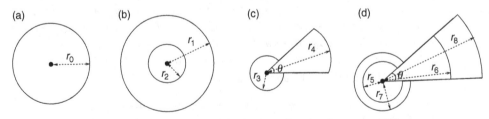

Figure 21.1 Exclusive regions for omnidirectional and directional antennas [11]. (a) Omnidirectional–omnidirectional; (b) directional–omnidirectional; (c) omnidirectional–directional; (d) directional–directional. Reproduced with permission from the IEEE.

[†] The necessary and sufficient condition to ensure that $R_i' \geq R_i$ is $\sum_{j \neq i} I_{j,i} \leq W N_0$, where flow j is scheduled to transmit concurrently with flow i. The sufficient condition given in the main text is more conservative, but it allows one to design much simpler and more practically feasible scheduling algorithms.

21.3.2   Case 2: Directional Antenna to Omni-antenna

In this case, the transmitter antennas are directional and the receiver antennas are omni-antennas ($G_R(i) = 1$). The pattern of a directional antenna consists of a mainlobe of gain G_{T_M} with beamwidth θ and a side lobe of gain G_{T_S} with beamwidth $2\pi - \theta$.

As shown in Figure 21.1(b), if the receiver is inside the radiation angle of an interferer, the interference is $I_{j,i} = k_1 G_0 G_{T_M} P d_{j,i}^{-\alpha}$. Thus, an interferer should be outside a circle centered on the receiver a radius r_1, where

$$r_1 = \left(\frac{k_1 G_0 G_{T_M} P}{N_0 W} \right)^{1/\alpha}. \tag{21.7}$$

If the receiver is outside the radiation angle of an interferer, we have $I_{j,i} = k_1 G_0 G_{T_S} P d_{j,i}^{-\alpha}$, and the ER is a circle a radius r_2, where

$$r_2 = \left(\frac{k_1 G_0 G_{T_S} P}{N_0 W} \right)^{1/\alpha}. \tag{21.8}$$

21.3.3   Case 3: Omni-antenna to Directional Antenna

When the receiver antennas are directional and the transmitter antennas are omnidirectional, the ER in this case is a sector of a circle centered on the receiver, with a radius r_4, plus a sector with a radius r_3 and angle $2\pi - \theta$, as shown in Figure 21.1(c), where θ is the beamwidth of the directional antenna of the receiver.

Let G_{R_M} be the antenna gain of the receiver within the beamwidth θ, and G_{R_S} the gain outside the beamwidth. If an interferer is located within the beamwidth of the receiver's antenna, $I_{j,i} = k_1 G_0 G_{R_M} P d_{j,i}^{-\alpha}$, and the interferer should be at least r_3 away from the receiver, where

$$r_3 = \left(\frac{k_1 G_0 G_{R_M} P}{N_0 W} \right)^{1/\alpha}. \tag{21.9}$$

Otherwise, $I_{j,i} = k_1 G_0 G_{R_S} P d_{j,i}^{-\alpha}$, and the interferer should be at least r_4 away from the receiver, where

$$r_4 = \left(\frac{k_1 G_0 G_{R_S} P}{N_0 W} \right)^{1/\alpha}. \tag{21.10}$$

21.3.4   Case 4: Directional Antenna to Directional Antenna

When both the transmitter and the receiver antennas are directional, the ER zones have four different types.

If both the interferer and the receiver are outside each other's radiation beamwidth, the ER zone is a sector with an angle $2\pi - \theta$ and radius r_5, where

$$r_5 = \left(\frac{k_1 G_0 G_{T_S} G_{R_S} P}{N_0 W} \right)^{1/\alpha}. \tag{21.11}$$

If an interferer is within the radiation angle of the receiver, but the receiver is outside the radiation angle of the interferer, the ER zone is a sector with an angle θ and radius r_6, where

$$r_6 = \left(\frac{k_1 G_0 G_{T_S} G_{R_M} P}{N_0 W}\right)^{1/\alpha}.$$

(21.12)

If an interferer is outside the radiation angle of the receiver with its radiation beamwidth toward the receiver, the ER zone is a sector with an angle $2\pi - \theta$ and radius r_7, where

$$r_7 = \left(\frac{k_1 G_0 G_{T_M} G_{R_S} P}{N_0 W}\right)^{1/\alpha}.$$

(21.13)

If an interferer is located within the beamwidth of the receiver, and the receiver is also within the beamwidth of the interferer, the ER zone is a sector with an angle θ and radius r_8, where

$$r_8 = \left(\frac{k_1 G_0 G_{T_M} G_{R_M} P}{N_0 W}\right)^{1/\alpha}.$$

(21.14)

The four types of ER zones for this case are shown in Figure 21.1(d).

21.4   REX: Randomized Exclusive Region Based Scheduler

It was shown in Section 21.3 that concurrent transmissions are more favorable than round-robin TDMA transmissions if all interfering devices are sufficiently far apart, i.e., outside the ERs of the other receivers. In other words, the network throughput can be improved by exploiting spatial reuse of the wireless channel with concurrent transmissions. Unlike traditional scheduling problems, the flow throughput per time slot depends on the network topology, the user deployment, the transmission power, the cross-correlations of interfering signals, and the scheduling decision itself, so it is unknown before the scheduling decision. With a random network topology, the optimal scheduling problem for concurrent transmissions is NP-hard [9, 10].

In [11], a Randomized EXclusive Region Based (REX) scheduling algorithm was proposed. This considers a femtocell with N active flows requesting transmissions, where the AP has global information about the femtocell, for example the number of active flows, and location information for all devices, etc., based on which the AP schedules peer-to-peer transmissions for active flows.[‡] We denote the set of all active flows by $S\{N\}$, with N elements. A subset of the flows $\gamma_t \subset S\{N\}$ contains the flows scheduled in slot t that satisfy the conditions favoring concurrent transmissions, as derived in (21.6)–(21.14). We denote by FS the set of scheduled flows in $S\{N\}$, and denote by $T_a(j)$ the number of slots allocated to flow j.

‡ In femtocells, the mobility is typically low, for example, ≤ 1 m/s, and the superframe duration is less than 100 ms. Thus, the node movement is normally less than 0.1 m during the superframe duration. Such a small change in location will not significantly affect the received power and interference power level, and it is acceptable to ignore mobility in the scheduling decision.

Initially, $FS = \text{NULL}$, $\gamma_t = \text{NULL}$ for each slot t, and $T_a(j) = 0$ for each flow j. The REX scheduling algorithm consists of the following steps, starting from the first slot $(t = 1)$:

- *Step 1.* Randomly choose one flow with the minimum number of slots allocated, labeled flow j, and schedule it in slot t. Increase $T_a(j)$ by one. Add flow j to γ_t. If flow j is not included in FS, add it to FS.
- *Step 2.* Check all the remaining active flows in the set $S\{N\} - \gamma_t$ for concurrent transmission conditions as derived in (21.6)–(21.14), starting from the flow with the minimum slots allocated. If any flow k satisfies the condition for concurrent transmission, i.e., flow k and each flow in the set γ_t are mutually outside each other's ERs, add flow k to γ_t and increase $T_a(k)$ by one. If flow k is not included in FS, add it to FS.
- *Step 3.* Increase the slot number t by one, and sort the flows according to their allocated number of slots in ascending order.
- *Step 4.* Repeat Steps 1–3 until all flows are scheduled, i.e., $FS = S\{N\}$.

It is worth noting that although sorting flows according to their allocated number of slots in step 3 increases the computational complexity by $O(N \log N)$, it is essential for maintaining fairness among flows. If we search the flows in a deterministic sequence for slot allocation, those flows with smaller sequence numbers are more likely to be scheduled. This will cause a serious unfairness problem. With the searching sequence used in step 2, the maximum access delay of all flows can be bounded.

The results for whether two flows are mutually exclusive can be saved in a lookup table to reduce the execution time of REX. Owing to the low mobility in femtocells, the frequency of updating of this table is low.

21.5   Estimating the Average Number of Concurrent Transmissions Using REX

Given the number of active users in an area, what is the number of flows that can be transmitted simultaneously under the constraint of the ER condition? Since the network topology and user deployment drastically affect the network performance, we focus on the expected number of concurrent transmissions, which is general and independent of the network topology and user deployment.

Consider an $L \times L$ square room containing N active flows, with N transmitters and N receivers randomly deployed. We define $P(k,n)$ as the probability that only k flows satisfy the ER condition and can be scheduled for concurrent transmission, after checking the first $n \leq N$ flows one by one. Without loss of generality, we check the flows in ascending order $1, 2, \ldots, N$. For slot t, the first flow, f_1, is scheduled for transmission in the set γ_t, and we have $P(1,1) = 1$. Flow f_2 is added to γ_t if it does not conflict with flow f_1. We define Q as the probability of a transmitter lying outside the ER of a receiver. The probability that a flow does not conflict with another flow is Q^2, because both transmitters need to be outside the ERs of the other receivers. Accordingly, the probability that two flows do not satisfy the ER condition is $1 - Q^2$. Therefore, in the two-flow case, we have $P(2,2) = Q^2$ and $P(1,2) = 1 - Q^2$. After we have checked the first n flows, there are k flows in γ_t if (1) there are $k-1$ flows in γ_t when we check the first $n-1$ flows, and the nth flow does not conflict with the other $k-1$ flows in γ_t; or (2) there are k flows in the set when we check the first $n-1$ flows, and the nth flow conflicts with one of the k flows in γ_t. The probability that a flow does not conflict with any of the other $k-1$ flows is $Q^{2(k-1)}$:

$$P(k,n) = P(k-1,n-1)Q^{2(k-1)} \qquad (21.15)$$
$$+ P(k,n-1)(1-Q^{2k}) \qquad \text{for } k < n.$$

If, from among the n flows, only the first flow can be added to γ_t, implying that the following $n-1$ flows do not satisfy the ER condition, we have

$$P(1,n) = (1-Q^2)^{n-1} \qquad \text{for } k = 1. \qquad (21.16)$$

Another extreme case is that in which all n flows can be scheduled concurrently, which means that none of the flows conflicts with the remaining $n-1$ flows:

$$P(n,n) = (Q^{n-1})^n \qquad \text{for } k = 1. \qquad (21.17)$$

Given the initial values of $P(1,1)$, $P(1,2)$, and $P(2,2)$, we can iteratively obtain $P(k,N)$ as a function of Q for $\forall k, 1 \le k \le N$. The expected number of concurrent transmissions is

$$E[CT] = \sum_{k=1}^{N} kP(k,N). \qquad (21.18)$$

To obtain $E[CT]$, we need to know Q. Let the size of the ER of a receiver be A, and the total area $S = L^2$. With each device randomly deployed in the room, an interferer of one flow is outside the ER of the receiver of another flow with probability $Q = 1 - A/S$. Since the ER and Q are related to the types of antennas used, in the following we derive Q by considering the four cases shown in Figure 21.1.

21.5.1   Case 1: Omni-antenna to Omni-antenna

In case 1, the ER is a circle of radius r_0, and $A_0 = \pi r_0^2$, as shown in Figure 21.1(a). The probability that an interferer is outside the ER of the receiver is given by

$$Q_1 = 1 - \frac{A_0}{S} = 1 - \frac{\pi r_0^2}{S} \qquad \text{for } r_0 \ll L. \qquad (21.19)$$

21.5.2   Case 2: Directional Antenna to Omni-antenna

Owing to the omnidirectional receivers and directional transmitters, the ER in case 2 contains two zones, a circle of radius r_1 and another circle of radius r_2, as shown in Figure 21.1(b). Accordingly, the areas of the two zones are $A_1 = \pi r_1^2$ and $A_2 = \pi r_2^2$. If a receiver is within the radiation angle of an interferer with probability $\theta/(2\pi)$, the interferer is outside the first ER zone (A_1) with probability $1 - A_1/S$. Similarly, if a receiver is outside the radiation angle of an interferer with probability $1 - \theta/(2\pi)$, the interferer is outside the second ER zone (A_2) with probability $1 - A_2/S$. Therefore, the probability that an interferer is outside the ER of a receiver is given by

$$Q_2 = 1 - \frac{\pi r_2^2}{S} + \frac{r_2^2 \theta}{2S} - \frac{r_1^2 \theta}{2S} \qquad \text{for } r_1, r_2 \ll L. \tag{21.20}$$

21.5.3 Case 3: Omni-antenna to Directional Antenna

The ER in case 3 contains two exclusive zones, one sector with radius r_3 and angle $2\pi - \theta$, and another sector with radius r_4 and angle θ, as shown in Figure 21.1(c). The areas of the two sectors are $A_3 = \pi r_3^2(1 - \theta/(2\pi))$ and $A_4 = \pi r_4^2 \theta/(2\pi) = \theta r_4^2/2$. Note that the two areas are exclusive to each other, i.e., $A_3 \cap A_4 = $ NULL. Thus, an omnidirectional interferer is outside the ER of a directional receiver if it is in neither A_3 nor A_4, and the probability is given by

$$Q_3 = 1 - \frac{\pi r_3^2}{S} + \frac{r_3^2 \theta}{2S} - \frac{r_4^2 \theta}{2S} \qquad \text{for } r_3, r_4 \ll L. \tag{21.21}$$

If the antenna gain of the directional transmitter in case 2 equals that of the directional receiver in case 3 ($G_{TM} = G_{RM}$), we have $r_2 = r_3$ and $r_1 = r_4$ from (21.7)–(21.10), and thus $Q_2 = Q_3$ from (21.20) and (21.21). Accordingly, from (21.18), the expected number of concurrent transmissions in case 2 equals that in case 3 if $G_{TM} = G_{RM}$.

21.5.4 Case 4: Directional Antenna to Directional Antenna

The ER in this case contains four zones, as shown in Figure 21.1(d). The first zone, A_5, is a sector with radius r_5 and angle $2\pi - \theta$, and the second zone, A_6, is a sector with radius r_6 and angle θ. The areas of these two exclusive zones are $A_5 = \pi r_5^2[1 - \theta/(2\pi)]$ and $A_6 = \pi r_6^2 \theta/(2\pi)$. If a receiver is outside the radiation angle of an interferer with probability $1 - \theta/(2\pi)$, the interferer is outside the ER zones of the receiver (A_5 and A_6) with probability $1 - (A_5 + A_6)/S$. Similarly, if a receiver is within the radiation angle of an interferer with probability $\theta/(2\pi)$, the interferer is outside the ER zones of the receiver (A_7 and A_8) with probability $1 - (A_7 + A_8)/S$, where $A_7 = \pi r_7^2[1 - \theta/(2\pi)]$ and $A_8 = \pi r_8^2 \theta/(2\pi)$. Therefore, we have

$$Q_4 = \left(1 - \frac{A_7 + A_8}{S}\right)\frac{\theta}{2\pi} + \left(1 - \frac{A_5 + A_6}{S}\right)\left(1 - \frac{\theta}{2\pi}\right), \tag{21.22}$$

for $r_5, r_6, r_7, r_8 \ll L$.

21.5.5 Edge Effect

If the areas of the ER zones are relatively large compared with the area of the room or the device is located near the edge of the room, it is likely that some parts of the ER zones will be outside the room, and this is referred to as the "edge effect." Using case 1 as an example, if the receiver is at the corner of the square room, we have $A' = \pi r_0^2/4 < A$.

The actual probability that a random interferer is within the ER of a receiver should be $A'/S \leq A/S$. Thus, the analytical model developed without considering the edge effect may result in a conservative estimate of $E[CT]$. The simulation results presented in the next section show that the edge effect is limited if $A_i \leq S$ and $r_i \leq \sqrt{2}L$, for $i = 0, 1, \ldots, 8$.

21.6   Performance Evaluation

We ran simulations to compare the performance of REX and round-robin TDMA. The network was set up in a 10×10 m^2 area, with 20–80 active flows deployed in the area. Such very dense scenarios can occur in conference rooms, shopping malls, and exhibition areas, or in the Internet of Things. All flows used a transmission power of 10 mW, and k_1 was -51 dB. We set $N_0 = -114$ dBm/MHz, $W = 500$ MHz, and $G_0 = 10^{-2}$. k_2 was set to 1 for calculational simplicity. We repeated the simulation 500 times with different random seeds and calculated the average values.

21.6.1   Spatial Multiplexing Gain

We applied the ideal flat-top model for the directional antenna, i.e., $\eta = 1$, so that the antenna gain within the beamwidth θ (mainlobe) was $2\pi/\theta$ and the gain outside the beamwidth was 0. As shown in Figures 21.2(a) and (b), as the ER size increases with the ER radius r and the beamwidth θ, fewer flows can be scheduled for concurrent transmission. The relationship between the number of concurrent transmissions and the path loss exponent α is shown in Figure 21.2(c). According to (21.6)–(21.14), a higher path loss exponent results in a smaller r and ER size, and thus more aggressive spatial reuse can be achieved. When α or θ is small, (e.g., $\alpha \leq 3$ or $\theta \leq 30°$), the ER radius can be very large and the edge effect becomes significant, so the analysis is more conservative owing to the edge effect. Overall, the simulation results validate the analysis.

The spatial multiplexing gain, defined as the ratio of the network throughput obtained with REX to that obtained with round-robin TDMA, is shown in Figure 21.3. For case 1 (omni-antenna to omni-antenna), the sufficient condition to favor concurrent transmissions is to set the ERs to circles of radius $r_0 = 4.47$ m centered on each receiver, as derived in (21.6). Figure 21.3(a) shows that with the ER size set to 4.47 m, the network throughput can be enhanced by more than four times, and the highest network throughput can be achieved if the ER size is around 3 m. Although more flows satisfy the concurrent-transmission condition with a smaller ER size, higher interference among these flows may limit the network throughput. By setting the ER size to r_0, we can ensure that the spatial multiplexing gain is always greater than 1. When the ER size is large enough to forbid any concurrent transmission, the performance of REX is the same as that of round-robin TDMA and the spatial multiplexing gain equals 1. The results demonstrate the effectiveness and efficiency of REX.

Figure 21.3(b) shows the spatial multiplexing gain obtained in cases 2–4. The ER size was set according to the values r_1–r_8 derived from (21.7)–(21.14). With $\theta = 6°$,

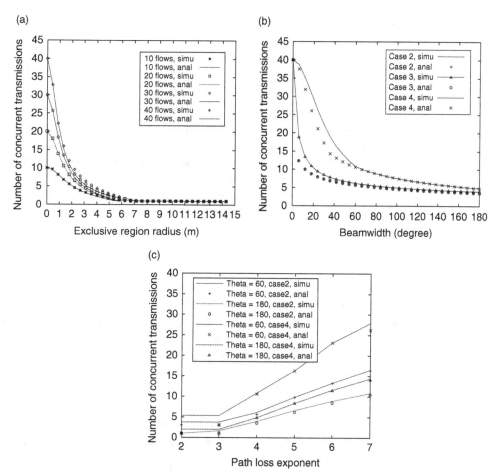

Figure 21.2 Number of concurrent transmissions [11]. (a) Case 1, $\alpha = 4$; (b) cases 2–4, $\alpha = 4$; (c) cases 2–4, $\theta = 60°$. Reproduced with permission from the IEEE.

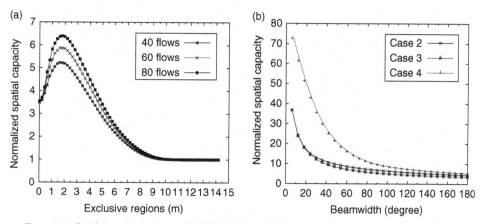

Figure 21.3 Spatial multiplexing gain [11]. (a) Case 1; (b) cases 2–4. Reproduced with permission from the IEEE.

the ER was a circle and a sector of radius 12.4 m in cases 2 and 3, respectively. Most flows can be transmitted concurrently with such a small beamwidth, and the spatial multiplexing gain is as high as 38 in cases 2 and 3, and 73 in case 4. The results show that a higher spatial multiplexing gain can be achieved with a smaller beamwidth, because more flows satisfy the concurrent-transmission conditions. We also observe that the spatial multiplexing gain obtained in case 2 is slightly higher than that in case 3, although the expected numbers of concurrent transmissions in these two cases are the same, as shown in Figure 21.2(b). This is because directional transmitters cause less interference to other concurrent transmissions than omnidirectional transmitters. The spatial multiplexing gain is improved further when both the receiver and the transmitter use directional antennas. In all cases, REX can achieve substantial spatial multiplexing gains.

21.6.2   Fairness

Obviously, in terms of the time slots allocated to each flow, round-robin TDMA can allocate time slots equally to each flow and is fair. On the other hand, for fair resource allocation, it is desirable to maximize the minimum flow throughput among all competing flows. We therefore compared the per-flow throughput obtained with REX and with round-robin TDMA, and the maximum and minimum per-flow throughputs are shown in Figure 21.4. Among the 40 flows, when the beamwidth is less than 80°, the minimum per-flow throughput with REX is even higher than the maximum per-flow throughput with round-robin TDMA; when the beamwidth is larger than 80°, the minimum per-flow throughput with REX is still much higher than that with round-robin TDMA. Thus, REX achieves better max–min fairness and is very desirable for improving the user experience.

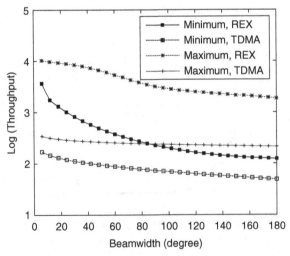

Figure 21.4 Comparison of per-flow throughput fairness [11]. Reproduced with permission from the IEEE.

21.7   Further Discussion

21.7.1   Fast Fading

Wireless channels may suffer from fast fading, which affects the instantaneous received signal and the interference power. Given the low mobility in femtocells, to minimize the complexity and overhead of channel measurement, the REX scheduler uses the path loss model to represent the average channel rate to simplify the scheduling process. In a mobile environment, it is possible to leverage opportunistic scheduling [12] and REX to devise a throughput-optimal scheduling solution, although this remains a further research issue.

21.7.2   Shadowing Effect

The shadowing effect occurs when a certain range of angle of arrival (AOA) of the signal is blocked. This effect depends on the angular distribution of the incident power and the range of AOA being blocked. A model has been developed [13] to quantify the body-shadowing effect. For mmWave, with a short wavelength and a directional antenna,the body-shadowing effect may penalize the link budget by 20–30 dB [14]. In an indoor environment with multiple moving people, mmWave links become intermittent if the antenna directions of the transmitter and receiver are fixed. Several approaches can help to mitigate the negative effects of shadowing. In addition to installing more APs, we can exploit non-line-of-sight (NLOS) paths such as reflected paths from walls, ceilings, floors, or other surfaces to steer around obstacles. Once an obstacle is blocking the LOS, the most favorable NLOS direction can be discovered and selected, or relay nodes can be used to deliver the data in a multihop fashion [15]. When a multihop path is found to be more favorable than direct transmission, we can reschedule the multihop transmission using the REX scheduler.

Furthermore, the shadowing effect may sometimes be beneficial as it may block interference as well. The REX scheduler has the option to reduce the ER when shadowing of interference is detected, to further enlarge the spatial multiplexing gain.

21.7.3   Three-Dimensional Networks

A real wireless network is three-dimensional. A three-dimensional cone-plus-sphere model can take the effects of the mainlobe and sidelobes into consideration [16]. In this model, the antenna gain consists of a mainlobe of beamwidth θ and aggregated spherical sidelobes of beamwidth $2\pi - \theta$ at the base of the mainlobe cone. Uniform gain is also assumed for simplicity in the cone-plus-sphere model. Consequently, the ERs for the omnidirectional-to-omnidirectional and directional-to-omnidirectional cases will be spheres instead of circles, and cones instead of sectors for the omnidirectional-to-directional and directional-to-directional cases. Once the ERs are defined, the REX scheduler is ready to work in a three-dimensional network.

21.7.4 Distributed Medium Access

In a system without a central scheduler to allocate resources, a distributed medium access control (MAC) protocol is needed. For distributed MAC design, the key issue is to maximize the resource utilization and limit the channel waste due to collisions. In a small-size network, such as a wireless local area network (WLAN), a wireless personal area network (WPAN), or a femtocell, since the propagation delay of the wireless signal is typically much smaller than the transmission time of a frame, carrier sense multiple access with collision avoidance (CSMA/CA) is the fundamental strategy to reduce the collision probability and improve the network throughput. Based on CSMA/CA, the distributed coordination function (DCF) protocol adopted in the IEEE 802.11 standard has contributed to the overwhelming success of WLANs, where wireless devices typically use omnidirectional antennas.

Fully distributed resource management for mmWave networks with directional antennas is much more complicated than the case with omni-antennas only, owing to the deafness problem (a receiver fails to drive the mainlobe of the antenna to the correct direction, so it cannot detect and receive packets) and the exaggerated hidden terminal problem (an interferer fails to sense an ongoing transmission, as it is outside the mainlobe of other senders). To mitigate the negative impact of these problems, nodes need to discover neighbors, drive the antenna to the right direction, and exchange directional request-to-send/clear-to-send (RTS/CTS) messages, as designed in the distributed, asynchronous directional-to-directional MAC (DtD MAC) protocol [17]. Using the DtD MAC protocol, multiple copies of directional RTS messages need to be transmitted sequentially in each direction to address the hidden terminal problem, and idle nodes (potential receivers) need to continuously scan different directions to emulate a receiver with an omnidirectional antenna to solve the deafness problem.

Given the high complexity and overhead of DtD MAC, another distributed approach, called dual sensing directional MAC (DSDMAC), has been proposed [18]. This uses two well-separated wireless channels, i.e., a data channel and a busy-tone channel, for coordinating transmissions in a distributed manner, resulting in lower overhead and higher spectrum efficiency. The DSDMAC protocol uses a noninterfering out-of-band busy-tone signal combined with sensing the activity on the actual data channel to identify deafness situations, and to avoid hidden terminals and unnecessary blocking.

With the above distributed approach based on CSMA, a key issue is to set the carrier sensing range. The ER introduced in this chapter can be applied to setting the sensing range appropriately to optimize distributed MAC protocols, considering both the omni-antenna and directional-antenna cases [19].

21.7.5 Hybrid Medium Access

Resource reservation is most beneficial if the traffic has a constant arrival pattern and if the network is saturated. Random access is desirable for bursty traffic and if the traffic intensity in the network is low, so fewer resources are wasted owing to collisions. Given that heterogeneous traffic may coexist in a femtocell and the traffic volume may change substantially over time, hybrid reservation/contention-based medium access control is desirable, i.e., some of the channel resources are reserved, and the rest are subject to free contention using a random access, CSMA-based MAC protocol [20, 21]. Using a hybrid MAC, during the resource reservation period, REX scheduling can be used to ensure high spatial reuse, and, similarly to the distributed-MAC case, the ER derived can be used to design the carrier-sensing range.

21.7.6   Optimal Scheduling

It is well known that optimal scheduling of concurrent transmissions in rate-nonadaptive wireless networks is NP-hard. Optimal scheduling in rate-adaptive mmWave networks (or other broadband wireless networks) is even more difficult as discussed earlier. Given the mutual interference, each flow's throughput in a time slot is unknown before the scheduling decision for that slot is finalized. The capacity bound derived for rate-nonadaptive networks is no longer applicable either. Given the hardness of these problems and the fact that the scheduling decisions need to be made within a few milliseconds, various heuristic global-search algorithms have been proposed [22–24]; these can outperform REX scheduling, especially when the number of flows in the network is large. Furthermore, when the transmission power of each node is adjustable, the throughput-maximization scheduling problem can be formulated as a mixed integer linear programming (LP) problem [25], and an LP-relaxation and rounding approach can be used to design an approximation algorithm. How to improve the scheduling design further to approach the optimum calls for further research.

21.8   Conclusion

In summary, this chapter has studied the sufficient conditions in terms of ERs to ensure that concurrent transmissions are beneficial, considering both omnidirectional and directional antennas. By applying the concept of ER, the REX scheduler can achieve a substantial spatial multiplexing gain, especially for mmWave femtocells with directional antennas. The analysis that we performed of the average number of concurrent transmissions obtained with REX not only is important for mmWave-based femtocells but also reveals the important relationship between the ER condition and the spatial multiplexing properties of wireless networks in general. The analytical framework should be helpful for revealing the fundamental theoretical bounds of general wireless networks, and provide important guidelines for both centralized scheduling and distributed MAC protocol design.

References

[1] S. Hara and R. Prasad, "Overview of multicarrier CDMA," *IEEE Commun. Mag.*, vol. 35, no. 12, pp. 126–133, Dec. 1997.

[2] C. Garnier, L. Clavier, Y. Delignon, M. Loosvelt, and D. Boulinguez, "Multiple access for 60 GHz mobile ad hoc network," in *Proc. of IEEE Vehicular Technology Conf. (VTC)*, May 2002.

[3] S. Roy, Y. C. Hu, D. Peroulis, and X. Y. Li, "Minimum energy broadcast using practical directional antennas in all-wireless networks," in *Proc. of IEEE INFOCOM*, Apr. 2006.

[4] C. Balanis, *Antenna Theory, Analysis and Design*, Wiley, 1997.

[5] J. E. Wieselthier, G. D. Nguyen, and A. Ephremides, "Energy-limited wireless networking with directional antennas: The case of session-based multicasting," in *Proc. of IEEE INFOCOM*, Jul. 2002.

[6] I. Kang and R. Poovendran, "Power-efficient broadcast routing in adhoc networks using directional antennas: Technology dependence and convergence issues," Technical report, UWEETR-2003-0015, Jul. 2003.

[7] IEEE, "Wireless medium access control (MAC) and physical layer (PHY) specifications for high rate wireless personal area networks (WPANs)," IEEE 802.15.3 TG, Sep. 2003.

[8] J. S. Davis, "Indoor wireless RF channels." Available at http://wireless.per.nl/reference/chaptr03/indoor.htm.

[9] S. Ramanathan, "A unified framework and algorithms for (T/F/C)DMA channel assignment in wireless networks," in *Proc. of IEEE INFOCOM*, Apr. 1997.

[10] K.H. Liu, L. Cai, and X. Shen, "Exclusive-region based scheduling algorithms for UWB WPAN," *IEEE Trans. Wireless Commun.*, vol. 7, no. 3, pp. 933–942, Mar. 2008.

[11] L. X. Cai, L. Cai, X. Shen, and J. W. Mark, "REX: A randomized exclusive region based scheduling scheme for mmWave WPANs with directional antenna," *IEEE Trans. Wireless Commun.*, vol. 9, no. 1, pp. 113–121, Jan. 2010.

[12] Y. Chen, X. Wang, and L. Cai, "HOL delay based scheduling in wireless networks with flow-level dynamics," in *Proc. of IEEE Global Communications Conf. Workshops (GLOBECOM)*, Dec. 2014.

[13] R. Zhang, L. Cai, S. He, X. Dong, and J. Pan, "Modeling, validation and performance evaluation of body shadowing effect in ultra-wideband networks," *Phys. Commun.*, vol. 2, no. 4, pp. 237–247, Jun. 2009.

[14] T. Manabe, Y. Miura, and T. Ihara, "Effects of antenna directivity and polarization on indoor multipath propagation characteristics at 60 GHz," *IEEE J. Sel. Areas Commun.*, vol. 14, no. 3. pp. 441–448, Apr. 1996.

[15] G. Zheng, C. Hua, R. Zheng, and Q. Wang, "Toward robust relay placement in 60 GHz mmWave wireless personal area networks with directional antenna," *IEEE Trans. Mobile Comput.*, vol. 15, no. 3, pp. 762–773, Mar. 2016.

[16] R. Ramanathan, "On the performance of ad hoc networks with beamforming antennas," in *Proc. of ACM International Symposium on Mobile Ad Hoc Networking and Computing (MobiHoc)*, Oct. 2001.

[17] E. Shihab, L. Cai, and J. Pan, "A distributed, asynchronous directional-to-directional MAC protocol for wireless ad hoc networks," *IEEE Trans. Veh. Technol.*, vol. 58, no. 9, pp. 5124–5134, Nov. 2009.

[18] A. Abdullah, L. Cai, and F. Gebali, "DSDMAC: Dual sensing directional MAC protocol for ad hoc networks with directional antennas," *IEEE Trans. Veh. Technol.*, vol. 61, no. 3, pp. 1266–1275, Mar. 2012.

[19] L. X. Cai, L. Cai, X. Shen, J. W. Mark, and Q. Zhang, "MAC protocol design and optimization for multi-hop ultra-wideband networks," *IEEE Trans. Wireless Commun.*, vol. 8, no. 8, pp. 4056–4065, Aug. 2009.

[20] R. Zhang, R. Ruby, J. Pan, L. Cai, and X. Shen, "A hybrid reservation/contention-based MAC for video streaming over wireless networks," *IEEE J. Sel. Areas Commun.*, vol. 28, no. 3, pp. 389–398, Apr. 2010.

[21] K. R. Malekshan, W. Zhuang, and Y. Lostanlen, "Coordination-based medium access control with space-reservation for wireless ad hoc networks," *IEEE Trans. Wireless Commun.*, vol. 15, no. 2, pp. 1617–1628, Feb. 2016.

[22] Z. Yang, L. Cai, and W. Lu, "Practical concurrent transmission scheduling algorithms for rate-adaptive wireless networks," in *Proc. of IEEE INFOCOM*, Mar. 2010.

[23] J. Qiao, X. Shen, J. W. Mark, and Y. He, "MAC-layer concurrent beamforming protocol for indoor millimeter wave networks," *IEEE Trans. Veh. Technol.*, vol. 64, no. 1, pp. 327–338, Jan. 2015.

[24] J. Qiao, X. Shen, J. W. Mark, Q. Shen, Y. He, and L. Lei, "Enabling device-to-device communications in millimeter wave 5G cellular networks," *IEEE Commun. Mag.*, vol. 53, no. 1, pp. 209–215, Jan. 2015.

[25] M. X. Cheng, Q. Ye, and L. Cai, "Rate-adaptive concurrent transmission scheduling schemes for WPANs with directional antennas," *IEEE Trans. Veh. Technol.*, vol. 64, no. 9, pp. 4113–4123, Sep. 2015.

22   Smart Data Pricing in 5G Systems

Carlee Joe-Wong, Liang Zheng, Sangtae Ha, Soumya Sen, Chee Wei Tan, and Mung Chiang

22.1   Introduction

The rapid deployment of fourth generation (4G) Long Term Evolution (LTE) standards in wireless networks has taken place against the backdrop of an enormous increase in user demand for mobile data. Indeed, Cisco estimates that demand for mobile data grew by 74% in 2015, and predicts an eightfold increase in demand by 2020 [1]. This increase in demand has strained the capacity of even LTE networks, and has made the development of fifth generation (5G) wireless networks both difficult and important. While the LTE and IEEE 802.11 standards continue to evolve, promising an expansion in cellular network capacity, purely technological solutions may not be sufficient for meeting 5G's stated goals of supporting 1–10 Gbps connections in the field and a 10- to 100-fold increase in the number of connected devices [2, 3].

Internet service providers (ISPs) looking to meet these 5G standards are exploring a number of possible solutions that aim to integrate and co-optimize networks, devices, and applications [2]. In this chapter, we investigate smart data pricing (SDP), an approach that addresses 5G challenges as well as broader changes in the ISP ecosystem. Smart data pricing focuses on the prices that end users pay for consuming data on cellular and other types of networks. By changing these prices, we can influence users' demands for different types of mobile data, thus allowing ISPs to improve their network performance by offloading or shifting traffic away from congested times.

SDP can help ISPs achieve their 5G performance goals while avoiding some of the costs of upgrading to new wireless standards. Deploying 5G networks will require significant upfront expenses [4], with the US Federal Communications Commission (FCC) chairman Tom Wheeler admitting that

While the FCC has taken many steps over the years and is still working to promote competition among network service providers, the fact remains that the financial barriers to building these networks are formidable, and most American consumers have few or no choices when it comes to this service.

SDP has a direct effect on ISPs' revenues, which can help to offset these development costs. Moreover, different types of pricing can induce more robust ISP competition for users, with ISPs using prices to attract users away from their competitors. Thus, SDP allows ISPs to achieve three key benefits:

- *Attracting users.* Most ISPs today offer a version of usage-based pricing for data, with users expected to pay a monthly fee for a fixed usage quota and extra overage fees for data above the quota [5]. While such a pricing plan would not significantly affect low-usage users, it would likely be unattractive to high-volume users, forcing them to limit their usage in order to remain below their monthly data cap. ISPs could thus attract such users by offering variations on usage-based data plans that make them more attractive to high-volume users, for example by allowing these users to share their data caps with other, lower-volume users.
- *Involving content providers (CPs).* By allowing content and application providers to pay ISPs for users' data transfers, ISPs can establish an economic link with them [6, 7]. They might even take this link one step further and tie such payments to network performance, allowing CPs to work with ISPs in order to customize users' experience of different types of apps. However, net neutrality concerns, which require that the treatment of Internet traffic be content-agnostic, may prohibit some types of customization, as these could easily lead to some CPs unfairly receiving preferential treatment. SDP, which emphasizes the economics over the technological aspect of ISP–CP relationships, offers a way for CPs to interact with ISPs while respecting net neutrality concerns.
- *Reducing network congestion.* ISPs can use data prices to incentivize users to shift their data away from congested cellular networks. For instance, users might shift their data usage from congested to uncongested times, given lower prices during those uncongested times [8]. Similarly, if less congested, supplementary networks were available, users might offload their data usage onto those networks, easing congestion in cellular networks. However, ISPs must carefully choose the price incentives to offer users: if they offer excessive incentives, for instance, users may shift too much data out of the congested times, creating congestion on other networks or at other times.

These potential benefits have fueled a revival of SDP research in the past few years [5, 9–11]. At the same time, many ISPs have begun to offer different types of pricing plans, as shown in Figure 22.1. Most of these changes fall under one of three themes, which echo SDP's three potential benefits above:

- *ISPs adapt their usage-based data plans to attract users, but avoid truly unlimited data plans.* While ISPs can simply lower their data plan prices or increase users' monthly quotas to attract more users, doing so risks lowering ISP revenue and increasing user demand, which may increase network congestion. Thus, while ISPs have introduced alternatives to charging overage fees on usage over a monthly data cap, these alternatives still penalize users for excessive data usage. Figure 22.1 shows that many ISPs, such as T-Mobile, Sprint, and Verizon, have introduced "throttle-above-a-cap" plans in which users have unlimited data usage, but can expect reduced speeds if their monthly usage exceeds a certain cap. Verizon has also tried to shift users away from their grandfathered-in unlimited data plans by raising the price of these plans. However, these penalties can prove unpopular with users; for example, Three's elimination of its unlimited data plan has been controversial [12]. Such moves can therefore drive users to competitor ISPs with more favorable data

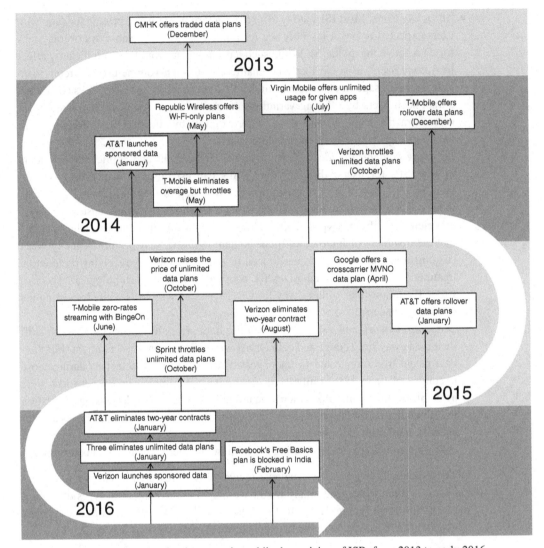

Figure 22.1 Major developments in mobile data pricing of ISPs from 2013 to early 2016.

plans. To cope with this concern, Verizon has introduced a time-dependent version of unlimited data plans that allows users to pay for a short session of unlimited data whenever the network is not congested.

Other variants of capped data plans can give users more flexibility and effectively allow them to increase their monthly usage quotas, without permitting unlimited data usage. For instance, T-Mobile introduced rollover data plans in December 2014, allowing users to carry their unused data quotas over to the next month; AT&T soon followed, in an effort to prevent their users from gravitating to T-Mobile's data plans. AT&T and Verizon, the dominant ISPs in the US market, have also taken steps to eliminate two-year data plan contracts, which had previously locked users into a two-year commitment to one data plan and one smartphone. Outside the US, China Mobile Hong Kong (CMHK) has introduced a traded data plan marketplace, in which users can purchase or sell leftover data quotas from and to each other [13]. Users are thus assured of neither wasting leftover data nor exceeding their data cap, and CMHK can collect additional revenue from users through administrative or other fees for running the data-trading marketplace.

- *CPs become involved in data pricing.* Another way for ISPs to attract users is to exempt certain types of traffic from data quotas; for example, T-Mobile's BingeOn service, introduced in November 2015, exempts certain types of video and music streaming from users' data caps [7]. Virgin Mobile introduced a similar feature in July 2014. CPs may also benefit from such an arrangement, with users more likely to consume content that does not count toward their data quota. Facebook has even introduced a "closed" version of sponsored data, in which it provides a free data plan to users but only allow access Facebook-selected own content [14]. This Free Basics plan, however, was blocked in India in February 2016 over net neutrality concerns.

  AT&T and Verizon have both sought to exploit CP interests by offering an "open toll-free" version of sponsored data, i.e., allowing CPs to actively sponsor, or subsidize, users' data consumption [6, 15]. This traffic would not count towards users' data plans, but ISPs would still gain revenue through CPs' sponsorship payments. Open toll-free data does not preclude users from accessing other applications, and is therefore less likely than the closed version to raise net neutrality concerns. Although an open toll-free version of sponsored data has been deployed globally with millions of users, net neutrality concerns in the US have generally prevented ISPs from allowing CPs to pay for prioritizing their traffic over that of others [16].

- *ISPs explore new ways to reduce cellular network congestion while supporting increases in mobile data usage.* Meeting 5G's network performance targets will likely require ISPs to integrate heterogeneous network technologies into their cellular service [2]. Mobile virtual network operators (MVNOs) have already begun to do so: since MVNOs must rent cellular network capacity from other ISPs, they and their users bear the brunt of increases in cellular data costs due to cellular network congestion. Republic Wireless has thus introduced a data plan that relies on public Wi-Fi hotspots, with supplemental cellular service if Wi-Fi is not available [17]. Google Fi has introduced a similar data plan that combines Sprint, T-Mobile, and Wi-Fi networks [18]. By leveraging and integrating access to supplementary networks, ISPs can support a greater volume of network traffic.

In this chapter, we first review recent SDP research in Section 22.2. Following the three trends in SDP practice above, we outline recent work on how ISPs can price data to attract users, who ISPs should charge for data, and what networks should be included in ISPs' data plans. We then provide a more detailed discussion of one example of each type of SDP practice, considering how traded data plans can benefit users in Section 22.3, how sponsored data can benefit users and CPs in Section 22.4, and how

ISP pricing affects traffic offloading in Section 22.5. We discuss future directions in Section 22.6 and, finally, conclude the chaper in Section 22.7.

22.2   Smart Data Pricing

Research into Smart data pricing dates back to the 1990s, when many alternatives to flat-fee unlimited data plans were put forth for wireline networks. An overview of SDP's history can be found in [5, 10]; for brevity, we concentrate here on recent developments in SDP for wireless networks. Such pricing proposals include congestion- or time-dependent pricing, in which ISPs charge users a lower price at uncongested times [19]; shared data plans, which allow users to share one monthly data cap across multiple devices [20]; and sponsored data. Given that many of these pricing ideas have been offered in practice (Figure 22.1), much of this SDP research aims to understand the effects of offering such data plans on users and providers, as well as helping users and providers determine their "best" responses to these types of data pricing. We can divide recent SDP research into three concrete questions.

22.2.1   How Should ISPs Charge for Data?

What prices should users be charged? How well can variants of capped data plans cope with heterogeneity in users' demands and traffic characteristics?

Usage-based and capped data plans do not explicitly help to alleviate network congestion, though they do penalize large amounts of data usage. Time-dependent pricing addresses congestion more directly by incentivizing users to time-shift their data to less congested times: for instance, [21] studied the economic viability of time-dependent variants of usage-based and flat-rate pricing. Other work has focused on users' reactions to time-dependent prices, for example investigating how to model users' willingness to time-shift data and their understanding of dynamic prices [19, 22]. While some users tend to be receptive to time-dependent prices, even increasing their overall usage in response to lower prices at off-peak times, others are less responsive to the prices offered.

Similarly, different users will respond to capped data plans in different ways. For instance, studies of users' data usage over one month have revealed two types of user behavior: price-sensitive users are more likely to limit their usage in order to avoid exceeding their data cap, while price-insensitive users pay less attention to prices and consume more data [23]. Applying these results to rollover data plans, we have found that moderately price-sensitive users benefit most from the option to roll over their unused data cap to the next month [24]. Different types of users will also gain different benefits from variants of capped data plans such as shared or traded data plans. Shared data plans allow heavy- and light-usage users to share a data cap, ensuring that they do not exceed a single combined data cap [25, 26]. However, users' benefits from shared data plans depend on their willingness to reduce their usage upon reaching the cap [20].

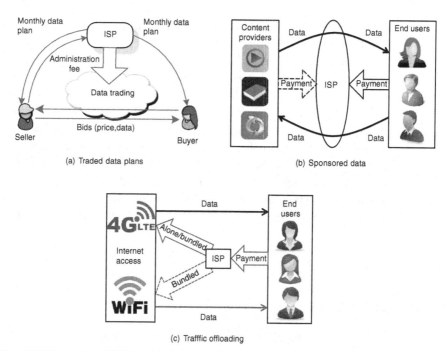

(a) Traded data plans

(b) Sponsored data

(c) Trafffic offloading

Figure 22.2 Examples of SDP and their data and payment flows.

Traded data plans allow light users to sell their leftover data to heavy users, creating an auction in which users submit bids to buy or sell data [13]. Figure 22.2(a) shows the flow of payments in this scenario: the ISP helps buyers and sellers match each other, collecting administrative fees in exchange for this service. The authors of [27, 28] considered the types of users likely to become buyers and sellers, as well as the bids they will make. Moreover, [28] showed that the ISP can also benefit from this market through its collection of administrative fees, even as both buyers and sellers benefit from the increased flexibility of traded data plans. We discuss these results in more detail in Section 22.3.

22.2.2   Whom Should ISPs Charge for Data?

How should CPs sponsor data? Who will eventually benefit from the opportunity to sponsor data?

Allowing CPs to sponsor data usage raises some implementation challenges, as discussed in [29]. Yet even designing sponsored data plans poses some challenges. Early work in this area assumed that ISPs could choose the amount of data to be sponsored [30, 31], but in today's deployments ISPs have generally allowed CPs to choose the amount of content that they wish to sponsor. Figure 22.2(b) illustrates the data and payment flows between end users, CPs, and ISPs. Data flows between the CP and the end users through the ISP's network, with users deciding how much content (i.e., data) to consume and CPs deciding how much content to sponsor. Thus, we can ask what prices the ISP charges users and CPs, as well as how much data CPs should sponsor. We consider these questions, as well as users' and CPs' benefits from sponsored data, in Section 22.4, which is adapted from [32].

Similar questions were considered in [33]. There can also be a temporal element to sponsored data [34], with CPs deciding to sponsor different amounts of data at different times, for example as network congestion varies. The authors of [35] derived optimal pricing policies for an ISP and evaluated the truthfulness of an auction mechanism for CPs to bid to sponsor data [36, 37]. Other work has considered CPs paying for content prioritization and improved quality of service (QoS) in addition to subsidizing data transfers [38, 39]. While such types of data pricing raise net neutrality concerns, these can be alleviated with competition between different content and service providers [40].

22.2.3   What Should ISPs Charge For?

How much data traffic can be offloaded to supplementary networks? What should the cost of accessing these networks be?

Many ISPs have begun offering supplementary network technologies such as Wi-Fi and femtocells to their users, encouraging users to offload data from cellular networks. Figure 22.2(c) shows the data and payment flows in the offloading mechanism: users can choose whether or not to purchase access to cellular data or to a bundle of cellular and supplementary networks. Orange and Google offer such data plans in the UK and the US, respectively [18, 41]. However, supplementary networks can also become congested if too many users subscribe to them and offload their data at the same time. Each user's subscription decision thus depends not only on the access prices for these networks but also on other users' subscription decisions and the coverage area of the supplementary network. Thus, ISPs must be careful about choosing the access prices for the two types of network, as well as the supplementary network's coverage area. We address these questions in Section 22.5, as discussed in [42].

The adoption of network technologies was also considered in [43, 44], including the relative network congestion externalities of these two technologies; [45] considered the specific case of femtocells providing a supplementary network. ISPs might also lease network capacity from access point owners, instead of offering supplementary networks themselves, as considered in [46, 47]. On a shorter timescale, other work has considered whether users should delay their data consumption in anticipation of future access to supplementary networks, with [48, 49] using prices to incentivize users to do so. If access to the supplementary network is free, users might even wait for Wi-Fi without explicit extra incentives, though they must trade off the cost savings from not using cellular data against the delay resulting from waiting for access to Wi-Fi [50]. Users' Wi-Fi access patterns are generally sufficient for such schemes to reduce cellular network congestion [51].

22.3   Trading Mobile Data

A major disadvantage of usage-based data plans is that they primarily impact higher-volume users, who are more likely to exceed their monthly data caps. The discrepancy between heterogeneous data usage and fixed data caps has been somewhat mitigated by shared data plans [20, 25]. Yet most users only share these data plans with their immediate family. If all the family members use similar amounts of data, they may still use significantly less or significantly more data than their shared data cap. The majority of users are understandably reluctant to share their leftover data caps with strangers, but they could be more amenable to *selling* their leftover data. Users with larger demands for data could then purchase additional data directly from other users, avoiding ISPs' high overage fees or exceeding their monthly caps.

At first glance, traded data plans appear to harm ISPs: by trading data caps, users can dramatically increase their ability to avoid overage fees, hurting ISP revenue. However, if ISPs involve themselves in the traded data market, they may be able to mitigate the damage and even increase their revenue overall. Indeed, ISP cooperation may be necessary for an effective traded data market, as the ISPs must enforce the traded data caps on users' bills (e.g., ensuring that buyers are not charged overage fees for using their purchased data). An ISP may also help buyers and sellers locate one another by running its own data exchange platform. We show in this section that the ISP's role as a middleman between buyers and sellers of data allows it to extract revenue from the secondary market, benefiting both users and the ISP.

Over time, the opportunity to trade data may induce some users to buy lower data caps from the ISP since they can buy data from other users; conversely, others may buy high data caps and resell them to other users. There may also be long-term branding and marketing benefits to an ISP for offering a secondary market. We do not consider these long-term effects in this section, and instead focus on user and ISP behavior within a month.

As shown in Figure 22.1, CMHK recently introduced a secondary market for trading data [13]. Their 2cm ("2nd exchange market") data exchange platform allows users to submit bids to buy and sell data, with CMHK acting as a middleman both to match buyers and sellers and to ensure that the sellers' trading revenue and buyers' purchased data are reflected in customers' monthly bills. In this section, we address two natural questions raised by traded data plans: *how do users choose bids to submit*, and *how can an ISP match buyers to sellers?* More fundamentally, *why would ISPs offer such data plans?* We answer these questions by assuming that users and the ISP behave optimally, i.e., so as to maximize their own benefits. We shall show that all three parties can benefit from the option of a secondary market, in results adapted from [28].

22.3.1   Related Work on Data Auctions

Since traded data plans take the form of auctions between buyers and sellers, we can draw on previously studied data auctions to analyze users' and ISPs' actions. However, most such auctions are designed to reduce real-time network congestion. For example,

[52] considered a scheme in which users make bids for each transmitted data packet and the ISP admits packets in order of decreasing bid. The study in [53] allocated different bandwidths to users based on QoS requirements specified in their bids, while [54] required users to send demand functions to an auctioneer who calculated the users' prices and their capacity allocation. The users' optimal bids in these auctions thus depend on their immediate valuations of a particular piece of data, rather than the overall value of a larger data cap.

Secondary markets can also occur in spectrum auctions, with secondary spectrum holders purchasing temporary spectrum from the primary spectrum holders [55]. The spectrum capacity, however, is only held on a temporary basis, and the buyers' and sellers' incentives are different from those in data trading. Spectrum auctions also do not have a third-party middleman, a feature shared by more generic studies of double auctions in the fields of electrical power and electronic commerce [56, 57].

22.3.2   Modeling User and ISP Behavior

We suppose that each seller (or buyer) can submit a bid to the secondary market consisting of the volume of data that the user wishes to sell (or buy) and the unit price that the user will accept (or pay) for this data. The ISP then matches buyers and sellers to each other. While the ISP determines the amount of data that users can buy or sell, a buyer always pays the bid price for any data bought, and similarly a seller always receives his or her bid price. Any difference between the amounts paid and received goes to the ISP; we call this money "bid revenue." This setup helps to ensure that users have no incentive to lie about the prices that they are willing to accept (as sellers) or pay (as buyers).

22.3.2.1   Choosing Optimal Bids

When choosing how much data to bid for, users must account for its effect on their usage in the rest of the month, which also depends on their unknown future usage preferences. For instance, buyers may use more data if they can buy data in the secondary market, but they may or may not value this potential data usage in the future. However, users might not be able to trade their entire bid amount in the market, which may give them an incentive to strategically overbid for data, in order to make sure that they can buy or sell their desired amount of data. In [28], we have shown that these strategies do not benefit users: it is optimal for users to assume they can trade their entire bid. We can derive the resulting amount of data to bid for as a function of the user's bid price. Users are assumed to maximize a utility function that accounts for their unknown desire for future usage and the revenue or cost from selling or buying data to or from others, as a function of their bid amounts and prices.

The prices that users bid affect whether their bids can be fully matched: for instance, some buyers may not pay a high price set by a seller. However, users do not know a priori how much of their bid can be matched, as they do not have information about the ISP's matching algorithm and the other users' bids. The user must therefore guess his or her chance of being matched at a given price. In [28], we gave an algorithm for users to make these guesses, and showed that these guesses were effective in simulations.

22.3.2.2 Matching Buyers and Sellers

The ISP matches users so as to optimize its revenue, including volume-based administration fees and "bid revenue," or the price difference between buyers who pay high and sellers willing to accept low prices. Since buyers will buy more data in the secondary market owing to its low prices compared with ISP overage fees, the ISP can collect substantial administration fees, which can exceed its primary market revenue. We can formulate the ISP's matching problem as a linear program, with an adjustable parameter allowing the ISP to put more or less emphasis on either administration or bid revenue. The constraints of this optimization problem ensure that buyers never pay more, and sellers never receive less, than their bid prices; at the same time, neither buyers nor sellers purchase or sell more data than their bids.

22.3.2.3 Market Dynamics

Users can bid for data at any time over the month, and the ISP must therefore run several matchings in each month. As buyers and sellers participate in more matchings, users can estimate more accurately the amount of data they can buy or sell at different prices. Some users, however, may not use these estimates to choose their bid prices; for example, optimistic sellers may always try to sell data at high prices, even if they are unlikely to be matched. In [28], we proposed an algorithm for users to adjust their expectations of being matched and to change their bid prices accordingly.

22.3.3 User and ISP Benefits

We have evaluated our user and ISP behavior models both analytically and by simulation. Users clearly benefit from the opportunity to trade data, since they can simply decide not to participate in the traded data plan if it does not benefit them. However, *which* users benefit and how much the ISP benefits is less clear. Thus, our analysis in [28] answers the following two questions:

- *Who can trade data?* We compared the users that were matched when the ISP optimized weighted combinations of its bid and administration revenue. Emphasizing the administration revenue results in more matchings: the more data that users successfully trade, the more administration revenue the ISP can collect. Bid revenue, however, depends on the relative difference in prices between buyers and sellers: emphasizing bid revenue might favor matches between sellers with very high and buyers with very low bids, and could reduce the total amount of data traded. No matter which type of revenue the ISP chooses to emphasize, however, buyers paying higher prices and sellers accepting lower prices are more likely to be matched, as we would intuitively expect.

- *Does ISP revenue increase with traded data?* Intuitively, users may purchase and consume more data with a traded data plan, since buyers in the traded data market

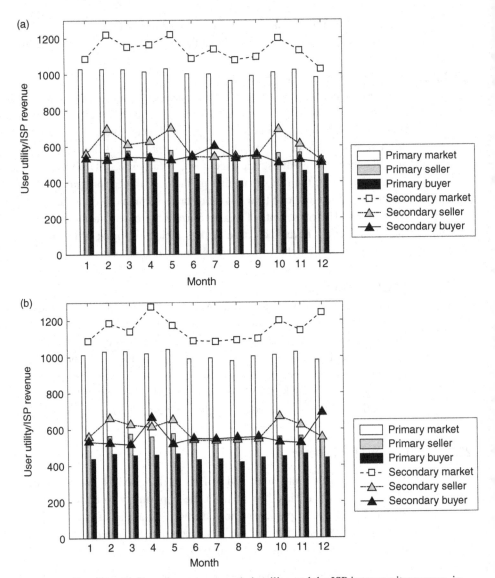

Figure 22.3 Buyers and sellers always increase their utility, and the ISP increases its revenue, in the secondary market [28]. (a) Optimizing bid revenue; (b) optimizing administration revenue. Reproduced with permission from the IEEE.

will likely offer lower prices than the ISP's data overage charges. We gave sufficient conditions under which the ISP could increase its revenue owing to this increase in data consumption. We then simulated the day-to-day user bidding over a one-year dataset of monthly usage for 100 US ISP customers. Figure 22.3 shows that the ISP increases its revenue and users increase their utilities with the traded data plan when the ISP emphasizes either bid or administration revenue.

22.4   Sponsoring Mobile Data

As shown in Figure 22.1, variations of sponsored data have recently become popular worldwide, with AT&T, T-Mobile, and Verizon offering sponsored data in the USA [6, 7, 15]. Despite the fact that sponsored data does not actively prioritize sponsoring CPs' traffic, thus arguably complying with net neutrality, these moves have not been without controversy. For instance, Facebook's FreeBasics [14] data plan would allow free access to Facebook-selected content, but would not include access to other content on the Internet. This plan was blocked in India over net neutrality concerns in February 2016.

An open version of sponsored data has the potential to benefit users, CPs, and ISPs: users are charged lower prices due to CP subsidies, CPs can attract more traffic as users increase their demand, and ISPs can generate more revenue. Yet such plans have raised concerns over the possible advantage they give to larger CPs that can better afford to sponsor data, echoing the controversy over CPs paying for higher QoS for their users. In this section, we do not consider QoS: we show that even *without* a QoS component, sponsored data can disproportionately benefit larger CPs, but can also even out demand among different types of users and benefit users more than CPs. In our model of sponsored data, CPs decide how much data to sponsor for each user. While ISPs can influence CPs by changing their data prices, sponsored data's effect on the mobile data market will be determined by *how much data CPs decide to sponsor.* It is this decision, and the impact on users and CPs, that we examine. More details of our results can be found in [32].

22.4.1   Modeling Content Provider Behavior

Since CPs decide how much data to sponsor for different users, an accurate model of their actions must include the full heterogeneity of CP and user behavior. We therefore explicitly model not only differences in user price sensitivity, as was done in [33], but also the different benefits that CPs receive from greater user demand, as shown in Table 22.1. While some CPs take in ad revenue, as considered in [35], others rely on subscription revenue, and still other CPs benefit from user goodwill or usage itself. We thus consider two different types of CPs: "revenue CPs," which benefit from usage as it contributes to revenue, and "promotion CPs," which benefit directly from usage.

Most revenue CPs rely on either ads or freemium subscriptions to make money. For apps that rely on ads, such as Pandora or Facebook, revenue grows approximately linearly with usage, as the number of ads shown is often proportional to the amount of content consumed (e.g., ads at regular intervals between songs or news stories). Other apps charge users per unit of content consumed, for example Vimeo's per-video viewing fees. Still others, such as Netflix and freemium apps, offer flat-fee subscriptions that do not depend on usage volume. These CPs are arguably less likely to sponsor data to increase subscriptions, since nonsubscribers likely derive little utility from the content itself and would require high subsidies. We therefore suppose that the benefit to revenue CPs from sponsored data is proportional to the amount of data that their users consume.

Table 22.1 Types of benefit to CPs from increased demand.

CP type	Benefit source	Benefit from usage	Example
Revenue	Ad revenue	Linear in usage	Pandora
	Subscriptions	Linear in usage	Vimeo
	Subscriptions	Linear in number of users	Netflix
Promotion	Goodwill	Concave in usage	Promotions
	Usage	Concave in usage	Enterprise

Promotion CPs benefit directly from increased usage. For instance, a new photo-sharing or social network app may sponsor data to attract users and usage in the early stages of its release. Enterprises can also fall into this category: they might subsidize employee usage of company apps, encouraging them to work more while out of the office and thus increasing worker productivity. In both cases, the CP's benefit from increased demand is concave rather than linear, mirroring users' diminishing marginal utility from more data usage.

22.4.2   Implications of Sponsored Data

We suppose that users, CPs, and an ISP make their decisions about sponsored data so as to selfishly maximize their utilities in a three-stage Stackelberg game. Users first decide how much data they will consume from each CP, given an amount of data sponsored by the CP and a price charged by the ISP. Each user chooses his or her demand level to maximize a utility function with two components: the satisfaction from using the data, and the cost of consuming this data. Users may be heterogeneous; for example, some users may be more cost-sensitive, with utilities that depend heavily on the cost of consuming data.

Given these user demands and the prices charged by the ISP, CPs then decide how much data to sponsor for each user; different CPs will likely sponsor different amounts of data. In the final stage of the game, the ISP optimizes its revenue from sponsored data by choosing the prices to charge users and CPs. We suppose that the ISP charges one price for all users and another price for all CPs.

By solving this model, we find that content sponsorship can:

- *Reverse intuitive relationships between user demand and different user and CP characteristics.* Figure 22.4(a) shows the distribution of user demand for different CPs with and without sponsored data. We see that the demand generally increases with sponsored data, as we would intuitively expect. However, as other parameters in the system are varied, users' demands can change in unexpected ways. For instance, while we would expect user utility to decrease as CPs show more ads, revenue CPs may sponsor more data if they show more ads, since doing so increases their ad revenue. We showed that this sponsorship can actually increase users' utilities overall as more ads are shown.

- *Disproportionately benefit more cost-aware users and less cost-aware CPs.** Figure 22.4(b) shows the distributions of user and CP utilities with and without sponsored data; while both users and CPs generally increase their utility with sponsored data, we see that the most dramatic increases for users occur in the low-utility region. Thus, the benefit of sponsored data to more cost-conscious users implies that it evens out the distributions of demand and utility across users. However, our findings for CPs justify the concern that sponsored data will exacerbate the advantage of larger, less cost-aware CPs.
- *Benefit users more than CPs.* We can show analytically that user utility increases proportionally more than a CP's utility when the CP chooses the amount of data sponsored so as to maximize its utility. Figure 22.4(c) shows the ratio of CP to user utilities in our simulations; we see that this ratio, as expected, decreases with sponsored data. This result suggests that the benefit of sponsored data to users may outweigh the exacerbation in competition that it induces in CPs.

22.5   Offloading Mobile Data

Offloading user demand onto supplementary technologies is not unique to data networks: successful technologies such as cellular data are often followed by other supplementary technologies, which when combined with the original enhance its features and quality. Yet the adoption of these supplementary and base technologies depends not just on their access prices but also on the externalities that users of each technology impose on others. The presence of negative congestion externalities in data networks complicates ISPs' decisions to offer users access to supplementary technologies such as Wi-Fi or femtocells. While users may be attracted to higher-quality networks (e.g., Wi-Fi generally offers higher speeds than cellular connections), if too many users begin to offload their data, these networks may become congested and deliver worse performance than cellular networks. Thus, to analyze the effects of data offloading, we have developed a model of users' adoption of a base (e.g., cellular) and supplementary (e.g., Wi-Fi) network. Solving this model allows us to analyze some example scenarios and ISPs' optimal behavior.

22.5.1   User Adoption and Example Scenarios

We suppose that users have three choices: subscribe to neither network, subscribe only to the base network, or subscribe to a bundle of the base and supplementary networks. Users act so as to selfishly maximize their utilities, which are the sum of the intrinsic satisfaction that users receive from subscribing to each network (e.g., users receive more satisfaction from Wi-Fi owing to its higher QoS), negative externalities from network congestion, and the access prices of each network. We consider heterogeneous

---

* We call CPs less "cost-aware" if they experience more utility from user demand relative to the cost of sponsorship.

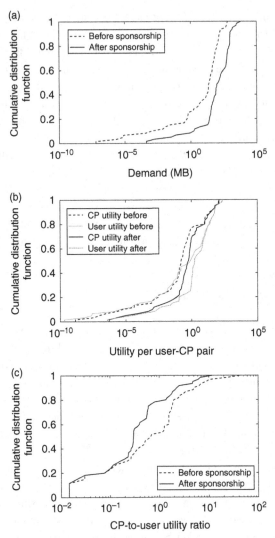

Figure 22.4 Simulation results for sponsored data [32]. (a) User demand; (b) CP and user utilities; (c) CP-to-user utility ratios. Reproduced with permission from the IEEE.

users, who may have different levels of satisfaction from subscribing to each network; for instance, some users may value Internet access highly, while others may find it convenient but unnecessary. We suppose that users change their subscription levels over time, and can show that the adoption levels for the base and bundled networks converge to an equilibrium. Given these results, we can then analyze some example ISP behaviors before considering an ISP's optimal operating point.

We have explored two example scenarios, which show that negative congestion externalities can lead to unexpected results:

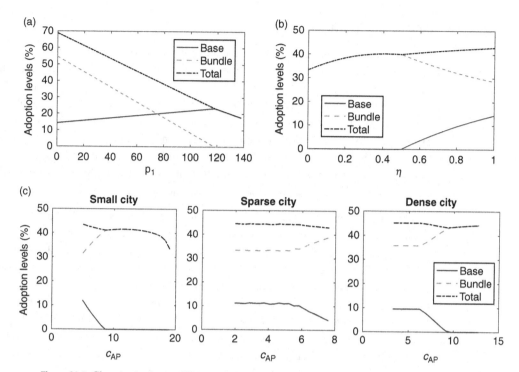

Figure 22.5 Changes in the equilibrium technology adoption levels as the access price of the base technology, the coverage area of the supplementary technology, and the deployment costs are varied. (a) As the base technology's access price p_1 increases, its adoption may increase. (b) As the supplementary coverage area η increases, total adoption increases. (c) Adoption levels for the profit-maximizing prices and coverage area in different types of cities, as the cost of deployment c_{AP} increases. Reproduced with permission from the IEEE.

- *Can increasing the base technology's access price reduce its adoption?* Consider an ISP trying to induce some users to leave its base technology by increasing the access price. In some scenarios, this move can increase the base technology's adoption: increasing the price of the base and the base + supplementary bundle by the same amount can lead some users dropping from a bundled to a base technology subscription. Figure 22.5(a) illustrates this scenario; in [42], we give precise conditions under which it occurs.
- *Will increasing the coverage area of a supplementary network always lead to more users adopting it?* Suppose that an ISP wishes to expand its supplementary network to offload more traffic from the base technology, but cannot change its pricing structure owing to exogenous factors, for example the presence of a major competitor. In [42], we derived conditions under which increased supplementary-network coverage will *decrease* its adoption: at the new equilibrium, each user offloads more traffic, which can increase overall congestion in the supplementary network and induce some users to drop the bundled service. This decrease can also occur when the ISP offers revenue-maximizing instead of fixed prices, as illustrated in Figure 22.5(b).

22.5.2   Optimal ISP Behavior

Given the adoption behaviors discussed above, we can now consider an ISP's optimal operating point. In our framework, the ISP may influence user adoption with three variables: the access prices of the two technologies, and the coverage area of the supplementary technology (e.g., supplementary Wi-Fi access may not be available at all locations). We first considered a generic model of ISP revenue and derived analytical expressions for the revenue-maximizing prices and coverage of the supplementary technology. We then focused on the scenario of offloading traffic from the base to the supplementary technology, estimating the benefits and deployment costs of offloading using empirical usage data. We considered three questions:

- *What is the optimal coverage area of a supplementary network?* Suppose that an ISP is seeking to maximize its revenue by optimizing the coverage area of the supplementary technology and the access prices of the base technology and the bundle. We showed that revenue is maximized with full supplementary-network coverage (i.e., it should be available everywhere).
- *Will users always adopt both the base and the bundled technologies at the revenue-maximizing prices?* Suppose that the ISP cannot expand its supplementary-network coverage, for example due to investment costs. We showed that the base and bundled offers will both have users unless the coverage area of the supplementary network is sufficiently small and the base technology experiences sufficiently severe congestion externalities. Figure 22.5(b) illustrates this phenomenon: as the coverage area increases, users begin to adopt the base technology.
- *Will deployment costs affect the optimal coverage and the resulting adoption of the base plus supplementary bundle?* While increasing the deployment costs of the supplementary technology decreases its optimal coverage area, it also decreases the potential congestion in this supplementary network. Consequently, we find that the adoption of the supplementary technology can increase with its deployment costs, even when the profit-maximizing prices and coverage area are chosen (Figure 22.5(c)). We considered three different deployment scenarios: a smaller city with less area to cover, and two cities with sparser and denser populations, who would require more concentrated Wi-Fi deployments to cover the same physical area. We would thus expect denser, larger cities to experience higher deployment costs; however, the lack of congestion due to a smaller deployment induces users to continue adopting the supplementary technology as the deployment cost rises.

22.6   Future Directions

Given current trends in the demand for mobile data and the diversity in mobile applications, we expect ISPs to continue experimenting with new data plans in the coming years. Below, we outline three promising trends in SDP research, which are motivated by existing ISP actions, new wireless technologies like 5G becoming available, and new applications for these technologies.

22.6.1   Capacity Expansion and New Supplementary Networks

To help ease cellular network congestion, the US FCC has taken steps allowing ISPs and users to access to a broader range of available spectrum. For instance, the FCC opened TV white space bands, i.e., unused broadcast television spectrum, to the public in 2011. Recent research on TV white space has shifted from focusing on interference management to considering the economics and pricing of access to white space [58, 59]. Other research has focused on utilizing millimeter wave frequencies (30–300 GHz), which are currently unlicensed. Measurements in New York City have confirmed the viability of millimeter wave wireless communications for supporting 5G systems [60]. Indeed, Starry has begun to sell millimeter wave connectivity for home Internet use [61], though the economics and long-term viability of its business model remain unknown.

The availability of new frequencies for wireless communication makes the problem of offloading data from traditional cellular networks more difficult. While most research on wireless traffic offloading considers one type of supplementary technology, devices today might have access to multiple heterogeneous networks with different transmission characteristics – for example, transmission in millimeter wave wireless systems incurs a larger path loss. Thus, pricing access to these supplementary networks or frequencies must take into account both the presence of multiple networks and the effect of their heterogeneity on users' preferences for accessing one network over another.

22.6.2   Two-Year Contracts Versus Usage-Based Pricing

The four largest ISPs in the US (AT&T, Sprint, T-Mobile, and Verizon) have all taken steps to eliminate two-year data plan contracts in favor of more flexible month-to-month pricing. Although this flexibility would appear to benefit users, Sprint reinstated the contract option for their customers just weeks after eliminating it [62]. Two-year data plan contracts generally come with a discounted purchase of a smartphone or tablet, allowing users to effectively purchase a discounted device in exchange for committing to one ISP's data plan for two years. Sprint's move raises the question of whether users would prefer this contract model over the more flexible option, and, if so, which users prefer which option.

The elimination of two-year contracts raises broader questions about the relationship between users changing their data plans and purchasing new devices for consuming data. Users who prefer to frequently update their smartphones as new models come out, for instance, would likely prefer a more flexible data plan. However, these users would run the risk of ISPs raising the price of their data plans, for example as network deployment becomes more expensive. They would also lose the phone subsidies included in two-year contracts, though these might be recovered as a result of rival ISPs offering rewards if users switch to their data plans from another ISP [63]. Such moves raise the question of how different ISPs might optimally design their data plans to maximize their revenue. Smartphone manufacturers would also watch user behavior closely, perhaps timing the release of new models to maximize users' desire for a new smartphone and discourage them from signing two-year contracts.

22.6.3   Incentivizing Fog Computing

5G wireless networks can encompass many different types of devices, each of which may need some form of cellular data plan. Newly connected devices such as connected cars, for instance, may not be able to rely on public Wi-Fi or short-range technologies such as Bluetooth to communicate; and it is unlikely that car manufacturers will set up their own communication network for vehicles to exchange data. Thus, we can consider the design of data plans for such devices, and whether these plans would be shared with users' smartphones or stand alone.

More broadly, SDP methods can also be used to consider non-price incentives in connected wireless systems. As more devices gain Internet connectivity as well as increased storage and computing capabilities, we anticipate that they will be able to handle a substantial amount of storage, communication, and system management services. This idea is embodied in the "fog computing" architecture [64], which devolves substantial functionality and control of computing and storage systems to the end devices. However, participating in fog systems may come at a cost for users. For instance, in a crowdsensing system, users' devices incur costs in sending data to each other, such as energy or battery consumption, or even the monetary cost of paying for data transfers [65, 66]. Thus, we can consider the problem of designing incentives that will induce users to exchange data despite these costs.

Since many users can participate in fog computing systems, we can also consider the design of incentives to overcome possible conflicts between users. For instance, users may compete for access to a storage or network resource, which may wish to prioritize access for those users with higher resource valuations. We can thus set up an auction framework that leverages computing power on users' devices to automate bidding for these shared resources according to users' individual preferences at different times [67].

22.7   Conclusion

ISPs today are beginning to experiment with different mobile data plans in an effort to address changes in the broader mobile ecosystem. In this chapter, we have reviewed three types of smart data pricing (traded data plans, sponsored data, and data offloading to supplementary networks) and outlined their consequences for users' demand patterns and the potential benefits for users and providers. We then pointed to some future directions for SDP research, driven by new developments in wireless network technology and applications. In particular, we anticipate the development of new types of data plans for connected devices, and perhaps even the use of pricing and incentives as a network management tool. Thus, if 5G networks aim to integrate networks, devices, and applications [2], we expect that SDP will have an important part to play in these wireless systems.

References

[1] Cisco Systems, "Cisco visual networking index: Global mobile data traffic forecast update, 2015–2020," Feb. 2016. Available at www.cisco.com/c/en/us/solutions/collateral/service-provider/visual-networking-index-vni/mobile-white-paper-c11-520862.pdf.

[2] B. Bangerter, S. Talwar, R. Arefi, and K. Stewart, "Networks and devices for the 5G era," *IEEE Commun. Mag.*, vol. 52, no. 2, pp. 90–96, Feb. 2014.

[3] GSMA Intelligence, "Understanding 5G: Perspectives on future technological advancements in mobile," Dec. 2014. Available at https://gsmaintelligence.com/research/?file=141208-5g.pdf&download.

[4] T. Wheeler, "Empowering small businesses to innovate in today's digital economy," Mar. 2016. Available at www.fcc.gov/news-events/blog/2016/03/25/empowering-small-businesses-innovate-today's-digital-economy.

[5] S. Sen, C. Joe-Wong, S. Ha, and M. Chiang, "Pricing data: A look at past proposals, current plans, and future trends," *ACM Comput. Surv.*, vol. 46, no. 2, p. 15, Feb. 2013.

[6] AT&T Inc., "AT&T sponsored data," Jan. 2014. Available at www.att.com/att/ sponsored-data/en/index.html#fbid=fNjHshoHkg.

[7] T-Mobile International AG, "Binge on," Nov. 2015. Available at www.t-mobile.com/offer/binge-on-streaming-video.html.

[8] S. Sen, C. Joe-Wong, S. Ha, and M. Chiang, "Incentivizing time-shifting of data: A survey of time-dependent pricing for Internet access," *IEEE Commun. Mag.*, vol. 50, no. 11, pp. 91–99, Oct. 2013.

[9] S. Sen, C. Joe-Wong, S. Ha, and M. Chiang, eds., *Smart Data Pricing*. Wiley, 2014.

[10] S. Sen, C. Joe-Wong, S. Ha, and M. Chiang, "Smart data pricing (SDP): Economic solutions to network congestion," in *Recent Advances in Networking*, H. Haddadi and O. Bonaventure, eds., ACM SIGCOMM, pp. 221–274, 2013.

[11] S. Sen, C. Joe-Wong, S. Ha, and M. Chiang, "Smart data pricing: Using economics to manage network congestion," *Commun. of the ACM*, vol. 58, no. 12, pp. 86–93, Dec. 2015.

[12] Z. Kleinman, "Three defends mobile tariff 'Price Hike'," Jan. 2016. Available at www.bbc.com/news/technology-35441452.

[13] China Mobile Hong Kong, "2cm (2nd exchange market)," Nov. 2013. Available at www.hk.chinamobile.com/en.

[14] Facebook, "Free basics platform," May 2015. Available at https://info.internet.org/en.

[15] Verizon, "FreeBee data," Jan. 2016. Available at http://freebee.verizonwireless.com/business/freebeedata.

[16] Federal Communications Commission, "In the matter of protecting and promoting the open Internet," Mar. 2015. Available at https://apps.fcc.gov/edocs_public/attachmatch/FCC-15-24A1_Rcd.pdf.

[17] D. Bean, "I spent a week with the Republic Wireless WiFi phone network," Jul. 2014. Available at www.yahoo.com/tech/i-spent-a-week-with-the-republic-wireless-wifi-phone-92437040379.html.

[18] Google, "Project Fi," Apr. 2015. Available at https://fi.google.com/.

[19] S. Ha, S. Sen, C. Joe-Wong, Y. Im, and M. Chiang, "TUBE: Time-dependent pricing for mobile data," *SIGCOMM Comput. Commun. Rev.*, vol. 42, no. 4, pp. 247–258, Oct. 2012.

[20] S. Sen, C. Joe-Wong, and S. Ha, "The economics of shared data plans," in *Proc. of Workshop on Information Technologies and Systems (WITS)*, Dec. 2012.

[21] L. Zhang, W. Wu, and D. Wang, "Time dependent pricing in wireless data networks: Flat-rate vs. usage-based schemes," in *Proc. of IEEE International Conf. on Computer Communications (INFOCOM)*, Apr. 2014.

[22] S. Sen, C. Joe-Wong, S. Ha, J. Bawa, and M. Chiang, "When the price is right: Enabling time-dependent pricing for mobile data," in *Proc. of ACM SIGCHI Conf. on Human Factors in Computing Systems*, Apr. 2013.

[23] M. Andrews, G. Bruns, M. Doğru, and H. Lee, "Understanding quota dynamics in wireless networks," *ACM Trans. Internet Technol.*, vol. 14, no. 2–3, p. 14, Oct. 2014.

[24] L. Zheng and C. Joe-Wong, "Understanding rollover data," in *Proc. of IEEE INFOCOM Workshop on Smart Data Pricing (SDP)*, Apr. 2016.

[25] Y. Jin and Z. Pang, "Smart data pricing: To share or not to share?" in *Proc. of IEEE INFOCOM Workshop on Smart Data Pricing (SDP)*, Apr. 2014.

[26] T. Yu, Z. Zhou, D. Zhang, X. Wang, Y. Liu, and S. Lu "INDAPSON: An incentive data plan sharing system based on self-organizing network," in *Proc. of IEEE International Conf. on Computer Communications (INFOCOM)*, Apr. 2014.

[27] J. Yu, M. H. Cheung, J. Huang, and H. V. Poor, "Mobile data trading: A behavioral economics perspective," in *Proc. of IEEE International Symposium on Modeling and Optimization in Mobile, Ad Hoc, and Wireless Networks (WiOpt)*, May 2015.

[28] L. Zheng, C. Joe-Wong, C. W. Tan, S. Ha, and M. Chiang, "Secondary markets for mobile data: Feasibility and benefits of traded data plans," in *Proc. of IEEE International Conf. on Computer Communications (INFOCOM)*, Apr. 2015.

[29] M. Andrews, "Implementing sponsored content in wireless data networks," in *Proc. of Allerton Conf.*, Sep. 2013.

[30] P. Hande, M. Chiang, R. Calderbank, and S. Rangan, "Network pricing and rate allocation with content provider participation," in *Proc. of IEEE International Conf. on Computer Communications (INFOCOM)*, Apr. 2009.

[31] Y. Wu, H. Kim, P. H. Hande, M. Chiang, and D. H. Tsang, "Revenue sharing among ISPs in two-sided markets," in *Proc. of IEEE International Conf. on Computer Communications (INFOCOM)*, Apr. 2011.

[32] C. Joe-Wong, S. Ha, and M. Chiang, "Sponsoring mobile data: An economic analysis of the impact on users and content providers," in *Proc. of IEEE International Conf. on Computer Communications (INFOCOM)*, Apr. 2015.

[33] L. Zhang, W. Wu, and D. Wang, "Sponsored data plan: A two-class service model in wireless data networks," in *Proc. of ACM International Conf. on Measurement and Modeling of Computer Systems (SIGMETRICS)*, Jun. 2015.

[34] L. Zhang, W. Wu, and D. Wang, "TDS: Time-dependent sponsored data plan for wireless data traffic market," in *Proc. of IEEE International Conf. on Computer Communications (INFOCOM)*, Apr. 2016.

[35] M. Andrews, U. Ozen, M. Reiman, and Q. Wang, "Economic models of sponsored content in wireless networks with uncertain demand," in *Proc. of IEEE INFOCOM Workshop on Smart Data Pricing (SDP)*, Apr. 2013.

[36] M. Andrews, G. Bruns, and H. Lee, "Calculating the benefits of sponsored data for an individual content provider," in *Proc. of IEEE Conf. on Information Sciences and Systems (CISS)*, Mar. 2014.

[37] M. Andrews, Y. Jin, and M. Reiman, "A truthful pricing mechanism for sponsored content in wireless networks," in *Proc. of IEEE International Conf. on Computer Communications (INFOCOM)*, Apr. 2016.

[38] N. Economides and J. Tåg, "Network neutrality on the Internet: A two-sided market analysis," *Inf. Econ. Policy*, vol. 24, no. 2, pp. 91–104, Jun. 2012.

[39] M. H. Lotfi, K. Sundaresan, M. A. Khojastepour, and S. Rangarajan, "The economics of quality sponsored data in wireless networks," in *Proc. of International Symposium on Modeling and Optimization in Mobile, Ad Hoc, and Wireless Networks (WiOpt)*, May 2015.

[40] R. T. B. Ma, "Subsidization competition: Vitalizing the neutral Internet," in *Proc. of ACM CoNEXT*, Dec. 2014.

[41] Orange, "Orange pay monthly pounds 2 WiFi bundle terms and conditions," Apr. 2011. Available at http://ee.co.uk/help/accounts-billing-and-topping-up/terms-and-conditions/orange-terms-and-conditions/broadband-and-email-services-terms/orange-pay-monthly-pounds2-wifi-bundle-terms-and-conditions.

[42] C. Joe-Wong, S. Sen, and S. Ha, "Offering supplementary network technologies: Adoption behavior and offloading benefits," *IEEE/ACM Trans. Netw.*, vol. 23, no. 2, pp. 355–368, Apr. 2015.

[43] S. Sen, Y. Jin, R. Guérin, and K. Hosanagar, "Modeling the dynamics of network technology adoption and the role of converters," *IEEE/ACM Trans. Netw.*, vol. 18, no. 6, pp. 1793–1805, Dec. 2010.

[44] K. Wang, F. Lau, L. Chen, and R. Schober, "Pricing mobile data offloading: A distributed market framework," *IEEE Trans. Wireless Commun.*, vol. 15, no. 2, pp. 913–927, Feb. 2016.

[45] S. Ren, J. Park, and M. van der Schaar, "Entry and spectrum sharing scheme selection in femtocell communications markets," *IEEE/ACM Trans. Netw.*, vol. 21, no. 1, pp. 218–232, Apr. 2013.

[46] W. Dong, S. Rallapalli, R. Jana, L. Qiu, K. K. Ramakrishnan, L. Razoumov, Y. Zhang, and T. W. Cho, "iDEAL: Incentivized dynamic cellular offloading via auctions," *IEEE/ACM Trans. Netw.*, vol. 22, no. 4, pp. 1271–1284, Aug. 2014.

[47] S. Paris, F. Martignon, I. Filippini, and L. Chen, "A bandwidth trading marketplace for mobile data offloading," in *Proc. of IEEE International Conf. on Computer Communications (INFOCOM)*, Apr. 2013.

[48] J. Lee, Y. Yi, S. Chong, and Y. Jin, "Economics of WiFi offloading: Trading delay for cellular capacity," *IEEE Trans. Wireless Commun.*, vol. 13, no. 3, pp. 1540–1554, Mar. 2014.

[49] X. Zhuo, W. Gao, G. Cao, and S. Hua, "An incentive framework for cellular traffic offloading," *IEEE Trans. Mobile Comput.*, vol. 13, no. 3, pp. 541–555, Feb. 2014.

[50] Y. Im, C. Joe-Wong, S. Ha, S. Sen, T.-Y. Kwon, and M. Chiang, "AMUSE: Empowering users for cost-aware offloading with throughput–delay tradeoffs," in *Proc. of IEEE Internatioal Conf. on Computer Communications (INFOCOM)*, Apr. 2013.

[51] K. Lee, J. Lee, Y. Yi, I. Rhee, and S. Chong, "Mobile data offloading: How much can WiFi deliver?" *IEEE/ACM Trans. Netw*, vol. 21, no. 2, pp. 536–550, Apr. 2013.

[52] J. MacKie-Mason and H. Varian, "Pricing the Internet," in *Public Access to the Internet*, B. Kahin and J. Keller, eds., Prentice-Hall, pp. 269–314, 1995.

[53] A. A. Lazar and N. Semret, "Design, analysis and simulation of the progressive second price auction for network bandwidth sharing," Technical Report CU/CTR/TR 487-98-21, Columbia University, Apr. 1998.

[54] P. Maillé and B. Tuffin, "Multibid auctions for bandwidth allocation in communication networks," in *Proc. of IEEE International Conf. on Computer Communications (INFOCOM)*, Apr. 2004.

[55] J. Jia, Q. Zhang, Q. Zhang, and M. Liu, "Revenue generation for truthful spectrum auction in dynamic spectrum access," in *Proc. of ACM International Symposium on Mobile Ad Hoc Networking and Computing (MobiHoc)*, May 2009.

[56] S. Hao, "A study of basic bidding strategy in clearing pricing auctions," *IEEE Trans. Power Syst.*, vol. 15, no. 3, pp. 975–980, Aug. 2000.

[57] P. R. Wurman, W. E. Walsh, and M. P. Wellman, "Flexible double auctions for electronic commerce: Theory and implementation," *Decision Support Syst.*, vol. 24, no. 1, pp. 17–27, 1998.

[58] X. Feng, Q. Zhang, and J. Zhang, "Hybrid pricing for TV white space database," in *Proc. of IEEE International Conf. on Computer Communications (INFOCOM)*, Apr. 2013.

[59] Y. Luo, L. Gao, and J. Huang, "Price and inventory competition in oligopoly TV white space markets," *IEEE J. Sel. Areas Commun.*, vol. 33, no. 5, pp. 1002–1013, Apr. 2015.

[60] G. R. Maccartney and T. S. Rappaport, "73 GHz millimeter wave propagation measurements for outdoor urban mobile and backhaul communications in New York City," in *Proc. of IEEE International Conf. on Communications (ICC)*, Jun. 2014.

[61] J. J. Roberts, "Meet 'Starry,' a radical new Internet service from the founder of Aereo," Jan. 2016. Available at http://fortune.com/2016/01/27/starry-wireless-internet/.

[62] M. Dano, "Sprint resurrects two-year wireless service contracts to give customers more choices," Feb. 2016. Available at www.fiercewireless.com/story/sprint-resurrects-two-year-wireless-service-contracts-give-customers-more-c/2016-02-26.

[63] D. Goldman, "Verizon will give you up to $650 to switch," Dec. 2015. Available at http://money.cnn.com/2015/12/29/technology/verizon-switch/index.html?iid=EL.

[64] OpenFog Consortium Architecture Working Group, "OpenFog architecture overview," Feb. 2016.

[65] T. Luo, H. P. Tan, and L. Xia, "Profit-maximizing incentive for participatory sensing," in *Proc. of IEEE International Conf. on Computer Communications (INFOCOM)*, Apr. 2014.

[66] D. Yang, G. Xue, X. Fang, and J. Tang, "Crowdsourcing to smartphones: Incentive mechanism design for mobile phone sensing," in *Proc. of ACM International Conf. on Mobile Computing and Networking (MobiCom)*, Aug. 2012.

[67] F. M. F. Wong, C. Joe-Wong, S. Ha, Z. Liu, and M. Chiang, "Improving user QoE for residential broadband: Adaptive traffic management at the network edge," in *Proc. of IEEE International Symposium on Quality of Service (IWQoS)*, Jun. 2015.

Index

4G, *see* fourth generation
5G, *see* fifth generation
5G Infrastructure Public Private Partnership
(5G-PPP), 77

access class barring (ACB), 410
access grant time interval (AGTI), 404
Access Network Discovery and Selection Function
(ANDSF), 359
access reservation protocol (ARP), 390
achievable-rate region, 177
ACK, *see* acknowledgment
acknowledgment, 11, 30, 62, 419
adaptive modulation and coding, 318, 319, 405, 463
ADC, *see* analog-to-digital converter
aggregator, 276
AMC, *see* adaptive modulation and coding
analog beamforming, 168, 196
analog-to-digital converter, 10, 165
application layer, 337, 338
asymptotic interference alignment, 226
asynchronous access, 243
attocell, 20, 290

backhaul, 13, 17, 81, 165, 180, 181, 208,
217, 280
base station (BS), 92, 110, 215, 271, 423
baseband unit (BBU), 3, 29, 48, 219, 406
beam alignment, 199, 205
beam pattern design, 204, 205
beamform–compress–forward strategy, 56
beamforming, 49, 121, 165, 192, 245, 282
blockage channel condition, 194
BS-based CS algorithm (BSCSA), 93

cache content placement, 238
cache model, 237, 245
cached interference channel, 242, 251
cached wireless network, 245
carrier aggregation, 40, 439
CDMA, *see* code division multiple access
cell association and power control, 217

cell-edge user, 221
cell range expansion, 216
cellular access protocols, 387, 390
central processor (CP), 48
centralized decoder, 228
channel acquisition, 202
channel estimation, 144, 158
channel gain, 296, 298
channel impulse response, 295
channel model, 189, 196, 292
channel quality indicator, 223
channel segregation, 92
channel state information (CSI), 3, 123, 154
channel state information at the transmitter, 123,
242
cloud computing, 4, 30, 48, 77, 133
cloud radio access network (C-RAN), 1, 28, 48, 217
cloudlets, 78
co-channel interference, 92, 172, 290
code division multiple access, 109, 461
coexistence, 441, 442
cognitive radio (CR), 112
commercial-of-the-shelf (COTS), 31
Common Public Radio Interface (CPRI), 30
compress–forward strategy, 52
compressed channel estimation, 205
computation offloading, 83
computational complexity, 6, 98, 233, 263, 303,
366, 467
compute-and-forward (CoF), 228
congestion control, 343, 416, 419
content provider (CP), 479
cooperative communication, 19, 74, 111, 181
cooperative interference network, 227
cooperative MIMO broadcast topology, 239
cooperative relay topology, 239
cooperative transmission and reception, 49
coordinated beamforming, 221
coordinated multipoint (CoMP), 3, 218, 236, 272,
280
coordinated scheduling, 221
cost-aware cellular network, 280
cross-correlation, 143

cross-layer, 78, 349, 353, 355
CSIT, *see* channel state information at the
    transmitter
cyclic prefix (CP), 134
cyclic suffix, 134

data aggregation, 81
degrees of freedom (DoF), 237
dense heterogeneous wireless network, 92
device-to-device (D2D), 15, 423
directional antenna, 462, 472, 475
distributed resource allocation, 438, 444
downlink, 50, 121, 153, 184, 203, 214, 293, 335
drift-plus-penalty (DPP), 342
duplexing, 173
dynamic channel assignment (DCA), 92
dynamic point selection (DPS), 221
dynamic programming (DP), 364
dynamic scheduling, 338

edge-based retransmission, 66
energy-aware network selection and resource
    allocation (ENSRA), 375
energy cooperation, 271
energy–delay trade-off, 370
energy efficiency, 1, 28, 92, 151, 403
energy harvesting, 16, 215, 271, 274, 423
energy harvesting region, 426
energy trading, 18, 276
enhanced intercell interference coordination
    (eICIC), 3
enhanced mobile broadband (eMBB), 27
eNodeB, 218
exclusive region, 462
Extended Coverage GSM (EC-GSM), 398

F-NOMA, *see* NOMA with fixed power allocation
fairness, 472
favorable propagation, 154
FDMA, *see* frequency division multiple access
Federal Communications Commission (FCC), 188
femtocell, 208, 290, 325, 460
fifth generation (5G), 1, 27, 48, 109, 172, 188, 214,
    236, 289, 383, 423, 460, 478
filtered OFDM (F-OFDM), 134
finite impulse response, 232
flexible physical layer, 133
fog computing, 4, 78
fourth generation (4G), 5, 27, 109, 133, 151, 179,
    214, 289, 369, 478
frequency division duplex (FDD), 173
frequency division multiple access, 109
frequency-selective channels, 144, 335
fronthaul, 28, 45, 48, 406
full-duplex, 10, 161, 172, 173

full-duplex relaying, 161
full-duplex transceiver, 175

Gale–Shapley (GS) algorithm, 445
generalized frequency division multiplexing
    (GFDM), 7, 134
global CSI, 17, 126, 242, 249
greedy maximization, 344

half-duplex, 9, 163, 172, 173
handoff, 80, 216
hard-core point process (HCPP), 324
heterogeneous cloud radio access network
    (H-CRAN), 405
heterogeneous network (HetNet), 16
hierarchical subcodebooks, 203
hospital resident (HR) model, 445
hybrid automatic repeat request (HARQ), 49, 397
hybrid beamforming, 196
hybrid medium access, 474

IA, *see* interference alignment
IEEE 802.11, 478
imperfect CSI, 9, 232
indoor propagation channel, 297
intercell interference coordination (ICIC), 41, 219
interchannel cooperation, 446, 450
interference alignment, 126, 214, 223, 236
interference alignment and cancellation (IAC), 227
interference-aware detection, 220
interference cancellation, 10, 48, 145, 162, 393
interference management, 216
interference mitigation, 17, 217
interference topology, 239
Internet of Things (IoT), 1, 27, 92, 109, 215, 380,
    403
Internet service provider (ISP), 478, 479

joint energy and spectrum sharing, 281, 284
joint scheduling across multiple cells, 220
joint signal processing across multiple cells, 220
joint transmission, 17, 50, 221

lens-based analog beamforming, 199
Licensed Assisted Access (LAA), 439
light-emitting diode (LED), 290
Light-Fidelity (LiFi), 19, 290
limited backhaul, 249
line-of-sight (LOS), 193, 217
linear precoder, 157
linear receiver, 154
link budget, 194, 217, 397
Long Term Evolution (LTE), 214, 358, 478
low latency, 66, 80, 86, 217, 278, 417

low-resolution receiver architecture, 201
LTE-Advanced (LTE-A), 218
LTE for M2M, 392
LTE-M, *see* LTE for M2M
LTE-Unlicensed (LTE-U), 13, 438, 441, 446
LTE Wi-Fi link aggregation (LWA), 439

machine-to-machine (M2M), 1, 380, 402
machine-type communications (MTC), 215, 404
massive antenna array, 162
massive machine-type communication, 15, 27
massive MIMO, 7, 151, 335
matching theory, 112, 444, 449
maximum-ratio combining (MRC), 155
MDS-coded PHY caching, 255
medium access control (MAC), 31, 148, 390, 403, 474
micro cloud center, 78
microgrid, 18, 276
millimeter wave, 1, 167, 188
MIMO, *see* multiple-input multiple-output (MIMO)
minimum mean squared error, 156, 231
mixed strategy, 96
MMSE, *see* minimum mean squared error
mMTC, *see* massive machine-type communication
mmWave, *see* millimeter wave
mobile cloud computing (MCC), 77
mobile data offloading, 13, 359
mobile edge application, 79
mobile edge computing, 4, 77
mobile edge host, 78
mobile virtual network operator (MVNO), 481
mobility management, 85
multilevel centralized and distributed (MCD) protocol stack, 42
multiple access, 109, 112, 290
multiple-input multiple-output (MIMO), 120, 134, 151, 237, 335
multiple-input multiple-output generalized frequency division multiplexing (MIMO-GFDM), 145
multi-resource scheduling, 84
multiuser superposition transmission (MUST), 110
multiuser MIMO (MU-MIMO), 336

NACK, *see* negative acknowledgment
narrowband IoT (NB-IoT), 15, 418
negative acknowledgment, 30, 62, 419
network congestion, 479, 484
network efficiency, 216
network functions virtualization (NFV), 29, 77, 133
network selection, 358
network utility, 340
network utility maximization (NUM), 337
Next Generation Fronthaul Interface (NGFI), 30
Next Generation Hotspot (NGH), 361

NOMA, *see* non-orthogonal multiple access (NOMA)
NOMA with fixed power allocation, 112
nonconvex objective function, 261
non-orthogonal multiple access (NOMA), 5, 109
NP-hard, 466, 475

OFDM, *see* orthogonal frequency division multiplexing (OFDM)
OFDMA, *see* orthogonal frequency division multiple access (OFDMA)
offloading, 358, 482, 496
on/off switching, 92
operating expenditure, 3, 271, 371
OPEX, *see* operating expenditure
opportunistic cooperative spatial diversity, 243
opportunistic cooperative spatial multiplexing, 267
optical attocell network, 304
optical OFDM, 294
orthogonal frequency division multiple access (OFDMA), 94, 109, 418
orthogonal frequency division multiplexing (OFDM), 134, 182, 290
orthogonal multiple access (OMA), 4, 109
outage probability, 429, 432

P2P, *see* peer-to-peer
partial MIMO cooperation, 244
path loss, 95, 188
peer-to-peer, 78, 237, 424
periodic reporting access, 395
phase noise, 5, 7, 461
phase shifter, 196
physical layer (PHY), 31, 133, 237, 336
physical layer caching, 16, 236
physical layer network coding, 228
physical random access channel (PRACH), 390
PNC, *see* physical layer network coding
Poisson point process, 312
power allocation, 65, 110, 272, 372
power line communication, 18, 293
precoding, 122, 157, 244, 347
precoding matrix indicator (PMI), 222
pricing, 85, 363, 478
privacy, 87

QoE, *see* quality of experience
QoS, *see* quality of service
quality of experience, 337, 358
quality of service, 1, 109, 215, 236, 278, 441

radio access network (RAN), 29, 77, 396
radio access technologies, 215
radio resource management, 16, 92, 359
rainfall channel condition, 194

random access opportunities (RAOs), 390
random backoff, 411
random matrix theory, 348
randomized exclusive region, 466
RAT, *see* radio access technologies
rate region, 339, 349
receiver design, 135
relay, 48, 111, 161, 173, 229, 402, 424
remote radio head (RRH), 3, 35, 48, 123, 219
renewable energy, 18, 271
resource allocation, 5, 84, 85, 92, 110, 286, 295, 338, 371, 405
retransmission, 67, 70, 73, 389

scheduling, 335, 462, 466
SDP, *see* smart data pricing
security, 6, 10, 79, 81, 179, 360, 381
self-backhauling, 181
self-interference, 10, 174
sequential decision, 364
shared data plan, 482, 485
SI, *see* self-interference
SIC, *see* successive interference cancellation
side information, 237
signal clipping, 303
signal-to-interference-plus-noise ratio (SINR), 54, 93, 112, 175, 239, 335
signal-to-noise ratio (SNR), 63, 112, 144, 153, 175, 196, 226, 236, 412
simplex, 174
simultaneous transmission and reception (STAR), 174
single-input single-output (SISO), 7, 112, 253
SINR, *see* signal-to-interference-plus-noise ratio (SINR)
small cell, 13, 76, 86, 123, 214, 218, 289, 358, 402, 441
smart data pricing, 14, 478, 479, 494
smart grid, 18, 271, 409
software-defined air interface (SDAI), 28
software-defined waveform, 137
sparse code multiple access (SCMA), 6
spatial degrees of freedom, 9, 128
spatial diversity, 18, 242
spatial multiplexing, 242, 335, 405, 461
spatial multiplexing gain, 461, 470, 472
spatial reuse, 304, 460
spectral efficiency, 10, 63, 93, 109, 142, 151, 160, 172, 236
spectrum sharing, 84, 272, 279, 444
sponsored data, 481, 483
state transition probability, 365
strong law of large numbers, 263

student–project allocation (SPA) model, 445
sub-6 GHz, 12, 208
subchannel allocation, 372
subspace sampling, 202, 203
successive interference cancellation, 110, 393
sum rate, 61, 111, 177, 345
synchronization, 134

Tactile Internet, 133
TDMA, *see* time division multiple access
Third Generation Partnership Project (3GPP), 110, 380
threshold policy, 368
time division duplex (TDD), 173
time division multiple access, 109, 113
traded data plan, 481, 485
traffic offloading, 278, 281, 439, 495
transmission mode selection, 425, 431
transmission point (TP), 220
transmission scheduling, 345, 355

UE-based CS algorithm (UECSA), 93
UE relay (UER), 423
ultra-reliable low-latency communication, 15, 27
unfavorable topology, 239
universal-filtered OFDM (UF-OFDM), 134
uplink, 50, 121, 153, 184, 202, 214, 293, 335
URLLC, *see* ultra-reliable low-latency communication
usage-based data plan, 479, 485
user-centric network (UCN), 29
user equipment (UE), 29, 79, 93, 215, 404, 423
user pairing, 112
user terminal (UT), 152

virtual machine (VM), 36, 78
virtualization, 30, 35, 48, 79, 133
virtualization infrastructure, 79
visible light communication (VLC), 1, 290

wideband fading channel, 182
Wireless Fidelity (Wi-Fi), 358
wireless information transfer, 165
wireless local area network (WLAN), 442, 474

X-channel, 225
X2, 218

zero-forcing (ZF), 8, 51, 155
zero-forcing beamforming, 222, 282, 347
ZFBF, *see* zero-forcing beamforming

.

Printed in the United States
By Bookmasters